Lecture Notes in Mathematics

Edited by A. Dold and B. Eckmann

886

Fixed Point Theory

Proceedings of a Conference Held at Sherbrooke, Québec, Canada, June 2–21, 1980

Edited by E. Fadell and G. Fournier

Springer-Verlag
Berlin Heidelberg New York 1981

Editors
Edward Fadell
Department of Mathematics, University of Wisconsin
Madison, WI 53706, USA

Gilles Fournier
Département de Mathématique et Informatique
Université de Sherbrooke
Sherbrooke, Québec, Canada, J1K 2R1

AMS Subject Classifications (1980): 47Hxx, 54H25, 55M20, 58C30, 58G10

ISBN 3-540-11152-2 Springer-Verlag Berlin Heidelberg New York
ISBN 0-387-11152-2 Springer-Verlag New York Heidelberg Berlin

Printing and binding: Beltz Offsetdruck, Hemsbach/Bergstr.
2141/3140-543210

FOREWORD

Fixed Point Theory has always played a central role in the problems of Nonlinear Functional Analysis; and Topology has certainly played a useful role in both areas. More recently, advanced techniques of Algebraic Topology have proved fruitful when applied to problems in these areas. In 1977, a conference at Oberwolfach (organized by Prof. A. Dold and E. Fadell) brought together mathematicians in all of the aforementioned areas to discuss and attack problems of mutual interest and also to learn about research directions in each of these fields.

The success of this conference suggested the need for establishing a tradition along these lines. The basic objective of this Sherbrooke Conference was to contribute another step toward the establishment of the tradition.

We gratefully acknowledge the assistance of many people who helped make this conference a success. In particular, Roger D. Nussbaum who was one of the organizers of this meeting and Reine Fournier who supervised the countless day-to-day details that are a necessary part of every conference.

Funding for this conference was provided by the Université de Sherbrooke, the Québec Ministry of Education and a Regional Development grant from the Natural Sciences and Engineering Research Council of Canada. Other funds for the preparation of these Proceedings was provided by the ACFAS.

In addition we are indebted to the Université de Sherbrooke for providing excellent facilities.

Finally, we thank Mrs Michèle Vallée for her careful typing of the manuscript of these Proceedings.

E. Fadell and G. Fournier

PREFACE

La théorie des points fixes a toujours joué un rôle de premier plan en analyse fonctionnelle non-linéaire, et la topologie fut certainement très utile aux deux précédents domaines. En 1977, le congrès organisé à Oberwolfach par les professeurs A. Dold et E. Fadell réunit des mathématiciens appartenant à chacune de ces trois spécialités afin de leur permettre d'échanger leurs points de vue sur des problèmes d'intérêt commun et d'être informés sur les différents développements dans chacun des autres domaines de recherche.

Le succès de cette rencontre a démontré le besoin de régulariser ce type de rencontre. Le premier objectif du congrès de Sherbrooke fut de faire un autre pas dans cette direction.

Nous remercions chaleureusement toutes les personnes qui contribuèrent à faire de ce congrès le succès qu'il fut. Nous tenons tout particulièrement à exprimer notre gratitude à Roger D. Nussbaum un des deux organisateurs de cette rencontre et à Reine Fournier qui s'est occupée avec efficacité de tous les petits détails inhérents à l'organisation matérielle de ce congrès.

Ce congrès fut financé par des subventions provenant de l'Université de Sherbrooke, du Ministère de l'Education du Québec et du programme de développement régional du Conseil de recherches en sciences naturelles et en génie du Canada. La préparation des comptes rendus fut subventionnée par l'ACFAS.

En outre, nous sommes obligés envers l'Université de Sherbrooke pour les services qu'elle a mis à notre disposition lors de cette rencontre.

Finalement, nous remercions Mme Michèle Vallée pour avoir consciencieusement dactylographié le manuscrit des comptes rendus de ce congrès.

E. Fadell et G. Fournier

TABLE OF CONTENTS

ALEXANDER, J.C., FITZPATRICK, P.M.	GLOBAL BIFURCATION FOR SOLUTIONS OF EQUATIONS INVOLVING SEVERAL PARAMETER MULTIVALUED CONDENSING MAPPINGS	1
ALLIGOOD, K.T.	TOPOLOGICAL CONDITIONS FOR THE CONTINUATION OF FIXED POINTS	20
BELLEY, J-M.	A MEASURE THEORETIC APPROACH TO FIXED POINTS IN ERGODIC THEORY	33
BROWN, R.F.	ON THE PRODUCT THEOREM FOR THE FIXED POINT INDEX ...	44
FADELL, E., HUSSEINI, S.	A FIXED POINT THEORY FOR FIBER-PRESERVING MAPS	49
FOURNIER, G.	A SIMPLICIAL APPROACH TO THE FIXED POINT INDEX	73
GAUTHIER, G.	FIXED POINT THEOREMS FOR APPROXIMATIVE ANR'S	103
GÓRNIEWICZ, L.	ON THE LEFSCHETZ COINCIDENCE THEOREM	116
HUSSEINI, S.Y.	COBORDISMS OF MAPS	140
JAWOROWSKI, J.	FIBRE PRESERVING MAPS OF SPHERE BUNDLES INTO VECTOR SPACE BUNDLES	154
JIANG, B.	FIXED POINT CLASSES FROM A DIFFERENTIAL VIEWPOINT ..	163
JIANG, B. SCHIRMER, H.	FIXED POINT SETS OF CONTINUOUS SELFMAPS ON POLYHEDRA	171
KIRK, W.A.	LOCALLY NONEXPANSIVE MAPPINGS IN BANACH SPACES	178
LAMI DOZO, E.	ASYMPTOTIC CENTERS IN PARTICULAR SPACES	199
LIN, T.Y.	WU-LIKE CLASSES AND GENERALIZED PETERSON-STEIN CLASSES	208

MARTELLI, M.	SEMI-FREDHOLM OPERATORS AND HYPERBOLIC PROBLEMS	249
MASSA, S.	MULTI-APPLICATIONS DU TYPE DE KANNAN	265
MASSABO, I., NISTRI, P., PEJSACHOWICZ, J.	ON THE SOLVABILITY OF NONLINEAR EQUATIONS IN BANACH SPACES	270
MORALES, P.	CONTRACTION PRINCIPLE IN PSEUDO-UNIFORM SPACES	300
NUSSBAUM, R.D.	EIGENVECTORS OF NONLINEAR POSITIVE OPERATORS AND THE LINEAR KREIN-RUTMAN THEOREM	309
PEITGEN, H-O., SIEBERG, H-W.	AN $\bar{\varepsilon}$ - PERTURBATION OF BROUWER'S DEFINITION OF DEGREE	331
PETRYSHYN, W.V.	FIXED POINTS AND SURJECTIVITY THEOREMS VIA THE A-PROPER MAPPING THEORY WITH APPLICATION TO DIFFERENTIAL EQUATIONS	367
POTTER, A.J.B.	AN EXISTENCE THEOREM AND APPLICATION TO A NON-LINEAR ELLIPTIC BOUNDARY VALUE PROBLEM	398
RAY, W.O., SINE, R.C.	NONEXPANSIVE MAPPINGS WITH PRECOMPACT ORBITS	409
SCHIRMER, H.	FIXED POINT SETS OF CONTINUOUS SELFMAPS	417
STEINLEIN, H.	WHAT IS THE RIGHT ESTIMATE FOR THE LJUSTERNIK-SCHNIRELMANN COVERING PROPERTY ?	429
WILLE, F.	ON A CONJECTURE OF HOPF FOR α-SEPARATING MAPS FROM MANIFOLDS INTO SPHERES	435
WILLIAMSON, T.E. JR.	THE LERAY-SCHAUDER CONDITION IS NECESSARY FOR THE EXISTENCE OF SOLUTIONS	447
ALEXANDER, J.C.	A PRIMER ON CONNECTIVITY	455
KIRK, W.A.	FIXED POINT THEORY FOR NONEXPANSIVE MAPPINGS	484
	OPEN PROBLEMS	506

LISTE DES PARTICIPANTS

LIST OF PARTICIPANTS

ALEXANDER, J.	University of Maryland College Park, Maryland 20742, U.S.A.
ALLIGOOD, Kathleen	College of Charleston Charleston, South Carolina 29401, U.S.A.
BELLEY, Jean-Marc	Université de Sherbrooke Sherbrooke, Qué., Canada J1K 2R1
BROWDER, Felix	University of Chicago 5734 University Ave, Chicago, Illinois 60637, U.S.A.
BROWN, Robert F.	University of California Los Angeles, California 90024, U.S.A.
CLAPP DE PRIETO, Mónica	Instituto de Mathemáticas de la U.N.A.M. México 20, D.F., Mexique
CRAMER, Walter	Iowa State University 400 Carver Hall, Ames, Iowa 50011, U.S.A.
DOLD, Albrecht	Universität Heidelberg 6900 Heidelberg 1, Deutschland
DOWNING, David	Oakland University Rochester, Michigan 48063, U.S.A.
DUBOIS, Jacques	Université de Sherbrooke Sherbrooke, Qué., Canada J1K 2R1
FADELL, Edward	University of Wisconsin Madison, Wisconsin 53706, U.S.A.
FESHBACH, Mark	University of Minnesota Minneapolis, Minnesota 55410, U.S.A.
FINBOW, Arthur	Dalhousie University Halifax, Nova Scotia, Canada B3J 2K9
FORSTER, Walter	University of Southampton Southampton, S09 5HN, England
FOURNIER, Gilles	Université de Sherbrooke Sherbrooke, Qué., Canada J1K 2R1
FOURNIER, Reine	Université de Sherbrooke Sherbrooke, Qué., Canada J1K 2R1

GAUTHIER, Gilles — Université du Québec à Chicoutimi
Chicoutimi, Qué., Canada G7H 2B1

GIROLO, Jack — California Polytechnic State University
San Luis Obispo, California 93407, U.S.A.

GÓRNIEWICZ, Lech — University of Gdańsk
Gdańsk, Poland

GOSSEZ, Jean-Pierre — Université Libre de Bruxelles, Campus de la Plaine
1050 Bruxelles, Belgique

GUÉNARD, François — Ecole Normale Supérieure de l'Enseignement Technique
75015 Paris, France

GUZZARDI, Renato — Università Della Calabria
C.P. Box 9, Roges, Cosenza 87030, Italia

HALPERN, Benjamin — Indiana University
Bloomington, Indiana 47401, U.S.A.

HEATH, Philip — Memorial University of Newfoundland
St. John's, Newfoundland, Canada A1B 3X7

HUSSEINI, Sufian Y. — University of Wisconsin
Madison, Wisconsin 53706, U.S.A.

ISAC, George — Collège Militaire Royal
St-Jean, Qué., Canada J0J 1R0

JAWOROWSKI, Jan — Indiana University
Bloomington, Indiana 47401, U.S.A.

et/and Forschungsinstitut für Mathematik
ETH Zentrum, CH-8092 Zürich, Schweiz

JERRARD, Richard — University of Illinois
Urbana, Illinois 61801, U.S.A.

JIANG, Boju (CHIANG, Po-Chu) — University of California
Los Angeles, California 90024, U.S.A.

et/and Beijing University
Beijing, China

KIRK, William A. — University of Iowa
Iowa City, Iowa 52242, U.S.A.

KNILL, Ronald — Tulane University
New Orleans, Louisiana 70118, U.S.A.

LALLI, Bikkar Sing — University of Saskatchewan
Saskatoon, Saskatchewan, Canada S7N 0W0

LAMI DOZO, Enrique	Université Libre de Bruxelles 1050 Bruxelles, Belgique
LIN, T.Y.	Louisiana State University Baton Rouge, Louisiana 70803, U.S.A.
et/and	University of South Carolina at Aiken Aiken, South Carolina 29801, U.S.A.
MARTELLI, Mario	Bryn Mawr College Bryn Mawr, Pennsylvania 19010, U.S.A.
MARTIN, John	University of Saskatchewan Saskatoon, Saskatchewan, Canada S7N 0W0
MASSA, Silvio	Università Degli Studi di Milano Via Saldini 50, 20133 Milano, Italia
MEADE, Barbara	Memorial University of Newfoundland St. John's, NFLD, Canada A1B 3X7
MESSANO, Basilio	Instituto Universitario Navale Via Acton 38, Napoli, Italia
MORALES, Pedro	Université de Sherbrooke Sherbrooke, Qué., Canada J1K 2R1
NORRIS, Carl	Memorial University of Newfoundland St. John's, NFLD, Canada A1B 3X7
NUSSBAUM, Roger D.	Rutgers University New Brunswick, New Jersey 08903, U.S.A.
PEITGEN, Heinz-Otto	Universität Bremen Postfach 330440, 2800 Bremen 33, Deutschland
PEJSACHOWICZ, Jacobo	Università Nazionale della Calabria C.P. Box 9, Roges, Cosenza, Italia
PETRYSHYN, Walter V.	Rutgers Unviersity New Brunswick, New Jersey 08903, U.S.A.
POTTER, Anthony, J.B.	Aberdeen and Rutgers University New Brunswick, New Jersey 08903, U.S.A.
PRIETO, Carlos	Instituto de Mathemáticas de la U.N.A.M. México 20, D.F., Mexique
RHOADES, B.E.	Indiana University Bloomington, Indiana 47405, U.S.A.
SCHIRMER, Helga	Carleton University Ottawa, Ontario, Canada K1S 5B6

SINE, Robert	University of Rhode Island Kingston, Rhode Island 02881, U.S.A.
SINGH, S.P.	Memorial University of Newfoundland St.John's, NFLD, Canada A1B 3X7
STANKIEWICZ, Jan	Université de Montréal C.P. 6128, succ. "A", Montréal, Qué., Canada H3C 3J7
STEINLEIN, Heinrich	Ludwig-Maximilians-Universität D-8000 München 2, Theresienstrasse 39, Deutschland
VIOLETTE, Donald	Université de Sherbrooke Sherbrooke, Qué., Canada J1K 2R1
VOON, Shu-Nan	Dalhousie University Halifax, Nova Scotia, Canada B3J 2K9
WILLE, Friedrich	University of Kassel 3500 Kassel, West Germany
WILLIAMSON, Thomas E., Jr.	Montclair State College Upper Montclair, New Jersey 07043, U.S.A.
ZANCO, Clemente	Università Degli Studi Via Saldini 50, 20133 Milano, Italia

GLOBAL BIFURCATION FOR SOLUTIONS OF EQUATIONS INVOLVING SEVERAL PARAMETER MULTIVALUED CONDENSING MAPPINGS

BY

J.C. ALEXANDER* AND P.M. FITZPATRICK**

Department of Mathematics
University of Maryland
College Park, Maryland 20742
U.S.A.

§1.

It is our purpose in the present paper to prove a global bifurcation result for solutions of an equation of the form

$$x \in F(\lambda,x), \qquad (\lambda,x) \in \mathcal{O}$$

where F is a *multivalued* mapping of $\mathcal{O} \subseteq \mathbb{R}^n \times X$ into 2^X, where X is a Banach space. Hence our equation is very general, in that the parameter λ is not restricted to being a scalar and we allow $F(\lambda,x)$ to be a compact, convex subset of X for each (λ,x) in $\mathcal{O}$. The global topological assumption we impose on F is that it be condensing. So in particular our results cover compact mappings; that is, when $F(\Omega)$ is relatively compact for each $\Omega \subseteq \mathcal{O}$, with Ω bounded. In fact, our proof for the general condensing case is carried through by using an inductive procedure to reduce to the case when F is compact.

Let us introduce some definitions in order to state our theorem.

For $x \in X$, and $A \subseteq X$ we let $d(x,A) = \inf\{\|x - a\| \mid a \in X\}$, and if $B \subseteq X$ we let the Hausdorff metric, $d(A,B)$, be defined by

$$d(A,B) = \max\{\sup\{d(a,B) \mid a \in A\},\ \sup\{d(b,A) \mid b \in B\}\}.$$

Also, if $\varepsilon > 0$, we let $N_\varepsilon(A) = \{x \mid d(x,A) < \varepsilon\}$. We denote by $K(X)$ the family

Partially supported by NSF Grants MCS 7609668 A01 and 7607461 A01**

of closed, convex subsets of X.

A mapping $F : M \to 2^X$, where M is a topological space, is called upper-semicontinuous provided that if $m \in M$ and $\varepsilon > 0$ there exists a neighborhood U of m such that

$$F(U) \subseteq N_\varepsilon(F(m)).$$

If $\Omega \subseteq X$ is bounded, the set-measure of noncompactness of Ω, $\gamma(\Omega)$, is defined by

$$\gamma(\Omega) = \inf\left\{ r > 0 \;\middle|\; \begin{array}{l} \Omega \text{ is contained in a finite union of sets,} \\ \text{each of which has diameter} < r \end{array} \right\}.$$

Given $\mathcal{O} \subseteq \mathbb{R}^n \times X$ and $F : \mathcal{O} \to K(X)$ we call F *condensing* provided that F is upper-semicontinuous and $\gamma(\mathcal{O} \cap (A \times \Omega)) < \gamma(\Omega)$ when $\Omega \subseteq \mathcal{O}$ is bounded, $A \subseteq \mathbb{R}^n$ is bounded and $\gamma(\Omega) > 0$.

(1.1) *DEFINITION*

Let $F : \mathcal{O} \subseteq \mathbb{R}^n \times X \to K(X)$ be condensing. Suppose $(0,0) \in \mathcal{O}$ and there exists a neighborhood V of 0 in $\mathbb{R}^n$ such that $0 \in F(\lambda,0)$ if $\lambda \in V$. We let $S = \{(\lambda,x) \in \mathcal{O} : x \in F(\lambda,x)\}$ and say the solutions of

$$x \in F(\lambda,x), \quad (\lambda,x) \in \mathcal{O}$$

bifurcate globally from $V \times \{0\}$, provided that there exists a connected subset C of $S \setminus \{V \times \{0\}\}$ with $(0,0) \in \overline{C}$ and at least one of the following occurs:

(i) C is unbounded.

(ii) $\overline{C} \cap \partial\mathcal{O} \neq \phi$.

(iii) $(\lambda^*,0) \in \overline{C}$ for some $\lambda^* \neq 0$.

For further amplification of this definition see Remark (2.1).

Given $\eta > 0$ let $D_\eta = \{x \in X \mid \|x\| \leq \eta\}$, and let

$$\hat{G}_\eta \text{ cond} = \left\{ T : D_\eta \to K(X) \;\middle|\; \begin{array}{l} T \text{ is condensing and } x \notin T(x) \\ \text{if } x \in \partial D_\eta \end{array} \right\}.$$

$\hat{G}_\eta$ cond becomes a metric space under $d(T,S) = \sup_{x \in D} d(T(x),S(x))$, for $T,S \in \hat{G}$ cond. Clearly the dependence on η is topologically unimportant, so we drop the superscript η. We let $\hat{G}$ comp $= \{T \in \hat{G}$ cond : T is compact$\}$, and G cond and G comp be the subsets of $\hat{G}$ cond and $\hat{G}$ comp, respectively, whose members consist of single-valued mappings. The homotopy properties of G comp and G cond, have been investigated in [3].

We can now formulate the hypothesis of our Theorem 1: for $a > 0$, let $D_a^n = \{\lambda \in \mathbb{R}^n : |d| \le a\}$.

(H1) Suppose n is a positive integer, X is a Banach space and $\mathcal{O} \subseteq \mathbb{R}^n \times X$ is open, with $T : \mathcal{O} \to K(X)$ condensing.

(H2) Suppose there exists $a > 0$, $b > 0$ such that $D_a^n \times D_b \subseteq \mathcal{O}$ and

$$F : D_a^n \times D_b \to K(X)$$

is such that if $\varepsilon > 0$ there exists a $\delta > 0$ such that $d(F(\lambda,x), F(\lambda',x)) < \varepsilon$ if $|\lambda - \lambda'| < \delta$, and $(\lambda,x), (\lambda',x) \in D_a^n \times D_b$.

(H3) For each compact $K \subseteq D_a^n \setminus \{0\}$, there exists $\delta(K) > 0$ such that $x \notin F(\lambda,x)$ when $\lambda \in K$, $0 < \|x\| \le \delta(K)$.

Under the above hypotheses it is clear that if $0 < \xi < a$,

$$S^{n-1} = \{\lambda \in \mathbb{R}^n : |\lambda| = \xi\}$$

and $0 < \eta < \delta(S^{n-1})$ then F induces a mapping of S^{n-1} into $\hat{G}_\eta$ cond. We may denote by γ the homotopy type of this mapping.

(1.1) *THEOREM*

Let F, $\mathcal{O}$ and X be as defined above and suppose assumptions (H1), (H2) and (H3) are satisfied. Suppose γ, as defined above, is nontrivial. Then there is global bifurcation of the solutions of equation (1.1) from the trivial solutions.

In case $n = 1$, and F is compact and single-valued the above theorem, under some additional differentiability assumptions which allow one to determine the nontriviality of γ and a slightly weaker definition of global bifurcation, was proved by Rabinowitz [7]. For $n = 1$, and F a single-valued k-set contraction, $0 < k < 1$, a result similar to the above was proven by Stuart [10].

For general n, with F single-valued and compact, and some additional differentiability assumptions, the above result appears in Alexander [1]. For general n, and F single-valued and condensing, the above result was proven in Alexander and Fitzpatrick [3].

The proof of the above Theorem proceeds in three steps. We first prove that the inclusion $G \text{ cond} \subseteq \hat{G} \text{ cond}$ is a weak homotopy equivalence; this is Theorem (2.1). Next we prove an approximation theorem for multivalued mappings, which guarantees F can be approximated in a suitable sense by single-valued mappings; this is Theorem (2.2). We then apply a refinement of the main result of [1] together with the previous two results to complete the proof.

We find it convenient to first prove Theorem (1.1) for compact F, and then in the last section prove the general case.

§2.

Recall that if W is a topological space, with $U \subseteq W$ a subspace then the inclusion $U \subseteq W$ is a *weak homotopy equivalence* provided that whenever K is a compact complex, of which K_0 is a subcomplex, and $\alpha : (K,K_0) \to (Y,U)$ then there exists $H : [0,1] \times K \to W$ such that $H([0,1] \times K_0) \subseteq U$, $H(0,\cdot) = \alpha$ and $H(\{1\} \times K) \subseteq U$.

Using the definition of G cond and G comp introduced in the preceeding section we have

(2.1) *THEOREM*

The inclusion $G \text{ comp} \subseteq \hat{G} \text{ comp}$ is a weak homotopy equivalence.

Proof

Let K be a compact complex, of which K_0 is a subcomplex, and let $\alpha : (K,K_0) \to (\hat{G} \text{ comp}, G \text{ comp})$. For each $k \in K$ let $\alpha(k) = f_k$. By a standard compactness argument we may choose $\varepsilon > 0$ such that

$$d(x, f_k(x)) \geq \varepsilon \quad \text{when} \quad k \in K, \quad x \in \partial D.$$

Since K is a compact, metric space, α is uniformly continuous, so we may choose $\delta > 0$ such that

$$d(f_k(x), f_{k'}(x)) < \frac{\varepsilon}{5} \quad \text{when} \quad d'(k,k') < \delta, \quad x \in D,$$

where d' is the metric on K. Choose $\{k_i : 1 \leq i \leq m\} \subseteq K$ such that

$$\{V_i : 1 \leq i \leq m\} \text{ cover } K,$$

where $V_i = \{k \in K : d'(k,k_i) < \delta\}$, for each $i \in \{1,\ldots,m\}$.

By the upper semicontinuity of f_{k_i}, for $1 \leq i \leq m$, we may, given $x \in \partial D$, choose $\delta(x) > 0$, $\delta(x) < \varepsilon/5$, so that $f_{k_i}(y) \subseteq N_{\varepsilon/5}(f_{k_i}(x))$, when $\|x - y\| < \delta(x)$, $1 \leq i \leq m$. Since ∂D is a metric space we may choose a locally finite star refinement $\{\mathcal{O}_\gamma\}_{\gamma\in\Gamma}$, of the cover $\{B(x,\delta(x)) \cap \partial D\}_{x\in\partial D}$, which is also a cover of ∂D, and a corresponding subordinate partition of unity, $\{\varphi_\gamma\}_{\gamma\in\Gamma}$. Finally, for each $\gamma \in \Gamma$ choose $y^i_\gamma \in f_{k_i}(\mathcal{O}_\gamma)$, and define $f_i : D \to X$ by $f_i(x) = \sum_{\gamma\in\Gamma} \varphi_\gamma(x) y^i_\gamma$, when $x \in \partial D$, and extend f_i radially to all of D. Clearly each f_i is compact.

By our choice of $\{\mathcal{O}_\gamma\}_{\gamma\in\Gamma}$ it follows that for any $x \in \partial D$ there exists $\hat{x} \in \partial D$ with

$$\bigcup_{\substack{\gamma\in\Gamma \\ x\in\mathcal{O}_\gamma}} \mathcal{O}_\gamma \subseteq B(\hat{x}, \delta(\hat{x})),$$

so that by the convexity of $f_{k_i}(\hat{x})$ and our choice of $\delta(\hat{x})$

$$f_i(x) \in N_{\varepsilon/5}(f_{k_i}(\hat{x})) \quad \text{and} \quad f_{k_i}(x) \subseteq N_{\varepsilon/5}(f_{k_i}(\hat{x})).$$

Now let $\{\psi_i\}_{i=1}^m$ be a partition of unity for K subordinate to $\{V_i\}_{i=1}^m$ and define $\hat{f} : K \to G$ comp by

$$\hat{f}(k) = \sum_{i=1}^{m} \psi_i(k) f_i ,$$

for $k \in K$. Since each f_i is compact and bounded and each ψ_i is uniformly continuous, $\hat{f}$ is continuous.

Let $H : [0,1] \times K \to \hat{G}$ comp be defined by $H(t,k) = tf_k + (1-t)\hat{f}_k$, for $k \in K$, $0 \le t \le 1$. The continuity of H is clear, as is the fact that $H([0,1] \times K \times D)$ is relatively compact, and, of course, $H(t,k)$ is single-valued if $k \in K_0$. What remains to verify in order to prove the weak homotopy equivalence is that $x \notin H(t,k)(x)$ if $x \in \partial D$, $k \in K$, and $0 \le t \le 1$. So suppose the contrary; that is, there exist such x, k and t with

$$x \in tf_k(x) + (1-t)\hat{f}_k(x) = \sum_{i=1}^{m} \psi_i(k)[tf_k(x) + (1-t)f_i(x)].$$

We may choose $\hat{x} \in \partial D$ such that $\|x - \hat{x}\| < \delta(\hat{x})$, and both $f_i(x) \in N_{\varepsilon/5}(f_{k_i}(\hat{x}))$, $f_{k_i}(x) \subseteq N_{\varepsilon/5}(f_{k_i}(\hat{x}))$, for $1 \le i \le m$.

Suppose $i \in \{1,\dots,m\}$ and $\psi_i(k) \ne 0$; then $k \in V_i$. Thus

$$f_i(x) \in N_{\varepsilon/5}(f_{k_i}(\hat{x})) \subset N_{2\varepsilon/5}(f_k(\hat{x})),$$

and

$$f_k(x) \subseteq N_{\varepsilon/5}(f_{k_i}(x)) \subset N_{2\varepsilon/5}(f_{k_i}(\hat{x})) \subseteq N_{3\varepsilon/5}(f_k(\hat{x})).$$

By the convexity of $f_k(\hat{x})$ we see that

$$x \in N_{3\varepsilon/5}(f_k(\hat{x})),$$

and since $\|x - \hat{x}\| < \varepsilon/5$ we conclude that $d(\hat{x}, f_k(\hat{x})) < \varepsilon$. This contradiction proves $H : [0,1] \times K \to \hat{G}$ comp, and so proves the Theorem. ■

Our next theorem guarantees the solutions of equation (1.1) can be approximated arbitrarily closely by solutions of a similar single-valued equation.

(2.2) *THEOREM*

Let $\mathcal{O} \subseteq \mathbb{R}^n \times X$ be open, with $a, b > 0$ such that $D^n_a \times D_b \subseteq \mathcal{O}$. Suppose $F : \mathcal{O} \to K(X)$ is compact and such that if $\varepsilon > 0$ there exists a $\delta > 0$ with $d(F(\lambda,x), F(\lambda',x)) < \varepsilon$ if $|\lambda - \lambda'| < \delta$, and $(\lambda,x), (\lambda',x) \in D^n_a \times D_b$.

Then corresponding to each $\eta > 0$ we may choose a compact single-valued mapping $F_\eta : \mathcal{O} \to X$ such that

(i) for each $(\lambda,x) \in \mathcal{O}$ there exists $(\lambda',x') \in \mathcal{O}$ with $|\lambda - \lambda'| < \eta$, $\|x - x'\| < \eta$ and $F_\eta(\lambda,x) \in N_\eta(F(\lambda',x'))$;

(ii) given $\varepsilon > 0$ there exists $\delta > 0$ such that

$$\|F_\eta(\lambda,x) - F_\eta(\lambda',x)\| < \varepsilon$$

if

$$|\lambda - \lambda'| < \delta \quad \text{and} \quad (\lambda,x), (\lambda',x) \in D^n_a \times D_b.$$

<u>*Proof*</u>

Choose $\delta \in (0, \eta/2)$ such that if $\lambda, \lambda' \in D^n_a$ and $|\lambda - \lambda'| < 2\delta$ then $d(F(\lambda,x), F(\lambda',x)) < \eta/2$ when $\|x\| \leq b$. Choose $\{\lambda_i : 1 \leq i \leq m\} \subseteq D^n_a$ such that $\{B(\lambda_i,\delta) : 1 \leq i \leq m\}$ cover D^n_a, and let $\{\psi_i : 1 \leq i \leq m\}$ be a partition of unity subordinate to this cover.

By the upper semicontinuity of $F(\lambda_i,\cdot)$, $1 \leq i \leq m$, for each $x \in D_b$ we may choose $\delta(x) \in (0,\eta)$ such that if $\|x - \hat{x}\| < \delta(x)$ then

$$F(\lambda_i,\hat{x}) \subset N_{\frac{1}{2}\eta}(F(\lambda_i,x)), \quad \text{for } 1 \leq i \leq m.$$

Then $\{B(x,\delta(x)) : x \in D_b\}$ is an open cover of D_b, so we may choose a locally finite star refinement, $\{\mathcal{O}_\alpha\}_{\alpha\in\Lambda}$, which also covers D_b. Let $\{P_\alpha\}_{\alpha\in\Lambda}$ be a partition of unity subordinate to $\{\mathcal{O}_\alpha\}_{\alpha\in\Lambda}$. For each $i \in \{1,\ldots,m\}$ and $\alpha \in \Lambda$ let $y^i_\alpha \in F(\lambda_i,\mathcal{O}_\alpha)$.

We define $h : D^n_a \times D_b \to X$ by

$$h(\lambda,x) = \sum_{i=1}^{m} \psi_i(\lambda)[\sum_{\alpha\in\Lambda} P_\alpha(x)y_\alpha^i].$$

Clearly h is continuous; h is compact since $h(D_a^n \times D_b) \subseteq \overline{co}(F(D_a^n \times D_b))$; and h is uniformly continuous in λ since each ψ_i is uniformly continuous and $\sum_{\alpha\in\Lambda} P_\alpha(x)y_\alpha^i$ is bounded, independent of $x \in D_b$.

For each $(\lambda,x) \in \mathcal{O} \setminus (D_a^n \times D_b)$ we select a neighborhood of (λ,x), $W(\lambda,x)$, such that if $(\lambda',x') \in W(\lambda,x)$ then $F(\lambda',x') \subseteq N_{\frac{1}{2}\varepsilon}(F(\lambda,x))$, such that diam $(W(\lambda,x)) < \eta/2$ and also such that $W(\lambda,x) \cap (D^n_{\frac{2}{3}a} \times D_{\frac{2}{3}b}) = \phi$. For each $(\lambda,x) \in (D_a^n \times D_b) \setminus (D^n_{\frac{1}{2}a} \times D_{\frac{1}{2}b})$ we select $i \in \{1,\ldots,m\}$ and $\alpha \in \Lambda$ such that $|\lambda - \lambda_i| < \delta$ and $x \in \mathcal{O}_\alpha$, and let $W(\lambda,x) = B(\lambda_i,\delta) \times \mathcal{O}_\alpha$. Then $\{W(\lambda,x) : (\lambda,x) \in \mathcal{O}\setminus (D^n_{\frac{1}{2}a} \times D_{\frac{1}{2}b})\}$ forms an open cover of $\mathcal{O}\setminus (D^n_{\frac{1}{2}a} \times D_{\frac{1}{2}b})$. So we may select a locally finite star refinement of this cover, $\{V_\beta\}_{\beta\in B}$. Let $y_\beta \in F(V_\beta)$, for $\beta \in B$, and let $\{q_\beta\}_{\beta\in B}$ be a partition of unity subordinate to this cover.

Next choose a continuous $g : \mathcal{O} \to [0,1]$ such that $g = 1$ on $D^n_{\frac{1}{2}a} \times D_{\frac{1}{2}b}$ and $g = 0$ on $\mathcal{O} \setminus (D^n_{\frac{2}{3}a} \times D_{\frac{2}{3}b})$. Let $F_\eta : \mathcal{O} \to X$ be defined by

$$F_\eta(\lambda,x) = g(\lambda,x)\left(\sum_{i=1}^{m} \psi_i(\lambda) \sum_{\alpha\in\Lambda} P_\alpha(x)y_\alpha^i\right) + (1 - g(\lambda,x))\left(\sum_{\beta\in B} q_\beta(\lambda,x)y_\beta\right).$$

Clearly F_η is continuous and since $F_\eta(\Omega) \subseteq \overline{co}(F(N_\eta(\Omega)))$ for each $\Omega \subseteq \mathcal{O}$, it follows that F_η is compact.

Since $F_\eta = h$ on $D^n_{\frac{1}{2}a} \times D_{\frac{1}{2}b}$, F_η has the required uniform continuity properties. It remains to check the approximation properties, and to do so we consider three cases.

Let $(\lambda,x) \in D^n_{\frac{1}{2}a} \times D_{\frac{1}{2}b}$. Choose some i_0 with $|\lambda - \lambda_{i_0}| < \delta$. Choose $\hat{x}$ such that $\bigcup_{x\in\mathcal{O}_\alpha} \mathcal{O}_\alpha \subset B(\hat{x},\delta(\hat{x}))$. Then if $\psi_i(\lambda)P_\alpha(x) \neq 0$ it follows that $|\lambda_i - \lambda_{i_0}| < 2\delta$ and $\mathcal{O}_\alpha \subseteq B(x,\delta(\tilde{x}))$. Then $F(\lambda_i,x') \subseteq N_{\frac{1}{2}\eta}(F(\lambda_{i_0},x)) \subseteq N_\eta(F(\lambda_{i_0},\tilde{x})$ if $x' \in \mathcal{O}_\alpha$, so that $y_\alpha^i \in N_\eta(F(\lambda_{i_0},\tilde{x}))$. But $F(\lambda_{i_0},\tilde{x})$ is convex, implying

$$F_\eta(\lambda,x) \in N_\eta(F(\lambda_{i_0},\tilde{x})), \quad |\lambda-\lambda_{i_0}| < \eta, \quad \|x-\tilde{x}\| < \eta.$$

Since $F_\eta(\lambda,x) = \sum_{\beta\in B} q_\beta(\lambda,x)y_\beta$ when $(\lambda,x) \in \mathcal{O} \setminus (D^n_{\frac{2}{3}a} \times D_{\frac{2}{3}b})$, an argument similar to the above yields the approximation property for such (λ,x).

Finally consider $(\lambda,x) \in (D^n_{\frac{2}{3}a} \times D_{\frac{2}{3}b}) \setminus (D^n_{\frac{1}{2}a} \times D_{\frac{1}{2}b})$. Since $(\lambda,x) \notin W(\lambda',x')$ when $(\lambda',x') \notin D^n_a \times D_b$, it follows that there exist $i_0 \in \{1,\ldots,m\}$, $\alpha_0 \in \Lambda$ such that $\bigcup_{(\lambda,x)\in V_\beta} V_\beta \subseteq B(\lambda_{i_0},\delta) \times \mathcal{O}_{\alpha_0}$. Choose x^* such that

$$\bigcup_{x\in\mathcal{O}_\alpha} \mathcal{O}_\alpha \subseteq B(x^*,\delta(x^*)).$$

Then if $\psi_i(\lambda)P_\alpha(x) \neq 0$, $|\lambda_i - \lambda_{i_0}| < 2\delta$, $\mathcal{O}_\alpha \subseteq B(x^*,\delta(x^*))$. Thus $|\lambda-\lambda_{i_0}| < \eta$, $\|x-\tilde{x}\| < \eta$ and $F_\eta(\lambda,x) \in N_\eta(F(\lambda_{i_0},x^*))$. ■

In [1] the first-named author proved a global bifurcation result for fixed-points of certain single-valued compact mappings. A careful reading of [1] reveals that the arguments presented there suffice to prove the following.

(2.3) *THEOREM*

Let X be a Banach space with k a positive integer. Let $\mathcal{O} \subseteq \mathbb{R}^k \times X$ be open, with $f:\mathcal{O} \to X$ compact and there exist $a,b > 0$ with $D^n_a \times D_b \subseteq \mathcal{O}$ and given $\varepsilon > 0$ there exists a $\delta > 0$ such that if $|\lambda-\lambda'| < \varepsilon$ and $(\lambda,x), (\lambda',x) \in D^n_a \times D^n_b$ then $\|f(\lambda,x) - f(\lambda',x)\| < \varepsilon$. Suppose there exists $0 < \alpha < a$, $0 \leq \beta < b$, such that $x \neq f(\lambda,x)$ if $|\lambda| = \alpha$ and $\beta \leq \|x\| \leq b$. Moreover suppose that the element $\Pi_{n-1}(G\ \text{comp})$ induced by $f\big|_{D^n_\alpha \times D_\beta}$ is nontrivial. Then letting

$$\Sigma = \{(\lambda,x) : x = f(\lambda,x)\} \cup \{\infty\},$$

with the one-point compactification topology, there exists a connected subset Σ^0 of $\Sigma \setminus \{(\lambda,x) : |\lambda| = \alpha,\ \beta \leq \|x\| \leq b\}$ which intersects both

$$\{(\lambda,x) : |\lambda| < \alpha,\ \|x\| \leq \beta\} \text{ and } \{(\lambda,x) : |\lambda| > \alpha,\ \|x\| \leq \beta\}.$$

(2.1) *REMARK*

It is not difficult to see that the definition of global bifurcation given in Definition (1.1) is implied by the global conclusion of Theorem (2.3) provided that in addition f satisfies the hypothesis that for every compact $K \subseteq D_b \setminus \{0\}$ there exists $\delta = \delta(K) > 0$ such that $f(\lambda,x) \neq x$ when $\lambda \in K$, $0 < \|x\| \leq \delta$. Indeed, let $\tilde{\Sigma}_0$ be the space obtained by collapsing

$$\{(\lambda,x) : |\lambda| < a, \quad x = 0\} \cap \Sigma_0 \quad \text{and} \quad \{(\lambda,x) : |\lambda| > \varepsilon_2, \quad x = 0\}$$

to points p and q, respectively, in Σ_0. Then $\tilde{\Sigma}_0$ is a connected set which contains p and q. Thus, by Lemma (5.2.4) of Kuratowski [5] there exists a connected Ω in $\hat{\Sigma}_0 \setminus \{p,q\}$ whose closure intersects both p and q. It is clear that $(0,0) \in \overline{\Omega}$, and $q \in \overline{\Omega}$ means either Ω is unbounded, $(\lambda^*,0) \in \overline{\Omega}$ for some $|\lambda^*| \geq \varepsilon_2$, or $\overline{\Omega} \cap \partial\mathcal{O} \neq \phi$. Recall that two subsets A and B of a topological space W are said to be separated in W if there exists $V \subseteq W$, V both open and closed, with $A \subseteq V$, $B \cap V = \phi$. The following result is proven in [4].

(2.1) *LEMMA*

Suppose $\{F_n\}$ is a family of closed subsets of a normal topological space, with $T = \cap F_n$. Assume that for each neighborhood U of T, there exists n such that $F_n \subseteq U$. Suppose A and B are closed subsets of T which cannot be separated in any F_n. Then A and B cannot be separated in T.

<u>*Proof of Theorem* (1.1)</u>

Fix $\varepsilon > 0$ such that $B^n_\varepsilon \leq V$ and let $\delta > 0$ be such that $x \notin F(\lambda,x)$ if $|\lambda| = \varepsilon$ and $0 < \|x\| \leq \delta$. Let $D = \{x : \|x\| \leq \delta\}$ and let $S^{n-1} = \{\lambda : |\lambda| = \varepsilon\}$. Now let $\Sigma = \{(\lambda, x) \in \mathcal{O}, \quad x \in F(\lambda,x)\} \cup \{\infty\}$, with the usual one-point compactification topology. We will show that there exists a connected subset Σ^0 of $\Sigma\setminus(S^{n-1} \times D)$ which intersects $D_0 = \{(\lambda,0) : |\lambda| < \varepsilon\}$ and $D_1 = \{(\lambda,0) : |\lambda| > \varepsilon\}$. Hence by Remark (2.1) we get global bifurcation in the sense of Definition (1.1)

Let $k \in N$ be such that $\frac{1}{k} \leq \delta$. By compactness of F we may choose $\eta_k > 0$ and $\beta_k > 0$ such that $d(x,F(\lambda,x)) \geq \beta_k$ if

$$d((\lambda,x), \quad \{(\lambda,x) : |\lambda| = \varepsilon, \quad k^{-1} \leq \|x\| \leq \delta\}) \leq \eta_k.$$

Let $0 < \gamma_k < \min\{k^{-1}, \eta_k, \frac{1}{2}\beta_k\}$.

By Theorem (2.2) we may choose $F_k : \mathcal{O} \to X$ compact, uniformly continuous in λ, in a neighborhood of $(0,0)$, and such that if $(\lambda,x) \in \mathcal{O}$ with $|\lambda| \le k$, $\|x\| \le k$, then there exists $(\lambda',x') \in \mathcal{O}$ such that $d((\lambda,x), (\lambda',x')) < \gamma_k$ and $d(F_k(\lambda,x), F(\lambda',x')) < \gamma_k$.

By our choice of γ_k it is easy to check that

$$x \notin t\, F_k(x) + (1-t)\, F(x)$$

when $0 \le t \le 1$, $|\lambda| = \varepsilon$ and $k^{-1} \le \|x\| \le \delta$. So we may apply Theorem (2.1) to conclude that the element which F_k induces in G comp is nontrivial. Thus Theorem (2.3) may be invoked for the fixed-points of F_k; that is letting

$$\Sigma_k = \{(\lambda,x) : x = F_k(\lambda,x)\} \cup \{\infty\},$$

with the one point compactification topology, there is a connected subset Σ_k^0 of $\Sigma_k \setminus (S^{n-1} \times D)$ which intersects both

$$\{(\lambda,x) : |\lambda| < \varepsilon, \ \|x\| \le k^{-1}\} \text{ and } \{(\lambda,x) : |\lambda| > \varepsilon, \ \|x\| \le k^{-1}\}.$$

We will apply the Lemma (2.1). We take as ambient space $(\mathbb{R}^n \times X)^+ \setminus (S^{n-1} \times D)$, and let

$$T_k = \big((\Sigma' \cup \Sigma_k) \setminus (S^{n-1} \times D)\big) \cup \{(\lambda,x) \in \mathcal{O} : |\lambda| \ne \varepsilon, \ \|x\| \le k^{-1}\}.$$

We claim that D_0 and D_1 cannot be separated in T_k. Indeed if this were not so, one could select W, an open and closed subset of T_k, with $D_0 \subseteq W$ and $D_1 \cap W = \phi$. Since both $\{(\lambda,x) : |\lambda| < \varepsilon, \ \|x\| \le k^{-1}\}$ and $\{(\lambda,x) : |\lambda| > \varepsilon, \ \|x\| \le k^{-1}\}$ are connected but not open in T_k it follows that

$$\{(\lambda,x) : |\lambda| < \varepsilon, \ \|x\| \le k^{-1}\} \subseteq W \text{ and } W \cap \{(\lambda,x) : |\lambda| > \varepsilon, \ \|x\| \le k^{-1}\} = \phi.$$

Thus $W \cap \Sigma_k^0$ is a proper, nonempty subset of Σ_k^0, which is both open and closed. This contradicts the connectedness of Σ_k^0, and so our claim is justified.

To conclude the proof we must show

$$\cap T_n = (\Sigma \setminus (S^{n-1} \times D)) \cup (D_0 \cup D_1) \equiv T,$$

and that if U is a neighborhood of T, then there exists some n with $T_n \subseteq U$.

So suppose $(\lambda,x) \in \cap T_n$, and $x \neq 0$. Then for sufficiently large n, $x = F_n(\lambda,x)$ and there exists (λ_n,x_n) such that $|\lambda - \lambda_n| + \|x - x_n\| < n^{-1}$ and $d(x,F(\lambda_n,x_n)) < n^{-1}$. By the upper-semicontinuity of F, together with the compactness of $F(\lambda,x)$ it follows that $x \in F(\lambda,x)$. So $(\lambda,x) \in T$.

Now let U be a neighborhood of T. Since $\infty \in T$ we may choose an n_0 such that it will suffice to find n' with the property that if $(\lambda,x) \in T_{n'}$ and $|\lambda| \leq n_0$, $\|x\| \leq n_0$ then $(\lambda,x) \in U$. So suppose there exists no such n'. Then we may choose $\{(\lambda_n,x_n)\}$ with $(\lambda_n,x_n) \in T_n$, with $|\lambda_n| \leq n_0$, $\|x\| \leq n_0$, and $(\lambda_n,x_n) \notin U$, for each n. It follows that $x_n = F_n(\lambda_n,x_n)$, for each n. Hence, if n is sufficiently large we may choose (λ_n',x_n') such that $\|x_n - x_n'\| \leq n^{-1}$, $|\lambda_n-\lambda_n'| < n^{-1}$ and $d(x_n,F(\lambda_n',x_n')) < n^{-1}$. But $\{(\lambda_n',x_n') : n \geq n_0\}$ is bounded, and thus $\bigcup_{n\geq n_0} F(\lambda_n',x_n')$ is relatively compact. Thus a subsequence of (λ_n,x_n) converges to $(\lambda^*,x^*) \in \mathcal{O}$ or $(\lambda^*,x^*) \to \infty$. The second possibility is precluded by the fact that $\infty \in T$ and the upper-semicontinuity of F implies $x^* = F(\lambda^*,x^*)$, if the first possibility occurs, and again we have a contradiction. ∎

§3.

In this section we will prove Theorem (1.1) for condensing F. Since the generalization from compact F to condensing F, when F is multivalued, parallels the similar single-valued extension which we carried out in [3] we will omit some of the details.

Recall that if $T : D \subseteq \mathbb{R}^n \times X \to X$ is condensing and V is closed, bounded and convex then one has the following reduction procedure (see [6] and [9]): let $K_0 = \overline{\mathrm{co}}F(D)$. If η is an ordinal such that K_β has been defined for $\beta < \eta$, let $K_\eta = \overline{\mathrm{co}}F(D \cap (\mathbb{R}^n \times K_{\eta-1}))$ if η has a predecessor, while let $K_\eta = \bigcap_{\beta<\eta} K_\beta$ if η has no predecessor. Then $K_{\beta+1} = \overline{\mathrm{co}}F(D \cap (\mathbb{R}^n \times K_\beta) \subseteq K_\beta$, for each ordinal β. So there exists some β with $K_\alpha = K_\beta$ if $\alpha \geq \beta$. Denoting this stationary set by K_∞, we see that $\overline{\mathrm{co}}F(D \cap (\mathbb{R}^n \times K_\infty) = K_\infty$ and that K_∞ is compact, since T is condensing. Using the retraction theorem of Dugundji to choose a retraction p of X onto $K_\infty \cap D$ an argument similar to that used in [7] shows that if $(\lambda,x) \in D$, $0 \leq t \leq 1$, and $x \in t\,F(\lambda,x) + (1-t)F(\lambda,p(x))$, then $x \in F(\lambda,x)$. This is basic idea behind carrying through an argument similar to the single-valued argument of [3] to prove the following.

(3.1) *THEOREM*

$\hat{G}$ comp $\subseteq \hat{G}$ cond is a weak homotopy equivalence.

The above construction also shows that the element Π_{n-1}(G comp) induced by $(\lambda,x) \to F(\lambda,p(x))$ is the same as that induced in $\Pi_{n-1}(\hat{G}$ cond) by F.

The proof of Theorem (1.1) now proceeds as follows. Choose $\{V_j\}_{j=1}^{\infty}$ to be a family of increasing closed, bounded convex subset of $\mathcal{O}$ each of which contains $D_a^n \times D_b$ and $\cup V_j = \mathcal{O}$. Apply the above construction to F on V_j. Now use the compact version of Theorem (1.1) for the map $(\lambda,x) \to F(\lambda,p_k(x))$, to obtain global bifurcation for the fixed points of $F(\lambda,p(x))$ on V_j, which coincides with the fixed points of F on V_j. Finally, one uses Lemma (2.1) to obtain the global bifurcation result for the fixed points of F on all of $\mathcal{O}$.

(3.1) *REMARK*

In case $n = 1$ the condition that $\gamma \neq 0$ means simply that the topological degree of $(I - T(\lambda,\cdot))|_{D_b}$ changes as λ crosses 0; the degree theory for multi-valued condensing mapping has been developed in [7]. It would be very interesting to give specific criteria which would allow one to check that the degree changes. More generally, can one, when $n \geq 1$, give reasonable criteria to guarantee $\gamma \neq 0$?

Appendix

The Compact Single-Valued Case

The bifurcation result for compact single-valued operators, analogous to the one established in the present paper, was proved in [1]. Generalizations to other classes of operators use the compact result as a starting point. However the result in [1] is stated in a restricted, technical form that although acceptable for applications, is awkward to use for extending the results to more general classes of operators. Moreover, computational methods were mixed in with existence results. A more conceptual approach to the general type of result was formulated in [3]; however that article presumes [1]. The proof in [1] is more general that the stated result, and several papers referred to the proof rather than the result.

The purpose here is to explicitly state the bifurcation result of [1] in a general and abstract enough form that it can be quoted directly.

The general setup

Let B be a Banach space, R^n n-dimensional Euclidean space and consider

$$F(\lambda,x) : R^n \times B \to B.$$

We are interested in fixed points of F; that is (λ,x) with $x = F(\lambda,x)$. If the domain of definition $\mathcal{O}$ of F is not all $R^n \times B$, results can still be proved. For example if $\mathcal{O}$ is convex, an R^n- and zero-preserving homeomorphism $\varphi : \mathcal{O} \to R^n \times B$ can be constructed to reduce the problem to the case $\mathcal{O} = R^n \times B$. For simplicity, we suppose F is defined on all of $R^n \times B$.

Let S^{n-1} be a small sphere around the origin in R^n and D a small disk in B around the origin. Let S be the boundary of D. Suppose the fixed points of F are bounded away from $S^{n-1} \times S$. Then we can choose a small annulus A^n in R^n which is S^{n-1} "thickened up". That is, suppose

$$S^{n-1} = \{\lambda \in R^n : |\lambda| = \varepsilon\},$$

$$A^n = \{\lambda \in R^n : \varepsilon - \eta < |\lambda| < \varepsilon + \eta\}, \quad \eta < \varepsilon.$$

Let S_0^{n-1} (respectively S_1^{n-1}) be the interior (exterior) boundary of A^n in R^n that is the set of λ with $|\lambda| = \varepsilon - \eta$ ($|\lambda| = \varepsilon + \eta$). Let $(R^n \times B)^+$ be $R^n \times B$ with a point ∞ adjoined. A neighborhood basis for ∞ consists of the complements of bounded sets.

Let $\mathrm{Fix}(F)$ be the set of fixed points of F. Let

$$\Sigma = (\mathrm{Fix}(F) \cup \{\infty\} \cup S_0^{n-1} \times D) \cup (S_1^{n-1} \times D))\backslash((A^n \times D) \cap \mathrm{Fix}(F)).$$

In Σ there are two distinguished subsets

$$S_0^{n-1} \times D, \quad S_1^{n-1} \times D.$$

DEFINITION

Global bifurcation of $\mathrm{Fix}(F)$ is said to occur if $S_0^{n-1} \times D$ and $S_1^{n-1} \times D$ cannot be separated in Σ.

This means there is no disjoint division of Σ into two open sets U, V with $S_0^{n-1} \times D \subset U$, $S_1^{n-1} \times D \subset V$. The reader is referred to [2] for a general discussion of these topological concepts.

This definition is well-suited to proving results by approximating F. In applications, it is usually known that $F(\lambda,0) = 0$ (at least for λ near 0) and that bifurcation from this zero solution cannot occur for small non-zero λ. With this assumption, the above subset formulation of global bifurcation immediately implies the standard version.

Heuristically, the definition means that the set of fixed point stretches from the inner wall of the cylinder $A^n \times D$ around to the outer wall without going through the cylinder itself (but possibly passing through ∞, which gives an unbounded branch).

APPROXIMATIONS

Let $F_N : \mathcal{O}_N \to B$ be a sequence of operators defined on some increasing sequence of open sets $\mathcal{O}_N \subset R^n \times B$ whose union exhaust all of $R^n \times B$. Define $\mathrm{Fix}(F_N)$

and

$$\Sigma_N = (\mathrm{Fix}(F_N) \cup \{\infty\} \cup (\text{complement of } \mathcal{O}_n) \cup (S_0^{n-1} \times D)) \cup (S_1^{n-1} \times D) \setminus ((A^n \times D) \cap \mathrm{Fix}(F_N)).$$

Suppose all $\mathrm{Fix}(F_N)$ are bounded away from $S^{n-1} \times X$.

PROPOSITION

Suppose for any bounded set C, there exists N such that $C \subset \mathcal{O}_N$, and $\mathrm{Fix}(F_N) \cap C$ is contained in the ε-neighborhood of $\mathrm{Fix}(F) \cap C$. Moreover, suppose $S_0^{n-1} \times D$, $S_1^{n-1} \times D$ cannot be separated in Σ_N for any N. Then $S_0^{n-1} \times D$, $S_1^{n-1} \times D$ cannot be separated in Σ.

This result is immediate from the results of [2].

THE COMPACT CASE

Suppose F is single-valued and compact, that is takes bounded sets to sets with compact closure. Suppose F is uniformly continuous in $\lambda \in S^{n-1}$ for $x \in D$; that is given $\varepsilon > 0$, there exists $\delta > 0$ such that

$$|F(\lambda,x) - F(\lambda',x)| < \varepsilon$$

if $|\lambda - \lambda'| < \delta$ $\lambda, \lambda' \in S^{n-1}$, $x \in D$. Recall from [3] that G comp is the space of compact operators $f : D \to B$ with no fixed points on S with the topology of uniform convergence. Then F induces a map

$$\Gamma_F : S^{n-1} \to G\,\text{comp}$$

and hence an element in the homotopy group

$$\gamma_F \in \pi_{n-1}(G\,\text{comp}).$$

The structure of these groups was determined in [3]. Suppose F satisfies these conditions as well as those of the preceeding section.

THEOREM

If $\gamma_F \neq 0$, then global bifurcation of the fixed point set of F occurs.

Sketch of proof

Suppose $x - F(\lambda,x)$ is bounded away from 0 by ε for $(\lambda,x) \in A^n \times D$. For each integer $N > \varepsilon^{-1}$, approximate F by a finite dimensional mapping $\bar{F}_N$ on $\mathcal{O}_N = \{(\lambda,x) : |\lambda,x| \leq N\}$ to within N^{-1} of F. This reduces the problem to a finite-dimensional one, except that we must take more care of the approximation on $S^{n-1} \times D$. Let $G\,\mathrm{fin}$ be the subspace of $G\,\mathrm{comp}$ consisting of operators with finite-dimensional image. In Proposition (3.1) of [3], it was shown

$$G\,\mathrm{fin} \to G\,\mathrm{comp}$$

is a weak homotopy equivalence. This was done by an approximation argument; we have to make sure our approximation agrees on $S^{n-1} \times D$ with the one prescribed by Proposition (3.1) of [3]. So we make such an approximation $\hat{F}_N$ of F on $S^{n-1} \times D$ to within N^{-1}. It extends to a N^{-1}-approximation $\hat{F}_N$ on some neighborhood U of $S^{n-1} \times D$. Let $\varphi : R^n \times B \to [0,1]$ be a continuous function which is 1 on $S^{n-1} \times D$ and 0 out of U. Define

$$F_N(\lambda,x) = \begin{cases} \varphi(\lambda,x)\hat{F}_N(\lambda,x) + (1 - \varphi(\lambda,x))\bar{F}_N(\lambda,x), & (\lambda,x) \in U \\ \bar{F}_N(\lambda,x) & (\lambda,x) \notin U. \end{cases}$$

Then $F_N(\lambda,x)$ is a suitable finite-dimensional approximation. If the result is true for each F_N, it is true for F.

Thus the problem is reduced to a finite-dimensional one, and the element

$$\gamma_{F_N} \in \pi_{n-1}(G\,\mathrm{fin})$$

is non-zero. Here the proof in [1] can be used with a couple of modifications. Because we are not using the linearization of F (it need not even exist) all discussion of the J-homomorphism can be dropped. Because $\mathrm{Fix}(F_N)$ is less restricted than in [1], the meaning of the notation has to be altered slightly as follows:

[1]	present
N	closure of $A^n \times D$
V	$\mathrm{Fix}(F_N) \cup \{\infty\} \cup$ (complement $\mathcal{O}_N$) $\cup (S_0^{n-1} \times D) \cup (S_1^{n-1} \times D)$
$\hat{S}^{n-1}$	$A^n \times D$
$\hat{D}_0$	$S_0^{n-1} \times D$
$\hat{D}_1$	$S_1^{n-1} \times D$
$V \setminus \hat{S}^{n-1}$	Σ_N
$N \setminus \hat{S}^{n-1}$	$(S_0^{n-1} \times D) \cup (S_1^{n-1} \times D)$

The definition of $\theta : S^{m+n} \setminus (A^n \times D) \to [0,1]$ should be altered to the following:

$$\theta(\lambda,x) = \min(\eta,\ \mathrm{dist}((\lambda,x),\ S_0^{n-1} \times D))$$

(recall η enters the definition of A^n).

We need only Lemma (3.2) and the general discussion immediately following. Finally, we need the formula analogous to, but easier then, Lemma (3.3), which reads (in the notation of [1]):

$$F_*(\Sigma^{n-1} \times \Sigma^{m-1}) = \gamma_F\, \Sigma^{m-1}.$$

With these modifications, the arguments of [1] complete the present proof.

REFERENCES

[1] ALEXANDER, J.C.: Bifurcation of zeros of parametrized functions, J. Funct. Anal. 29 (1978), 37-53.

[2] ALEXANDER, J.C.: A primer on connectivity, to be published.

[3] ALEXANDER, J.C. and FITZPATRICK, P.M.: The homotopy of certain spaces of nonlinear operators, and its relation to global bifurcation of the fixed points of parametrized condensing operators, J. Funct. Anal, 34 (1979), 87-106.

[4] ALEXANDER, J.C. and FITZPATRICK, P.M.: Galerkin approximations in several parameter bifurcation problems, Math. Proc. Comb. Phil. Soc., 87 (1980), 489-500.

[5] KURATOWSKI, K.: Topology, vol II; New York, Academic Press, (1968).

[6] NUSSBAUM, R.D.: The fixed point index for local condensing mappings, Ann. Mat. Pura Appl., 89 (1971), 217-258.

[7] PETRYSHYN, W.V. and FITZPATRICK, P.M.: A degree theory, fixed point theorems and mapping theorems for multivalued noncompact mapping, Trans. Am. Math. Soc. 194 (1974), 1-25.

[8] RABINOWITZ, P.H.: Some global results for nonlinear eigenvalue problems, J. Funct. Anal. 7 (1971), 487-513.

[9] SADOVSKII, B.N.: Ultimately compact and condensing mappings, Uspehi Mat. Nauk, 27 (1972), 81-146.

[10] STUART, C.A.: Some bifurcation theory for k-set contractions, Proc. London Math. Soc., 27 (1973), 531-550.

TOPOLOGICAL CONDITIONS FOR THE CONTINUATION OF FIXED POINTS

By

KATHLEEN T. ALLIGOOD

Department of Mathematics
College of Charleston
Charleston, S.C. 29401

0. Introduction

The Lefschetz Fixed Point Theorem is one of many theorems in fixed point theory which prove the existence of solutions but provide no information as to how to find them. One method currently employed, the Homotopy Continuation Method (see [1] and [2]), involves numerically following paths of solutions through homotopies to find previously unknown solutions. Suppose, for example, we are interested in locating fixed point of a function f on the disk. The method can be described as follows: start with a function f_0 with a known fixed point (such as a constant function), homotop f_0 to f, and then follow a path of fixed points from the known solution through the homotopy until a fixed point of f is reached. There are examples of problems (see [1]) whose solutions cannot be found by this method due to the fact that no connecting path exists in the solution space of the homotopy.

In this paper we investigate topological conditions which will ensure the existence of such connecting components for appropriate homotopies. More formally, consider a manifold M with boundary (possibly empty), a manifold N, and a map $f : M \to N$. Let P be a closed submanifold of N without boundary, and let $F : M \times I \to N$ be any homotopy of f. When do we know that $F^{-1}(P)$ will connect $M \times \{0\}$ and $M \times \{1\}$? Alexander and Yorke [2] have provided the following general algebraic condition on f:

Given $f : (M, \partial M) \to (N, N - P)$ and $F : (M \times I, \partial M \times I) \to (N, N - P)$, a homotopy of f. If $f_* : h_*(M, \partial M) \to h_*(N, N - P)$ is non-zero for some (generalized) homology h (or the corresponding dual condition for cohomology), then $F^{-1}(P)$ contains a component which connects $M \times \{0\}$ and $M \times \{1\}$.

Solutions which possess this property will be called *C-essential*.

In Section 1, we examine fixed points of multi-valued maps and prove that solutions of the Kakutani Fixed Point Theorem for multi-valued maps on the disk are C-essential. In Section 2, we give an example of a bundle map on $S^1 \times S^1$ whose fixed points are detected by the Dold (transfer) index and prove that a non-zero index implies such solutions are C-essential. In Section 3, we use fiber bundle techniques to study equivariant problems and, in particular, to show that $\mathbb{Z}_p$-coincidences of maps $f : S^n \to \mathbb{R}^m$ for certain m, n, and p are C-essential. (Given a free $\mathbb{Z}_p$-action μ on S^n, n odd, a $\mathbb{Z}_p$-coincide is a point $x \in S^n$ such that

$$f(x) = f(\mu x) = f(\mu^2 x) = \ldots = f(\mu^{p-1} x).)$$

I would like to express my sincere thanks to James C. Alexander for his guidance in this work.

1. Multi-Valued Maps and the Kakutani Fixed Point Theorem.

We begin this section with an intersection class condition for continuation. A classical example of the use of intersection classes is the Lefschetz Fixed Point Theorem. We shall combine this result with the theory of multi-valued maps to prove that solutions of the Kakutani Fixed Point Theorem are C-essential.

(The following definition is due to Dold [7].)

Let M^n be an oriented manifolds; (V,S) and (W,T), arbitrary open pairs in M; $o \in H_n(M)$, the fundamental class of M; $\xi \in H_i(V,S)$ and $\eta \in H_j(W,T)$; $x \in \check{H}^{n-i}(M-S, M-V)$ and $y \in \check{H}^{n-j}(M-T, M-W)$, the Poincaré duals of ξ and η, (i.e., $\xi = x \frown o$ and $\eta = y \frown o$). Then the *intersection class of* ξ *and* η is

$$\xi \cdot \eta = (x \smile y) \frown o.$$

$\xi \cdot \eta \in H_{i+j-n}(V \cap W, (S \cap W) \cup (V \cap T))$.

(Note: if $M-S$ or $M-T$ is not compact, then Čech chohomology with compact supports must be used.)

(1.1) *THEOREM*

Let $f : M^m \to N^n$; P, a closed subset of N and $i : P \to N$; $\sigma \in H_j(M)$ and

$\tau \in H_k(P)$ for some k such that $j+k \geq n$. If $f_*(\sigma)\cdot i_*(\tau) \neq 0$ (in N), then $f^{-1}(P)$ is C-essential.

Proof

Let $j : N \to (N, N-P)$ be the natural map, and let $\bar{f} = j \circ f$. We will verify that $H_j(M) \xrightarrow{\bar{f}_*} H_j(N, N-P)$ is non-zero. $f_*(\sigma) \neq 0$ in $H_j(N)$ and $i_*(\tau) \neq 0$ in $H_k(N)$, since their intersection product is non-zero. Look at a portion of the homology sequence of $(N, N-P)$:

$$\cdots \longrightarrow H_j(N-P) \xrightarrow{i_*} H_j(N) \xrightarrow{j_*} H_j(N, N-P) \longrightarrow \cdots$$

Suppose that $\bar{f}_*(\sigma) = j_* f_*(\sigma) = 0$. Then there is some $v \in H_j(N-P)$ such that $i_*(v) = f_*(\sigma)$. $N-P$ is open. v is represented by some closed set $V \subset N-P$. Let U be an open set such that $V \subset U \subset \bar{U} \subset N-P$. Then $v \neq 0$ in $H_j(U)$, since

$$H_j(U) \xrightarrow{k_*} H_j(N-P)$$

and $k_*(v) \neq 0$. Let $\mathcal{W}$ be the open set $N \setminus \bar{U}$. Then $\tau \neq 0$ in $H_k(\mathcal{W})$, since $P \subset \mathcal{W} \subset N$. Now $v \cdot t$ lies in $H_*(U \cap \mathcal{W}) = 0$, since $U \cap \mathcal{W} = \emptyset$. But, for $i : U \cup \mathcal{W} \to N$ the injection,

$$0 = v \cdot \tau \quad \Rightarrow 0 \;=\; i_*(v \cdot \tau) \;=\; i_*(v) \cdot i_*(\tau) \;=\; f_*(\sigma) \cdot i_*(\tau) \;\neq 0.$$

Contradiction. Thus, $H_j(M) \xrightarrow{\bar{f}_*} H_j(N, N-P)$ is non-zero. ∎

The treatment of multi-valued maps we use in this section follows that of Montgomery and Eilenberg [8] and Górniewicz [9]. Greek letters will be used for multi-valued maps; Roman, for single-valued maps.

A map $\varphi : M \to N$ is called *admissible* if for some $p, q : M \xleftarrow{p} K \xrightarrow{q} N$,

(i) $q \circ p^{-1}(x) \subseteq \varphi(x)$, $\quad \forall\, x \in M$, and

(ii) p is a Vietoris map; i.e., p is proper and $p^{-1}(x)$ is acyclic, $\forall\, x \in M$.

(For details of the definition of continuity of φ, see [9].)

Consider, for example, an acyclic map φ, (i.e., $\varphi(x)$ is connected and $H_i(\varphi(x))$ is trivial for $i > 0$, $\forall x \in M$). Let p and q be the projections $M \xleftarrow{p} \Gamma_\varphi \xrightarrow{q} N$, where $\Gamma_\varphi = \{(x_1,x_2) \in M\times N : x_2 \in \varphi(x_1)\}$ is the graph of φ. Then $\varphi(x) = q\circ p^{-1}(x)$. We know that when p is a Vietoris map, P_* is an isomorphism (see [12]). Thus we can define $\varphi_* = q_* \circ p_*^{-1}$. In the following, we deal only with maps φ in which $\varphi = q\circ p^{-1}$, although the proofs can be easily modified for the more general case of all admissible maps. Notice that if $\varphi = q\circ p^{-1}$, where p is Vietoris, then φ is acyclic.

For continuation results, we shall allow only homotopies which can be similarly decomposed; i.e., we consider homotopies $\Phi : M\times I \to N$ between φ_0 and $\varphi_1 : M\to N$ such that

(i) Φ is acyclic, and

(ii) $\Phi(x,0) \subset \varphi_0(x)$ and $\Phi(x,1) \subset \varphi_1(x)$, $\forall\, x \in M$.

(1.2) *THEOREM (Continuation Condition for Multi-Valued Maps).*

Given an acyclic map $\varphi : M \to N$, and P, a closed set in N, let $\Phi : M\times I \to N$ be a homotopy of Φ, such that $\Phi^{-1}(P) \cap \partial(M\times I) = \phi$. If $\varphi_* : H_*(M,\partial M) \to H_*(N,N-P)$ is non-zero (or the corresponding cohomological condition), then some component of $\Phi^{-1}(P)$ connects $M\times\{0\}$ and $M\times\{1\}$.

<u>Proof</u> *(Adapted from the Alexander-Yorke proof for single-valued maps [2]).*

Suppose $\Phi^{-1}(P)$ is the disjoint union of two closed subsets Q_0 and Q_1 in $M\times I$, where $Q_0 \cap (M\times\{0\}) = \phi$ and $Q_1 \cap (M\times\{1\}) = \phi$. Let $j_i : M \to M\times I$ be the inclusion of M as $M\times\{i\}$, for $i = 0,1$, and let $\overline{\varphi} = \Phi\circ j_0$. Thus

$$\overline{\varphi}(x) \subseteq \varphi(x), \quad \forall\, x \in M.$$

Now $j_{i*} : H_*(M,\partial M) \to H_*(M\times I, M\times I - Q_i)$ is the zero map since $j_i(M) \subset M\times I - Q_i$. Also, $j_{0*} : H_*(M,\partial M) \to H_*(M\times I, M\times I - Q_1)$ is the zero map since

$$j_0, j_1 : (M,\partial M) \to (M\times I, M\times I - Q_1)$$

are homotopic.

$$H_*(M\times I, M\times I - \Phi^{-1}(P)) \xrightarrow{(i_{1*},\, i_{2*})} H_*(M\times I, M\times I - Q_0) \oplus H_*(M\times I, M\times I - Q_1)$$

is an isomorphism; therefore, $j_{0*} : H_*(M,\partial M) \to H_*(M\times I, M\times I - \Phi^{-1}(P))$ is zero.

Hence, $\bar\varphi_* = \bar\varphi_* j_{0*} = 0$.

Now $\varphi = q \circ p^{-1}$ and $\bar\varphi = \bar q \circ \bar p^{-1}$, where $M \xleftarrow{p} \Gamma_\varphi \xrightarrow{q} N$ and $M \xleftarrow{\bar p} \Gamma_{\bar\varphi} \xrightarrow{\bar q} N$. Since p_* and $\bar p_*$ are isomorphisms, $i_* : H_*(\Gamma_{\bar\varphi})\ \ H_*(\Gamma_\varphi)$ is an isomorphism, where $i : \Gamma_{\bar\varphi} \to \Gamma_\varphi$ is the inclusion. Since $\bar q_* = q_* \circ i$, $\bar q_* = 0$ if and only if $q_* = 0$. Also, $\bar p_* = 0$ if and only if $p_* = 0$. Therefore, if $\bar\varphi_* = 0$, then either $H_*(M,\partial M) = 0$ (if $\bar p_* = 0$) or $\bar q_* = 0$. Thus, $\varphi_* = q_* \circ p_*^{-1} = 0$. ■

Now we may use homological (or cohomological) results based on the Alexander-Yorke condition (such as Theorem (1.1)) for acyclic maps. For the following arguments, assume that homology coefficients lie in some ring R. (Use $\mathbb{Z}_2$ coefficients if M is not orientable.)

(1.3) *COROLLARY (Kakutani Fixed Point Theorem).*

Given $\varphi : D^n \to D^n$ such that $\varphi(x)$ is convex, $\forall\, x \in D^n$. Then φ has a C-essential fixed point. (A fixed point of φ is a point $x \in D^n$ such that $x \in \varphi(x)$.)

Proof

Let $d : M \to M \times M$ be the diagonal map, $d(m) = (m,m)$; $0 \in H_n(M\,;R)$, the fundamental class of M; and $\delta = d_*(0)$, the diagonal class. If

$$\Delta = \{(m_1,m_2) \in M \times M : m_1 = m_2\}$$

is the diagonal, then $\delta \in H_n(\Delta)$ and $(\mathrm{id} \times \varphi)^{-1}(\Delta)$ is the set of fixed points of φ.

$\varphi(x)$ is acyclic, hence φ_* and $(\mathrm{id} \times \varphi)_*$ are well-defined. Let γ_φ denote $(\mathrm{id} \times \varphi)_*(\delta)$, the class of the graph of φ. By a standard argument (see, for example, Roitberg [11]), we have

$$(\mathrm{id} \times \varphi)_*(\delta) \cdot i_*(\delta) = \gamma_\varphi \cdot i_*(\delta) = \lambda_\varphi .$$

Here $\lambda_\varphi = \sum_{i=0}^{n} (-1)^i$ trace $(q_* p_*^{-1})_i$, where $\varphi = q \circ p^{-1}$, and p is Vietoris. Since $\lambda_\varphi \neq 0$, φ has a C-essential fixed point. ■

Given two maps $p,q : N \to M$, the theory of multi-valued maps can also be used to prove the existence of coincidence points of p and q, (i.e., points $x \in N$ such that $p(x) = q(x)$). Thus we obtain similarly a continuation result for coincidences:

(1.4) *COROLLARY*

Given $M \xleftarrow{p} N \xrightarrow{q} M$, where p is Vietoris, let $\varphi = q \circ p^{-1}$. If λ_φ is non-zero, then p and q have a C-essential coincidence.

Proof

A fixed point of φ is a coincidence of p and q. ∎

2. Fiber-Preserving Maps and the Dold Index

As an illustration of continuation of fixed points for fiber-preserving maps, we consider the function $\eta : S^1 \times S^1 \to S^1 \times S^1$, given by $\eta(g,b) = (gb,b)$. Notice that although $\mathrm{Fix}(\eta)$ is $S^1 \times \{1\}$, η can be homotoped to a fixed-point-free function provided we do not require that maps be fiber-preserving. If we do impose such a restriction, however, we find that any map homotopic to η will have fixed points. In fact, the fixed point set (generically) will be the continuation of the circle $\mathrm{Fix}(\eta)$.

For example, let $\eta_t : S^1 \times S^1 \to S^1 \times S^1$ be the homotopy defined by

$$\eta_t(g,b) = (gbe^{-it\theta},b),$$

for $t \in [0,1]$ and some $\theta \in (0,2\pi)$. For each t, $\mathrm{Fix}(\eta_t)$ is $S^1 \times \{e^{it\theta}\}$. Thus the entire fixed point set of the homotopy is homeomorphic to $S^1 \times I$. This essential fixed point set of the bundle map η is detected by a non-zero Dold (transfer) index $I(\eta)$. [For the cohomology theory π^*_{st} (stable cohomotopy), $I(\eta)$ is the element $(1,0)$ in $\pi^0_{st}(S^1 \oplus pt) \approx \mathbb{Z}_2 \oplus \mathbb{Z}$. (See Dold [4] for details of the calculation of this element .)] In this section we show that a non-zero index implies the existence of C-essential fixed points.

Following Dold [5], our discussion of the fixed point transfer will be set in the category of Euclidean neighborhood retracts over B (denotes ENR_B):

$E \xrightarrow{p} B$ is an ENR_B if $E \xrightarrow{i} U \xrightarrow{r} E$ (i,r maps over B) is the identity, where U is an open subset of $\mathbb{R}^n \times B$, for some n. Let $f : E \to E$ be a map over B, (i.e., fiber-preserving), such that $p|\mathrm{Fix}(f)$ is proper (called compactly fixed). In this case, there exists $\rho : B \to (0,\infty)$ such that

$$i(\mathrm{Fix}(f)) \subset \text{interior of } U_\rho = \{(y,b) \in \mathbb{R}^n \times B : \|y\| \leq \rho(b)\}.$$

Let $F : E \times I \to E$ (or $\{f_t\} : E \to E$, maps over B, $\forall\, t \in I$) be a homotopy of f. $\{f_t\}$ will be called a *compactly fixed homotopy* providing

$$i(\mathrm{Fix}(f_t)) \subset \text{interior of } U_\rho, \quad \forall\, t \in I.$$

In the special case where $E \subseteq \mathbb{R}^n \times B$, consider the following sequence:

$$\left\{\begin{array}{l} (\mathbb{R}^n,\mathbb{R}^n - 0) \times X = ((\mathbb{R},\mathbb{R}^n - 0) \times B) \times_B X \xleftarrow{(id-f,id)} (X, X - \mathrm{Fix}\ f) \\ \qquad \xrightarrow{EXC} (E, E - \mathrm{Fix}\ f) \xleftarrow{incl.} (E, E - E_\rho) \sim (\mathbb{R}^n,\mathbb{R}^n - 0) \times B \end{array}\right\} \qquad (2.1)$$

where $(id - f)(b,y) = (b, y - \varphi(b,y))$, (i.e., $f(b,y) = (b,\varphi(b,y))$); X is a neighborhood of the fixed points; and U_ρ is denoted by E_ρ for this case where $E = U$.

Apply cohomology to (2.1) and consider the composite map

$$k^* : h^{i+n}((\mathbb{R}^n,\mathbb{R}^n - 0) \times X) \to h^{i+n}((\mathbb{R}^n,\mathbb{R}^n - 0) \times B).$$

We obtain the trace map $t_f^{\times} : h^i X \to h^i B$, by "de-suspending" k^*. Passing to $\varinjlim \{hX\}$ (for neighborhoods X of $\mathrm{Fix}\ f$), we get the transfer $t_f : \check{h}(\mathrm{Fix}\ f) \to h(B)$.

In the general case, look at $E \xrightarrow{i} U \xrightarrow{fr} E$. $\mathrm{Fix}(f) \underset{r}{\overset{i}{\rightleftarrows}} \mathrm{Fix}(ifr)$ are easily shown to be inverse homeomorphisms.

The *transfer* t_f is defined as the composite

$$\check{h}(\mathrm{Fix}\ f) \overset{r^*}{\simeq} \check{h}(\mathrm{Fix}(ifr)) \xrightarrow{t_{ifr}} hB$$

and is independent of the choice of i and r [5]. The Fixed-Point Index of f is the element $I(f) = t_f(1) \in h^0(B)$.

(2.1) *THEOREM*

Given $E \xrightarrow{p} B$, an ENR_B; i and r, as before, and $f : E \to E$, all maps over B. If $t_f \neq 0$, then f has a C-essential fixed point, for any compactly fixed homotopy of f.

Proof

Suppose $t_f \neq 0$. Then $t_{ifr} \neq 0$. Denote ifr by g. Consider the following diagram, which induces a commutative diagram in cohomology:

$$
\begin{array}{ccccccc}
(U,U-\overset{\circ}{U}_\rho) & \xrightarrow{\text{incl.}} & (U,U\text{-Fix } g) & \xrightarrow{(id-g,id)} & (\mathbb{R}^n \times U, (\mathbb{R}^n-0)\ U) & \xleftarrow{\text{EXC}} & (X\times\mathbb{R}^n \cup ((U-X)\times\{0\}), X\times(\mathbb{R}^n-0)) \\
\| & & & & & & \downarrow p \\
\| & & & & & & (X\times\mathbb{R}^n \cup ((U-X)\times\{0\}), X\times(\mathbb{R}^n-0)\cup((U-X)\times\{0\})) \\
\| & & & & & & \downarrow \text{EXC} \\
(U,U-\overset{\circ}{U}_\rho) & \xrightarrow{\text{incl.}} & (U,U\text{-Fix } g) & \xleftarrow{\text{EXC}} & (X,X\text{-Fix } g) & \xrightarrow{(id-g,id)} & ((\mathbb{R}^n,\mathbb{R}^n-0)\times B)\times_B X \approx (\mathbb{R}^n,\mathbb{R}^n-0)\times X \\
\uparrow \text{incl.} & & & & & & \\
(U,U-U_\rho) \sim (\mathbb{R}^n,\mathbb{R}^n-0)\times B. & & & & & &
\end{array}
$$

Along the top row, we excise first $(U-X)\times(\mathbb{R}^n-0)$, then $(U-X)\times\{0\}$. The bottom row is non-zero in cohomology, since $t_{ifr} \neq 0$. Thus the top row is also non-zero in cohomology, and

$$
\begin{array}{ccc}
h^n(U\times\mathbb{R}^n, U\times(\mathbb{R}^n-0)) & \xrightarrow{i^*(id-g,id)^*} & h^n(U,U-\overset{\circ}{U}_\rho) \\
& & \downarrow \text{EXC} \\
& & h^n(U_\rho,\partial U_\rho)
\end{array}
$$

is non-zero.

Restrict $(id\text{-}g,id)$ to U_ρ. Then $(id\text{-}g,id)^{-1}(U\times 0)$ Fix g. Thus we have shown that $g = ifr$ has an essential fixed point. For $\{f_t\}: E \to E$, $t \in I$, a compactly fixed homotopy of f, let $Q \subset U \times I$ be a component of $\mathrm{Fix}(if_t r)$ connecting $U\times\{0\}$ and $U\times\{1\}$. Then $R: Q \to E\times I$, defined by $R(u,t)$ $(r(u),t)$, maps Q onto a component of Fix(f) connecting $E\times\{0\}$ and $E\times\{1\}$. ■

__Note:__ A similar argument shows that a non-zero coincidence/fixed point index (defined by Dold in [6]) implies the existence of C-essential solutions, (see [3]).

3. Equivariant Maps and $\mathbb{Z}_p$-Coincidences

Certain basic problems cannot be solved directly by the cohomological condition $h^*(N,N-P) \to h^*(M,\partial M)$ due to the fact that M is closed, (i.e., $\partial M = 0$), and N carries no cohomology. Obvious examples of this situation are problems in which $N \to \mathbb{R}^n$. Unfortunately, these are the most likely candidates for the continuation method because of the ready existence of a homotopy between any two maps $f,g : M \to \mathbb{R}^n$. Some of these spaces, however, either directly or indirectly possess group actions and can be put into a vector bundle formulation which introduces more cohomology into the problem. In the following, let G be a group; M and N, right G-spaces, where ∂M is G-invariant; P, a closed G-invariant submanifold of N; and $f : (M,\partial M) \to (N,N-P)$, a G-equivariant map. First we look at an equivariant formulation of the Alexander-Yorke cohomological condition:

Let $F : M \times I \to N$ be an equivariant homotopy of f, (i.e., $F(mg,t) = F(m,t)\cdot g$, $\forall\, g \in G$). If $F^{-1}(P)$ connects $M \times \{0\}$ and $M \times \{1\}$, then $f^{-1}(P)$ is called G-*essential*.

(3.1) *PROPOSITION*

Let M, N, P, and f be as above; and let X be a left G-space. If

$$(f \times id)^* : h^*(N \times_G X,\ (N-P) \times_G X) \to h^*(M \times_G X,\ \partial M \times_G X)$$

is non-zero, then $f^{-1}(P)$ is G-essential.

Proof

Notice that $(f \times id) : (M \times_G X,\ M \times_G X) \to (N \times_G X,\ (N \times_G X) - (P \times_G X))$. Thus, if $(f \times id)^*$ is non-zero, $(f \times id)^{-1}(P \times_G X)$ is G-essential. (Of course, the homotopy of $f \times id$ we consider here is $(F \times id) : (M \times I) \times_G X \to N \times_G X$, where $F : M \times I \to N$ is a G-equivariant homotopy of f, and the G action on $M \times I$ is defined as: $(m,t)g = (mg,t)$.) Thus $(M \times I) \times_G X \cong (M \times_G X) \times I$, and we consider

$$(F \times id) : (M \times_G X) \times I \to N \times_G X.$$

Therefore, there is a component $Q \subset (F \times id)^{-1}(P \times_G X) \subset (M \times_G X) \times I$ which connects $M \times_G X \times \{0\}$ and $M \times_G X \times \{1\}$. Let $\pi : M \times X \times I \to M \times I$. Then $\pi(Q)$ connects

$M \times \{0\}$ and $M \times \{1\}$. ■

As an example of continuation for an equivariant problem, we shall consider $\mathbb{Z}_p$-coincidences of maps $f : S^n \to \mathbb{R}^m$. Given a finite cyclic group G acting on S^n, a point $x \in S^n$ will be called a *coincidence point of* (f,G) if

$$f(xg_i) = f(xg_j), \quad \forall\, g_i, g_j \in G.$$

(3.2) *PROPOSITION*

Consider $f : S^n \to \mathbb{R}^m$ and a free G action on S^n, (n odd). (f,G) has a G-essential coincidence in the following cases:

$G = \mathbb{Z}_q$ and $n \geq (q-1)m$, where q is odd.

$G = \mathbb{Z}_4$, $m = 1$, and $n \geq 3$.

We shall need the following result, due to H. Munkolm, to prove the proposition:

THEOREM (Munkholm [10]*).*

Given a free G-action μ on S^n, (G a finite cyclic group) and $f : S^n \to \mathbb{R}$. Let $\mathbb{R}G$ be the product of $|G|$ copies of $\mathbb{R}$; writing its elements as $\sum_g r(g)g$, it is a G-space under the action $(\sum r(g)g)h = \sum r(g)gh = \sum r(gh^{-1})g$. In $\mathbb{R}G$ there is the G-invariant subspace $IG = \{\sum r(g)g \in \mathbb{R}G : \sum r(g)\} = 0$. Consider the bundle $\xi_\mu : S^n \times_G IG \to S^n/G$. ξ_μ is $\mathbb{Z}_q$-orientable, hence has a (mod q) Euler class, $e_q(\xi_\mu) \in H^{|G|-1}(S^n/G)$. $e_q(\xi_\mu)^m \neq 0$ in the following cases:

(1) $G = \mathbb{Z}_q$ and $n \geq (q-1)m$, for q odd,

(2) $G = \mathbb{Z}_4$, $m = 1$, and $n \geq 3$.

<u>*Proof of Proposition* (3.2).</u>

Let $\mu : S^n \times \mathbb{Z}_p \to S^n$ be a free $\mathbb{Z}_p$ action. $f : S^n \to \mathbb{R}^m$ induces the map $\overline{f} : S^n \to \mathbb{R}^m G = \mathbb{R}^{mp}$, by $\overline{f}(x) = (f(x), f(\mu x), f(\mu^2 x), \ldots, f(\mu^{p-1}x))$.

Let

$$\Delta_m = \{\underbrace{(a,a,\dots,a)}_{p} \in \mathbb{R}^{mp} : a \in \mathbb{R}^m\} = \underbrace{\Delta \times \Delta \times \dots \times \Delta}_{m},$$

the m-fold cross product of the diagonal of $\mathbb{R}^p$. Now $\overline{f}^{-1}(\Delta_m)$ is the set of coincidences of $(f, \mathbb{Z}_p)$. In the special case where $m = 1$, we have maps

$$S^n \times_{\mathbb{Z}_p} S^n \xrightarrow{\overline{f} \times id} \mathbb{R}^p \times_{\mathbb{Z}_p} S^n \xrightarrow{j} (\mathbb{R}^p \times_{\mathbb{Z}_p} S^n, (\mathbb{R}^p - \Delta) \times_{\mathbb{Z}_p} S^n),$$

where j is the inclusion. Let Z be the $p-1$ dimensional hyperplane in $\mathbb{R}^p$ consisting of points $(x_1,\dots,x_p) \in \mathbb{R}^p$ such that $x_1 + \dots + x_p = 0$. Since Z is perpendicular to Δ, let $\pi : \mathbb{R}^p \to Z \approx \mathbb{R}^{p-1}$ be the projection along Δ. Composing π with $\overline{f}$ yields:

$$(S^n \times_{\mathbb{Z}_p} S^n) \xrightarrow{h} \mathbb{R}^{p-1} \times_{\mathbb{Z}_p} S^n \xrightarrow{j} (\mathbb{R}^{p-1} \times_{\mathbb{Z}_p} S^n, (\mathbb{R}^{p-1} - 0) \times_{\mathbb{Z}_p} S^n),$$

where $h = \pi\overline{f} \times id$. Here $(\pi\overline{f})^{-1}(0)$ is the set of coincidences. Let ξ_μ be the vector bundle $(\mathbb{R}^{p-1} \times_{\mathbb{Z}_p} S^n \to S^n/\mathbb{Z}_p)$. Then ξ_μ is the bundle of the previous theorem since here $IG = I\mathbb{Z}_p = Z$. By Proposition (3.1), we want to show that

$$h^*j^* : H^{p-1}(\mathbb{R}^{p-1} \times_{\mathbb{Z}_p} S^n, (\mathbb{R}^{p-1} - 0) \times_{\mathbb{Z}_p} S^n; \mathbb{Z}_p) \to H^{p-1}(S^n \times_{\mathbb{Z}_p} S^n; \mathbb{Z}_p)$$

is non-zero. Look at the Serre spectral sequence (with $\mathbb{Z}_p$ coefficients) for the bundles $(S^n \times_{\mathbb{Z}_p} S^n \to L^n_p)$, and $(\mathbb{R}^{p-1} \times_{\mathbb{Z}_p} S^n \to L^n_p)$, where $L^n_p = S^n/\mathbb{Z}_p$ si the n-dimensional lens space with fundamental group $\mathbb{Z}_p$:

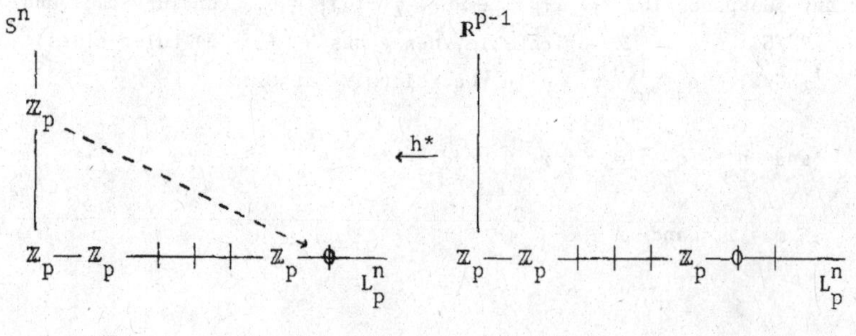

All differentials are trivial. Thus, if $u \in H^i(\mathbb{R}^{p-1} \times_{\mathbb{Z}_p} S^n; \mathbb{Z}_p) = \mathbb{Z}_p$ (for $i \le n$), then $h^*(u)$ is non-zero in $H^i(S^n \times_{\mathbb{Z}_p} S^n; \mathbb{Z}_p)$ since h acts as the identity on L^n_p.

In the following let E stand for $S^n \times_{\mathbb{Z}_p} \mathbb{R}^{p-1}$ and E_0, for $S^n \times_{\mathbb{Z}_p} (\mathbb{R}^{p-1} - 0)$, the non-zero vectors of E. Consider the cohomology exact sequence for (E, E_0):

$$\cdots \longrightarrow H^k(E,E_0) \xrightarrow{j^*} H^k(E) \xrightarrow{i^*} H^k(E_0) \xrightarrow{\delta} H^{k+1}(E,E_0) \longrightarrow \cdots$$

Using the Thom Isomorphism: $- \smile e(\xi_\mu) : H^{k-n}(E) \to H^k(E,E_0)$, for

$$e(\xi_\mu) \in H^n(L_p^n) \cong H^n(E), \quad n = p-1,$$

we obtain the Gysin sequence for (E,E_0):

$$\cdots \longrightarrow H^{k-n}(E) \xrightarrow{g} H^k(E) \xrightarrow{i^*} H^k(E_0) \longrightarrow H^{k-n+1}(E) \longrightarrow \cdots$$

where $g(x) = (x \smile e(\xi))|E = x \smile (e(\xi)|E)$. Thus $H^{p-1}(E,E_0) \xrightarrow{j^*} H^{p-1}(E)$ is non-zero if and only if $1 \smile e(\xi_\mu)$ is non-zero (where $1 \in H^0(E) \cong H^0(L_p^n)$) id and only if $e(\xi_\mu)$ is non-zero. Therefore, the problem reduces to calculating $e(\xi_\mu)$.

In the more general case, where $m \geq 1$, we have

$$(S^n \times_{\mathbb{Z}_p} S^n) \xrightarrow{\overline{f} \times id} (\mathbb{R}^{mp} \times_{\mathbb{Z}_p} S^n) \longrightarrow (\mathbb{R}^{mp} \times_{\mathbb{Z}_p} S^n, (\mathbb{R}^{mp} - \Delta_m) \times_{\mathbb{Z}_p} S^n).$$

Composing with the projection along Δ_m yields:

$$(S^n \times_{\mathbb{Z}_p} S^n) \xrightarrow{\pi\overline{f} \times id} (\mathbb{R}^{m(p-1)} \times_{\mathbb{Z}_p} S^n) \longrightarrow (\mathbb{R}^{m(p-1)} \times_{\mathbb{Z}_p} S^n, (\mathbb{R}^{m(p-1)} - 0) \times_{\mathbb{Z}_p} S^n).$$

Clearly, the vector bundle $\zeta = (\mathbb{R}^{m(p-1)} \times_{\mathbb{Z}_p} S^n) \to L_p^n$ is $m\xi_\mu = \xi_\mu \oplus \ldots \oplus \xi_\mu$. Again, the problem reduces to calculating $e(\zeta) = (e(\xi_\mu))^m$. Thus, Munkholm's Theorem provides the result. ■

REFERENCES

[1] ALEXANDER, J.C.: Topological theory of an embedding method, Continuation Methods, ed. H. Wacker. (New York: Academic Press), (1978).

[2] ALEXANDER, J.C. and YORKE, J.A.: The homotopy Continuation method: numerically implementable topological procedures, Trans. Amer. Math. Soc. 242 (1978), 271-284.

[3] ALLIGOOD, K.T.: Homological indices and homotopy continuation, Ph. D. thesis, University of Maryland, (1979).

[4] DOLD, A.: The fixed point transfer of fibre-preserving maps, Invent. Math. 25 (1974), 281-297.

[5] DOLD, A.: The fixed point transfer of fibre-preserving maps, Math. Zeit. 148 (1976), 215-244.

[6] DOLD, A.: A coincidence-fixed-point index, Enseign. Math. 24 (1978). 41-53.

[7] DOLD, A.: Lectures on Algebraic Topology. (Berlin-Heidelberg-New-York: Springer-Verlag), (1972).

[8] EILENBERG, S. and MONTGOMERY, D.: Fixed point theorems for multi-valued transformations, Amer. J. Math. (1946), 214-222.

[9] GÓRNIEWICZ, L.: Homological methods in fixed point theory of multi-valued maps, Diss. Math. 129 (1976), 1-71.

[10] MUNKHOLM, H.J.: Borsuk-Ulam type theorems for proper $\mathbb{Z}_p$-actions on (mod p homology) n-spheres, Math. Scand. 24 (1969), 167-185.

[11] ROITBERG, J.: On the Lefschetz fixed point formula, Comm. Pure and Appl. Math. 20 (1967), 139-143.

[12] VIETORIS, L.: Über den höheren zusammenhang kompakter räume und ein klasse von zusammenhangstreuen abbildungen, Math. Ann. 97 (1927), 454-472.

A MEASURE THEORETIC APPROACH TO FIXED POINTS IN ERGODIC THEORY

By

J-M. BELLEY*

Département de mathématiques et d'informatique
Université de Sherbrooke
Sherbrooke, Québec, Canada, J1K 2R1

1. INTRODUCTION

Let $S = \{0,1,2...\}$ or $[0,\infty)$. We will show the existence of i) an algebra α of Borel subsets of S and ii) a bounded nonnegative charge $\overline{m}_S : \alpha \to [0,1]$ (that is a finitely additive set function) which is, in some sense, the restriction to S of the normalized Haar measure on the Bohr compactification on the ordered group $G = S \cup (-S)$ (with dual group $G^\wedge$). We will then take a directed set D and a net $\{(S,\alpha_\delta,\mu_\delta) : \delta \in D\}$ of charged spaces for which 1) α_δ is an algebra of subsets of S containing α and $\mu_\delta : \alpha_\delta \to [0,\infty)$ $(\delta \in D)$, 2) $\{\mu_\delta(S) : \delta \in D\}$ is bounded, and 3) if $<z,\hat{z}>$ denotes $\hat{z} \in G^\wedge$ evaluated at $z \in G$, then $\hat{\mu}_\delta(\hat{z}) = \int_S <z,\hat{z}> d\mu_\delta(z)$ (in the sense of Moore-Smith convergence) converges to $\hat{\overline{m}}_S(\hat{z}) = \int_S <z,\hat{z}> d\overline{m}_S(z)$, which is 1 if $\hat{z}$ is the identity in $G^\wedge$ and 0 for all other $\hat{z}$ in $G^\wedge$ (that is $\lim_\delta \mu_\delta = \overline{m}_S$ weakly). Note that, for $S = \{0,1,2,...\}$ such a net of charged spaces generalizes the notion of a strongly regular matrix with nonnegative entries. To obtain our ergodic result, a concept of integrability with respect to a finitely additive set function will be introduced.

Taking x_o in a Hilbert space H with inner product $(\cdot,\cdot)$, we consider a semigroup $\{\varphi_\sigma : \sigma \in S\}$ of transformations on H for which i) the complex-valued function $\sigma \to (\varphi_\sigma(x_o),y)$ on S is continuous and μ_δ-integrable $(\delta \in D, y \in H)$, and ii) $\lim_{\sigma\to\infty}(\varphi_{\sigma+z}(x_o), \varphi_\sigma(x_o))$ exists uniformly in $z \in S$. If we let x_δ and x' designate those elements of H (which, as we show, exist) given uniquely by

$$(x_\delta,y) = \int_S (\varphi_\sigma(x_o),y) \;\; d\mu_\delta(\sigma)$$

*Research supported by NSERC grant A8858 and the ministry of education of Québec.

and

$$(x',y) = \int_S (\varphi_\sigma(x_o),y) \quad d\bar{m}_S(\sigma)$$

respectively, for all $y \in H$ and $\delta \in D$, then we show that $\{x_\delta : \delta \in D\}$ converges strongly to x'. Furthermore, we show that x' is a fixed point for $\{\varphi_\sigma : \sigma \in S\}$ if a) $\{\varphi_\sigma : \sigma \in S\}$ is a semigroup of linear transformations on H, or b) H is a real Hilbert space, x_o lies in a closed bounded convex subset C of H for which $\varphi_\sigma ; C \to C$ $(\sigma \in S)$, and $\{\varphi_\sigma : \sigma \in S\}$ is a continuous nonexpansive semigroup on C. This generalizes results proved by Blum and the others [7] and S. Reich [12]; the latter being in turn generalizations of results of Baillon [1, 2, 3], Baillon and Brézis [5], and Brézis and Browder [8].

2. Measure Theoretic Preliminaries

(2.1) Let β be an algebra of subsets of a set X. A *charge* μ on β is a complex-valued bounded (in total variation) finitely additive set function defined on β; and we call the triple (X,β,μ) a *charged space*. A function $f : X \to \mathbb{C}$ (the complex numbers) is said to be β-*continuous with respect to* μ if, given $\varepsilon > 0$, there exists A_ε in β for which $\mu(A_\varepsilon) = 0$, and there exists a β-measurable step function $\varphi : X \to \mathbb{C}$ such that

$$\sup \{|f(x) - \varphi(x)| : x \in X \setminus A_\varepsilon\} < \varepsilon \ .$$

Let $C(X,\beta,\mu)$ denote the space of such functions. Now, given $f : S \to \mathbb{C}$, we define the integral $\int_X f \, d\mu$ by the usual Moore-Smith convergence method as follows. Without loss of generality, let f and μ be nonnegative. Given a finite partition π of X consisting of disjoint sets $E_i \in \beta$ $(i = 1,\dots,n)$, let $S(f,\mu,\pi; z_1,\dots,z_n)$ designate the sum $\sum_{i=1}^{n} f(z_i)\,\mu(E_i)$ where $z_i \in E_i$. Let

$$\overline{S}(f,\mu,\pi) = \sup \{S(f,\mu,\pi \,;\, z_1,\dots,z_n) : z_1 \in E_1,\dots,z_n \in E_n\}$$

and

$$\underline{S}(f,\mu,\pi) = \inf \{S(f,\mu,\pi \,;\, z_1,\dots,z_n) : z_1 \in E_1,\dots,z_n \in E_n\} \ .$$

Writing $\overline{\int} f \, d\mu$ and $\underline{\int} f \, d\mu$ for the lim inf and lim sup of $\overline{S}(f,\mu,\pi)$ and $\underline{S}(f,\mu,\pi)$

respectively over the net of all finite partitions $\pi \subset \beta$ of X (ordered as usual by refinements), we say that f is *μ-integrable* (in the sense of Moore-Smith convergence) if $\overline{\int} f\, d\mu = \underline{\int} f\, d\mu$; in which case we write $\int_X f\, d\mu$ for the common limit. Clearly any $f \in C(X,\beta,\mu)$ is μ-integrable.

(2.2) Henceforth G will denote the locally compact abelian group $\{0,\pm 1,\pm 2,\pm 3,\ldots\}$ or $(-\infty,\infty)$ and S will denote the semigroup $\{0,1,2,\ldots\}$ or $[0,\infty)$ respectively. Let $G\hat{}$ and $\overline{G}$ designate the dual group and the Bohr compactification of G respectively. The value of $\hat{z} \in G\hat{}$ at $z \in G$ will be denoted by $\langle z,\hat{z}\rangle$. Denote by m the normalized Haar measure on the Borel subsets $\beta(\overline{G})$ of $\overline{G}$ and let

$$\theta = \{A \in \beta(\overline{G}) : m(\overline{A} \setminus \mathring{A}) = 0\}$$

where $\overline{A}$ and $\mathring{A}$ is the closure and interior of A in $\overline{G}$, respectively. Clearly, θ is an algebra of sets in $\overline{G}$ for which $A + z \in \theta$ for all $A \in \theta$ and all $z \in \overline{G}$. Now, if Y is any dense subset of $\overline{G}$, the trace $\theta_Y = \{A \cap Y : A \in \theta\}$ of θ over Y is an algebra of sets contained in the Borel subsets $\beta(Y)$ of Y with topology induced by that of $\overline{G}$. As shown in [6, section (2.1)], the set function m_Y on θ_Y given by $m_Y(A \cap Y) = m(A)$ $(A \in \theta)$ is a well defined charge on θ_Y. So, for $Y = S$ which is dense in $\overline{G}$, we have that $m_S((A+z) \cap S) = m(A+z) = m_S(A \cap S)$ for all $A \in \theta$ and all $z \in G$. Let α be the algebra generated by the sets $\{A + z : A \in \theta_S, z \in S\}$ and let $\overline{m}_S$ be the well defined charge on α given by $\overline{m}_S(A) = m_S((A-z) \cap S)$ for all $z \in S$ such that $(A-z) \cap S \in \theta_S$.

(2.3) *REMARKS*

i) If $f \in C(S,\alpha,\overline{m}_S)$ and if $z \in S$, then the function $f_z(w) = f(w+z)$ $(w \in S)$ is also in $C(S,\alpha,\overline{m}_S)$. Taking into account that $A + z \in \alpha$ and $\overline{m}_S(A+z) = \overline{m}_S(A)$ for all $A \in \alpha$, and that, outside of a set of $\overline{m}_S$-measure zero, f is approximated by an α-measurable step function, it follows that $\int_S f\, d\overline{m}_S = \int_S f_z\, d\overline{m}_S$.

ii) Any element in the space $B^+(G)$ of positive definite functions on G is integrable with respect to any charge on $\theta_G = \{A \cap G : A \in \theta\}$ (see [6, section (4.6)]). Hence, as shown in [9, p. 293], each element of $B^+(G)$ is θ_G-continuous with respect to any charge on θ_G, and so $C(G,\theta_G,\mu)$ contains $B^+(G)$ for all charges μ on θ_G. So, for all charges μ on α, $\{f|S : f \in B^+(G)\} \subset C(S,\alpha,\overline{m}_S)$ and, since $G\hat{} \subset B^+(G)$, $\{\hat{z}|S : \hat{z} \in G\hat{}\} \subset C(S,\alpha,\mu)$.

iii) In [6, section (4.6)] it is shown that, given any charge $\mu : \alpha \to \mathbb{C}$, and given any measure $\nu : \beta(G^\wedge) \to \mathbb{C}$ which is bounded in total variation, the functions $\hat{\nu}(z) = \int_{G^\wedge} \langle z,\hat{z}\rangle\, d\nu(\hat{z})$ and $\hat{\mu}(\hat{z}) = \int_S \langle z,\hat{z}\rangle\, d\mu(z)$ on S and $G^\wedge$ are integrable with respect to μ and ν respectively, and

$$\int_S \int_{G^\wedge} \langle z,\hat{z}\rangle\, d\nu(\hat{z})\, d\mu(z) = \int_{G^\wedge} \int_S \langle z,\hat{z}\rangle\, d\mu(z)\, d\nu(\hat{z}).$$

iv) Let $AP(G)$ designate the space of continuous almost periodic functions on G with supremum norm. By a well known theorem of harmonic analysis (see [10, p. 168]) every element of $AP(G)$ is the restriction on G of a unique function in the space $C(\bar{G})$ of continuous functions on $\bar{G}$. So, for all charges μ on α, by the Stone-Weierstrass theorem and since $\{\hat{z}|S : \hat{z} \in G^\wedge\} \subset C(S,\alpha,\mu)$, it follows that $\{f|S : f \in AP(G)\} \subset C(S,\alpha,\mu)$. Furthermore, for all $f \in C(\bar{G})$ and all charges μ on α, $f|S \in C(S,\alpha,\mu)$ and $\int_{\bar{G}} f\, dm = \int_S f|S\, d\bar{m}_S$. Since $(\bar{G})^\wedge = (G^\wedge)_d$ (that is $G^\wedge$ with discrete topology), it follows (see [10, p.154]) that the Fourier-Stieltjes transform $\hat{m}_S(\hat{z}) = \int_S \langle z,\hat{z}\rangle\, dm_S(z)$ equals 1 if $\hat{z}$ is the identity and 0 for all other $\hat{z}$ in $G^\wedge$, as is the case with $\hat{m}_G(\hat{z})$ also.

v) If D is a directed set and if $\{(S,\alpha_\delta,\mu_\delta) : \delta \in D\}$ is a net of charged spaces for which a) $\alpha \subset \alpha_\delta$ for all $\delta \in D$ and b) the Fourier-Stieltjes transforms $\hat{\mu}_\delta(\hat{z}) = \int_S \hat{z}|S\, d\mu_\delta$ converge to 1 if $\hat{z}$ is the identity 0 in $G^\wedge$ and to 0 if $\hat{z} \in G^\wedge \setminus \{0\}$ (that is $\lim_\delta \mu_\delta(\hat{z}) = \hat{m}_S(\hat{z})$, or equivalently, $\lim_\delta \mu_\delta = m_S$ weakly), then $\lim_\delta \mu_\delta(S \setminus (S+\sigma)) = 0$ for all $\sigma \in S$. To show this, note that for all $t \in (0,\infty)$, the set

$$E_t = \bigcup \{[nt, nt+\sigma) \cap G : n \in Z\}$$

lies in θ_G since, as shown in [6, Lemma (2.5)], θ_G can be identified with the class of all sets $E \subset G$ for which

$$\inf\{\int_G (h-g)\, dm_G : g \le I_E \le f\,;\, f,\, g \in AP(G)\} = 0$$

(where I_E is the characteristic function of E). So,

$$\mu_\delta(S \setminus (S+\sigma)) \le \mu_\delta(E_t \cap S)$$
$$\le \inf\{\int_S f\, d\mu_\delta : I_{E_t} \le f \le 1,\ f \in AP(G)\}$$

for all $\delta \in D$ and all $t \in (0,\infty)$. But, given $t \in (\sigma,\infty)$ and given $\varepsilon > 0$ small enough, there clearly exists a periodic (and so almost periodic) function f_ε such

that $I_{E_t} \leq f_\varepsilon \leq 1$ and $f_\varepsilon(z) = 0$ for all $z \notin \cup\{[w-\varepsilon, w+\varepsilon) : w \in E_t\} = E'_t$.
Hence $\int_G (f_\varepsilon - I_{E_t})\, dm_G \leq m_G(E'_t \setminus E_t) \leq 2\varepsilon/t$ and $\mu_\delta(S\setminus(S+\sigma)) \leq \int_S f_\varepsilon|S\, d\mu_\delta \to \int_G f_\varepsilon\, dm_G$ (by the Stone-Weierstrass theorem and the hypothesis $\hat{\mu}_\delta(\hat{z}) \to \hat{m}_G(\hat{z})$ for all $\hat{z} \in G^\wedge$).
Thus, for all $t > \sigma$,

$$\lim_\delta \mu_\delta(S\setminus(S+\sigma)) \leq \int_G f_\varepsilon\, dm_G = \int_G (f_\varepsilon - I_{E_t})dm_G + \int_G I_{E_t}\, dm_G$$

$$\leq 2\varepsilon/t + m_G(E_t) = 2\varepsilon/t + \sigma/t.$$

Since t is arbitrarily large, it follows that $\lim_\delta \mu_\delta(S\setminus(S+\sigma)) = 0$. To evaluate the measure of the sets above, we use the fact that $m_G(G) = 1$ in conjunction with the translation invariance of m_G.

3. Fixed Points in Ergodic Theory

(3.1) If $S = \{0,1,2,\ldots\}$ and if, for $n = 1,2,3,\ldots$, α_n is the algebra of all subsets of S and $\mu_n : \alpha_n \to [0,1]$ is a charge on α_n given by

$$\mu_n(\{k\}) = \begin{cases} n^{-1} & \text{if } k = 0,\ldots, n-1 \\ 0 & \text{if } k = n, n+1, n+2,\ldots \end{cases}$$

then $\{(S,\alpha_n,\mu_n) : n = 1,2,3,\ldots\}$ is an example of a sequence of charged spaces for which the Fourier-Stieltjes transforms $\hat{\mu}_n(t) = \int_S e^{ikt}\, d\mu_n(k)$ $(t \in G^\wedge = [0,2\pi))$ converges as $n \to \infty$ to 1 if $t = 0$ and to 0 otherwise. Similarly, if $S = [0,\infty)$ and if, for $T > 0$, α_T is the algebra of Borel subsets of S and

$$\mu_T(E) = \begin{cases} |E|/T & \text{if } E \subset [0,T],\ E \in \alpha_T \\ 0 & \text{if } E \subset (T,\infty),\ E \in \alpha_T, \end{cases}$$

where $|E|$ is the Lebesgue measure of E, then $\{(S,\alpha_T,\mu_T) : T > 0\}$ is an example of a generalized sequence of charged spaces for which the Fourier-Stieltjes transforms $\hat{\mu}_T(s) = \int_S e^{ist}\, d\mu_T(t)$ $(s \in G^\wedge = (-\infty,\infty))$ converge, as $T \to \infty$, to 1 if $s = 0$ and to 0 otherwise.

(3.2) *REMARK*

It is known (see, for example, [1], [4] and [12, p. 270]) that if

$\{\varphi_\sigma : \sigma \in S\}$ is a continuous semigroup of nonexpansive transformations of a bounded closed convex subset C of a real Hilbert space, then for all $x \in C$, the Cesaro means

$$\frac{1}{n}\sum_{k=0}^{n-1} \varphi_k(x) = \int_S \varphi_k(x)\, d\mu_n(k)$$

when $S = \{0,1,2,\ldots\}$ and the time averages

$$\frac{1}{T}\int_0^T \varphi_\sigma(x)\, d\sigma = \int_S \varphi_\sigma(x)\, d\mu_T(\sigma)$$

when $S = [0,\infty)$ converge weakly in C to a fixed point of the semigroup. Note that, for any $x,y \in H$, the function $\sigma \to (\varphi_\sigma(x),y)$ is μ_δ-integrable ($\delta = n$ or T). We will use this in the proof of the following generalization of a result of S. Reich [12, p. 269] for strongly regular matrices.

(3.3) *THEOREM*

Given a directed set D, let $\{(S,\alpha_\delta,\mu_\delta) : \delta \in D\}$ be a net of charged spaces for which 1) $\alpha \subset \alpha_\delta$ for all $\delta \in D$, 2) $\{\mu_\delta : \delta \in D\}$ is a uniformly bounded (in total variation) net of non-negative charges, and 3) $\hat{\mu}_\delta(\hat{z})$ converges to 1 if $\hat{z} = 0$ and to 0 otherwise. If $\{\varphi_\sigma : \sigma \in S\}$ are transformations on a Hilbert space H such that, for a given $x_0 \in H$, the complex-valued function $\sigma \to (\varphi_\sigma(x_0),y)$ on S is continuous and μ_δ-integrable ($\delta \in D$, $y \in H$) and $\lim_\sigma (\varphi_{\sigma+z}(x_0),\varphi_\sigma(x_0))$ exists uniformly in $z \in S$, then

i) for all $y \in H$, the complex-valued function $\sigma \to (\varphi_\sigma(x_0),y)$ on S is $\overline{m}_S$-integrable,

ii) the net $\{\int_S \varphi_\sigma(x_0)\, d\mu_\delta(\sigma) : \delta \in D\}$ of elements in H converges strongly to $\int_S \varphi_\sigma(x_0)\, d\overline{m}_S(\sigma)$, and

iii) $\int_S \varphi_\sigma(x_0)\, d\overline{m}_S(\sigma)$ is a fixed point for $\{\varphi_\sigma : \sigma \in S\}$ in each of the following two cases : a) $\{\varphi_\sigma : \sigma \in S\}$ is a semigroup of linear transformations on H, b) H is a real Hilbert space, x_0 lies in a bounded closed convex subset C of H for which $\varphi_\sigma : C \to C$ ($\sigma \in S$), and $\{\varphi_\sigma : \sigma \in S\}$ is a continuous nonexpansive semigroup on C.

(3.4) *REMARK*

By $\int_S \varphi_\sigma(x_0)\, d\mu_\delta(\sigma)$ and $\int_S \varphi_\sigma(x_0)\, d\overline{m}_S(\sigma)$ we mean those elements x_δ and x' of H given uniquely by

$$(x_\delta, y) = \int_S (\varphi_\sigma(x_0), y)\, d\mu_\delta(\sigma)$$

and

$$(x', y) = \int_S (\varphi_\sigma(x_0), y)\, d\overline{m}_S(\sigma)$$

respectively, for all $y \in H$ and $\delta \in D$.

(3.5) *PROOF OF THEOREM* (3.3)

Let x_σ denote $\varphi_\sigma(x_0)$. If z, w, and $w - z$ lie in S, then $\lim_\sigma (x_{\sigma+w}, x_{\sigma+z})$ and $\lim_\sigma (x_{\sigma+(w-z)}, x_\sigma)$ exist and are equal. So, if we write $p(z)$ for the continuous complex-valued function on S given by $\lim_\sigma (x_{\sigma+z}, x_\sigma)$, then

$$p(w - z) = \lim_\sigma (x_{\sigma+w}, x_{\sigma+z}).$$

Let

$$\lambda(z) = \begin{cases} p(z) & \text{if } z \in S \\ \overline{p(-z)} & \text{if } z \in -S \end{cases}$$

where $\overline{p(-z)}$ denotes the complex conjugate of $p(-z)$. Clearly, $\lambda(z)$ is a continuous complex-valued positive definite function on G. Hence, by remark (2.3) ii) $p(z)$ lies in $C(S, \alpha, \overline{m}_S)$ and in $C(S, \alpha, \mu_\delta) \subset C(S, \alpha_\delta, \mu_\delta)$ for all $\delta \in D$. Assume part i) for the moment.

We now show the strong convergence of the net $\{\int_S x_\sigma\, d\mu_\delta(\sigma) : \delta \in D\}$ to $\int_S x_\sigma\, d\overline{m}_S(\sigma)$. Given $\delta \in D$, if $\|\cdot\|$ is the norm on H, then

$$\|\int_S x_\sigma\, d\mu_\delta(\sigma) - \int_S x_\sigma\, d\overline{m}_S(\sigma)\|^2$$

$$= \int_S\int_S ((x_\sigma, x_\gamma) - \lambda(\sigma - \gamma))\, d\mu_\delta(\sigma)\, d\mu_\delta(\gamma)$$

$$-2\mathrm{Re}\int_S\int_S ((x_\sigma, x_\gamma) - \lambda(\sigma - \gamma))\, d\mu_\delta(\sigma)\, d\overline{m}_S(\gamma)$$

$$+\int_S\int_S((x_\sigma,x_\gamma) - \lambda(\sigma-\gamma))\, d\bar{m}_S(\sigma)\, d\bar{m}_S(\gamma)$$

$$+\int_S\int_S\lambda(\sigma-\gamma)\, d\mu_\delta(\sigma)\, d\mu_\delta(\gamma)$$

$$-2\mathrm{Re}\int_S\int_S\lambda(\sigma-\gamma)\, d\mu_\delta(\sigma)\, d\bar{m}_S(\gamma)$$

$$+\int_S\int_S\lambda(\sigma-\gamma)\, d\bar{m}_S(\sigma)\, d\bar{m}_S(\gamma).$$

Using the fact that $\lim_\sigma (x_{\sigma+z},x_\sigma) = p(z)$ uniformly in $z \in S$, and that for all $\sigma \in S$, the net $\{\mu_\delta(S\setminus(S+\sigma)) : \delta \in D\}$ converges to 0, it follows that the first three integrals can be made arbitrarily small by choosing $\delta > \delta_0$ for some $\delta_0 \in D$. The sum of the last three integrals can also be made arbitrarily small by using Bochner's integral representation for $\lambda(z)$ (see [13, p. 19]) in conjunction with remark (2.3) iii), hypothesis 3), and Lebesgue's dominated convergence theorem.

The legitimacy of iii) part a) is clear. To show part b), we can use remark (3.2) along with part ii) above which shows that the strong limit $\int_S x_\sigma\, d\bar{m}_S(\sigma)$ is independent of the net of charged spaces with the stated properties.

To prove i) it is enough to show that any net $\{\sum_{E\in\pi} x_\sigma\, \bar{m}_S(E) : \pi \in P\}$ $(\sigma \in E)$ is strongly Cauchy in H; where P is the directed set of all finite α-measurable partitions of S ordered by refinements.

$$\|\sum_{E\in\pi} x_\sigma\, \bar{m}_S(E) - \sum_{E'\in\pi'} x_{\sigma'}\, \bar{m}_S(E')\|^2$$

$$= \sum_{E,F\in\pi} (x_\sigma,x_\gamma)\, \bar{m}_S(E)\, \bar{m}_S(F)$$

$$-2\mathrm{Re} \sum_{E\in\pi}\sum_{E'\in\pi'} (x_\sigma,x_{\sigma'})\, \bar{m}_S(E)\, \bar{m}_S(E')$$

$$+ \sum_{E',F'\in\pi'} (x_{\sigma'},x_{\gamma'})\, \bar{m}_S(E')\, \bar{m}_S(F')$$

and so, by remark (2.3) v),

$$\lim_{\pi,\pi'} \|\sum_{E\in\pi} x_\sigma\, \bar{m}_S(E) - \sum_{E'\in\pi'} x_{\sigma'}\, \bar{m}_S(E')\|^2$$

$$= \int_S\int_S\lambda(\sigma-\gamma)\, d\bar{m}_S(\sigma)\, d\bar{m}_S(\gamma)$$

$$-2\mathrm{Re}\int_S\int_S\lambda(\sigma-\gamma)\, d\bar{m}_S(\sigma)\, d\bar{m}_S(\gamma)$$

$$+\int_S\int_S \lambda(\sigma-\gamma)\, d\bar{m}_S(\sigma)\, d\bar{m}_S(\gamma)$$

which vanishes, by Bochner's integral representation for λ in conjunction with remark (2.3) iii) and the fact that $\hat{\bar{m}}_S(\hat{z}) = 1$ if $\hat{z}$ is the identity and 0 otherwise.

The theorem is proved. ■

(3.6) *REMARKS*

i) Blum and the others [7, p. 18] considered the case of a sequence $\{\mu_n : n = 0,1,2,\ldots\}$ of probability measures on G and a continuous unitary representation $\varphi_z = U_z$ $(z \in G)$ of G with pure point spectrum so as to have a continuous complex-valued function $z \to (U_z x, y)$ on G for all $x,y \in H$, which can be extended to a continuous complex-valued function on $\bar{G}$. Our approach has the advantage that, instead of trying to pass to $\bar{G}$, we stay on our original group G. In so doing, we obtain a generalization of their ergodic result in that the convergence is strong and the transformations need not be unitary with pure point spectrum.

ii) A regular matrix $\{a_{n,k} : k,n = 1,2,3,\ldots\}$ with $a_{n,k} \geq 0$ is said to be strongly regular if $\lim_{n\to\infty} \sum_{k=1}^{\infty} |a_{n,k+1} - a_{n,k}| = 0$. If $S = \{0,1,2,\ldots\}$ and if, for $n = 1,2,3,\ldots$, α_n is the algebra of all subsets of S and $\mu_n : \alpha_n \to [0,1]$ is that charge on α_n given by a strongly regular matrix $\mu_n(\{k\}) = a_{n,k} \geq 0$, then $\{(S,\alpha_n,\mu_n) : n = 1,2,3,\ldots\}$ is a sequence of charged spaces for which $\hat{\mu}_n(t)$ converges to 1 if $t = 0$ and to 0 for all $t \in (0,2\pi)$. To see this note that, by definition of a strongly regular matrix, $\hat{\mu}_n(0) = \sum_{k=1}^{\infty} a_{n,k} = 1$, while

$$(1-e^{it}) \sum_{k=1}^{\infty} a_{n,k} e^{ikt} = a_{n,1} + \sum_{k=2}^{\infty} (a_{n,k} - a_{n,k-1}) e^{ikt}$$

and so, for $t \in (0,2\pi)$,

$$\left|(1-e^{it}) \sum_{k=1}^{\infty} a_{n,k} e^{ikt}\right| \leq |a_{n,2}| + \sum_{k=2}^{\infty} |a_{n,k} - a_{n,k-1}|$$

which converges to 0 as $n \to \infty$; that is $\hat{\mu}_n(t) = \sum_{k=1}^{\infty} a_{n,k} e^{ikt}$ converges to 0 as $n \to \infty$. Since W.O. Ray [11, p. 531] has shown that nonexpansive mappings on a closed convex subset of a real Hilbert space have a fixed point if and only if it is bounded, theorem (3.3) generalizes the result of S. Reich [12, p. 269] not only in the summability method but also in that the strong convergence does not depend on having a semigroup of transformations induced by a nonexpansive mapping on a closed convex subset of a real Hilbert space.

REFERENCES

[1] BAILLON, J.-B.: Un théorème de type ergodique pour les contractions nonlinéaires dans un espace de Hilbert, C. R. Acad. Sci. Paris 280 (1975), 1511-1514.

[2] BAILLON, J.-B.: Quelques propriétés de convergence asymptotique pour les semigroupes de contractions impaires, C. R. Acad. Sci. Paris 283 (1976), 75-78.

[3] BAILLON, J.-B.: Quelques propriétés de convergence asymptotique pour les contractions impaïres, C. R. Acad. Sci. Paris 283 (1976), 587-590.

[4] BAILLON, J.-B.: Comportement asymptotique des itérés de contractions nonlinéaires dans les espaces L^p, C. R. Acad. Sci. Paris 286 (1978), 157-159.

[5] BAILLON, J.-B. and BREZIS, H.: Une remarque sur le comportement asymptotique des semigroupes nonlinéaires, Houston J. Math. 2 (1976), 5-7.

[6] BELLEY, J.-M.: A Representation Theorem and Applications to Topological groups, Trans. Amer. Math. Soc. (to appear).

[7] BLUM, J.R., EISENBERG, B. and HAHN, L.-S.: Ergodic theory and the measure of sets in the Bohr group, Acta Sci. Math. 34 (1973), 17-24.

[8] BREZIS, H. and BROWDER, F.E.: Nonlinear ergodic theorems, Bull. Amer. Math. Soc. 82 (1976), 959-961.

[9] DARST, R.B.: A Note on Abstract Integration, Trans. Amer. Math. Soc. 99 (1961), 292-297.

[10] LOOMIS, L.H.: An introduction to Abstract Harmonic Analysis, Van Nostrand, New York, (1953).

[11] RAY, W.O.: A Fixed Point Property and Unbounded Sets in Hilbert Space,

Trans. Amer. Math. Soc. 258 (1980), 531-537.

[12] REICH, S.: Almost Convergence and Nonlinear Ergodic Theorems, J. Approximation Theory, 24 (1978), 269-272.

[13] RUDIN, W.: Fourier Analysis on Groups, Interscience, New York, (1967).

ON THE PRODUCT THEOREM FOR THE FIXED POINT INDEX

By

ROBERT F. BROWN

Department of Mathematics
University of California
Los Angeles CA90024 U.S.A.

1. INTRODUCTION

For $T : X \to X$ a map of a finite polyhedron and x_0 an isolated fixed point of T, we denote by $i(X,T,x_0)$ the local (Hopf) index of T on any neighborhood U of x_0 such that $\overline{U} \cap \mathrm{Fix}(T) = \{x_0\}$. The result to which the title of this note refers is:

PRODUCT THEOREM

Let $p : E \to B$ be a locally trivial fibre space with fibre F, where E, B and F are finite polyhedra. Let $f : E \to E$ be a fibre-preserving map inducing $\overline{f} : B \to B$ and suppose e and $b = p(e)$ are isolated fixed points of f and $\overline{f}$ respectively, in the interiors of maximal simplices. Denote by f_b the restriction of f to $F_b = p^{-1}(b)$. Then

$$i(E,f,e) = i(B,\overline{f},b) \cdot i(F_b,f_b,e).$$

For trivial fibre spaces, the theorem was proved by Leray in [5] using techniques of analysis (see pages 21-25 of [1] for the details). The only proof of the Product Theorem in the literature, as Lemma (5.2) of [3], consists of using local triviality to reduce the theorem to the case considered by Leray. It then refers to Leray's proof to complete the argument.

The Product Theorem is a necessary tool in the fixed point theory of fibre-preserving maps. Note its role in the proof of the main result of [4] (page 80). So it is convenient to have an easier and more accessible proof of the Product Theorem, and that is what this paper presents.

2. THE PROOF

In this section, we give a proof of the Product Theorem that depends only on elementary facts from algebraic topology.

Let U be a neighborhood of b such that $\overline{U} \cap \mathrm{Fix}(\overline{f}) = \{b\}$ and $\overline{U}$ is homeomorphic to a cell. Let V be a Euclidean neighborhood of b contained in U such that $\overline{f}(V) \subseteq U$. By local triviality, we may identify $f : p^{-1}(V) \to p^{-1}(U)$ with $\varphi : V \times F \to U \times F$ where $\varphi(v,y) = (\overline{f}(v), f'(v,y))$, $f' : V \times F \to F$ (compare the proof of Proposition (4.1) of [4]). We may identify the restriction of f' to $\{b\} \times F$ with the restriction of f to F_b, so we use the same notation: $f_b : F \to F$.

Let the fixed point of φ corresponding to e be (b,y_0), which is in a euclidean neighborhood S in $V \times F$ containing no other fixed point. There exists $\varepsilon > 0$ such that if the distance from $(v,y) \in V \times F$ to (b,y_0) is less that 3ε, then (v,y) and $\varphi(v,y)$ are in S. Imposing the structure of a real vector space on S, we define $\psi : V \times F \to U \times F$ by $\psi(v,y) = (\overline{f}(v), g'(v,y))$, where $g' : V \times F \to F$ is as follows (d denotes distance)

(1) if $d(v,b) \leq \varepsilon$ and $d(y,y_0) \leq \varepsilon$, then $g'(v,y) = f_b(y)$

(2) if $d(v,b) \leq \varepsilon$ and $d(y,y_0) = (1+t)\varepsilon$, $0 \leq t \leq 1$, then $g'(v,y) = tf'(v,y) + (1-t)f_b(y)$

(3) if $d(v,b) = (1+s)\varepsilon$, $0 \leq s \leq 1$, and $d(y,y_0) \leq \varepsilon$, then $g'(v,y) = sf'(v,y) + (1-s)f_b(y)$

(4) if $d(v,b) = (1+s)\varepsilon$, $0 \leq s \leq 1$, and $d(y,y_0) = (1+t)\varepsilon$, $0 \leq t \leq 1$, then $g'(v,y) = (s+t)f'(v,y) + (1-s-t)f_b(y)$ if $0 \leq s \leq 1-t$ and $g'(v,y) = f'(v,y)$ if $1-t \leq s \leq 1$

(5) otherwise, let $g'(v,y) = f'(v,y)$.

Define $H : (V \times F) \times I \to U \times F$ by

$$H((v,y),r) = (\overline{f}(v), rg'(v,y) + (1-r)f'(v,y)).$$

We see that φ and ψ are homotopic by a homotopy with the property that in a 3ε-neighborhood of (b,y_0), $H((v,y),r) = (v,y)$ only if $(v,y) = (b,y_0)$ because

$\overline{f}(v) \neq v$ for $v \neq b$ and $\varphi = \psi$ on $\{b\} \times F$. Thus, by the homotopy property of the index, $i(V \times F, \varphi, (b,y_0)) = i(V \times F, \psi, (b,y_0))$.

Let $W_V = \{v \in V : d(v,b) \leq \epsilon\}$, $W_F = \{y \in F : d(y,y_0) \leq \epsilon\}$ and $W = W_V \times W_F \subset V \times F$. Note that the restriction of ψ to W is of the form $\psi = \overline{f} \times f_b$. Suppose that W_V is a cell of dimension m and W_F is a cell of dimension k, then S may be identified with $\mathbb{R}^{m+k}$. The map $1 - \psi : (W, W - (b,y_0)) \to (\mathbb{R}^{m+k}, \mathbb{R}^{m+k} - 0)$ is given by

$$(1 - \psi)(v,y) = (v,y) - \psi(v,y) = (v,y) - (\overline{f}(v), f_b(y))$$

$$= (v - \overline{f}(v), y - f_b(y))$$

so we may identify

$$1 - \psi = (1 - \overline{f}) \times (1 - f_b) : (W_B, W_B - b) \times (W_F, W_F - y_0) \to (\mathbb{R}^m, \mathbb{R}^m - 0) \times (\mathbb{R}^k, \mathbb{R}^k - 0) .$$

The Künneth Theorem gives us the following commutative diagram, where μ_W and μ are isomorphisms [7, page 235]

$$\begin{array}{ccc} H_m(W_B, W_B - b) \otimes H_k(W_F, W_F - y_0) & \xrightarrow{(1-\overline{f})_* \otimes (1-f_b)_*} & H_m(\mathbb{R}^m, \mathbb{R}^m - 0) \otimes H_k(\mathbb{R}^k, \mathbb{R}^k - 0) \\ \downarrow \mu_W & & \downarrow \mu \\ H_{m+k}(W, W - (b,y_0)) & \xrightarrow[(1-\psi)_*]{} & H_{m+k}(\mathbb{R}^{m+k}, \mathbb{R}^{m+k} - 0) \end{array}$$

We may choose generators $\sigma'_m \in H_m(W_B, W_B - b)$, $\sigma'_k \in H_k(W_F, W_F - y_0)$, $\sigma_m \in H_m(\mathbb{R}^m, \mathbb{R}^m - 0)$, and $\sigma_k \in H_k(\mathbb{R}^k, \mathbb{R}^k - 0)$ so that $(1 - \overline{f})_*(\sigma'_m) = i(W_B, \overline{f}, b)\sigma_m$ and $(1 - f_b)_*(\sigma'_k) = i(W_F, f_b, y_0)\sigma_k$ (compare [2, page 121]). Furthermore, letting $\sigma'_{m+k} = \mu_W(\sigma'_m \otimes \sigma'_k)$ and $\sigma_{m+k} = \mu(\sigma_m \otimes \sigma_k)$ then $(1 - \psi)_*(\sigma'_{m+k}) = i(W, \psi, (b,y_0))\sigma_{m+k}$.

Commutativity of the diagram and the definition of tensor product then imply

$$i(W, \psi, (b,y_0))\sigma_{m+k} = (1 - \psi)_* \mu_W(\sigma'_m \otimes \sigma'_k)$$

$$= \mu((1 - \overline{f})_* \otimes (1 - f_b)_*)(\sigma'_m \otimes \sigma'_k)$$

$$= \mu(i(W_B, \overline{f}, b)\sigma_m \otimes i(W_F, f_b, y_0)\sigma_k)$$

$$= (i(W_B,\overline{f},b) \cdot i(W_F,f_b,y_0))\sigma_{m+k} .$$

Therefore, by the first part of the proof $i(V\times F,\varphi,(b,y_0)) = i(V,\overline{f},b) \cdot i(F,f_b,y_0)$ and so from the identifications

$$i(E,f,e) = i(B,\overline{f},b) \cdot i(F_b,f_b,e). \blacksquare$$

3. EULER CHARACTERISTICS

As an application of the Product Theorem, we will prove a "folk theorem" that seems to be widely known among topologists, but apparently has never been published because no one has been able to find an excuse to do it. The symbol $\chi(X)$ denotes the Euler characteristic of a space X.

PROPOSITION

Let $p : E \to B$ be a locally trivial fibre space with fibre F where E, B and F are compact triangulated manifolds (possibly with boundary). Then

$$\chi(E) = \chi(B) \cdot \chi(F).$$

Proof

By a theorem of Wecken [9], there is a map $\overline{f} : B \to B$ homotopic to the identity map with one fixed point b, in the interior of B. (See the paper of Chiang in these *Proceedings* for an elegant proof of Wecken's Theorem on manifolds.) The Covering Homotopy Theorem gives a fibre-preserving map $f' : E \to E$ homotopic to the identity map such that $pf' = \overline{f}p$. By Wecken's Theorem applied to the restriction f'_b of f' to $F_b = p^{-1}(b)$ and Proposition (4.1) of [4], there is a fibre-preserving map $f : E \to E$ with one fixed point e, in the interior of E, such that f is homotopic to f' and $pf = \overline{f}p$. Let f_b be the restriction of f to F_b. The Lefschetz-Hopf Theorem implies that if a map $T : X \to X$ has the properties

(i) T is homotopic to the identity map, and

(ii) T has exactly one fixed point,

then, for x_0 the fixed point of T, $i(X,T,x_0) = \chi(X)$. Since the maps f, $\overline{f}$ and f_b all have properties (i) and (ii), the Product Theorem implies

$$\chi(E) = i(E,f,e) = i(B,\overline{f},b) \cdot i(F_b,f_b,e) = \chi(B) \cdot \chi(F). \blacksquare$$

If the fibre space is orientable, the proposition is a very special case of a celebrated theorem of Serre [6]. The proposition could be generalized somewhat by fixed point methods without imposing any orientability conditions. However, there is a spectral sequence which requires only that $p : E \to B$ be a fibre space where B is the homotopy type of a finite polyhedron and E and the fibre F have the homology of such polyhedra. Using this spectral sequence in place of the Leray-Serre spectral sequence with Serre's original argument produces a far more general version of the proposition than one could hope to prove by fixed point theory [8].

References

[1] BROWN, R.: Notes on Leray's index theory, Advances in Math., 7 (1971), 1-28.

[2] BROWN, R.: The Lefschetz Fixed Point Theorem, Scott, Foresman and Co., Chicago, (1971).

[3] BROWN, R.: The Nielsen number of a fibre map, Ann. Math. 88 (1967), 483-493.

[4] FADELL, E.: Natural fibre splittings and Nielsen numbers, Houston J. Math., 2 (1976), 71-84.

[5] LERAY, J.: Sur les équations et les transformations, J. Math. Pures appl. (9ième série) 24 (1945), 201-248.

[6] SERRE, J.-P.: Homologie singulière des espaces fibrés, Ann. Math. 54 (1951), 425-505.

[7] SPANIER, E.: Algebraic Topology, McGraw-Hill, New York, (1966).

[8] SPANIER, E.: Private communication.

[9] WECKEN, F.: Fixpunktklassem, III, Math. Ann. 118 (1941-3), 544-577.

A FIXED POINT THEORY FOR FIBER-PRESERVING MAPS

BY

E. FADELL AND S. HUSSEINI

E. Fadell
Mathematisches Institut
Universität Heidelberg
6900 Heidelberg
West Germany

S. Husseini
Department of Mathematics
University of Wisconsin
Madison, WI 53706
U.S.A.

1. INTRODUCTION

The objective of this note is to begin the study of a fixed point index theory for fiber-preserving maps which will parallel the theory for self-maps of spaces, as set forth in [4] and [5], which emphasizes the role of the fundamental group. To review the global case, given a (smooth and compact, for simplicity) manifold M, one considers the inclusion map $i : M \times M - \Delta M \subset M \times M$, where ΔM is the diagonal in $M \times M$, and replaces it by a fiber map $q : E \to M \times M$. The fiber $F = q^{-1}(*)$ in this case in $(m-2)$-connected where $m = \dim M$ and, in particular, when $m \geq 3$, F is simply connected. Furthermore, one easily establishes isomorphisms

$$\pi_{m-1}(F) \approx \pi_m(M \times M, M \times M - \Delta M) \approx \pi_m(M, M - y_0) \approx \mathbb{Z}[\pi],$$

where $y_0 \in M$ and $\pi = \pi_1(M)$. If we designate by $\mathcal{B}$ the local system on $M \times M$ with local group $\pi_{m-1}(F) \approx \mathbb{Z}[\pi]$, the (right) action of $\pi \times \pi$ on $\mathbb{Z}[\pi]$ which determines $\mathcal{B}$ is given by the formula

$$\alpha \circ (\sigma, \tau) = (\operatorname{sgn} \sigma)\, \sigma^{-1} \alpha \tau$$

where $\operatorname{sgn} \sigma = \pm 1$ according as $\sigma \in \pi$ preserves or reverses a local orientation of M. Now, given a self map $f : M \to M$ one considers the diagram

$$\begin{array}{ccc} E(f) & \longrightarrow & E \\ \downarrow{\scriptstyle q_f} & & \downarrow{\scriptstyle q} \\ M & \xrightarrow{1 \times f} & M \times M \end{array}$$

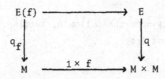

where $(E(f), q_f, M)$ is the fiber space induced from $(E,q, M \times M)$ by the map $1 \times f$. A geometric argument shows that q_f admits a cross section if, and only if, f is deformable to a fixed point free map $g : M \to M$. If we let $\mathcal{B}(f)$ denote the local system on M induced from $\mathcal{B}$ by $1 \times f$, then the local group of $\mathcal{B}(f)$ is $\mathbb{Z}[\pi]$ with π acting on $\mathbb{Z}[\pi]$ by

$$\alpha \circ \sigma = (\mathrm{sgn}\,\sigma)\,\sigma^{-1}\alpha\varphi(\sigma), \qquad \alpha, \sigma \in \pi \tag{*}$$

where $\varphi : \pi \to \pi$ is induced by f. Let $o(f) \in H^m(M; \mathcal{B}(f))$ denote the primary obstruction to finding a cross section to q_f. $o(f)$ is invariant under homotopy and consequently a necessary and sufficient condition that f is deformable to a fixed point free map $g : M \to M$ is that $o(f) = 0$. Thus, $o(f)$ serves as a (global) fixed point index for f. (There is, of course, the corresponding local version [5]). $o(f)$ is computable in terms of Nielsen classes as follows: First, one introduces the action $\alpha * \sigma = \varphi(\sigma^{-1})\alpha\sigma$ of π on $\mathbb{Z}[\pi]$ and then let $\mathcal{R}(f)$ denote the corresponding local system. If $\mathcal{T}(M)$ is the orientation sheaf on M, there is a dual pairing

$$\mathcal{B}(f) \otimes \mathcal{T}(M) \to \mathcal{R}(f)$$

given by $\alpha \otimes 1 \mapsto \alpha^{-1}$ which induces a cap product

$$<\ ,\ > : H^m(M; \mathcal{B}(f)) \otimes H_m(M; \mathcal{T}(M)) \to H_0(M, \mathcal{R}(f))$$

where $H_0(M, \mathcal{R}(f)) \approx \mathbb{Z}R[\varphi]$, and $R[\varphi]$ is the space of orbits (usually called Reidemeister classes) of π under the action $*$.

THEOREM

If μ is the twisted integral fundamental class of M, then

$$<o(f), \mu> = \sum_{\rho \in R} \lambda_\rho \rho$$

where $R = R[\varphi]$, and λ_ρ is the local numerical index of the Nielsen class corresponding to the Reidemeister class ρ. In particular, the Lefschetz number $L(f)$ of f is the sum of the coefficients λ_ρ.

We will consider in this note the following setting. Let $Y \xrightarrow{i} M \xrightarrow{p} B$ denote a smooth fiber bundle and let

$$\begin{array}{ccc} M & \xrightarrow{\quad f \quad} & M \\ & {\scriptstyle p}\searrow \quad \swarrow{\scriptstyle p} & \\ & B & \end{array}$$

denote a fiber-preserving map, that is $pf = p$. Now, let $M \times_B M$ denote the fiber square of M, that is

$$M \times_B M = \{(x,y) \in M \times M : p(x) = p(y)\}.$$

Then, the inclusion map $M \times_B M - \Delta M \subset M \times_B M$ is replaced by a fiber map

$$q : E_B(M) \to M \times_B M$$

and in this case the fiber F has as homotopy

$$\pi_{j-1}(F) \approx \pi_j(M \times_B M, M \times_B M - \Delta M) \approx \pi_j(Y, Y - y_0).$$

Thus, if $k = \dim Y$, then F is $(k-2)$-connected and hence when $k \geq 3$, F is simply connected and $\pi_{k-1}(F) \approx \mathbb{Z}[\pi']$, where $\pi' = \pi_1(Y)$. If we let $\pi = \pi_1(M \times_B M)$, then because there is an obvious section for the bundle $M \times_B M \to M$, π admits a semi-direct product representation $\pi = \pi''\pi'$, where $\pi'' = \pi_1(M)$. Again designate by $\mathcal{B}_B$ the local system on $M \times_B M$ with local group $\pi_{k-1}(F) \approx \mathbb{Z}[\pi']$, as induced by the fibration $E_B(M) \to M \times_B M$. The action of π on $\mathbb{Z}[\pi']$ which determines $\mathcal{B}_B$ is computed in §3. Now, just as in the single space case above, the fiber-preserving map $f : M \to M$ induces a diagram

$$\begin{array}{ccc} E_B(f) & \longrightarrow & E_B(M) \\ {\scriptstyle q_f}\downarrow & & \downarrow{\scriptstyle q} \\ M & \xrightarrow{\quad 1 \times f \quad} & M \times_B M \end{array}$$

and q_f will admit a section if, and only if, f is fiberwise homotopic to a fixed point free map. The primary obstruction

$$o_B(f) \in H^k(M; \mathcal{B}_B(f)), \quad \mathcal{B}_B(f) = f^* \mathcal{B}_B$$

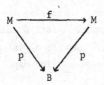

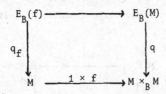

to finding a section to q_f over the k-skeleton of B is called the *primary* (obstruction) *fixed point index* of f. When B is a point, $o_B(f)$ coincides with $o(f)$, as previously defined. In addition to verifying the elementary properties of $o_B(f)$, we show the following as an application. Let G denote a compact manifold of $\dim k \geq 3$ and let $M = G \times G$. Suppose that G admits a continuous multiplication xy with right identity e. Then, the fiber-preserving map

$$\begin{array}{ccc} G \times G & \xrightarrow{f} & G \times G \\ & \searrow \quad \swarrow & \\ & G & \end{array} \qquad (x,y) \mapsto (x,xy)$$

always has a non-zero index $o_G(f)$. Thus, every map fiber homotopic to f has fixed points even though f may be deformable (not over G) to a fixed point free map. This example, when $G = S^1$ is given in Dold [2] and for this example the Dold index $\Lambda(f) \in \pi^0_{\text{stable}}(S^1) \approx \mathbb{Z}_2$ is the unique non-zero element. Thus, the index $o_B(f)$ can be viewed as an unstable alternative to the Dold index $\Lambda(f)$.

Finally, in Section 4 we exhibit an example of a fiber-preserving map $f : M \to M$ with $o_B(f) = 0$ but the existence of a non-zero secondary obstruction precludes f from being deformable to a fixed point free map using a fiberwise homotopy. Thus, the need for considering higher (obstruction) fixed point indices $o_B^j(f)$, $j \geq 2$ is indicated.

2. The Primary Obstruction Fixed Point Index of a Fiber Preserving Map.

We will consider first the following setting: $Y \xrightarrow{i} M \xrightarrow{p} B$ is a smooth fiber bundle with total space M, fiber Y, base B, inclusion i and projection p. Y, M and B are smooth, connected compact manifolds (without boundary). We consider next the fiber square

$$\begin{array}{ccc} M \times_B M & \xrightarrow{j_2} & M \\ {\scriptstyle j_1}\downarrow & & \downarrow{\scriptstyle p} \\ M & \xrightarrow{p} & B \end{array}$$

where

$$M \times_B M = \{(x,y) \in M \times M : p(x) = p(y)\}$$

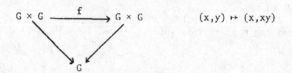

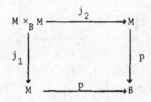

and $j_1(x,y) = x$, $j_2(x,y) = y$. Of course, the fiber $j_1^{-1}(x) = \{x\}\times Y \equiv Y$ and j_1 admits the diagonal section $\Delta_B : M \to M \times_B M$, where $\Delta_B(x) = (x,x)$. If we let $j : M \times_B M \to M \times M$ denote inclusion, $\Delta : M \to M \times M$ the diagonal map, then $j_1\Delta = \Delta_B$ and the diagonal $\Delta M = \Delta_B M$.

(2.1) *PROPOSITION*

$(M \times_B M, M \times_B M - \Delta(M), j_1, M)$ is a locally trivial fibered pair [3] with fiber $(Y, Y - y_0)$, $y_0 \in Y$.

Proof

A simple exercice following the proof of the corresponding result [3] when B is a point.■

Consider now the inclusion map $M \times_B M - \Delta M \hookrightarrow M \times_B M$ which we replace by the fiber map $q : E_B(M) \to M \times_B M$. Thus,

$$E_B(M) = \{(\alpha,\beta) \in (M \times_B M)^I : \alpha(0) \neq \beta(0)\}$$

and $q(\alpha,\beta) = (\alpha(1), \beta(1))$. Alternatively, $E_B(M)$ consists of pairs of paths (α,β) in M such that $p\alpha(t) = p\beta(t)$, $0 \le t \le 1$, and $\alpha(0) \neq \beta(0)$. Let F denote the fiber $q^{-1}(x,y)$, $(x,y) \in M \times_B M - \Delta M$. Then, it is easy to see that

$$\pi_{n-1}(F, (\bar{x},\bar{y})) \approx \pi_n(M \times_B M,\ M \times_B M - \Delta M,\ (x,y))$$

where $\bar{u}$ denotes the constant path at u. But using Proposition (2.1),

$$\pi_n(M \times_B M,\ M \times_B M - \Delta M,\ (x,y)) \approx \pi_n(Y, Y - x, y)$$

where $Y = p^{-1}(x)$. Thus, if $\dim Y = k$

$$\pi_{n-1}(F, (\bar{x},\bar{y})) \approx \begin{cases} 0 & n-1 \le k-2 \\ \mathbb{Z}[\pi'] & n-1 = k-1 \end{cases}$$

where $\pi' = \pi_1(Y)$. Thus, if $k \ge 3$, the fiber F is 1-connected and the first non-vanishing homotopy group is $\mathbb{Z}[\pi']$ and occurs in dimension $k-1$. In particular, F is then n-simple for every n. Thus, as F varies over $M \times_B M$, $\pi_{k-1}(F)$ generates a local system $\mathcal{B}_B$ on $M \times_B M$ with local group isomorphic to $\mathbb{Z}[\pi']$. We collect these facts in the following proposition.

(2.2) *PROPOSITION*

The fibration

$$F \longrightarrow E_B(M) \xrightarrow{q} M \times_B M$$

has (k-2)-connected fiber and when $k \geq 3$ the fiber F is simply connected so that $\pi_{k-1}(F)$ generates a local system $\mathcal{B}_B$ on $M \times_B M$, with local group isomorphic to $\mathbb{Z}[\pi']$, $\pi' = \pi_1(Y)$.

(2.3) *DEFINITION*

We call the fibration in Proposition (2.2) *the fixed point indicator of* M *(over* B*)*. $\mathcal{B}_B$ will be called the *primary local system for* M *(over* B*)*.

We will compute the action of $\pi = \pi_1(M \times_B M)$ on $\mathbb{Z}[\pi']$ which determines $\mathcal{B}_B$ in the next section.

Now, we continue under the assumption that $k = \dim Y \geq 3$. Suppose, we are given two fiber-preserving maps

$$\begin{array}{ccc} M & \xrightarrow{f,g} & M \\ & {\scriptstyle p}\searrow \quad \swarrow{\scriptstyle p} & \\ & B & \end{array} \qquad fp = p, \quad gp = p.$$

Then,

$$g \times f : M \to M \times_B M$$

and a lift

$$\begin{array}{ccc} & & E_B(M) \\ & \overset{\Gamma}{\nearrow} & \downarrow \\ M & \xrightarrow{g \times f} & M \times_B M \end{array}$$

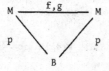

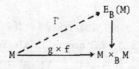

provides a homotopy

$$\widetilde{\Gamma} : M \times I \to M \times_B M$$

such that if we write,

$$\widetilde{\Gamma} : M \to M \times_B M, \quad \widetilde{\Gamma}_t = g_t \times f_t$$

we have $g_1 = g$, $f_1 = f$, g_0 and f_0 are coincidence free, that is

$$g_0(u) \neq f_0(u), \quad u \in M$$

and $pg_t(u) = pf_t(u)$, $u \in M$, $0 \leq t \leq 1$. In other words, the lift Γ provides deformations g_t and f_t of g_0 and f_0 respectively, such that g_1 and f_1 are coincidence free and g_t and f_t cover the same homotopy into the base B.

(2.4) *PROPOSITION*

The existence of the lift Γ above implies the existence of a fiber-preserving homotopy

$$\begin{array}{ccc} M & \xrightarrow{f_t} & M \\ & {\scriptstyle p}\searrow \quad \swarrow{\scriptstyle p} & \\ & M & \end{array}$$

such that $f_1 = f$ and f_0 and g are coincidence free.

Proof

The proof is a straightforward adaptation of the corresponding result for B = point, using the fact that $(M \times_B M, M \times_B M - \Delta M, j_1, M)$ is a Hurewicz fibered pair. ■

(2.5) *COROLLARY*

Suppose we are given a single fiber-preserving map $f : M \to M$ and

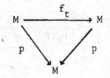

$$\begin{array}{ccc} E_B(f) & \longrightarrow & E_B(M) \\ {\scriptstyle q_f}\downarrow & & \downarrow{\scriptstyle q} \\ M & \xrightarrow{1\times f} & M \times_B M \end{array}$$

represents the pull-back of the fixed point indicator by $1 \times f$. Then, f is fiber homotopy equivalent to a fixed point free map if, and only if, q_f admits a section.

(2.6) *DEFINITION*

We call

$$F \longrightarrow E_B(f) \xrightarrow{q_f} M$$

the *fixed point indicator of* f *(over* B*)* and the bundle of coefficients (local system) $\mathcal{B}_B(f)$ induced from $\mathcal{B}_B$ by $1 \times f$, the *primary local system for* f.

(2.7) *DEFINITION*

Let $o_B(f) \in H^k(M; \mathcal{B}_B(f))$ denote the primary obstruction [6] for finding a cross section for the map q_f. We call $o_B(f)$ the *primary obstruction fixed point index for* f.

(2.9) *THEOREM (Fiber Homotopy Invariance)*

$o_B(f)$ is invariant under fiber homotopy equivalence.

Proof

If f and g are fiber homotopic, then $1 \times f$ and $1 \times g$ are homotopic as maps from M into $M \times_B M$. Then, standard obstruction theory applies and when the equivalent local systems $\mathcal{B}_B(f)$ and $\mathcal{B}_B(g)$ are identified, we have $o_B(f) = o_B(g)$. ■

(2.9) *FIBERWISE FIXED POINT THEOREM*

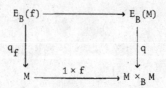

If $f : M \to M$ is a map over B, then $o_B(f) \neq 0$ implies every map fiber homotopic to f has fixed points.

(2.10) *REMARK*

Because $o_B(f)$ is only the primary obstruction, it can happen that the conclusion of (2.9) is valid when $o(f) = 0$ (because higher obstructions do not vanish, see Section 4). Also, $o(f) \neq 0$ does not preclude the possibility that f map be deformed into a fixed point free map, outside the class of maps over B (see Section 4).

(2.11) *REMARK*

The Fiber Homotopy Invariance of $o_B(f)$ is slighty more general. Namely, let $\alpha_t : M \to M$ denote a cyclic homotopy of the identity, and let $\beta_t : M \to M$ denote any homotopy such that

$$p(\alpha_t(u)) = p(\beta_t(u)), \quad u \in M, \quad 0 \leq t \leq 1.$$

Then, if $f = \beta_0$, $g = \beta_1$, $o_B(f) = o_B(g)$. Of course, then Theorem (2.9) has (ostensibly) a corresponding improvement.

We next formulate a naturality property of the primary obstruction fixed point index $o_B(f)$.

(2.11) *(NATURALITY) THEOREM*

Suppose C is a submanifold of B (above) and $N = p^{-1}(C)$. Then, if $f : M \to M$ is a fiber-preserving map (map over B) and $f_0 = f|N$, we have an inclusion induced homomorphism

$$\ell_* : H^k(M; \mathcal{B}_B(f)) \to H^k(N; \mathcal{B}_C(f_0))$$

which carries $o_B(f)$ to $o_C(f_0)$. In particular if C is a point, then

$$\ell_*(o_B(f)) = o(f_0),$$

the obstruction index for the single space Y as defined in [4] and [5].

(2.12) *REMARK*

The naturality property can be proved in more generality, for example C need not be a submanifold. However, (2.12) suffices for the purposes of this introductory work on the subject.

<u>*Proof of* (2.11).</u>

First, observe that there is a diagram

$$\begin{array}{ccc} E_C(N) & \xrightarrow{\lambda} & E_B(M) \\ \downarrow q_N & & \downarrow q_M \\ N \times_C N & \xrightarrow{\ell \times \ell} & M \times_B M \end{array}$$

where the horizontal maps are inclusions, $q_M = q$ above and q_N the corresponding map for N. Next, $M \times_B M$ is an ANR (compact metric) and hence the space of paths $(M \times_B M)^I$ is an ANR (sep. metric). If $w : (M \times_B M)^I \to M \times_B M$ is the initial-point map, then $w^{-1}(M \times_B M - \Delta M) = E_B(M)$. Thus, the open subset $E_B(M) \subset (M \times_B M)^I$ is also an ANR (sep. metric). Similarly $E_C(N)$ is an ANR (sep. metric). Now, let F_0 denote the fiber $q_N^{-1}(x,y)$ and F the fiber $q_M^{-1}(x,y)$, where $(x,y) \in N \times_C N - \Delta N$. Then, we have the diagram

$$\begin{array}{ccccc} \pi_{n-1}(F_0,\bar{x},\bar{y}) & & \xrightarrow{\lambda_{0*}} & & \pi_{n-1}(F,\bar{x},\bar{y}) \\ \updownarrow \approx & & & & \updownarrow \approx \\ \pi_n(N \times_C N, N \times N - \Delta N) & & \longrightarrow & & \pi_n(M \times_B M, M \times_B M - \Delta M) \\ & \approx \searrow & & \nearrow \approx & \\ & & \pi_n(Y, Y - x,y) & & \end{array}$$

where $\lambda_0 = \lambda | F_0$. Thus, λ_{0*} is an isomorphism for all n. Since, F_0 and F are neighborhood retracts in $E_C(N)$ and $E_B(M)$, respectively, λ_0 is a homotopy equivalence

$$\begin{array}{ccc} E_C(N) & \xrightarrow{\lambda} & q_M^{-1}(N \times_C N) \\ & \searrow \quad \swarrow & \\ & N \times_C N & \end{array}$$

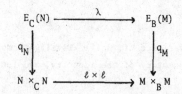

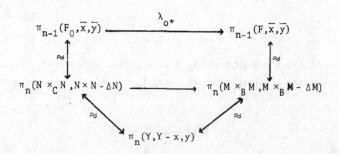

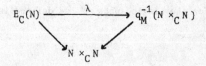

Now, standard obstruction theory [6] gives, the result that $\ell_*(o_B(f)) = o_C(f_0)$. ■

3. CALCULATING THE COEFFICIENT BUNDLES $\mathcal{B}_B$ AND $\mathcal{B}_B(f)$.

In order to describe the local system $\mathcal{B}_B$ on $M \times_B M$ in more detail we will need to set up some notation and preliminairies.

Consider the fibration

$$Y \xrightarrow{i} M \xrightarrow{p} B, \quad Y = p^{-1}(b)$$

with fiber Y, base B and total space M with all spaces 0-connected and in the category ANR (metric). We choose also a base point $y_1 \in Y \subset M$. Now, let $\tilde{M}$, $\tilde{B}$ denote the corresponding universal covers and the induced diagram

$$\begin{array}{ccccc} \eta_M^{-1}(Y) & \longrightarrow & \tilde{M} & \xrightarrow{\tilde{p}} & \tilde{B} \\ \downarrow & & \downarrow \eta_M & & \downarrow \eta_B \\ Y & \longrightarrow & M & \xrightarrow{p} & B \end{array}$$

where $\tilde{p}(\tilde{y}_1) = \tilde{b}_1$; $\tilde{y}_1$ and $\tilde{b}_1$ are base points chosen above y_1 and b_1, respectively; and η_M, η_B are the corresponding covering maps. The discrete fiber $\eta_B^{-1}(b)$ will be indexed by the covering group π'' of $\tilde{B}$, that is

$$\tilde{b}_\sigma = \tilde{b}_1\sigma, \quad \sigma \in \pi''.$$

(3.1) *REMARK*

In keeping with the convention used in [5], groups π of covering transformations (covering groups) will be treated as *right* operators on the corresponding covering spaces. Furthermore, the group operation in π, $\alpha\beta$ will be read as α followed by β. Thus, $x(\alpha\ \beta) = \beta(\alpha(x))$, $\alpha,\beta \in \pi$, x a point in the corresponding covering space.

The components $\tilde{Y}_\sigma$ of $\eta_M^{-1}(Y)$ may also be indexed by π'' with

$$\tilde{p}(\tilde{Y}_\sigma) = \tilde{b}_\sigma, \quad \sigma \in \pi''.$$

(3.2) *PROPOSITION*

Let $\eta_\sigma = \eta_M|\widetilde{Y}_\sigma : \widetilde{Y}_\sigma \to Y$. Then, η_σ is a covering map with

$$\pi_1(\widetilde{Y}_\sigma) \cong \ker i_* : \pi_1(Y) \to \pi_1(M).$$

Now, suppose we set π, π', π'' equal to the covering groups of $\widetilde{M}$, $\widetilde{Y}$, and $\widetilde{B}$, where $\widetilde{Y} = \widetilde{Y}_1$. Then, we have induced homorphisms

$$\pi' \xrightarrow{\ \bar{i}\ } \pi \xrightarrow{\ \bar{p}\ } \pi''$$

where for $\tau \in \pi'$, $\bar{i}(\tau)$ is the covering transformation of $\widetilde{M}$ which takes $\widetilde{y}_1$ to $\widetilde{y}_1\tau$, so that $\bar{i}(\tau)$ is the natural extension of τ. Furthermore, for $\alpha \in \pi$, $\bar{p}(\alpha)$ is determined by

$$\widetilde{b}_1\,\bar{p}(\alpha) = \widetilde{p}(\widetilde{y}_1\,\alpha)\,.$$

Suppose now that $p : M \to B$ admits a section $\theta : B \to M$ with $\theta(b_1) = y_1$, and $\widetilde{\theta} : \widetilde{B} \to \widetilde{M}$ the lift of θ determined by $\widetilde{\theta}(\widetilde{b}_1) = \widetilde{y}_1$. Then, we have an induced homomorphism $\widetilde{\theta} : \pi'' \to \pi'$ and a semi-direct product representation

$$1 \longrightarrow \pi' \xrightarrow{\ \bar{i}\ } \pi \underset{\bar{\theta}}{\overset{\bar{p}}{\rightleftarrows}} \pi'' \longrightarrow 1$$

where, as usual, $\bar{\theta}(\sigma)$, $\sigma \in \pi'$ is determined by

$$\widetilde{y}_1\ \bar{\theta}(\sigma) = \widetilde{\theta}(\widetilde{b}_1\,\sigma)\,.$$

Under the above conditions, therefore, each element $\gamma \in \pi$ can be uniquely written $\gamma = \sigma\tau$, $\sigma \in \pi''$, $\tau \in \pi'$. In short, we represent π as $\pi''\pi'$. Continuing under the assumption that θ is a cross section for p, we identify B and $\theta(B)$ and assume further that $(M, M - B, p, B)$ is a locally trivial fibered pair [3] with fiber $(Y, Y - y_1)$.

(3.3) *PROPOSITION*

Suppose in addition to the above conditions that M, Y, B are smooth manifolds of dim $n + k$, k and n, respectively, with $k \geq 3$, and fundamental groups π, π', π'' and $p : M \to B$ is a smooth bundle. Then, the action of $\pi = \pi''\pi'$ on

$$\pi_k(M,\ M-B) = \pi_k(Y,\ Y-y_1) = \mathbb{Z}[\pi']$$

is given by

$$\alpha \circ \sigma\tau = (\mathrm{sgn}_B\,\sigma)\,(\mathrm{sgn}_Y\,\sigma)\,\sigma^{-1}\alpha\sigma\tau, \quad \sigma \in \pi'',\ \alpha,\ \tau \in \pi'$$

with $\mathrm{sgn}_B\,\sigma = \pm 1$ $(\mathrm{sgn}_Y\,\sigma = \pm 1)$ according as σ preserves or reverses a local orientation of $B(Y)$.

Proof

First we agree that π, π', π'' are represented as covering transformations. Then, consider the diagram of fibered pairs

$$\begin{array}{ccccc}
\widetilde{B} & \longleftarrow & (\widetilde{M}, \widetilde{M}-\eta_M^{-1}(B)) & \longleftarrow & (\widetilde{Y}, \widetilde{Y}-\eta^{-1}(y_1)) \\
\downarrow{\scriptstyle \eta_B} & & \downarrow{\scriptstyle \eta_M} & & \downarrow{\scriptstyle \eta} \\
B & \longleftarrow & (M,\ M-B) & \longleftarrow & (Y, Y-y_1)
\end{array}$$

where $\eta : \widetilde{Y} \to Y$ is given by $\eta = \eta_M|\widetilde{Y}$. Note that $\widetilde{Y}$ is 1-connected under these circumstances. The set $\eta^{-1}(y_1)$ may be labelled

$$\eta^{-1}(y_1) = \{\widetilde{y}_1\alpha = \widetilde{y}_\alpha,\quad \alpha \in \pi'\}$$

and furthermore, we may identify

$$\pi_k(M, M-B) \equiv H_k(\widetilde{Y}, \widetilde{Y}-\eta^{-1}(y_1)) \equiv \mathbb{Z}[\pi']$$

and, in turn,

$$H_k(Y, Y-\eta^{-1}(y_1)) \equiv \sum_{\alpha\in\pi'} H_k(\widetilde{V}_\alpha, \widetilde{V}_\alpha - \widetilde{y}_\alpha)$$

where $\widetilde{V}_\alpha$, $\alpha \in \pi'$ is a Euclidean neighborhood of $\widetilde{y}_\alpha$ and $\widetilde{V}_\alpha = \alpha(\widetilde{V}_1)$. If $V = \eta(\widetilde{V}_1)$, we choose a local orientation of V at y_1, thereby determining a generator

$$\gamma_1 \in H_k(\widetilde{V}_1, \widetilde{V}_1 - \widetilde{y}_1).$$

Then, $\gamma_1\alpha$ generates $H_k(\widetilde{V}_\alpha, \widetilde{V}_\alpha - \widetilde{y}_\alpha)$ and if

$$j_1 : (\tilde{Y}, \tilde{Y} - \eta^{-1}(y_1)) \subset (\tilde{M}, \tilde{M} - \eta_M^{-1}(B))$$

we may identify $\mathbb{Z}[\pi']$ with the image of $H_k(\tilde{Y}, \tilde{Y} - \eta^{-1}(y_1))$ via the correspondence

$$\alpha \leftrightarrow j_{1*}(\gamma_\alpha) = j_{1*}(\gamma_1 \alpha).$$

For $\sigma \in \pi''$, we set $\operatorname{sgn} \sigma = \pm 1$ according as σ preserves or reverses a local orientation of Y. Observe, that $j_{1*}(\gamma_1)$ is represented by a k-cell D_1^k in $\tilde{M}$ transverse to $\tilde{B}$ identified with $\tilde{\theta}(\tilde{B}) \subset \tilde{M}$, see Fig. 1.

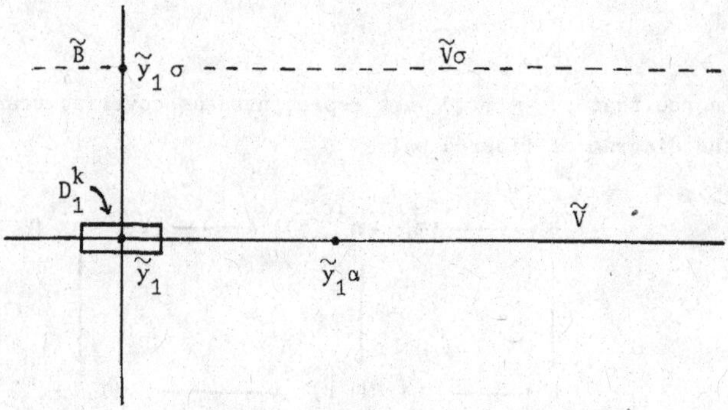

Figure 1.

We are now in a position to compute the action of $\pi = \pi''\pi'$ on $\mathbb{Z}[\pi']$ as follows. Take $\sigma \in \pi''$, $\tau \in \pi'$. Then,

$$\begin{aligned}
\alpha \circ \sigma\tau &= [j_{1*}(\gamma_1 \alpha)]\ \sigma\tau \\
&= [j_{1*}(\gamma_1)]\ \alpha\ \sigma\ \tau \\
&= [j_{1*}(\gamma_1)]\ \sigma\ \sigma^{-1}\ \alpha\ \sigma\ \tau \\
&= (\operatorname{sgn}_B \sigma)\ (\operatorname{sgn}_Y \sigma)\ [j_{1*}(\gamma_1)]\ \sigma^{-1}\ \alpha\ \sigma\ \tau \\
&= (\operatorname{sgn}_B \sigma)\ (\operatorname{sgn}_Y \sigma)\ (\sigma^{-1}\ \alpha\ \sigma)\ \tau
\end{aligned}$$

where we note that $(\sigma^{-1}\ \alpha\ \sigma)\ \tau \in \pi'$. ∎

We now apply Proposition (3.3) to the following special case. Let $Y \to M \to B$ denote a smooth fiber bundle and $Y \to M \times_B M \to M$ the associated fiber square. Then, applying Proposition (3.3) to the latter, we obtain

(3.4) *COROLLARY*

The action of $\pi = \pi'' \pi'$ on $\mathbb{Z}[\pi'] \equiv \pi_k(M \times_B M, M \times_B M - \Delta M)$, where π, π', π'' are the fundamental groups of $M \times_B M$, Y and M, respectively, is given by

$$\alpha \circ \sigma\tau = (\mathrm{sgn}_M \sigma)\,(\mathrm{sgn}_Y\, p_*(\sigma))\, \sigma^{-1} \alpha \sigma \tau, \quad \sigma \in \pi'',\ \alpha, \tau \in \pi'.$$

If, M is orientable and the fiber bundle $M \to B$ is also orientable, then

$$\alpha \circ \sigma\tau = \sigma^{-1} \alpha \sigma \tau, \quad \sigma \in \pi'',\ \alpha,\ \tau \in \pi'.$$

An important special case of this corollary is the case when the given fibration $Y \xrightarrow{i} M \xrightarrow{p} B$ has the special property that $i_* : \pi' \to \pi''$ is injective. Then, if $j : M \times_B M \to M \times M$ represents the inclusion map, the diagram

$$\begin{array}{ccccccccc}
1 & \longrightarrow & \pi' & \longrightarrow & \pi & \rightleftarrows & \pi'' & \longrightarrow & 1 \\
 & & \downarrow{\scriptstyle i_*} & & \downarrow{\scriptstyle j_*} & & \downarrow{\scriptstyle id} & & \\
1 & \longrightarrow & \pi'' & \longrightarrow & \pi'' \times \pi'' & \rightleftarrows & \pi'' & \longrightarrow & 1
\end{array}$$

tells us that $\ker i_* = \ker j_*$ and i_* injective forces j_* to be injective and we may identify the group π with image j_* so that

$$\pi = \{(\tau_1, \tau_2) \ni \pi'' \times \pi'' : \tau_1^{-1} \tau_2 \in \pi'\}.$$

Then, for $\alpha \in \pi'$

$$\begin{aligned}
\alpha \circ (\tau_1, \tau_2) &= (\mathrm{sgn}_M \tau_1)\,(\mathrm{sgn}_Y\, p_*(\tau_1))\,(\tau_1^{-1}, \tau_1^{-1})(1,\alpha)(\tau_1, \tau_1)(1, \tau_1^{-1}\tau_2) \\
&= (\mathrm{sgn}_M \tau_1)\,(\mathrm{sgn}_Y\, p_*(\tau_1))\, \tau_1^{-1} \alpha\, \tau_2
\end{aligned} \tag{1}$$

where, τ_1, τ_2 belong to π'' and $\tau_1^{-1}\tau_2$ belongs to π'.

When $Y \to B \times Y \to B$ is the trivial bundle, the element of $\pi = \pi_1(B \times Y \times Y)$ has the form (σ, τ_1, τ_2) where $\sigma \in \pi_1(B)$, $\tau_1, \tau_2 \in \pi'$, and the action takes the form

$$\begin{aligned}
\alpha \circ (\sigma, \tau_1, \tau_2) &= (\mathrm{sgn}_B \sigma)\,(\mathrm{sgn}_Y \sigma)\,(\sigma^{-1}, \tau_1^{-1})(1,\alpha)(\sigma, \tau_2) \\
&= (\mathrm{sgn}_B \sigma)\,(\mathrm{sgn}_Y \tau_1)\,(1, 1, \tau_1^{-1}\alpha, \tau_2) \qquad (2) \\
&= (\mathrm{sgn}_B \sigma)\,(\mathrm{sgn}_Y \tau_1)\, \tau_1^{-1} \alpha\, \tau_2
\end{aligned}$$

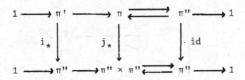

When B is a point (2) becomes

$$\alpha \circ (\tau_1,\tau_2) = (\mathrm{sgn}_Y \tau_1)\tau_1^{-1}\alpha\,\tau_2 \; ; \; \alpha, \tau_1, \tau_2 \in \pi', \tag{3}$$

which coincides with the action in [4] and [5] for the single space theory.

Suppose now that

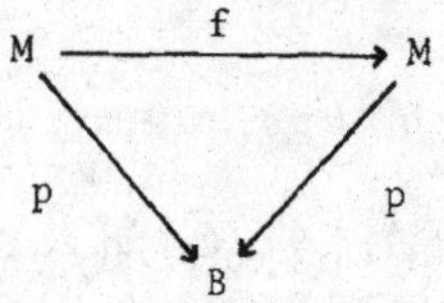

is a fiber preserving map and we wish to calculate the local system $\mathcal{B}_B(f)$. We consider the diagram

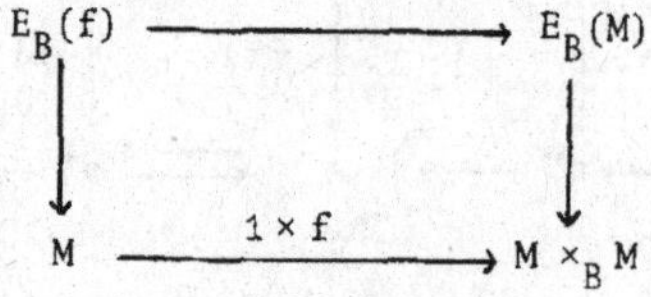

and observe again that $\mathcal{B}_B(f)$ is the pull-back of $\mathcal{B}_B$ by $1 \times f$. Thus initially the action of π'' on $\mathbb{Z}[\pi']$ is given by

$$\alpha \circ \rho = \alpha \circ (1\times f)_* (\rho), \quad \alpha \in \pi', \quad \rho \in \pi''.$$

Now, consider the diagram of self-explanatory maps

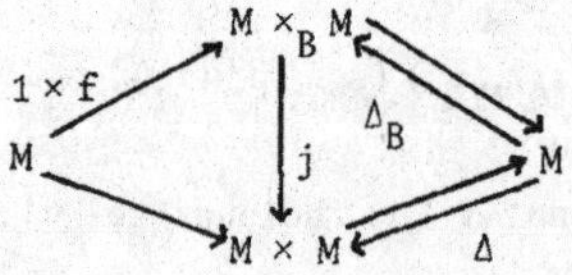

which allows us to write

$$(1\times f)_* (\rho) = \sigma\tau, \quad \sigma \in \pi'', \quad \tau \in \pi'$$

where

$$\sigma = (\Delta_B)_*(\rho) \equiv \rho, \quad \tau = [(\Delta_B)_*(\rho^{-1})]\,(1\times f)_*(\rho) \equiv \rho^{-1}(1\times f)_* (\rho).$$

Thus, we obtain

(3.5) *PROPOSITION*

The local system $\mathcal{B}_B(f)$ is given by the action

$$\alpha \circ \rho = (\mathrm{sgn}_M \rho)\,(\mathrm{sgn}_Y\, p_*(\rho))\; \rho^{-1} \alpha\, (1\times f)_*(\rho) \quad \alpha \in \pi', \quad \rho \in \pi''.$$

(3.6) *PROPOSITION*

In the special case, that the given fibration $Y \xrightarrow{i} M \xrightarrow{p} B$ has the property that $i_* : \pi' \to \pi''$ is injective then the local system $\mathcal{B}_B(f)$ is given by the action

$$\alpha \circ \rho = (\mathrm{sgn}_M \rho)\,(\mathrm{sgn}_Y\, p_*(\rho))\; \rho^{-1} \alpha\, f_*(\rho).$$

(3.7) *REMARK*

Note that in the special case that B is a point, the action in Proposition (3.6) becomes $\alpha \circ \rho = (\mathrm{sgn}_M \rho)\, \rho^{-1} \alpha \varphi(\rho)$, where $\varphi = f_*$, agreeing with the result in [4].

4. Examples.

Our first example is a generalization of one given by Dold in [2]. Dold's example is the following: Let $S^1 \times S^1 \xrightarrow{p} S^1$ denote the trivial fibration and $f : S^1 \times S^1 \to S^1$ the fiber preserving map given by $f(x,y) = (x,xy)$, using complex multiplication xy. Then, Dold's index $[f] \in \pi^0_{\text{stable}}(S^1) = \mathbb{Z}_2$ is the unique non-zero element so that f has the property that every fiberwise perturbation of f has fixed points.

Let G denote a compact connected manifold of dimension ≥ 3 with a multiplication $G \times G \to G$, written xy, which admits a right identity e. Thus, G may be a Lie group, an H-space or, in fact, we may take $G \times G \to G$ as projection on the first factor, for any G. Now, we consider the trivial bundle $G \times G \xrightarrow{p} G$,

$p(x,y) = x$ and the fiber-preserving map

$$f : G \times G \to G \times G, \quad (x,y) \mapsto (x,xy)$$

Note that, if G is a Lie groupe (for example) and $\alpha(t)$ is a simple path in G starting at e, then the homotopy $H_t(x,y) = (x\alpha(t),xy)$ is a *non* fiberwise perturbation of f with $g = H_1$ being fixed point free. However, as we now show our primary obstruction $o(f)$ is never zero so that any *fiberwise* perturbation of f necessarily has fixed points.

(4.1) *THEOREM*

$o_G(f) \in H^m(G \times G, \dot{B}_G(f))$ is never zero.

Proof

First let $e : G \to G$ denote the constant map. Then, considering G as a trivial bundle over a point we have $o_B(e)$ coinciding with the obstruction index

$$o(e) \in H^m(G, B(e))$$

as set forth in [4] and [5]. The constant map e has only one Nielsen class and its index is the Lefschetz number $L(e) = 1$. Also, there is only one Reidemesiter class [4] (the action $\alpha * \sigma = \alpha\sigma$). Thus

$$< o(e), \mu > = [1] \in \mathbb{Z}[1]$$

where μ is the (twisted) fundamental class of G. Thus, $o(e) \neq 0$.

Now, consider the maps

$$G \xrightarrow{1\times e} G \times G \xrightarrow{1\times f} G \times G \times G \equiv (G \times G) \times_G (G \times G)$$

given by

$$x \mapsto (x,e) \quad (x,y) \mapsto (x,y,xy).$$

The fixed point indicator $E_G(G \times G)$ consists of triples of paths in G (α,β,γ) such that $\beta(0) \neq \gamma(0)$. Then, if we consider the diagram of pull-backs

$$\begin{array}{ccc} E' & \longrightarrow & E_G(f) \\ \Big\downarrow & & \Big\downarrow q_f \\ G & \xrightarrow{\;1\times e\;} & G \times G \end{array}$$

where $E_G(f)$ is the pull-back of $E(G\times G)$ by $1\times f$ and E' the pull-back of $E_G(f)$ by $1\times e$, we notice that the fiber $F'_x = \eta^{-1}(x)$, $x \in G$ is given by

$$F'_x = \{(\alpha,\beta,\gamma),\ \alpha(1) = x,\ \ \beta(1) = e,\ \gamma(1) = x,\ \ \beta(0) \neq \gamma(0)\}$$

where again α,β,γ are paths in G. Since the first coordinate path α is "free" we see that E' may be fiberwise deformed to the subspace

$$E'(G) = \{(\bar{x},\beta,\gamma),\ \ \bar{x} = \text{constant path at } \ \gamma(1) = x\}$$

$$= \{(\beta,\gamma) : \beta(1) = e,\ \ \beta(0) \neq \gamma(0)\}.$$

But the obstruction $o(e) \in H^m(G, \mathcal{B}(e))$ is to finding a cross section to $E(G) \xrightarrow{q_e} G$ where

$$E(G) = \{(\gamma,\beta) : \beta(1) = e,\ \beta(0) \neq \gamma(0)\}.$$

Thus, since $E(G)$ and $E'(G)$ are fiberwise homeomorphic we have $o'(e) \neq 0$ where $o'(e)$ is the obstruction for finding a cross section for $E' \to G$. Thus,

$$(1\times e)^*\ (o_G(f)) = o'(e) \neq 0$$

and hence $o_G(f) \neq 0$. ■

(4.2) *REMARK*

While the fiberwise obstruction index $o_B(f)$ is not defined without dimensional restrictions, nevertheless the argument above can be made purely geometric to include the S^1 case.

(4.3) *COROLLARY*

Let $f : G\times G \to G\times G$ denote the above map where G has a given multiplication with right sided identity. Then,

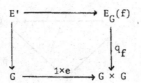

$$\begin{array}{ccc} G \times G & \xrightarrow{f} & G \times G \\ & {\scriptstyle p} \searrow \quad \swarrow {\scriptstyle p} & \\ & G & \end{array}$$

is not fiberwise homotopic to a fixed point free map. However, if G is a topological group or H-space, f is homotopic to a fixed point free map (using of course a non-fiberwise homotopy).

(4.4) *REMARK*

When G is H-space, the ordinary Lefschetz number $L(f) = 0$ and since G is also a Jiang space [1], the Nielsen number $n(f) = 0$. This justifies the last implication in the above corollary.

Our next example, will be of the form

$$\begin{array}{ccc} S^1 \times S^n & \xrightarrow{f} & S^1 \times S^n \\ & {\scriptstyle p} \searrow \quad \swarrow {\scriptstyle p} & \\ & S^1 & \end{array}$$

where S^n is a sphere of dimension $n \geq 3$ and p is projection, with the property that $o_B(f) = 0$, but a non-zero secondary obstruction to finding a cross section for the fixed point indicator $q_f : E_{S^1}(f) \to S^1 \times S^n$, precludes deforming f in a fiberwise fashion to a fixed point free map.

Let $M = S^1 \times S^n$ and let $L = S^1 \vee S^n$, the n-skeleton of M. If $g : S^1 \times S^n \to S^n$ is any map, g induces the map

$$f_0 = 1 \times g : M = S^1 \times S^n \to M \times_B M = S^1 \times S^n \times S^n$$

by

$$(t,u) \mapsto (t,u,g(t,u)).$$

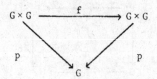

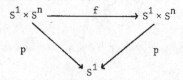

Let $q : S^1 \times S^n \times S^n \to S^1 \times S^n$ denote the projection $q(t,u,v) = (t,u)$ and consider $q_* : \pi_{n+1}(S^1 \times S^n \times S^n) \to \pi_{n+1}(S^1 \times S^n)$, viewed as projection $\mathbb{Z}_2 \oplus \mathbb{Z}_2 \to \mathbb{Z}_2$. Let $d \in C^{n+1}(M,L,\pi_{n+1}(S^1 \times S^n \times S^n))$ denote the $(n+1)$-cochain corresponding to the non-zero element in the kernel of q_*. Then, there is a map $G : M \times \{0,1\} \cup L \times I \to S^1 \times S^n \times S^n$ such that the difference cochain [6]

$$d^{n+1}(f_0, G, f_1) = d \neq 0$$

where f_1 corresponds to $G|M \times \{1\}$ and $G(x,t) = f_0(x)$, $x \in L$. Thus, the partial homotopy G is not extendable to $M \times I$. On the other hand, the partial homotopy

$$qG : M \times \{0,1\} \cup L \times I \to S^1 \times S^n$$

has difference cochain 0 and so is extendable to $M \times I$. Thus, qf_1 is homotopic to the identify and hence f_1 is homotopic to a map $f_2 : M \to S^1 \times S^n \times S^n$ of the form

$$f_2 : (t,u) \mapsto (t,u,\overline{f}_2(t,u))$$

and the homotopy $f_1 \sim f_2$ is relative to L. Clearly, f_2 is not homotopic to f_0, relative to L. The map we desire

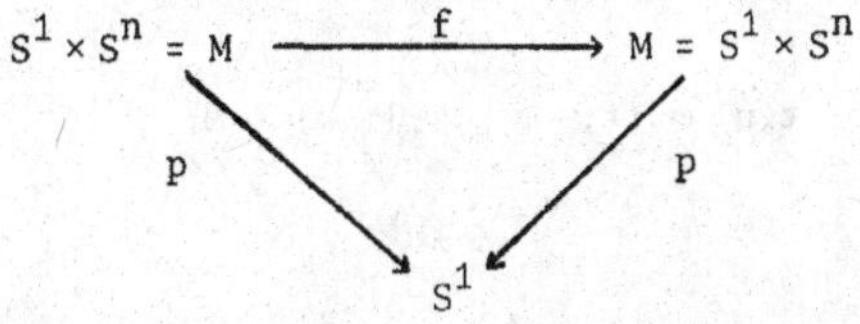

is given by $f : (t,u) \mapsto (t,\overline{f}_2(t,u))$ where we start with the special map $g(t,u) = -u$, that is

$$f_0 : (t,u) \mapsto (t,u, -u).$$

Note first that f is fixed point free on L. Thus, the primary fixed point obstruction $o_B(f) \in H^n(S^1 \times S^n, \mathcal{B}_B(f))$ vanishes. Notice, also using the results of Section 3, that the local group for $\mathcal{B}_B(f)$ is $\pi_n(S^n, S^n - y) \approx \pi_n(S^n) \approx \mathbb{Z}$ with trivial action. Next, consider the diagram

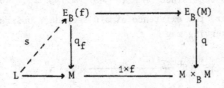

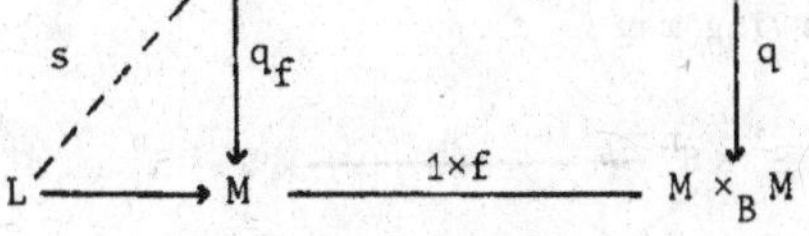

with partial section $s(x) = (\overline{x}, \overline{f(x)})$. If F is the fiber for q, note that

$$\pi_n(F) \approx \pi_{n+1}(S^n, S^n - y) \approx \mathbb{Z}_2$$

so that the fundamental group of $M \times_B M$ acts trivially in this example. Let

$$c^{n+1}(s) \in C^{n+1}(M, \mathbb{Z}_2)$$

denote the (secondary) obstruction to extending s to M. If $c^{n+1}(s) = 0$, then an extension of s to a full section implies that f is fiber homotopic to a fixed point free map h, relative to L. h would take the form

$$h: (t,u) \mapsto (t, \bar{h}(t,u)) \quad \bar{h}(t,u) \neq u$$

where

$$h(t,u) = (t, -u) \quad \text{for} \quad (t,u) \in L.$$

Then, by using the great circle arc joining $\bar{h}(t,u)$ and $-u$, h, and hence f, is homotopic (rel L) to the map

$$(t,u) \mapsto (t, -u), \qquad (t,u) \in M.$$

Thus,

$$f_0 : (t,u) \mapsto (t,u,-u)$$

and

$$f_2 : (t,u) \mapsto (t,u,\bar{f}_2(t,u))$$

are homotopic relative to L which is a contradiction. Thus, if $o^{n+1}(s)$ represents the secondary obstruction cohomology class containing $c^{n+1}(s)$, $o^{n+1}(s) \neq 0$. Thus, $o^{n+1}(s)$ is a non-zero secondary obstruction to finding a cross section for the fixed point indicator $q_f : E_{S^1}(f) \to S^1 \times S^n$. To see that $o^{n+1}(s)$ is the only secondary obstruction class observe that since $H^{n+1}(S^1 \times S^n; \mathbb{Z}_2) \approx \mathbb{Z}_2$, we have only to eliminate the 0 class. However, the existence of a 0 secondary obstruction would imply the existence of a fiber-preserving map

$$\begin{array}{ccc} S^1 \times S^n & \xrightarrow{h} & S^1 \times S^n \\ & \searrow \quad \swarrow & \\ & S^1 & \end{array}$$

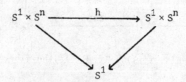

which is fixed point free and fiber homotopic to f. This induces a section for the fixed point indicator $q_f : E_{S^1}(f) \to S^1 \times S^n$, which, because the fiber F is at least 1-connected, since $n \geq 3$, we may assume coincides with the section s_0 on the 1-skeleton $S^1 \times e_0$. But, this would force $o^{n+1}(s) = 0$ which is a contradiction. We summarize with the following result.

(4.5) *PROPOSITION*

There is a fiber-preserving map

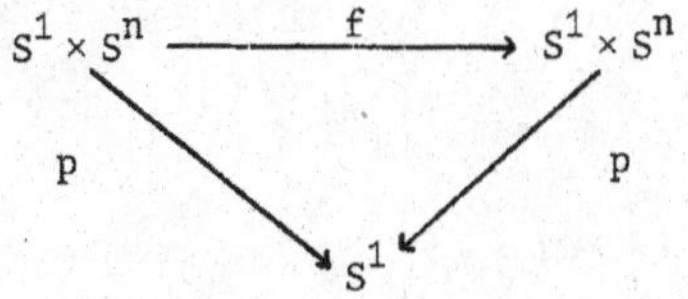

such that

1) The map $f|S^n$, the restriction of f to the fiber, is the antipodal map and hence the ordinary Lefschetz number $L(f|S^n) = 0$. The Nielsen number $n(f)$ of f is zero so that f is homotopic (not over S^1) to a fixed point free map.

2) The primary obstruction fixed point index $o_{S^1}(f) = 0$.

3) The secondary obstruction set for finding a cross section for the fixed point indicator $q_f : E_{S^1}(f) \to S^1 \times S^n$ is the single non-zero cohomology class and hence f is not fiber homotopy equivalent to a fixed point free map.

References

[1] BROWN, R.: The Lefschetz Fixed Point Theorem, Scott-Foresman (1971).

[2] DOLD, A.: The fixed point index of fiber-preserving maps, Inventiones Math. 25 (1974), 281-297.

[3] FADELL, E.: Generalized normal bundles for locally-flat imbeddings, Trans. Amer. Math. Soc. 114 (1965), 488-513.

[4] FADELL, E. and HUSSEINI, S.: Fixed point theory for non-simply connected manifolds, to appear shortly in Topology.

[5] FADELL, E. and HUSSEINI, S.: Local fixed point index theory for non-simply connected manifolds, to appear in the Illinois Journal of Math.

[6] WHITEHEAD, G.W.: Elements of Homotopy Theory, Springer Verlag (1978).

A SIMPLICIAL APPROACH TO THE FIXED POINT INDEX

By

GILLES FOURNIER*

Département de mathématique
Université de Sherbrooke
Sherbrooke, Québec, Canada J1K 2R1

0. Introduction

The fixed point index in the Euclidean spaces has been the object of many papers, H. Hopf [8], J. Leray [10], B. O'Neill [13], F.E. Browder [1] and A. Dold [3]. There has been many generalizations of these indices to infinite dimensional spaces: J. Leray [11], A. Granas [6,7], F.E. Browder [2], J. Eells and G. Fournier [4] and R.D. Nussabum [12]. Se also H.-O. Peitgen and H.-W. Siegberg [14].

The nicest exposition of fixed point index in the Euclidean space which has the broadest range and the most properties is the one given by A. Dold. In this paper, we shall give a new definition of this index which is almost the one given by O'Neill but which has in addition the commutativity property and the index modulo p property. This last property was first proved by H. Steinlein [16,17,18]. Other proofs are given in Krasnosel'skiǐ [9], Tromba [19] and Geoghegan [5]. We give a very short and elementary proof of this property.

We shall define the fixed point index of a map $f : V \to X$ to be the Lefschetz number or trace of a selfmap (not necessarily a chain map) on a simplicial chain complex. This provides an alternative way of computing the fixed point index (particularly in view of lemma (5.1)). This index can be generalized in the usual way (see A. Granas [7]) to some infinite dimensional spaces and their r-images.

A generalization of this paper to fiber preserving maps (except for the modulo p property) will be given in a forthcoming article by E. Fadell and S.Y. Husseini.

---

*This research was partially supported by the NSERC of Canada and a Bourse d'action concertée du ministère de l'éducation du Québec.

1. THE SUBDIVISION CHAIN MAP

In this paper we shall use the notations and definitions of Spanier [15].

There exists a natural way of comparing the orientation of a simplex

$$S = \langle v_0,\dots,v_n\rangle$$

of a simplicial complex K with the orientation of a simplex $\sigma = \langle v_0',\dots,v_n'\rangle$ of a subdivision K' of K for which $\sigma \subset |\overline{S}|$. Let us denote by $S_{K'}$ the subcomplex of K' which is a subdivision of $\overline{S}$, a subcomplex of K. We have that $v_0'-v_0,\dots,v_n'-v_0$ are in the convex hull of the vectors $0, v_1-v_0,\dots,v_n-v_0$ thus they can be expressed uniquely as a linear combination of the vectors $v_1-v_0,\dots,v_n-v_0$. Let

$$v_i-v_0 = \sum_{j=1}^{n} v_{ij}' \quad (v_j-v_0)$$

for all $i=0,\dots,n$. We shall say that σ has the same orientation as S, denoted $\sigma \in \mathcal{O}_S$, provided that

$$\det\begin{pmatrix} 1 & v_{01}' & \dots & v_{0n}' \\ \vdots & & & \\ 1 & v_{n1}' & \dots & v_{nn}' \end{pmatrix} = \det\begin{pmatrix} v_{11}'-v_{01}' & \dots & v_{1n}'-v_{0n}' \\ \vdots & & \\ v_{n1}'-v_{01}' & \dots & v_{nn}'-v_{0n}' \end{pmatrix} > 0. \qquad (1.1)$$

Let us note here that if the w_{ij}' (respectively v_{ij}) are the condinates of the vectors v_i' (respectively v_i) with respect to some other basis then $\sigma \in \mathcal{O}_S$ if and only if

$$\det\begin{pmatrix} 1 & w_{01}' & \dots & w_{0n}' \\ \vdots & & & \\ 1 & w_{n1}' & \dots & w_{nn}' \end{pmatrix} \times \det\begin{pmatrix} 1 & v_{01} & \dots & v_{0n} \\ \vdots & & & \\ 1 & v_{n1} & \dots & v_{nn} \end{pmatrix} > 0. \qquad (1.2)$$

Let $C(K) = \{C_q(K)\}_{q\in\mathbb{N}}$ denote the simplicial chain complex of the simplicial complex K, with boundary operator ∂. Note that if K is a subcomplex of L then $C_qK \subset C_qL$ for all q, that is $C(K) \subset C(L)$. The following is well known if the subdivision is a barycentric subdivision.

(1.3) *PROPOSITION*

Let K be a simplicial complex and let K' be a subdivision of K, then there exists a unique chain map $\tau\colon C(K) \to C(K')$ called the subdivision map, such that

$$\tau_0(\langle v\rangle) = \langle v\rangle, \text{ for any vertex } v \text{ of } K \tag{1.3.1}$$

$$\tau_q(S) \in C_q(S_{K'}), \text{ for any } q\text{-simplex } S \text{ of } K. \tag{1.3.2}$$

Furthermore if S is a q-simplex of K we have that

$$\tau_q(S) = \sum_{\sigma\in\mathcal{O}_S} \sigma. \tag{1.3.3}$$

Before proving this proposition we need the following lemmas and definition.

(1.4) *DEFINITION*

Let S be a q-simplex, and let K be a subdivision of $\bar{S}$. Then a (q-1)-simplex σ of K' is an *interior simplex* of S if and only if $\langle\sigma\rangle \subset \langle S\rangle$; it is a *frontier simplex* of S if and only if $|\bar{\sigma}| \subset |\dot{S}|$. Note that any (q-1)-simplex K is either a frontier or an interior simplex of S. The following three lemmas are needed for the proof of (1.3).

(1.5) *LEMMA*

Let K be a subdivision of $\bar{S}$, where $S = \langle v_0,\dots,v_q\rangle$ is a q-simplex. Let $\sigma = \langle a_0,\dots,a_{q-1}\rangle$ be a frontier simplex of K, then there exists exactly one simplex, say $\langle a_0,\dots,a_q\rangle$, which has σ as a face. Furthermore if $\sigma \in \mathcal{O}_{\langle v_0,\dots,v_{q-1}\rangle}$ then

$$\langle a_0,\dots,a_q\rangle \in \mathcal{O}_S.$$

<u>Proof</u>

Consider $|\bar{S}|$ included in $\mathbb{R}^q$. Take a point P in the interior $\langle\sigma\rangle$ of σ, take a straight line D intersecting the affine hyperplane Π containing $\{v_0,\dots,v_{q-1}\}$ hence containing $\{a_0,\dots,a_{q-1}\}$. Since S is only on one side of Π, the set of intervals $\{\langle\sigma'\rangle \cap D : \sigma' \in K\}$ is a finite partition of the convex $|S| \cap D$ which has P as an endpoint. Since $\langle\sigma\rangle \cap D = P$, there is exactly one σ' such that $P \notin \langle\sigma'\rangle \cap D$ which is an interval with endpoint P. Thus $P \in |\bar{\sigma'}|$ and so $\langle\sigma\rangle \in |\bar{\sigma'}|$ which implies that $\sigma \in |\bar{\sigma'}|$, that is $\sigma \subset \sigma'$. Finally $\{v_1-v_0,\dots,v_q-v_0\}$ is a basis of $\mathbb{R}^q$, and we may assume that $\sigma \in \mathcal{O}_{\langle v_0,\dots,v_{q-1}\rangle}$. Then if

$$a_i - v_0 = \sum_{j=1}^{q} a_{ij}\,(v_j - v_0),$$

for $i = 0,\dots,q$, we have that $a_{iq} = 0$ for $i = 0,\dots,q-1$ and $a_{qq} > 0$, since $v_0,\dots,v_{q-1}$, $a_0,\dots,a_{q-1} \in \Pi$ and a_q and v_q are on the same side of Π. That is

$$\det \begin{pmatrix} 1 & a_{01} & \dots & a_{0q} \\ \vdots & \vdots & & \vdots \\ 1 & a_{q1} & \dots & a_{qq} \end{pmatrix} = \det \begin{pmatrix} 1 & a_{01} & \dots & a_{0q-1} \\ \vdots & \vdots & & \vdots \\ 1 & a_{q11} & \dots & a_{q-1\ q-1} \end{pmatrix} > 0,$$

and we have the conclusion. ■

(1.6) *LEMMA*

Let K be a subdivision of $\overline{S}$, where $S = \langle v_0,\dots,v_q\rangle$ is a q-simplex. Let $\sigma = \langle a_0,\dots,a_{q-1}\rangle$ be an interior simplex of S, then there exists exactly two simplexes of K, say $\langle a_0,\dots,a_q\rangle$ and $\langle a_0,\dots,a_{q-1},b_q\rangle$ which have σ as a face. Furthermore exactly one of these has the same orientation as S.

Proof

It is similar to the proof of (1.5). ■

(1.7) *LEMMA*

Let K be a subdivision of $\overline{S}$, where S is a q-simplex. Take $\sigma_0 \in \mathcal{O}_S$, then there exists a sequence $\sigma_0,\dots,\sigma_m$ of elements of $\mathcal{O}_S$ such that

(i) σ_{i-1} and σ_i have a common face

(ii) σ_m has a frontier simplex as a face.

Proof

The proof is easier than (1.3) though it uses the same technique. ■

Proof of (1.3)

Let $\tau = \{\tau_q\colon C_q(K) \to C_q(L)\}$ be a sequence of linear maps. Since $\mathcal{O}_S$ is a basis of $C_q(S_{K'})$, τ satisfies (1.3.2) if and only if for any q-simplex S of K we can write

$$\tau_q(S) = \sum_{\sigma \in \mathcal{O}_S} \alpha_\sigma \, \sigma, \tag{1.3.4}$$

for a choice of $\alpha_\sigma \in \mathbb{R}$. Furthermore τ satisfies (1.3.1) if and only if $\alpha_{<v>} = 1$ for any vertex v of K.

It remains to show that τ is a chain map if and only if $\alpha_\sigma = 1$ for any $\sigma \in \mathcal{O}_S$, any q-simplex S of K and any $q > 0$. It is sufficient to show, by induction on q, that $\partial_q \circ \tau_q = \tau_{q-1} \circ \partial_q$ for all $0 < q \leq n$ if and only if $\alpha_\sigma = 1$ for any $\sigma \in \mathcal{O}_S$, any q-simplex S and any $q \leq n$.

The case $n = 0$ is evident from the fact that τ satisfies (1.3.1).

Let us suppose that the statement is true for (n-1) and let us prove it for n.

Take $S = \langle v_0,\dots,v_n\rangle$ an n-simplex of K. It is sufficient to prove that $\partial_q \circ \tau_q(S) = \tau_{q-1} \circ \partial_q(S)$ if and only if $\alpha_\sigma = 1$ for any $\sigma \in \mathcal{O}_S$. But

$$\partial_q(S) = \sum_{j=0}^{q} (-1)^j \, S_j,$$

where $S_j = \langle v_0,\dots,v_q\rangle$ is a face of S. Thus

$$\tau_{q-1} \circ \partial_q(S) = \sum_{j=0}^{q} (-1)^j \, \tau_{q-1}(S_j) = \sum_{j=0}^{q} \sum_{\sigma \in \mathcal{O}_{S_j}} (-1)^j \, \alpha_\sigma \, \sigma$$

and, since S_j is an (n-1)-simplex, we get by our induction hypothesis that $\alpha_\sigma = 1$ for all $\sigma \in \mathcal{O}_{S_j}$ and all j; that is

$$\tau_{q-1} \circ \partial_q(S) = \sum_{j=0}^{q} \sum_{\sigma \in \mathcal{O}_{S_j}} (-1)^j \, \sigma. \tag{1.3.5}$$

On the other hand, by (1.3.4)

$$\partial_q \circ \tau_q(S) = \sum_{\langle w_0,\dots,w_q\rangle \in \mathcal{O}_S} \sum_{j=0}^{q} (-1)^j \, \alpha_{\langle w_0,\dots,w_q\rangle} \langle w_0,\dots,\overset{j}{\check{\ }},\dots,w_q\rangle . \tag{1.3.6}$$

Now we will have equality between (1.3.5) and (1.3.6) if and only if for any (n-1)-simplex σ of $S_{K'}$, the subdivision of $\overline{S}$, the sum of its coefficients in (1.3.5) equals the sum of its coefficients in (1.3.6).

Take $\sigma' = \langle a_0,\ldots,a_{n-1}\rangle$ an interior simplex of S. Since all the simplexes occuring in the last member of (1.3.5) are frontier simplexes of S, the coefficient of σ' in (1.3.5) is zero. By (1.6), let $\sigma_a = \langle a_0,\ldots,a_q\rangle$ and $\sigma_b = -\langle a_0,\ldots,a_{q-1},b_q\rangle$ be the two elements of $\mathcal{O}_S$ having σ' as a face. Thus the two coefficients of σ' in (1.3.6) are obtained by deriving these two simplexes; they are $(-1)^q\,\alpha_{\sigma_a}$ and $(-1)^{q+1}\,\alpha_{\sigma_b}$. Thus we have equality for σ' if and only if $\alpha_{\sigma_a} = \alpha_{\sigma_b}$.

Take $\sigma' = \langle a_0,\ldots,a_{n-1}\rangle$ a frontier simplex of S. If $q > 1$, without loss of generality, we may assume that $\sigma' \in \mathcal{O}_{S_q}$. Thus the coefficient of σ' in (1.3.5) is $(-1)^q$. But, by (1.5), the unique q-simplex $\sigma = \langle a_0,\ldots,a_q\rangle$ having σ' as a face belongs to $\mathcal{O}_S$. Thus its coefficient in (1.3.6) is $(-1)^q\,\alpha_\sigma$. We have equality of the coefficients of σ' if and only if $\alpha_\sigma = 1$.

Thus we have that if $\alpha_\sigma = 1$ for any $\sigma \in \mathcal{O}_S$ then equation (1.3.5) equals equation (1.3.6). On the other hand if we have equality, for any $\sigma_0 \in \mathcal{O}_S$ let $\sigma_0,\ldots,\sigma_m$ be a sequence as in (1.7). By what preceeds we must have that $\alpha_{\sigma_i} = \alpha_{\sigma_{i-1}}$ for all $i = 1,\ldots,m$ and $\alpha_{\sigma_m} = 1$, that is $\alpha_{\sigma_0} = 1$ and thus we have the conclusion. ■

We have the following fact about the subdivision map.

(1.8) *PROPOSITION*

Let K' be any subdivision of K, then for any simplicial approximation $\varphi : K' \to K$ of the identity map $1_{|K|}$ we have that $C_q\varphi \circ \tau_q = 1_{C_q(K)}$, where $\tau : C(K) \to C(K')$ is the subdivision map.

Proof *(By induction on* q*)*

If $q = 0$, we have that $\tau_0(v) = v$ for any $v \in K$, but $C_0\varphi(v) = v$ for any $v \in K \cap K'$, thus $C_0\varphi \circ \tau_0 = 1_{C_0(K)}$. Now suppose that it is true for all $j < i$.

Let S be a q-simplex of K. We have that

$$\partial_q(C_q\varphi \circ \tau_q(S)) = C_{q-1}\varphi \circ \tau_{q-1}(\partial_q S) = \partial_q S,$$

since these are chain maps and $C_{q-1} \circ \tau_{q-1} = 1_{C_{q-1}}(K)$. Now take $\sigma \in \mathcal{O}_S$, then $\langle\sigma\rangle \subset \langle S\rangle$. Hence $\varphi(\sigma) \subset S$, since φ is a simplicial approximation of the identity. Thus $C_q\varphi(\sigma) = r_\sigma S$ where $r_\sigma \in \{-1,0,1\}$. This implies that

$$C_q\varphi \circ \tau(S) = C_q\varphi(\sum_{\sigma\in\mathcal{O}_S} \sigma) = \sum_{\sigma\in\mathcal{O}_S} C_q\varphi(\sigma) = \sum_{\sigma\in\mathcal{O}_S} r_\sigma S = r_S S$$

for some integer r_S. It remains to show that $r_S = 1$. But we already have that

$$\partial_q S = \partial_q(C_q\varphi \circ \tau(S)) = \partial_q(r_S S) = r_S \partial_q S,$$

which gives the conclusion. ■

(1.9) *COROLLARY (Spener's lemma, for the simplex)*

If $K = (V,\Sigma)$ is a subdivision of $\overline{S}$, where S is a q-simplex, and $i:V \to \{0,\ldots,q\}$ is a function such that

i is a bijection on the vertices of S (1.9.1)

if σ is a face of S then $i(V \cap |\overline{\sigma}|) \subset i(\sigma)$. (1.9.2)

Then there exists a q-simplex in K on which i is a bijection.

Proof

First notice that i define a simplicial approximation of the identity of $|\overline{S}|$ in the following may: we define $\varphi:K \to S$ by $\varphi = j^{-1} \circ i$ (since $j = i|_S$ is a bijection).

It is a simplicial approximation of the identity: since, if $\langle S'\rangle \cap \langle\sigma\rangle \neq \phi$ for some simplex S' of K and some face σ of S, then by (1.1) we have that $\langle S'\rangle \subset \langle\sigma\rangle$, thus $S' \subset |\overline{\sigma}|$ and

$$\varphi(S') \subset j^{-1} \circ i(V \cap |\overline{\sigma}|) \subset j^{-1} \circ i(\sigma) = \sigma.$$

By (1.8), we get that $C_q\varphi \circ \tau_q = 1_{C_q(\overline{S})}$. That is $\sum_{\sigma\in\mathcal{O}_S} C_q\varphi(\sigma) = S$. Since by definition $C_q\varphi(\sigma) \in \{-S,0,S\}$ we get the conclusion.

Let us now see the closest to the converse of (1.3) that we can get.

(1.10) *PROPOSITION*

If K' is a subdivision of K and τ is the subdivision map, then for any simplicial approximation $\varphi:K' \to K$ of the identity of $|K|$, there exists a chain homotopy $D:\tau \circ C\varphi \sim 1_{CK'}$ such that

$\sigma \in S_{K'}$ implies that $D(\sigma) \in C(S_{K'})$, for any simplex σ of K' and any simplex S of K. (1.10.1)

Proof

It follows the usual proof using the fact that $S_{K'}$ is acyclic. ■

2. THE LOCAL LEFSCHETZ NUMBER.

Let $\varphi:K' \to L$ be a simplicial map, where K' is a subdivision of K, a subcomplex of L. Let $\pi:CL \to CK$ be the linear map defined by

$$\pi_q(S) = \begin{cases} S & \text{if } S \text{ is a simplex of } K \\ 0 & \text{otherwise} \end{cases}$$

for any simplex S of L. Note that π is usually not a chain map.

(2.1) *DEFINITION*

The *relative Lefschetz number* $\lambda(\varphi)$ of φ is defined to be:

$$\lambda(\varphi) = \lambda(\pi \circ C\varphi \circ \tau) = \sum_q (-1)^q \ \mathrm{tr}(\pi_q \circ C_q\varphi \circ \tau_q)$$

where $\tau:CK \to CK'$ is the subdivision chain map and $C\varphi:CK' \to CL$ is the chain map induced by φ, and where tr denotes the ordinary trace of endomorphisms of finite demensional vector spaces.

Let us denote the scalar product in C_qL determined by the basis $\mathcal{B}$ of all the oriented q-simplex of L: that is if $a = \sum_{S\in\mathcal{B}} a_S S$ and $b = \sum_{S\in\mathcal{B}} b_S S$ then $a \cdot b = \sum_{S\in\mathcal{B}} a_s b_s$.

Denote by ∂K the subcomplex of $K = (V,\Sigma)$ whose set of vertexes is $V \cap \partial|K|$ and whose set of simplexes is $\{S \in \Sigma : |\bar{S}| \subset \partial|K|\}$, where $\partial|K|$ is the frontier of $|K|$ in $|L|$. Note that $|\partial K| = \partial|K|$.

Let S be a simplex of L; denote by L_S, the biggest subcomplex of L containing no vertexes of S. In order to get some properties of the relative Lefschetz number, we need the following general lemma.

(2.2) *LEMMA*

Let R,T: $CK' \to CL$ be chain maps, and let D: $R \sim T$ be a homotopy such that

$D(\sigma) \in C_{q+1}(L_S)$ for any $\sigma \in S_{K'}$ and any $S \in \partial K$, (2.2.1)

then $\lambda(\pi \circ R \circ \tau) = \lambda(\pi \circ T \circ \tau)$.

Proof

We have that, for all $q > 0$,

$$\partial_{q+1} \circ D_q + D_{q-1} \circ \partial_q = R_q - T_q.$$

Thus

$$\pi_q \circ R_q \circ \tau_q - \pi_q \circ T_q \circ \tau_q = \pi_q \circ \partial_{q+1} \circ D_q \circ \tau_q + \pi_q \circ D_{q-1} \circ \tau_{q-1} \circ \partial_q.$$

Let us split this last term as the sum of

$$F_q = \partial_{q+1} \circ \pi_{q+1} \circ D_q \circ \tau_q + \pi_q \circ D_{q-1} \circ \tau_{q-1} \circ \partial_q$$

and

$$G_q = [\pi_q \circ \partial_{q+1} - \partial_{q+1} \circ \pi_{q+1}] \circ D_q \circ \tau_q$$

It is sufficient to show that $\lambda(F) = 0$ and $\lambda(G) = 0$: in fact then

$$\lambda(\pi \circ R \circ \tau) - \lambda(\pi \circ T \circ \tau) = \lambda(\pi \circ R \circ \tau - \pi \circ T \circ \tau) = \lambda(F) + \lambda(G) = 0.$$

But evidently $\partial_q \circ F_q = F_{q-1} \circ \partial_q$ and so F is a chain map. Furthermore if $D'_q = \pi_{q+1} \circ D_q \circ \tau_q$, then D': $F \sim 0$ the zero chain map. By Spanier [15; Th 6 p.195 and

Th 2 p. 163], we get that

$$\lambda(F) = \lambda(F_*) = \lambda(0_*) = \lambda(0) = 0.$$

On the other hand $G_q(C_q K) \subset C_q(\partial K)$: in fact

$$[\pi_q \circ \partial_{q+1} - \partial_{q+1} \circ \pi_{q+1}] (C_{q+1}(L)) \subset C_q(\partial K),$$

since if σ in a simplex of L, then either σ in a simplex K and

$$[\pi_q \circ \partial_{q+1} - \partial_{q+1} \circ \pi_{q+1}] (\sigma) = \partial_{q+1}(\sigma) - \partial_{q+1}(\sigma) = 0$$

or σ is not a simplex of K and

$$[\pi_q \circ \partial_{q+1} - \partial_{q+1} \circ \pi_{q+1}] (\sigma) = \pi_q(\partial_{q+1}(\sigma)) \in C_q(K) \cap C_q(\dot{\sigma})$$

thus it is contained in $C_q(\partial K)$.

It is sufficient to prove that for any q-simplex S of K, we have that $S \cdot G_q(S) = 0$. But this is evident if $S \notin \partial K$. Furthermore, take S a q-simplex of ∂K and let us show that $S \cdot G_q(S) = 0$: in fact $\tau_q(S) = \sum_{\sigma \in O_S} \sigma$ and so by (2.2.1);

$$D_q \circ \tau_q(S) = \sum_{\sigma \in O_S} D_q(\sigma) \in C_{q+1}(L_S) ;$$

thus

$$G_q(S) \in (\pi_q \circ \partial_{q+1} - \partial_{q+1} \circ \pi_{q+1}) C_{q+1}(L_S) \subset C_q(K_S)$$

and so $S \cdot G_q(S) = 0$. ■

(2.3) *PROPOSITION*

Let $\varphi, \psi: K' \to L$ be contiguous simplicial maps such that

$\psi(v), \varphi(v) \notin S$ for any $v \in S_{K'}$ and any simplex S of ∂K, (2.3.1)

then there exists a homotopy $D: C\varphi \sim C\psi$ satisfying (2.2.1) and $\lambda(\varphi) = \lambda(\psi)$.

<u>*Proof*</u>

Since φ and ψ are contiguous, the homotopy $D_q: C_q K \to C_{q+1} L$ defined by

$$D_q(<v_0,\dots,v_q>) = \sum_{i=0}^{q} (-1)^i <\varphi(v_0,\dots,\varphi(v_i), \psi(v_i),\dots,\psi(v_q)> ,$$

for any simplex $<v_0,\dots,v_q>$ of K', is a homotopy between φ and ψ satisfying (2.2.1). ■

Other properties of λ can be deduced from for the following properties of the ordinary trace of endomorphisms of vector spaces:

(i) $tr(A\circ B) = tr(B\circ A)$

(ii) $tr(A^p) \equiv tr\,A \pmod p$ for any prime number p.

(iii) $tr(A\times B) = tr\,A + tr\,B$, where $A\times B\ (v\times w) = A(v) \times B(w)$

Note that since the matrices of τ, $C\varphi$ and π expressed in the canonical basis, have only integer entries, then $\lambda(\varphi)$ is an integer.

3. THE SIMPLICIAL INDEX.

Before defining the index of a continuous map $f:|K| \to |L|$, where K is a subcomplex of L, we need the following lemma.

(3.1) *LEMMA*

Let $f:|K| \to |L|$ be a continuous map without any fixed points on a subcomplex M of K, where K is a subcomplex of the simplicial complex L. Then there exists L' a subdivision of L such that

$$f(|\overline{\sigma}|) \cap st\,\sigma = \phi \text{ for any simplex } \sigma \text{ of } M_{L'}, \qquad (3.1.1)$$

where $st\,\sigma = \bigcup_{v\in\sigma} st\,v$ and st v is taken in L', and st v denotes the star of v.

Furthermore if $\varphi:K' \to L'$ is a simplicial approximation of f, where K' is a subdivision of $K_{L'}$, then φ satisfies (2.3.1) with $\partial K = M$ and so

$$S\cdot(\pi \circ C\varphi \circ \tau(S)) = 0$$

for any $S \in M_{L'}$.

Proof

Denote $d = \inf\{d(x,f(x) : x \in |M|\}$; since $|M|$ is compact, we have that $d > 0$. Choose L' a subdivision of L such that diam $L' < d/2$. We then have the desired conclusions. ∎

Let $f:|K| \to |L|$ be a continuous map without fixed points on $\partial|K|$. By (3.1) for $M = \partial K$ let L' be a subdivision of L satisfying (3.1.1). Let $\varphi:K' \to L'$ be a simplicial approximation of f, where K' is a subdivision of $K_{L'}$.

(3.2) *DEFINITION*

Define the *simplicial index* of $f:|K| \to |L|$ by putting $\mathrm{ind}(K,f,L) = \lambda(\varphi)$. Note that since $\lambda(\varphi)$ is an integer, the fixed point index of f is an integer.

(3.3) *LEMMA*

The definition (3.2) is well defined, that is $\mathrm{ind}(K,f,L)$ is independent of φ, K' and L'.

Proof

a) Independence from φ.

Let φ,ψ be two such simplicial approximations of f, then φ and ψ are contiguous and by (3.1) they satisfy (2.3.1). Thus, by (2.3), we have that $\lambda(\varphi)=\lambda(\psi)$.

b) Independence from K'.

Let K'' be another subdivision of $K_{L'}$, and let $\varphi':K'' \to L'$ be a simplicial approximation of f.

First assume that K'' is a subdivision of K'. Then by a), we may assume that $\varphi' = \varphi \circ \psi$ where $\psi:K'' \to K'$ is a simplicial approximation of the identity. Let $\tau:K_{L'} \to K'$ and $\tau':K' \to K''$ be the subdivision chain maps; by the uniqueness of the subdivision map we have that $\tau' \circ \tau:K_{L'} \to K''$ is the subdivision map. Now, by definition, $\lambda(\varphi) = \lambda(\pi \circ C\varphi \circ \tau)$ and $\lambda(\varphi') = \lambda(\pi \circ C\varphi' \circ \tau' \circ \tau)$, where $\pi: C(L') \to C(K_{L'})$ is the canonical projection. But we have that

$$\pi \circ C\varphi' \circ \tau' \circ \tau = \pi \circ C\varphi \circ C\psi \circ \tau' \circ \tau = \pi \circ C\varphi \circ \tau$$

and so $\lambda(\varphi') = \lambda(\varphi)$, since $C\psi \circ \tau' = 1_{CK'}$ by (1.8).

If K'' is not a subdivision of K', let K''' be a subdivision of both, then if $\varphi'': K''' \to L'$ is a simplicial approximation of f, we have proven that

$$\lambda(\varphi) = \lambda(\varphi'') = \lambda(\varphi').$$

c) Independence from L'.

Let L'' be another subdivision of L satisfying (3.1.1). Without loss of generality, as in part b), we may assume that L'' is a subdivision of L'. Then, by b), we may assume that K' is a subdivision of $K_{L''}$, and thus of $K_{L'}$. Let $\varphi: K' \to L''$ be a simplicial approximation of f and $\psi: L'' \to L'$ be a simplicial approximation of the identity. Since $\psi \circ \varphi: K' \to L'$ is a simplicial approximation of f, we must prove that $\lambda(\varphi) = \lambda(\psi\circ\varphi)$. But $\lambda(\varphi) = \lambda(\pi' \circ C\varphi \circ \tau')$ and

$$\lambda(\psi\circ\varphi) = \lambda(\pi \circ C\psi \circ C\varphi \circ \tau' \circ \tau),$$

where $\pi: C(L') \to C(K_{L''})$ and $\pi': C(L'') \to C(K_{L''})$ are the canonical maps and $\tau: CK_{L'} \to CK_{L''}$ and $\tau': CK_{L''} \to CK'$ are the subdivision chain maps. By the commutativity property of the ordinary trace, we have that

$$\tau(\psi\circ\varphi) = \lambda(\tau \circ \pi \circ C\psi \circ C\varphi \circ \tau').$$

Notice that the following diagram commutes

$$\begin{array}{ccc} CL' & \xrightarrow{\ \pi\ } & CK_{L'} \\ \downarrow{\scriptstyle\tau} & & \downarrow{\scriptstyle\tau} \\ CL'' & \xrightarrow[\ \pi\ ]{} & CK_{L''} \end{array}$$

thus $\lambda(\psi\circ\varphi) = \lambda(\pi' \circ \tau \circ C\psi \circ C\varphi \circ \tau')$.

By (1.2), it is now sufficient to show that there exists a chain homotopy $D: \tau \circ C\psi \circ C\varphi \sim C\varphi$ satisfying (2.2.1) that is $D(\sigma) \in C(L''_S)$ for any $\sigma \in S_{K'}$ and any $S \in \partial K_{L''}$. But, by (1.10), there exists a chain homotopy $D': \tau \circ C\psi \sim 1_{CL''}$ satisfying (1.10.1).

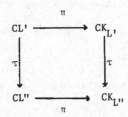

Define D by putting $D_q = D'_q \circ C_q\varphi$, then

$$D:\tau \circ C\psi \circ C\varphi \sim C\varphi.$$

If $S \in \partial K_{L'}$ and $\sigma \in S_{K'}$, then $C\varphi(\sigma) \in C\,((L'_S)_{L''})$: since, by (3.1.1) with $M_{L'} = \partial K_{L'}$ we have that

$$f(|\bar{\sigma}|) \subset f(|\bar{S}|) \subset |L'| \setminus \bigcup_{S \in stS} \langle S' \rangle = |L'_S| = |(L'_S)_{L''}|,$$

thus $\varphi(\sigma) \subset (L'_S)_{L''}$. Now take $S' \in L'_S$ such that $\varphi(\sigma) \in S'_{L''}$; this is possible since $\varphi(\sigma)$ is a simplex of L''. Then, by (1.10), we have that

$$D(\sigma) = D'(C\varphi(\sigma) \subset C(S'_{L''}) \subset C((L'_S)_{L''}) \subset C(L''_{S''})$$

for any $S'' \in S_{L''}$, That is the conclusion. ■

4. Some properties of the simplicial index

Let us denote $\varkappa_f = \{x \in |K| : f(x) = x\}$ then by our hypothesis $\varkappa_f \cap \partial|K| = \phi$.

(4.1) *PROPOSITION (Fixed points)*

If $\mathrm{ind}(K,f,L) \neq 0$ then $\varkappa_f \neq \phi$. Or equivalently: if $\varkappa_f = \phi$ then $\mathrm{ind}(K,f,L) = 0$.

Proof

By (3.1) with $M = K$, we have that $\varphi: K' \to L'$ satisfies (2.3.1), and

$$S \cdot (\pi \circ C\varphi \circ \tau)(S) = 0$$

for any $S \in K_{L'}$, that is

$$\lambda(\varphi) = \lambda(\pi \circ C\varphi \circ \tau) = 0. ■$$

(4.2) *PROPOSITION (Additivity)*

If $V = V_1 \cup \cdots \cup V_n$ and $K = (V,\Sigma)$ and $K_i = (V_i, P(V_i) \cap \Sigma)$, if $|K| = \bigcup_1^n |K_i|$ and if f has no fixed points on $|K_i| \cap |K_j|$ for all $i \neq j$ and no fixed points on $\partial|K|$, then

$$\mathrm{ind}(K,f,L) = \mathrm{ind}(K_1,f,L) + \cdots + \mathrm{ind}(K_n,f,L)$$

Proof

Let $\varphi: K' \to L'$ be as in (3.1) for the simplicial complex

$$M = \left\{ \bigcup_{i\neq j} [V_i \cap V_j] \cup \partial V, \bigcup_{i\neq j} [P(V_i) \cap P(V_j) \cap \Sigma] \cup \partial\Sigma \right\}$$

where $\partial K = (\partial V, \partial\Sigma)$, since then

$$|M| = \bigcup_{i\neq j} \left(|K_i| \cap |K_j|\right) \cup \partial K.$$

Since φ satisfies $S \cdot (\pi \circ C\varphi \circ \tau)(S) = 0$ for any $S \in M_{L'}$, then

$$\mathrm{tr}(\pi_q \circ C_q\varphi \circ \tau_q) = \sum_{i=1}^{n} \sum_{\substack{S\in K_{i_{L'}} \\ S\notin M}} S \cdot (\pi \circ C\varphi \circ \tau)(S) = 0.$$

But

$$\mathrm{tr}(\pi_i \circ C_q\varphi \circ \tau_q|K_i) = \sum_{\substack{S\in K_{i_{L'}} \\ S\notin M}} S \cdot (\pi_i \circ C\varphi \circ \tau)(S)$$

where $\pi_i : L' \to K_{i_{L'}}$ is the canonical projection. Since

$$S \cdot (\pi \circ C\varphi \circ \tau)(S) = S \cdot (\pi_i \circ C\varphi \circ \tau)(S)$$

for any $S \in K_{i_{L'}}$, we have that

$$\mathrm{tr}(\pi_q \circ C_q\varphi \circ \tau_q) = \sum_{i=1}^{n} \mathrm{tr}(\pi_i \circ C_q\varphi \circ \tau_q|K_i),$$

hence taking the alternate sum of the above, we get the conclusion. ■

(4.3) *COROLLARY (Excision)*

If K_1 is a subcomplex of K such that f has no fixed points on

$$|K| \setminus (|K_1| \setminus |\mathring{K}_1|),$$

then

$$\operatorname{ind}(K,f,L) = \operatorname{ind}(K_1,f,L).$$

<u>*Proof*</u>

Evident from (4.1) and (4.2). ■

(4.4) *PROPOSITION (Homotopy)*

If $h:|K| \times [0,1] \to |L|$ is such that h_t has no fixed points on $\partial|K|$, for any $t \in [0,1]$ then

$$\operatorname{ind}(K,h_1,L) = \operatorname{ind}(K,h_0,L).$$

<u>*Proof*</u>

Consider $d = \inf\{d(x,h(\{x\} \times [0,1])) : x \in \partial|K|\}$ we have that $d > 0$, since $\partial|K|$ is compact and $d(x,h(\{x\} \times [0,1]))$ is a continuous map different from zero on $\partial|K|$. Choose L' a subdivision of L such that diam $L' < d/2$. As in (3.1), L' satisfies (3.1.1) for $f = h_t$ for any t.

Consider $\{h^{-1}(\text{st } v) : v \in L'\}$, the open cover of $|K| \times [0,1]$ and take δ to be the Lebesgue number of this cover. Let K' be a subdivision of K such that diam $K' < \delta/4$.

Now choose $\varphi_t: K' \to L$ a simplicial approximation of h_t, for any t, such that

$$\overline{\text{st } v} \times [t-\delta/4,\ t+\delta/4] \subset h^{-1}(\text{st } \varphi_{v_t}(v)):$$

this is possible, since the first of these sets is of diameter less than δ. We have that φ_t is a simplicial approximation of h_r for any r such that $|r-t| \le \delta/4$. That is, by definition

$$\mathrm{ind}(K,h_r,L) = \lambda(\varphi_t) = \mathrm{ind}(K,h_t,L)$$

for any $(r-t) \le \delta$. Since $[0,1]$ is compact we get the conclusion. ■

(4.5) *PROPOSITION (Normalization)*

If $f:|K| \to |L|$ is a constant mapping $f(x) = c$ then

$$\mathrm{ind}(K,f,L) = \begin{cases} 1 & \text{if } c \in |K| \setminus \partial|K| \\ 0 & \text{if } c \in |L| \setminus |K| \end{cases}$$

<u>Proof</u>

Choose $v \notin \partial K$ such that $c \in \mathrm{st}\, v$. Define $\varphi:K \to L$ by $\varphi(w) = v$ for any $w \in K$. We have that φ is a simplicial approximation of f.

Now $C_q\varphi = 0$ if $q > 0$ and $C_0\varphi$ is the projection on the subspace generated by $\langle v\rangle$. Notice that $\tau:K \to K$ is the identity and $\pi(\langle v\rangle) = 0$ if $v \notin K$ and $\pi(\langle v\rangle) = \langle v\rangle$ if $v \in K$. Thus

$$\mathrm{tr}(\pi_q \circ C_q\varphi \circ \tau_q) = \begin{cases} 1 & \text{if } q = 0 \text{ and } v \in K \\ 0 & \text{otherwise} \end{cases}$$

and, we have the conclusion. ■

For the following property, we need to consider the singular homology functor H_S.

(4.6) *DEFINITION*

Let $f:|K| \to |K|$ be continuous, define the *Lefschetz number* of f to be

$$\lambda(f) = \lambda(H_S(f)) = \sum_q (-1)^q \ \mathrm{tr}(H_{S_q}(f)).$$

We then have the following property.

(4.7) *PROPOSITION*

Let $f:|K| \to |K|$ be continuous, then

$$\mathrm{ind}(K,f,L) = \lambda(f).$$

Proof

We have that $\mathrm{ind}(K,f,L) = \lambda(C\varphi \circ \tau)$ where $\varphi : K' \to K$ is a simplicial approximation of f and $\tau : K \to K'$ is the subdivision map. By Spanier [15; Th 6 p. 195]; $\lambda(C\varphi \circ \tau) = \lambda(C\varphi_* \circ \tau_*)$. It is thus sufficient to show that $\mathrm{tr}(H_{S_q}(f)) = \mathrm{tr}(C_q\varphi \circ \tau_q)$ for all q. But by Spanier [15; Th 8 p. 171 and Th 8 p. 191] there is a natural equivalence n between the functors $\varphi \mapsto C\varphi_*$ and $\varphi \mapsto H_S(|\varphi|)$. That is, we have the following commutative diagram

$$\begin{array}{ccc} H(K') & \xrightarrow{C\varphi_*} & H(K) \\ {\scriptstyle n_{K'}}\Big\downarrow\wr & & \wr\Big\downarrow{\scriptstyle n_K} \\ H_S(|K'|) & \xrightarrow{H_S(|\varphi|)} & H_S(|K|) \end{array}$$

Notice that $|\varphi|$ is homotopic to f and so $H_S(|\varphi|) = H_S(f)$. Moreover if $\psi : K' \to K$ is a simplicial approximation of the identity we get that $H_S(|\psi|) = H_S(1_{|K|}) = 1_{H_S(|K|)}$

Thus $C\varphi_* = n_K \circ H_S(f) \circ n_{K'}^{-1}$ and $C\psi_* = n_K \circ n_{K'}^{-1}$, since by (1.8) and (1.10), we get that τ_* is the inverse of $C\psi_*$, thus $\tau_* = n_{K'} \circ n_K^{-1}$ and so

$$C\varphi_* \circ \tau_* = n_K \circ H_S(f) \circ n_K^{-1} .$$

By the commutativity property of the trace we get the conclusion. ■

5. MORE PROPERTIES OF THE SIMPLICIAL INDEX

In order to obtain the commutativity and (mod p) properties, we need the following lemma. Denote $\mathrm{Int}|K| = |K| \setminus \partial|K|$.

(5.1) *LEMMA*

Let K_i be a subcomplex of L_i and $f_i : |K_i| \to |L_{i+1}|$ be continuous, for $i \in \mathbf{Z}$. Assume that, for one $n \geq 2$, $K_i = K_{i+n+1}$, $L_i = L_{i+n+1}$ and $f_i = f_{i+n+1}$ for all $i \in \mathbf{Z}$.

Denote

$$F_i = f_i^{-1}(f_{i+1}^{-1}(\cdots f_{i+n-1}^{-1}(|K_{i+n}|)\cdots))$$

and

$$g_i = f_{i+n} \circ \cdots \circ f_i : F_i \to L_i$$

If $\varkappa_{g_i} \subset \mathrm{Int}|K_i|$ for all i, then $f_i(\varkappa_{g_i}) = \varkappa_{g_{i+1}}$ which is compact and there exists L_i' a subdivision of L_i and M_i a subcomplex of $(K_i)_{L_i'}$ for all i, such that

1) $\varkappa_{g_i} \subset \mathrm{Int}|M_i|$

2) if $\tau_i : CM_i \to CM_i^1$ are the subdivision maps and $\pi_i : CL_i' \to CM_i$ are the canonical projections, and $\varphi_i : M_i \to L_{i+1}$ are simplicial approximations of $f_i : X\big|_{|M_i|}$, it follows that

$$\mathrm{ind}(M_i, g_i, L_i') = \lambda((\pi_{i+n+1} \circ C\varphi_{i+n} \circ \tau_{i+n}) \circ \cdots \circ (\pi_{i+1} \circ C\varphi_i \circ \tau_i)).$$

Proof

a) $f_i(\varkappa_{g_i}) = \varkappa_{g_{i+1}}$ which is compact.

Denote $\varkappa_i = \varkappa_{g_i}$. First $\varkappa$ is closed in the compact F_i so it is compact. Notice that $f_i(\varkappa_i) = \varkappa_{i+1}$: in fact $f_i(\varkappa_i) \subset \varkappa_{i+1}$ since if $x \in \varkappa_i$ then

$$f_{i+n} \circ \cdots \circ f_i(x) = x \in |K_i| = |K_{i+n+1}|$$

and so $f_i(x) \in F_{i+1}$ and

$$f_{i+n+1} \circ \cdots \circ f_i(x) = f_{i+n+1}(x) = f_i(x),$$

that is $f_i(x) \in \varkappa_{i+1}$; furthermore $f_i : \varkappa_i \to \varkappa_{i+1}$ is an isomorphism since it has as an inverse $f_{i+n} \circ \cdots \circ f_{i+1} : \varkappa_{i+1} \to \varkappa_i$ which is well defined, since $f_j(\varkappa_j) \subset \varkappa_{j+1}$ for all j.

b) The choices

Since $K_i \subset \mathrm{Int}|K_i|$ then

$$\varkappa_i \subset U_i = \bigcap_{j=0}^{n+1} f_i^{-1}(f_{i+1}^{-1}(\dots f_{i+j-1}^{-1}(\mathrm{Int}|K_{i+j}|\dots))$$

which is open. Now let L_i' be a subdivision of L_i such that there exists a subcomplex M_i of L_i' such that

$$K_i \subset \mathrm{Int}|M_i| \subset M_i \subset U_i \subset F_i.$$

Since $U_i \subset \mathrm{Int}|K_i|$, we have that M_i is a subcomplex of $(K_i)_{L_i'}$.

Assume that the L_i''s are fine enough so that there exist subcomplexes $N_{i,j}$ of M_{i+j} such that if $N_i = N_{i,0}$ then $\varkappa_i \subset \mathrm{Int}(N_i) \subset |M_i|$ and

$$f(|N_{i,j}|) \subset \mathrm{Int}(N_{i,j+1}) \tag{5.1.1}$$

for all $j = 0, \dots, n$.

Denote

$$d = \inf\{d(g_i(x),x) : x \in F_i \setminus \mathrm{Int}|N_i|,\ i \in \mathbb{Z}\};$$

since $F_i \setminus \mathrm{Int}|N_i|$ is compact for every $i \in \mathbb{Z}$ and the number of different g_i is n, we have that $d > 0$.

Without loss of generality we may assume that $\mathrm{diam}|\bar{\sigma}| < \varepsilon$ for any simplex σ of L_i', for any chosen $\varepsilon > 0$.

c) Reduction of domain.

Now if $\varphi_i : M_i' \to L_i'$ is a simplicial approximation of f_i, then for any $v \in M_i'$ we have that $f_i(v) \in \mathrm{st}\,\varphi_i(v)$ thus $\varphi_i(v) \in \overline{\mathrm{st}\, f_i(v)} \in N_\varepsilon(f_i(v))$. Hence for any $\sigma \in M_i$

$$\pi_{i+1} \circ C\varphi_i \circ \tau_i(\sigma) \subset C(M_{i+1} \cap N_\varepsilon(f_i(|\sigma|)),$$

where for any subset A of $|L_i|$ and for any i, $M_i \cap A$ denotes the biggest subcomplex of M_i contained in A. That is if B is a subcomplex of M_i,

$$\pi_{i+1} \circ C\varphi_i \circ \tau_i(C(B)) \subset C(M_{i+1} \cap N_\varepsilon(f_i(|B|))).$$

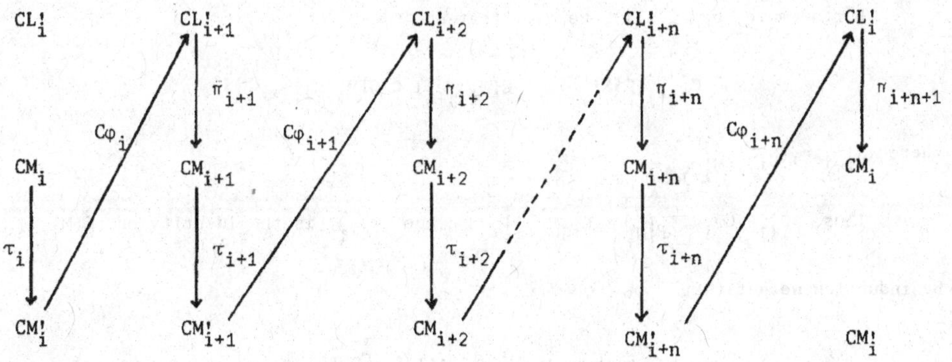

Figure 1.

Thus if $|\sigma| \in |M_i| \setminus \mathrm{Int}|N_i|$

$$(\pi_{i+n+1} \circ C\varphi_{i+n} \circ \tau_{i+n}) \circ \ldots \circ (\pi_{i+1} \circ C\varphi_i \circ \tau_i)\ (\sigma)$$

$$\subset C(M_{i+n+1} \cap N_\varepsilon(f_{i+n}(\ldots(M_{i+1} \cap N_\varepsilon(f_i(|\sigma|)))\ldots)))$$

$$\subset C(L\sigma),$$

provided ε is small enough: in fact since $f_i : |K_i| \to (L_{i+1})$ is uniformly continuous we may choose $\varepsilon_1, \ldots, \varepsilon_{n+1}$ such that $\varepsilon_{n+1} < d/2$ and $|x-y| < \varepsilon_i$ implies that

$$|f_j(x) - f_j(y)| < \varepsilon_{i+1} - \varepsilon_i$$

for $j,i = 1,\ldots,n$; this gives us that, since $|\bar{\sigma}| \cap F_i \neq \phi$, if $\varepsilon < \varepsilon_i$ then the image of σ is contained in $C(M_{i+n+1} \cap N_{\varepsilon_{n+1}}(f_{i+n} \circ \cdots \circ f_i(|\bar{\sigma}| \cap F_i)))$ but by definition of d, σ is not in this last complex. Thus the scalar product of σ and its image is zero, if $\sigma \notin N_i$ or if $\sigma \in \partial N_i$.

We now have that if

$$T_i = (\pi_{i+n+1} \circ C\varphi_{i+n} \circ \tau_{i+n}) \circ \ldots \circ (\pi_{i+1} \circ C\varphi_i \circ \tau_i)$$

then $\lambda(T_i) = \lambda(\pi'_i \circ T_i|_{N_i})$ where $\pi'_i : CM'_i \to CN_i$ is the canonical projection.

d) Relation to g_i.

Furthermore, by (5.1.1), we have that

$$C\varphi_{i+j}(C(N'_{i,j})) \subset C(N_{i,j+1}) \subset C(M_{i+j+1}),$$

where $N'_{i,j} = (N_{i,j})_{M'_{i+j}}$.

Thus $\pi_{i+1} \circ C\varphi_i \circ \tau_i|_{N_i} = C\varphi_i \circ \tau_i|_{N_i}$ since π_{i+1} is the identity on $C(M_{i+j})$.

By induction we get that

$$T_i|_{N_i} = (C\varphi_{i+n} \circ \tau_{i+n}) \circ \cdots \circ (C\varphi_i \circ \tau_i).$$

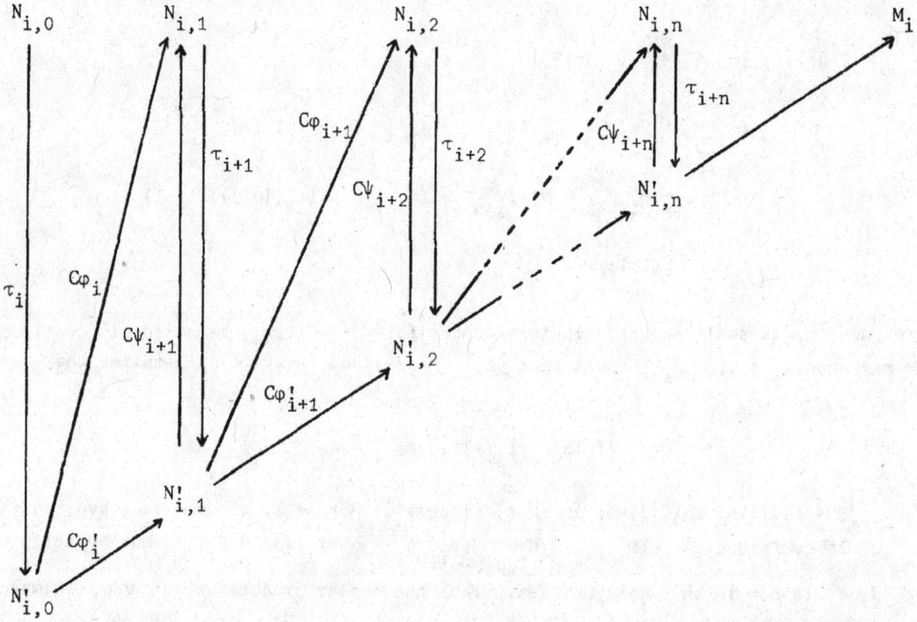

(The homotopies of the proof of (4.1), part c) go from the top route to the bottom route of this diagram.)

Figure 2.

We may assume that $N'_{i,j}$ are fine enough so that there exists

$$\varphi'_{i+j}: N'_{i+j} \to N'_{i+j+1}$$

a simplicial approximation of $f_{i+j}: |N_{i+j}| \to |N_{i+j+1}|$ and $\psi_{i+j+1}: N'_{i+j+1} \to N_{i+j+1}$ a simplicial approximation of the identity.

By (2.2), it is sufficient to show that there exists a homotopy D between

$$S_i = (C\varphi_{i+n} \circ \tau_{i+n}) \circ \ldots \circ (C\varphi_{i+1} \circ \tau_{i+1}) \circ C\varphi_i$$

and $C\varphi_{i+n} \circ \ldots \circ C\varphi'_i$ satisfying (2.2.1), since $C\varphi'_{i+n} \circ \cdots \circ C\varphi'_i$ is a simplicial approximation of g_i.

This will be done by steps. First we replace one by one the $C\varphi_j$ by the $C\psi_{j+1} \circ C\varphi'_j = C(\psi_{j+1} \circ \varphi'_j)$: since both are chain maps of simplicial approximations of f_j, by (2.3) there exists a homotopy $D_i: C\varphi_i \sim C(\psi_{j+1} \circ \varphi'_j)$ satisfying the condition that $D_j(\sigma)$ is a chain of the simplicial complex whose set of vertices is

$$\varphi_j(\sigma) \cup (\psi_{j+1} \circ \varphi'_j)(\sigma);$$

thus as in c), we get that $D_j(\sigma) \subset C(N_{j+1} \cap N_\varepsilon(f_j(|\sigma|)))$ and so the composite, replacing in S_i, the map $C\varphi_j$ by the homotopy D_j and any number of other $C\varphi_k$ by $C(\psi_{k+1} \circ \varphi'_k)$, sends any q-simplex σ of $\partial N'_i$ into $C_{q+1}(L_S)$ for any $S \in \partial N_i$ such that $\sigma \in S_{N'_i}$; by (2.2) this replacement does not change the Lefschetz number.

We thus get the intermediary map

$$C\varphi'_{i+n} \circ (\tau_{i+n} \circ C\psi_{i+n}) \circ C\varphi'_{i+n-1} \circ \ldots \circ C\varphi'_{i+1} \circ (\tau_{i+1} \circ C\psi_{i+1}) \circ C\varphi'_i \;;$$

it is sufficient to replace in this map the parentheses by the identity maps.

But, by (1.10), there exists $D'_j: \tau_j \circ C\psi_j \sim 1_{CN'_{i,j}}$ satisfying (1.10.1), thus as above we get the desired replacements and the conclusion. ■

(5.2) *PROPOSITION (Commutativity)*

If $f_i: |K_i| \to |L_i|$ $i = 1,2$ are continuous maps such that K_i is a subcomplex of L_j, where $i \neq j$. Consider $f_j \circ f_i: f_i^{-1}(|K_j|) \to L_j$, where $i \neq j$ if

$$\varkappa_{f_j \circ f_i} \subset \operatorname{Int}|K_i|$$

for $i = 1,2$, then there exist M_j neighbourhoods of $\varkappa_{f_j \circ f_i}$ and a subcomplex of $(K_i)_{L_j'}$ where L_j' is a subdivision of L_j such that

$$\mathrm{ind}(M_2,\ f_2 \circ f_1,\ L_2') = \mathrm{ind}(M_1,\ f_1 \circ f_2,\ L_1').$$

Proof

By (5.1) these indexes are defined for some M_i, $i = 1,2$, and

$$\mathrm{ind}(M_1,\ f_1 \circ f_2,\ L_1) = \lambda(\pi_2 \circ C\varphi_1 \circ \tau_1 \circ \pi_1 \circ C\varphi_2 \circ \tau_2)$$

and

$$\mathrm{ind}(M_2,\ f_2 \circ f_1,\ L_2) = \lambda(\pi_1 \circ C\varphi_2 \circ \tau_2 \circ \pi_2 \circ C\varphi_1 \circ \tau_1)$$

for some φ_i. By the commutativity property of the ordinary trace we get the conclusion. ■

(5.3) *PROPOSITION (Index mod p)*

Let p be a prime number and let $f\colon |K| \to |L|$ be continuous such that K is a subcomplex of L. If $\varkappa_{f^p} \subset \mathrm{Int}\,|K|$, where $f^p\colon f^{-p+1}(|K|) \to |L|$ is well defined, then there exists a neighbourhood M of $\varkappa_{f^p}$ and a subcomplex of a subdivision L' of L such that

$$\mathrm{ind}(M,\ f^p,\ L') \equiv \mathrm{ind}(K,\ f,\ L') \qquad (\mathrm{mod}\ p)$$

Proof

By lemma (4.1), this index is defined for some M and, for some simplicial approximation φ of f,

$$\begin{aligned}\mathrm{ind}(M,\ f^p,\ L) &= \lambda(\underbrace{(\pi \circ C\varphi \circ \tau) \circ \ldots \circ (\pi \circ C\varphi \circ \tau)}_{p\ \text{times}}) \\ &\equiv \lambda(\pi \circ C\varphi \circ \tau) \qquad (\mathrm{mod}\ p) \\ &= \mathrm{ind}(M,\ f,\ L)\end{aligned}$$

since $\operatorname{tr} A^p \equiv \operatorname{tr} A \pmod{p}$ for any linear map A. By excision we get

$$\operatorname{ind}(M, f, L) = \operatorname{ind}(K, f, L). \blacksquare$$

(5.4) *COROLLARY (Index mod p)*

Under the hypotheses of (5.3) if in addition f^p is defined on all of $|K|$ and

$$f^p(x) = x \in |K| \text{ implies that } f(x) \in |K| \tag{5.4.1}$$

then

$$\operatorname{ind}(K, f^p, L) = \operatorname{ind}(K, f, L).$$

Proof

All the fixed points of f^p in K are thus in M. By excision we get the conclusion. $\blacksquare$

6. THE FIXED POINT INDEX ON THE POLYHEDRA.

Let X be a finite polyhedron with triangulation (L, T). Let $f: U \to X$ be continuous where U is open in X. Assume that $\varkappa_f = \{x \in U : f(x) = x\}$ is compact. Consider the map $g = T^{-1} \circ f \circ T : T^{-1}(U) \to |L|$. Then there exists a subdivision L' of L and a subcomplex K of L' such that $\varkappa_g \subset |K| \subset T^{-1}(U)$ and g has no fixed points on $\partial|K|$.

(6.1) *DEFINITION*

Define the *fixed point index of* $f: U \to X$ by putting

$$\operatorname{ind}(U, f, X) = \operatorname{ind}(K, g, L').$$

(6.2) *LEMMA*

The definition (6.1) is independent of K, L' and (L,T).

Proof

a) Independence from K, L' if (L,T) is chosen.

Let K', L" be two other such complexes. Without loss of generality we may assume that L" is a subdivision of L', since otherwise L" and L' would have a common subdivision anyway.

We already have that ind(K, g' L') = ind($K_{L''}$, g, L"), by the definition of the simplicial index. But by excision,

$$\text{ind}(K_{L''}, g, L'') = \text{ind}(K', g, L''),$$

since $\varkappa_g \subset \text{Int}|K'| \cap \text{Int}|K|$.

b) Independence from (L,T)

Let (M, S) be an another triangulation of X, and K' be a subcomplex of M', a subdivision of M such that $\varkappa_{g'} \subset |K'| \subset S^{-1}(U)$ and $g' = S^{-1} \circ f \circ S$ has no fixed points on $\partial|K'|$.

We have that, by the commutativity property

$$\begin{aligned}\text{ind}(K_1', g', M'') &= \text{ind}(K_1', S^{-1} \circ f \circ T \circ T^{-1} \circ S, M'') \\ &= \text{ind}(K_1, T^{-1} \circ S \circ S^{-1} \circ f \circ T, L'') \\ &= \text{ind}(K_1, g, L'')\end{aligned}$$

for some subdivision K_1 (respectively K_1') of $K_{L''}$ (respectively $K'_{M''}$) where L" (respectively M") is a subdivision of L' (respectively M'); furthermore by the same property, we have that $\varkappa_g \subset \text{Int}|K_1|$ and $\varkappa_{g'} \subset \text{Int}|K_1'|$. Thus by excision

$$\text{ind}(K_1', g', M'') = \text{ind}(K'_{M''}, g', M'') = \text{ind}(K', g', M').$$

Likewise for g, thus we have the conclusion. ■

We have the following properties of the index on the polyhedra.

(6.3) *PROPOSITION (Fixed point)*

If $\operatorname{ind}(U, f, X) \neq 0$, then $\varkappa_f \neq 0$.

(6.4) *PROPOSITION (Additivity)*

If $U = \bigcup_{i=1}^{n} U_i$ and $\varkappa_f \cap U_i \cap U_j = \phi$ $i \neq j$ then

$$\operatorname{ind}(U, f, X) = \operatorname{ind}(U_1, f, X) + \dots + \operatorname{ind}(U_n, f, X).$$

(6.5) *PROPOSITION (Excision)*

If $\varkappa_f \subset V \subset U$ then $\operatorname{ind}(U, f, X) = \operatorname{ind}(V, f, X)$.

(6.6) *PROPOSITION (Homotopy)*

If $h: U \times [0,1] \to X$ is such that $\varkappa_h = \bigcup_{t \in [0,1]} \varkappa_{h_t}$ is compact then

$$\operatorname{ind}(U, h_0, X) = \operatorname{ind}(U, h_1, X).$$

(6.7) *PROPOSITION (Normalization)*

If f is the constant map c then

$$\operatorname{ind}(U, f, X) = \begin{cases} 0 & \text{if } c \notin U \\ 1 & \text{if } c \in U. \end{cases}$$

(6.8) *PROPOSITION (Normality)*

If $f: X \to X$ then $\operatorname{ind}(X, f, X) = \lambda(f)$.

(6.9) *PROPOSITION (Commutativity)*

Let $f_i : U_i \to X_i$ be continuous for $i = 1,2$, and $U_i \subset X_j$ for $j \neq i$. Consider $f_j \circ f_i : f_i^{-1}(U_j) \to X_j$; if $\varkappa_{f_2 \circ f_1}$ is compact then $\varkappa_{f_1 \circ f_2}$ is compact and

$$\mathrm{ind}(f_1^{-1}(U_2),\ f_2 \circ f_1,\ X_2) = \mathrm{ind}(f_2^{-1}(U_1),\ f_1 \circ f_2,\ X_1).$$

(6.10) *PROPOSITION (Modulo p)*

If p is prime and $\varkappa_{f^p}$ is compact then

$$\mathrm{ind}(U,\ f,\ X) \equiv \mathrm{ind}(f^{-p+1}(U),\ f,\ X) \qquad (\mathrm{mod}\ p).$$

(6.11) *PROPOSITION (Modulo p)*

If p is prime, $\varkappa_{f^p}$ is compact and, if f^p is defined on all of U, and

$$f^p(x) = x \in U \ \text{ implies that } \ f(x) \in U \tag{6.11.1}$$

then $\mathrm{ind}(U,\ f^p,\ X) \equiv \mathrm{ind}(U,\ f,\ X) \quad (\mathrm{mod}\ p)$.

Proof

All these properties follow from the corresponding ones for the simplicial index and from the independence of the definition on the choice of K. ■

(6.12) *REMARK*

It is now an easy task to get an index in the Euclidean space: if $f : V \to W$ is continuous $V \subset W \subset \mathbb{R}^n$, V and W are open and $\varkappa_f$ is compact, then there exists an open set U and a polyhedron X such that $\varkappa_f \subset U \subset X$ and $f(U) \subset X$; define $\mathrm{ind}(V, f, W) = \mathrm{ind}(U, f, X)$. This index is well defined and possesses the same properties as Dold's index, where for normality one uses the generalized Lefschetz number.

By the uniqueness property of such indexes, it must be exactly Dold's index.

Thus this index can be generalized in the same way as Dold's has been generalized.

REFERENCES

[1] BROWDER, F.E.: On the fixed point index for continuous mappings of locally connected spaces, Summa Brasil. Math. 4 (1960), 253-293.

[2] BROWDER, F.E.: Local and global properties of nonlinear mappings in Banach spaces, Institute Nag. di Alta Math. Simposia Math. Vol II (1968), 13-35.

[3] DOLD, A.: Fixed point index and fixed point theorem for Euclidean neighbourhood retracts, Topology, 4, (1965), 1-8.

[4] EELLS, J. and FOURNIER, G.: La théorie des points fixes des applications à itérée condensante, Bull. Soc. Math. France, Mémoire 46 (1976), 91-120.

[5] GEOGHEGAN, R.: The homomorphism on fundamental group induced by a homotopy idempotent having essential fixed points (preprint).

[6] GRANAS, A.: Some theorems in fixed point theory. The Leray-Schauder Index and the Lefschetz Number, Bull. Acad. Polon. Sci. 17 (1969), 131-137.

[7] GRANAS, A.: The Leray-Schauder index and fixed point theory for arbitrary ANR's, Bull. Soc. Math. France 100 (1972), 209-229.

[8] HOPF, H.: Über die algebraische Anzahl von Fixpunkten, Math. Z. 29 (1929), 493-524.

[9] KRASNOSEL'SKIǏ, M.A.: Fixed points of cone-compressing or cone-extending operators, Soviet. Math. Dokl. 1 (1960), 1285-1288.

[10] LERAY, J.: Sur les équations et les transformations, J. Math. Pures et appl. 24 (1945), 201-248.

[11] LERAY, J.: Théorie des points fixes: indice total et nombre de Lefschetz, Bull. Soc. Math. France 87 (1959), 221-233.

[12] NUSSBAUM, R.D.: Generalizing the fixed point index, Math. Ann. 228 (1977), no 3, 259-278.

[13] O'NEILL, B.: Essential sets and fixed points, Amer. J. Math. 75 (1953), 497-509.

[14] PEITGEN, H.-O. and SEIGBERG, H.-W.: An $\bar{\varepsilon}$-Permutation of Brouwer's Definition of Degree, these proceedings.

[15] SPANIER, E.H.: Algebraic Topology, McGraw-Hill, New-York, (1966).

[16] STEINLEIN, H.: Ein Satz über den Leray-Schauderschen Abbildungsgrad, Math. Z. 120 (1972), 176-208.

[17] STEINLEIN, H.: Über die verallgemeinerten Fixpunktindizes von Iterierten verdichtender abildungen, Manuscripta Math. 8, (1973), 251-266.

[18] STEINLEIN, H.: A new proof of the (mod p)-theorem in asymtotic fixed point theory, Proc. Conf. on Problems in Nonlinear Functional Analysis, Bonn (1974), ed Ber. Ges. Math. Datenverarbeitung Bonn 103 (1975), 29-42.

[19] TROMBA, A.J.: The Beer Barrel Theorem, Proc. Conf. Functional Differential Equations and Approximation of Fixed Points, Bonn (1978), Lecture Notes in Mathematics No 730, Springer-Verlag, Heidelberg (1979), 484-488.

FIXED POINT THEOREMS FOR APPROXIMATIVE ANR'S

By

GILLES GAUTHIER

Université du Québec à Chicoutimi
Chicoutimi, Québec, Canada.

0. Introduction

There have been many generalizations of the notions of approximative absolute neighborhood retracts of Noguchi [13] ($AANR_N$) and Clapp [3] ($AANR_C$). In this paper we are interested in those generalizations which use coverings: Powers [14] ($AANR_N(M)$) and Granas [12] ($AANR_N(C)$, $AANR_C(C)$; for various C). We recall the definitions, get properties and characterizations of these classes of spaces in Section 1. In Section 2 we prove a homological result which is used in Section 3 to obtain Lefschetz type fixed point theorems for the $AANR_C(K)$ and $AANR_C(M)$ spaces. These results appear with more details and generality in [7].

By a space we mean a Hausdorff topological space and by a map $f : X \to Y$ between spaces we mean a continuous function. We use Čech homology with compact carriers and coefficients in the field of rational numbers.

1. Approximative ANR-spaces and NES-spaces

Let C be one of the following classes of spaces: compact spaces (K), compact metrizable spaces (KM), metrizable spaces (M). We denote by $\mathrm{Cov}(X)$ the set of all open coverings of a space X. Let X,Y be spaces and $\alpha = (U_i)_{i\in I} \in \mathrm{Cov}(X)$; two maps $f,g : Y \to X$ are said to be α-near ($f \underset{\alpha}{=} g$) if for every $x \in Y$ there is an $i \in I$ such that $\{f(x),g(x)\} \subset U_i$.

A closed subspace A of a space X is said to be an approximative retract of X if for every $\alpha \in \mathrm{Cov}(A)$ there exists a map $r_\alpha : X \to A$ such that $r_\alpha|_A \underset{\alpha}{=} 1_A$. Consider again a space X and a closed subspace A of X. The subspace A is said to be an approximative neighborhood retract of X, in the weak sense, if for every $\alpha \in \mathrm{Cov}(A)$ there exist an open subset U_α of X containing A and a map $r_\alpha : U_\alpha \to A$ such that $r_\alpha|A \underset{\alpha}{=} 1_A$. The subspace A is said to be an approximative neighborhood retract of X, in the strong sense, if there exists an open subset U of X containing A such that for every $\alpha \in \mathrm{Cov}(A)$ there exists a map $r_\alpha : U \to A$ satisfying $r_\alpha|A \underset{\alpha}{=} 1_A$. A space X is said to be an approximative absolute neighborhood retract, in the weak [resp. strong] sense, with respect to C ($X \in \mathrm{AANR}_C(C)$ [resp. $\mathrm{AANR}_N(C)$]) if $X \in C$ and for any closed embedding of X in a space $Z \in C$, the image of X is an approximative neighborhood retract, in the weak [resp. strong] sense, of Z.

(1.1) *THEOREM*

$$\mathrm{AANR}_N = K \cap \mathrm{AANR}_N(M),$$
$$\mathrm{AANR}_C = K \cap \mathrm{AANR}_C(M).$$

The notion of approximative neighborhood extension space appears in [12]. We recall the definitions. A space Y is said to be an approximative neighborhood extension space, in the weak [resp. strong] sense, with respect to C ($X \in \mathrm{ANES}_C(C)$ [resp. $\mathrm{ANES}_N(C)$]) if for any pair (X,A), where $X \in C$ and A is closed in X, and any map $f : A \to Y$ the following condition is satisfied: for every $\alpha \in \mathrm{Cov}(Y)$ there exist a neighborhood U_α of A in X and a map $f_\alpha : U_\alpha \to Y$ [resp. there exists a neighborhood U of A in X such that for every $\alpha \in \mathrm{Cov}(Y)$ there exists a map $f_\alpha : U \to Y$] satisfying $f_\alpha|A \underset{\alpha}{=} f$.

Using various embedding theorems (Tychonoff, Wojdyslawski, Urysohn) we get,

(1.2) *THEOREM*

$$\mathrm{AANR}_N(C) = C \cap \mathrm{ANES}_N(C),$$
$$\mathrm{AANR}_C(C) = C \cap \mathrm{ANES}_C(C).$$

Let us remark that in the proofs of the following theorems we consider only the case where the various notions are in the weak sense; the proofs in the case where the notions are in the strong sense are similar.

Granas [12] showed that $\mathrm{ANR}(M) \subset \mathrm{NES}(K)$. We have the similar relations in the approximative case.

(1.3) *THEOREM*

$$\mathrm{AANR}_N(M) \subset \mathrm{ANES}_N(K),$$
$$\mathrm{AANR}_C(M) \subset \mathrm{ANES}_C(K).$$

<u>*Proof*</u>

Let $Y \in \mathrm{AANR}_C(M)$. Embed Y as a closed subspace of a normed space using the Arens-Eells embedding theorem [1] and then use the facts that $Y \in \mathrm{AANR}_C(M)$ and that a normed vector space is an $ES(K)$-space [12]. ■

(1.4) *LEMMA*

If $C \subset C'$ then $C \cap \mathrm{AANR}_N(C') \subset \mathrm{AANR}_N(C)$ and $C \cap \mathrm{AANR}_C(C') \subset \mathrm{AANR}_C(C)$.

(1.5) *THEOREM*

$$\mathrm{AANR}_N(KM) = K \cap \mathrm{AANR}_N(M),$$
$$\mathrm{AANR}_C(KM) = K \cap \mathrm{AANR}_C(M).$$

<u>*Proof*</u>

By Lemma (1.4) we have $K \cap \mathrm{AANR}_C(M) = KM \cap \mathrm{AANR}_C(M) \subset \mathrm{AANR}_C(KM)$, this complete the first part. Let $Y \in \mathrm{AANR}_C(KM)$. We have $Y \in KM$. Using the Urysohn embedding theorem we can show that $Y \in \mathrm{ANES}_C(M)$. The fact that $\mathrm{AANR}_C(KM) \subset K \cap \mathrm{AANR}_C(M)$ follows from Theorem (1.2). ■

There is yet another way to obtain the AANR_N and AANR_C spaces:

(1.6) *THEOREM*

$$\mathrm{AANR}_N = M \cap \mathrm{AANR}_N(K),$$
$$\mathrm{AANR}_C = M \cap \mathrm{AANR}_C(K).$$

Proof

By the Theorems (1.1), (1.3) and (1.2) we have

$$\mathrm{AANR}_C = K \cap \mathrm{AANR}_C(M) \subset K \cap \mathrm{ANES}_C(K) = \mathrm{AANR}_C(K)$$

and so $\mathrm{AANR}_C \subset M \cap \mathrm{AANR}_C(K)$. On the other hand, using the Theorems (1.2), (1.5) and (1.1) we have

$$M \cap \mathrm{AANR}_C(K) = KM \cap \mathrm{ANES}_C(K) \subset KM \cap \mathrm{ANES}_C(KM) = \mathrm{AANR}_C(KM) = K \cap \mathrm{AANR}_C(M) = \mathrm{AANR}_C. \blacksquare$$

(1.7) *DEFINITION*

Let H be a class of spaces. A map $f: X \to Y$ is nearly factorizable, in the strong sense, with respect to H if there exist a space $C \in H$ and a map $\varphi: X \to C$ such that: for every $\alpha \in \mathrm{Cov}(Y)$ there is a map $\psi_\alpha : C \to Y$ such that $f \underset{\alpha}{=} \psi_\alpha \varphi$.

(1.8) *DEFINITION*

Let H be a class of spaces. A map $f: X \to Y$ is nearly factorizable, in the weak sense, with respect to H if for every $\alpha \in \mathrm{Cov}(Y)$ there exist a space C_α in H and maps $\varphi_\alpha : X \to C_\alpha$, $\psi_\alpha : C_\alpha \to Y$ such that $f \underset{\alpha}{=} \psi_\alpha \varphi_\alpha$.

We have the following characterizations:

(1.9) *THEOREM*

a) Let $X \in K$. Then $X \in \mathrm{AANR}_C(K)$ [resp. $\mathrm{AANR}_N(K)$] if and only if 1_X is nearly factorizable in the weak [resp. strong] sense with respect to $H = \mathrm{NES}(K)$.

b) Let $X \in M$. Then $X \in \mathrm{AANR}_C(M)$ [resp. $\mathrm{AANR}_N(M)$] if and only if 1_X is nearly factorizable in the weak [resp. strong] sense with respect to $H = \mathrm{ANR}(M)$.

c) Let $X \in KM$. Then $X \in \mathrm{AANR}_C$ [resp. AANR_N] if and only if 1_X is nearly factorizable in the weak [resp. strong] sense with respect to $H = \mathrm{ANR}$ $(= K \cap \mathrm{ANR}(M))$.

Proof

a) Let $X \in AANR_C(K)$. Using the Tychonoff embedding theorem, embed X as a subspace of a Tychonoff cube. Open subsets of Tychonoff cube being $NES(K)$-spaces we obtain, since $X \in AANR_C(K)$, that 1_X is nearly factorizable in the weak sense with respect to $H = NES(K)$.

Conversely, let $X \in K$ and 1_X be nearly factorizable in the weak sense with respect to $H = NES(K)$. It is easy to show that $X \in ANES_C(K)$ and by Theorem (1.2) we obtain $X \in AANR_C(K)$.

b) Similar to a) using the Arens-Eells embedding theorem and the facts that a normed vector space is an $ES(M)$-space and that $NES(M) \cap M = ANR(M)$.

c) Similar to a) using Theorem (1.5), the Urysohn embedding theorem and the fact that in the Hilbert cube any neighborhood of a compact set A contains an ANR-space containing A. ■

Concerning part c) of Theorem (1.9) see also [9].

(1.10) *THEOREM*

a) $X \in AANR_C(K)$ [resp. $AANR_N(K)$] if and only if X is homeomorphic to an approximative neighborhood retract, in the weak [resp. strong] sense, of a Tychonoff cube.

b) $X \in AANR_C(M)$ [resp. $AANR_N(M)$] if and only if X is homeomorphic to an approximative neighborhood retract, in the weak [resp. strong] sense, of a normed vector space.

c) $X \in AANR_C$ [resp. $AANR_N$] if and only if X is homeomorphic to an approximative neighborhood retract, in the weak [resp. strong] sense of the Hilbert cube.

2. MAPS INTO COMPACT SPACES OF FINITE TYPE

A space X is said to be of finite type if the graded vector space

$$H_*(X) = \{H_q(X)\}$$

is of finite type (i.e. $\dim H_q(X) < \infty$ for all q and $H_q(X) = 0$ for almost all q). The set of all finite open coverings of a space X is denoted by $\mathrm{Cov}^f(X)$. If $\alpha, \beta \in \mathrm{Cov}(X)$ and β is a refinement of α, we write $\alpha \leq \beta$. For $\alpha \in \mathrm{Cov}^f(X)$ we denote by $N(\alpha)$ the nerve of α, it is a finite simplicial complex.

In the sequel the following theorem is of importance.

(2.1) *THEOREM*

Let X be a compact space of finite type. Then there exists $\alpha_0 \in \mathrm{Cov}^f(X)$ such that, for every compact space Y and maps $f,g : Y \to X$, if f and g are α_0-near then $H_*(f) = H_*(g)$.

Let H^* be the Čech cohomology functor with coefficients in the field of rational numbers from the category of compact pairs and maps to the category of graded vector spaces over $\mathbb{Q}$ and maps of degree zero. The functors $\mathrm{Hom}(H^*(-),\mathbb{Q})$ and H_* (restricted to the category of compact pairs and maps) are naturally isomorphic [10], so Theorem (2.1) is implied by

(2.2) *THEOREM*

Let X be a compact space of finite type. Then there exists $\alpha_0 \in \mathrm{Cov}^f(X)$ such that, for every compact space Y and maps $f,g : Y \to X$, if f and g are α_0-near then $H^*(f) = H^*(g)$.

Proof

Since X is of finite type, $H^*(X)$ is of finite type. Let q be such that $H^q(X) \neq 0$. Let $k = \dim H^q(X)$. For $\alpha, \beta \in \mathrm{Cov}^f(X)$, $\alpha \leq \beta$, we denote by

$$\varphi_\alpha^\beta : H^q(N(\alpha)) \to H^q(N(\beta))$$

the homorphism induced by any projection $p_{\alpha\beta} : N(\beta) \to N(\alpha)$ and for $\alpha \in \mathrm{Cov}^f(X)$ we denote by φ_α the canonical map of $H^q(N(\alpha))$ into $H^q(X) = \varinjlim\{H^q(N(\alpha)), \varphi_\alpha^\beta, \mathrm{Cov}^f(X)\}$ (see [5]).

Take $\alpha_i \in \mathrm{Cov}^f(X)$ and $u_{\alpha_i} \in H^q(N(\alpha_i))$, for $i = 1,2,\ldots,k$, such that $\varphi_{\alpha_1}(u_{\alpha_1}), \varphi_{\alpha_2}(u_{\alpha_2}), \ldots, \varphi_{\alpha_k}(u_{\alpha_k})$ is a basis of $H^q(X)$. Let $\alpha = (U_j)_{j=1}^{N} \in \mathrm{Cov}^f(X)$ $\alpha_i \leq \alpha$ for $i = 1,2,\ldots,k$. The space X being normal, let $\beta = (V_j)_{j=1}^{N} \in \mathrm{Cov}^f(X)$ be such that $\overline{V}_j \subset U_j$ for $j = 1,2,\ldots,N$. For $j = 1,2,\ldots N$, let

$$\lambda_j = (U_j, X\backslash\overline{V}_j) \in \mathrm{Cov}^f(X).$$

Take $\lambda_0 \in \mathrm{Cov}^f(X)$ such that $\lambda_j \leq \lambda_0$ for $j = 1,2,\ldots,N$.

We now show: if Y is a compact space and $f,g : Y \to X$ are two maps which are λ_0-near, then $H^q(f) = H^q(g)$.

Let Y be a compact space and let $f,g : Y \to X$ be two maps which are λ_0-near. By the choice of λ_0 we have $g^{-1}(V_j) \subset f^{-1}(U_j)$ for $j = 1,2,\ldots,N$. Set $\lambda = f^{-1}(\alpha)$ $\delta = g^{-1}(\beta)$, then $\lambda \leq \delta$.

The following diagram commutes (where α,β are as defined)

$$\begin{array}{cccc}
\alpha & N(\lambda) & \xrightarrow{\ f_\alpha\ } & N(\alpha) \\
\wedge\!| & \uparrow p'_{\lambda\delta} & & \uparrow p_{\alpha\beta} \\
\beta & N(\delta) & \xrightarrow{\ g_\beta\ } & N(\beta)
\end{array}$$

where f_α, g_β are inclusions and $p_{\alpha\beta} : N(\beta) \to N(\alpha) : j \to j$, $p'_{\lambda\delta} : N(\delta) \to N(\lambda) : j \to j$ are projections. This gives rise to the following commutative diagram.

$$\begin{array}{cccc}
\alpha & H^q(N(\alpha)) & \xrightarrow{\ H^q(f_\alpha)\ } & H^q(N(\lambda)) \\
\wedge\!| & \downarrow \varphi_\alpha^\beta & & \downarrow \psi_\lambda^\delta = H^q(p'_{\lambda\delta}) \\
\beta & H^q(N(\beta)) & \xrightarrow{\ H^q(g_\beta)\ } & H^q(N(\delta))
\end{array}$$

It is now easy to show that the two maps $H^q(f), H^q(g) : H^q(X) \to H^q(Y)$ coincide on the chosen basis for $H^q(X)$ and so $H^q(f) = H^q(g)$.

The space X being of finite type the proof is complete. ■

3. FIXED POINT THEOREMS

Let $f : E \to E$ be an endomorphism of a vector space Let $N(f) = \bigcup_{n\geq 1} \mathrm{Ker}(f^n)$ and $E_f = E/N(f)$. Since $f(N(f)) \subset N(f)$, f induces an endomorphism $\tilde{f} : E_f \to E_f$. Let $f = \{f_q\} : E \to E$ be an endomorphism of degree zero of a graded vector space $E = \{E_q\}$, f is said to be a Leray endomorphism if $E_f = \{(E_q)_{f_q}\}$ is of finite type. For such an f, we define the (generalized) Lefschetz number $\Lambda(f)$ of f by

$$\Lambda(f) = \sum_q (-1)^q \mathrm{tr}(\tilde{f}_q).$$

A map f from a space X into itself is said to be a Lefschetz map if $f_* = H_*(f)$ is a Leray endomorphism. For such an f we define the (generalized) Lefschetz number $\Lambda(f)$ of f by $\Lambda(f) = \Lambda(f_*)$. If the space X is of finite type, the generalized Lefschetz number coincide with the usual one.

We will make use of the following lemma.

(3.1) *LEMMA (See* [6]*).*

Assume that in the category of Hausdorff spaces the following diagram commutes.

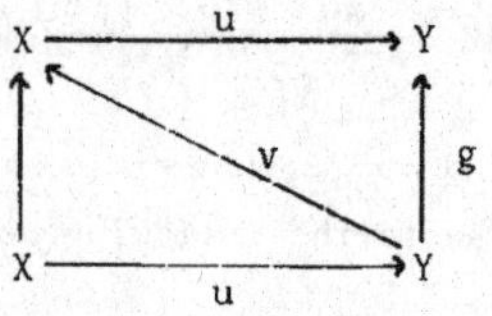

Then (i) if one of the maps f or g is a Lefschetz map, then so is the other and in that case $\Lambda(f) = \Lambda(g)$, (ii) f has a fixed point if and only if g does.

A space X is said to be a Lefschetz space if any map $f : X \to X$ is a Lefschetz map and $\Lambda(f) \neq 0$ implies that f has a fixed point. Let $\mathcal{P}$ be a class of maps, a space X is said to be a Lefschetz space for the class $\mathcal{P}$ if any map $f : X \to X$ which belongs to $\mathcal{P}$ is a Lefschetz map and $\Lambda(f) \neq 0$ implies that f has a fixed point.

(3.2) *THEOREM*

Let $\mathcal{H}$ be a class of Lefschetz spaces for the class of compact maps and let

X be a compact space of finite type. Let $f : X \to X$ be nearly factorizable, in the weak sense, with respect to H. If $\Lambda(f) \neq 0$ then f has a fixed point.

Proof

Let $f : X \to X$ be a nearly factorizable map, in the weak sense, with respect to H and be such that $\Lambda(f) \neq 0$. Using Theorem (2.1), let $\alpha_0 \in \mathrm{Cov}^f(X)$ be such that for any compact space Y and maps $g,h : Y \to X$, if g and h are α_0-near then $g_* = h_*$. Consider any $\alpha \in \mathrm{Cov}(X)$ such that $\alpha_0 \leq \alpha$. There exist $C_\alpha \in H$ and maps $\varphi_\alpha : X \to C_\alpha$, $\psi_\alpha : C_\alpha \to X$ such that $f \underset{\alpha}{=} \psi_\alpha\varphi_\alpha$. Since $\alpha_0 \leq \alpha$, we have $(\psi_\alpha\varphi_\alpha)_* = f_*$. Consider the compact map $\varphi_\alpha\psi_\alpha : C_\alpha \to C_\alpha$. Since C_α is a Lefschetz space for the class of compact maps and $\Lambda(\varphi_\alpha\psi_\alpha) = \Lambda(\psi_\alpha\varphi_\alpha) = \Lambda(f) \neq 0$ we obtain that $\varphi_\alpha\psi_\alpha$ has a fixed point and by Lemma (3.1) f has an α-fixed point.

The space X being Haudsdorff f has a fixed point. ■

(3.3) *COROLLARY (Granas [11]).*

If $X \in \mathrm{AANR}_N$ then X is a Lefschetz space.

Proof

Use Theorem (1.9) c) and the fact that ANR-spaces are Lefschetz spaces. ■

(3.4) *COROLLARY (Clapp [3]).*

If $X \in \mathrm{AANR}_C$ and is of finite type then X is a Lefschetz space.

Proof

Use Theorem (1.9) c) and the fact that ANR-spaces are Lefschetz spaces. ■

(3.5) *COROLLARY*

If $X \in \mathrm{AANR}_C(K)$ and is of finite type then X is a Lefschetz space.

Proof

Use theorem (1.9) a) and the fact that NES(K)-spaces are Lefschetz spaces for the class of compact maps [6], [12]. ∎

It follows from Theorem (1.6) that Corollary (3.5) is a generalization of the fixed point theorem of Clapp (Corollary (3.4)). Using the fact that if $X,Y \in AANR_C(K)$ then $X \times Y \in AANR_C(K)$ one can construct many examples of spaces which are in $AANR_C(K)$ but not in $AANR_C$: take any nonempty $X \in AANR_C$, then $X \times (\prod_{i \in \Gamma}[0,1])$ is in $AANR_C(K)$ but not in $AANR_C$ if Γ is uncountable.

(3.6) *DEFINITION*

Let X,Y be spaces, $f : X \to Y$ is a compact map of finite type if $f(X) \subset K$ for some compact subspace of finite type K of Y.

(3.7) *THEOREM*

If $Y \in ANES_C(K)$, then Y is a Lefschetz space for the class of compact maps of finite type.

Proof

Let $Y \in ANES_C(K)$ and $f : Y \to Y$ be a compact map of finite type. Let K be a compact subspace of finite type of Y such that $f(Y) \subset K$. Let $f_K : K \to K$, $f' : Y \to K$ be the maps induced by $f : Y \to Y$ and $i : K \to Y$ be the inclusion. By Lemma (3.1)(i), since $f_K = f'i$ is a Lefschetz map, $f = if'$ is also a Lefschetz map.

Suppose $\Lambda(f) \neq 0$. The space K being compact, there are a Tychonoff cube T and an embedding $h : K \to T$. Let $\tilde{K} = h(K)$ and $s : K \to \tilde{K}$ be the map induced by h. The following diagram is commutative.

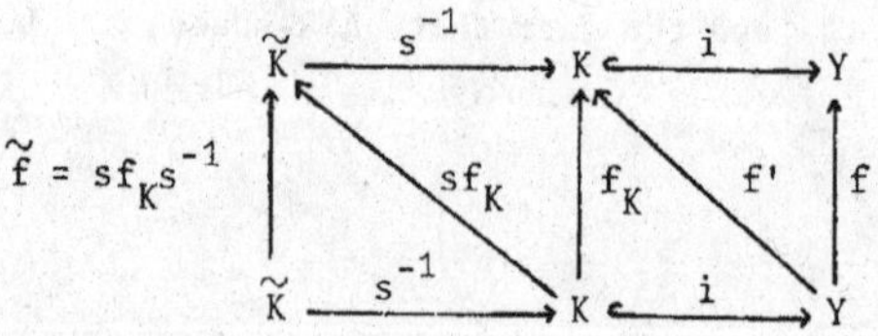

So $\Lambda(\tilde{f}) = \Lambda(f_K) = \Lambda(f) \neq 0$. The space $\tilde{K}$ is compact and of finite type, using Theorem (2.1) let $\beta_0 \in \mathrm{Cov}^f(\tilde{K})$ be such that for any compact space Z and maps $g,k : Z \to \tilde{K}$, if g and k are β_0-near then $g_* = k_*$.

Consider any $\beta \in \mathrm{Cov}(\tilde{K})$ such that $\beta_0 \leq \beta$. Let $\alpha \in \mathrm{Cov}(Y)$ be defined by $\alpha = (sf')^{-1}(\beta)$.

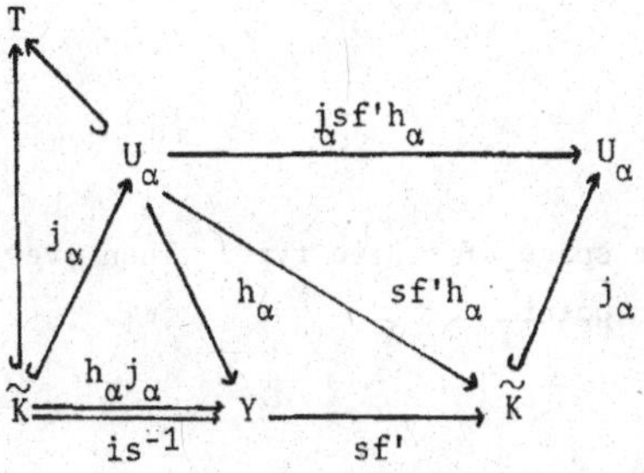

Since $Y \in \mathrm{ANES}_C(K)$, there are an open neighborhood U_α of $\tilde{K}$ in T and a map $h_\alpha : U_\alpha \to Y$ such that $h_\alpha j_\alpha \underset{\alpha}{=} is^{-1}$, where $j_\alpha : \tilde{K} \to U_\alpha$ is the inclusion. Then $\tilde{f} = sf'is^{-1} \underset{\beta}{=} sf'h_\alpha j_\alpha$. Since $\beta_0 \leq \beta$, we have $\tilde{f}_* = (sf'h_\alpha j_\alpha)_*$. Since U_α is open in $T \in ES(K)$ we have that $U_\alpha \in NES(K)$ and so U_α is a Lefschetz space for the class of compact maps. The map $j_\alpha sf'h_\alpha : U_\alpha \to U_\alpha$ is compact and

$$\Lambda(j_\alpha sf'h_\alpha) = \Lambda(sf'h_\alpha j_\alpha) = \Lambda(\tilde{f}) \neq 0,$$

so it has a fixed point. By Lemma (3.1) (ii) this implies that $sf'h_\alpha j_\alpha$ has also a fixed point and we get that $\tilde{f}$ has a β-fixed point.

The space $\tilde{K}$ being Hausdorff, $\tilde{f}$ has a fixed point and so f has a fixed point. ■

From Theorem (1.2) we have,

(3.8) *COROLLARY*

If $X \in \mathrm{AANR}_C(K)$, then X is a Lefschetz space for the class of compact maps of finite type.

From Theorem (1.3) we get this other generalization of the fixed point theorem of Clapp.

(3.9) *COROLLARY*

If $X \in AANR_C(M)$, then X is a Lefschetz space for the class of compact maps of finite type.

In [8] we define a notion of Borsuk presentation for compact spaces. This gives rise to the notions of NE-maps and NE-spaces. We obtain (compare with [2], [4]).

(3.10) *THEOREM*

Let X be a compact space of finite type. Then every NE-map $f : X \to X$ such that $\Lambda(f) \neq 0$ has a fixed point.

(3.11) *COROLLARY*

Let X be a NE-space of finite type, then X is a Lefschetz space.

(3.12) *THEOREM*

Let X be a NE-space. Then X is a Lefschetz space for the class of compact maps of finite type.

The $AANR_C(K)$-spaces being NE-spaces we get new, but much more tedious, proofs of Corollary (3.5) and Corollary (3.8).

It is not known whether all NE-spaces are $AANR_C(K)$-spaces.

REFERENCES

[1] ARENS, R.F. and EELLS, J.: On embedding uniform and topological spaces, Pacific J. Math., 6 (1956), 397-403.

[2] BORSUK, K.: On the Lefschetz-Hopf fixed point theorem for nearly extendable maps, Bull. Acad. Polon. Sci., Ser. Sci. Math. Astr. Phys., 23 (1975), 1273-1279.

[3] CLAPP, M.H.: On a generalization of absolute neighborhood retracts, Fund. Math., 70 (1971), 117-130.

[4] DUGUNDJI, J.: On Borsuk's extension of the Lefschetz-Hopf theorem, Bull. Acad. Polon. Sci., Ser. Sci. Math. Astr. Phys., 25 (1977), 805-811.

[5] EILENBERG, S. and STEENROD, N.: Foundation of algebraic topology, Princeton 1952.

[6] FOURNIER, G. and GRANAS A.: The Lefschetz fixed point theorem for some classes of non metrizable spaces, J. Math. Pures et Appl., 52 (1973), 271-284.

[7] GAUTHIER, G.: La théorie des rétracts approximatifs et le théorème des points fixes de Lefschetz, Ph. D. Thesis, Université de Montréal (1980).

[8] GAUTHIER, G. : Le théorème des points fixes de Lefschetz pour les NE-applications des espaces compacts non métrisables, Bull. Acad. Polon. Sci., Ser. Sci. Math. Astr. Phys., to appear.

[9] GAUTHIER, G. and GRANAS, A.: Notes sur le théorème de Lefschetz pour les ANR approximatifs, Coll. Math., to appear.

[10] GÓRNIEWICZ, L.: Homological methods in fixed point theory of multi-valued maps, Dissertationes Math., 129 (1976), 1-71.

[11] GRANAS, A.: Fixed point theorems for approximative ANR-s, Bull. Acad. Polon. Sci., Ser. Sci. Math. Astr. Phys., 16 (1968), 15-19.

[12] GRANAS, A.: Points fixes pour les applications compactes: espaces de Lefschetz et la théorie de l'indice, Les Presses de l'Université de Montréal, Montréal (1980).

[13] NOGUCHI, H.: A generalization of absolute neighborhood retracts, Kodai Math. Sem. Rep., 1 (1953), 20-22.

[14] POWERS, M.: Fixed point theorems for non-compact approximative ANR-s, Fund. Math., 75 (1972), 61-68.

ON THE LEFSCHETZ COINCIDENCE THEOREM

By

LECH GÓRNIEWICZ
Institute of Mathematics
University of Gdańsk
Wita Stwosza 57
80952 Gdańsk, Poland.

0. Introduction

A generalization of the fixed point problem is the coincidence problem. The coincidence problem is formulated as follows; given two continuous maps $p,q : \Gamma \to X$ of Hausdorff topological spaces the coincidence problem for (p,q) is concerned with conditions which guarantee that the pair (p,q) admits one or more coincidence points, that is points $y \in \Gamma$ such that $p(y) = q(y)$. The study of this problem (first treated in a topological setting in 1946 by Eilenberg and Montgomery [3]) was recently taken up in [8] and [10], where an extension of the Eilenberg-Montgomery coincidence theorem to non-compact ANR-s ([8]) and for NES(compact) spaces ([10]) was established. Note that the Eilenberg-Montgomery theorem is a natural generalization of the Lefschetz fixed point theorem. In this paper we will apply and develop results given in [10]. Moreover, we want to give several applications of our general coincidence theorem to fixed point theory of multivalued maps. We will use the Dold's fixed point theory methods ([2]) and its generalizations given in [8]. The present paper is strictly connected to the paper [10]. The main results of this paper are contained in section 5. The last three sections are devoted to applications of the main result to the fixed point theory of multivalued maps. The important parts of this paper are sections 3 and 4, where we consider the notion of morphisms and we prove the coincidence theorem for open subsets of euclidean spaces and for open subsets of the Tychnoff cube.

In what follows by a space we understand a Hausdorff topological space and by map a continuous transformation.

1. Homological Preliminaries

By H we shall denote the Čech homology functor with compact carriers and coefficients in the field of rationals Q from the category of Hausdorff topological spaces and continuous maps to the graded vector spaces over Q and linear maps of degree zero. A space X is called *acyclic* provided (i) $H_0(X) = Q$ and (ii) $H_n(X) = 0$ for every $n \geq 1$; X is called of *finite type* provide $\dim H_n(X) < +\infty$ for all $n \geq 0$ and $H_n(X) = 0$ for almost all n. Let X be a space by $\mathrm{Cov}(X)$ we will denote the family of all open coverings of X. Let X be a space and let $\alpha \in \mathrm{Cov}(X)$. For two given maps $f,g : Y \to X$ we will say that f and g are α*-close* if for each $y \in Y$ there is a member $U \in \alpha$ such that $f(y) \in U$ and $g(y) \in U$.

Recall the following theorem.

(1.1) *THEOREM*

Let X be a compact space of finite type. Then there exists $\alpha_0 \in \mathrm{Cov}(X)$ such that for every space Y and for every two maps $f,g : Y \to X$ the condition: f and g are α_0*-close*, implies that the induced linear maps (on homology) are the same, that is, $f_* = g_*$.

In the metric case Theorem (1.1) was proved in [8]; for non-metric case see [6] (compare also with [23]).

A continuous map $p : \Gamma \to X$ is said to be a *Vietoris map* provided p is proper and $p^{-1}(x)$ is acyclic for each $x \in X$ (written $p : \Gamma \Rightarrow X$). The Vietoris-Begle Mapping Theorem (compare [8]) implies that, if $p : \Gamma \Rightarrow X$ is a Vietoris map, then the induced map $p_* : H(\Gamma) \to H(X)$ is a linear isomorphism.

Similarly, a map of pairs $p : (\Gamma,\Gamma_0) \to (X,X_0)$ is said to be a Vietoris map provided (i) p is proper, (ii) $p^{-1}(X_0) = \Gamma_0$ and (iii) $p^{-1}(x)$ is acyclic for every $x \in X$.

From the classical Vietoris-Begle Mapping Theorem automatically follows

(1.2) *PROPOSITION*

If $p : (\Gamma,\Gamma_0) \to (X,X_0)$ is a Vietoris map, then the induced map

$$p_* : H(\Gamma,\Gamma_0) \to H(X,X_0)$$

is a linear isomorphism.

Let us consider the following two maps:

$$q : \Gamma \to X \quad \text{and} \quad p : \Gamma_1 \to X.$$

In this case we define a space $\Gamma \boxtimes \Gamma_1$ by putting:

$$\Gamma \boxtimes \Gamma_1 = \{(y,y_1) \,;\, q(y) = p(y_1)\}$$

and a map $\bar{p} : \Gamma \boxtimes \Gamma_1 \to X$ given as follows:

$$\bar{p}(y,y_1) = y.$$

The map $\bar{p} : \Gamma \boxtimes \Gamma_1 \to X$ is called *pull-back* of the map p (with respect to q).

Some important properties of Vietoris maps are summarized in the following.

(1.3) *PROPOSITION*

(i) If $p : (\Gamma,\Gamma_0) \Rightarrow (X,X_0)$ is a Vietoris map and $(A,A_0) \subset (X,X_0)$, then the map $\tilde{p} : (p^{-1}(A), p^{-1}(A_0)) \Rightarrow (A,A_0)$ is also Vietoris, where $\tilde{p}(y) = p(y)$ for each $y \in p^{-1}(A)$.

(ii) The composition of two Vietoris maps is Vietoris, too.

(iii) The pull-back of a Vietoris map is also a Vietoris map.

First, we remark that on the category of all pairs (U,V) and continuous maps the Čech homology functor with compact carriers and the singular homology functor are equivalent, where by U we understand an arbitrary open subset of some euclidean space or a finite polyhedron and by V we understand an open subset of U (for details see [8]). In what follows we will use the following notations. By R^{n+1} we will denote the $(n+1)$-dimensional euclidean space and by S^n the unit sphere in R^{n+1}. We identify S^n with $R^n \cup \{\infty\}$. Let $A \subset U \subset R^n$, where A is compact and U is open in R^n. Then from the excision axiom (for singular homology) we deduce:

(1.4) The inclusion $j : (U, U \setminus A) \to (S^n, S^n \setminus A)$ induces an isomorphism $j_* : H(U, U \setminus A) \to H(S^n, S^n \setminus A)$.

Let K be a finite polyhedron which is contained in U. Consider the diagram:

$$U \xleftarrow{p} \Gamma \xrightarrow{q} K .$$

With the above diagram we associate the following:

$$(U, U \setminus K) \xleftarrow{\tilde{p}} (\Gamma, \Gamma \setminus p^{-1}(K)) \xrightarrow{\tilde{q}} (R^n, R^n \setminus \{0\}) ,$$

where $\tilde{p}(y) = p(y)$ and $\tilde{q}(y) = p(y) - q(y)$ for each $y \in \Gamma$. We observe that $\tilde{p}$ is a Vietoris map. Let $\Delta : (U, U \setminus K) \to (U, U \setminus K) \times U$ and $d : (U, U \setminus K) \times K \to (R^n, R^n \setminus \{0\})$ be two maps given as follows: $\Delta(x) = (x,x)$, $d(x,x') = x - x'$, for each $x \in U$ and $x' \in K$.

(1.5) *LEMMA* ([8]).

The following diagram commutes

$$\begin{array}{ccccc} H(U,U \setminus K) & \xrightarrow{\Delta_*} & H(U,U \setminus K) \otimes H(U) & \xrightarrow{Id \otimes q_* p_*^{-1}} & H(U,U \setminus K) \otimes H(K) \\ & \searrow^{\tilde{q}_* \tilde{p}_*^{-1}} & & \swarrow_{d_*} & \\ & & H(R^n, R^n \setminus \{0\}) & & \end{array}$$

Let us fix for each n an orientation $1 \in H_n(S^n) \approx Q$ of the n-th sphere $S^n = R^n \cup \{\infty\}$. Consider the diagram

$$S^n \xrightarrow{i} (S^n, S^n \setminus A) \xrightarrow{j} (U, U \setminus A) ,$$

in which i,j are inclusions. From (1.4) we deduce that the map j_* is a linear isomorphism. We define the *fundamental class* O_A of the pair (U,A) by putting (compare with [2]):

$$O_A = j_{*n}^{-1} \circ i_{*n}(1).$$

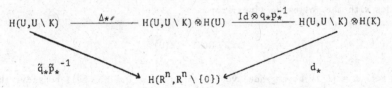

(1.6) *LEMMA (compare with [2] or [8]).*

Let $A \subset A_1 \subset V \subset U \subset R^n$, where A, A_1 are compact, U, V are open subsets of R^n and let $k : (V, V \setminus A_1) \to (U, U \setminus A)$ be the inclusion map. Then we have $k_{*n}(O_{A_1}) = O_A$.

Now, we shall formulate the Dold's Lemma (see [2]) in the terms of Čech homology with compact carriers (compare with [8]). For this consider the following maps: $t : U \times K \to K \times U$, $t(x,x') = (x',x)$ for each $x \in U$ and $x' \in K$, $O_K^{\times} : H(K) \to H(U, U \setminus K) \otimes H(K)$, $O_K^{\times}(u) = O_K \otimes u$ for each $u \in H(K)$, $\times : Q \otimes H(U) \to H(U)$, $\times(v \otimes u) = v \cdot u$ for each $u \in H(K)$ and $v \in Q$.

(1.7) *LEMMA*

The composite

$$1 = 1(K,U) : H(K) \xrightarrow{O_K^{\times}} H(U, U \setminus K) \otimes H(K) \xrightarrow{\Delta_* \otimes Id} H(U, U \setminus K) \otimes H(U) \otimes H(K) \longrightarrow$$

$$\xrightarrow{Id \otimes t_*} H(U, U \setminus K) \otimes H(K) \otimes H(U) \xrightarrow{d_* \otimes Id} Q \otimes H(U) \xrightarrow{\times} H(U)$$

coincides with the following linear map

$$i_* : H(K) \to H(U).$$

Let $E = \{E_n\}$ be a graded vector space over Q. We shall consider the following graded vector spaces:

(1) $E^* = \{E_n^*\}$, where $E_n^* = Hom(E_{-n}, Q)$,

(2) $Hom(E,E) = \{(Hom(E,E))_n\}$, where $(Hom(E,E))_n = \bigoplus_{-\ell+k=n} Hom(E_\ell, E_k)$,

(3) $E^* \otimes E = \{(E^* \otimes E)_n\}$, where $(E^* \otimes E)_n = \bigoplus_{\ell+k=n} E_\ell^* \otimes E_k$,

and the *evaluation map* $e : (E^* \otimes E)_0 \to Q$ given by putting

$$e(u_k \otimes v_\ell) = u_k(v_\ell), \quad \text{for } u_k \in Hom(E_k, Q), \ v_\ell \in E_\ell, \ k = \ell.$$

Now, we are able to give the definition of the coincidence index and to formulate some of its properties which we will use in the following.

For the diagram: $U \xleftarrow{p} \Gamma \xrightarrow{q} U$, when $q(\Gamma) \subset K \subset U$, by $\varkappa_{p,q}$ we will denote the following set $\varkappa_{p,q} = \{x \in U : x \in q(p^{-1}(x))\}$. It is easy to see that $\varkappa_{p,q}$ is a compact set. So as before, we can associate with the above diagram the following:

$$(U, U \setminus \varkappa_{p,q}) \xleftarrow{\tilde{p}} (\Gamma, \Gamma \setminus p^{-1}(\varkappa_{p,q})) \xrightarrow{\tilde{q}} (R^n, R^n \setminus \{0\}).$$

We define the index of coincidence of (p,q) by putting (compare with [2] or [8]):

$$I(p,q) = q_* \circ (p_*)^{-1}(0_{\varkappa_{p,q}}).$$

From (1.5), (1.6) and the definition of index is not difficult to obtain the following (for details see [8])

(1.8) *PROPOSITION*

(i) If $I(p,q) \neq 0$, then the pair (p,q) has a coincidence point.

(ii) If A is a compact set such that $\varkappa_{p,q} \subset A \subset U$, then

$$I(p,q) = q_*(p_*)^{-1}(0_A).$$

(iii) Let K be a finite polyhedron such that $q(\Gamma) \subset K$ and let

$$a \in (H(K))^* \otimes H(K)$$

be define as follows $a = (\hat{d} \otimes \mathrm{Id}) \circ (\mathrm{Id} \otimes q_{1*}\, p_*^{-1}) \circ \Delta_*(0_K)$, where $q_1 : U \to K$ the contraction of q to the pair (Γ, K) and $\hat{d} : H(U, U \setminus K) \to (H(K))^*$ is given by the formula $\hat{d}(u)(v) = d_*(u \otimes v)$, for $u \in H(U, U \setminus K)$ and $v \in H(K)$. Then we have: $I(p,q) = e(a)$.

2. The Lefschetz number

Let $f : E \to E$ be an endomorphism of a finite dimensional vector space E over Q. By $\mathrm{tr}(f)$ we will denote the (ordinary) trace of f. Let $E = \{E_n\}$ be a graded vector space over Q of finite type, that is $E_n = 0$ for almost all n

and $\dim E_n < +\infty$ for all n. For an endomorphism $f: E \to E$ of degree zero of such space we define the *(ordinary) Lefschetz number* $\lambda(f)$ by putting

$$\lambda(f) = \sum_n (-1)^n \operatorname{tr}(f_n).$$

Now, we want to define the Lefschetz number in the terms of linear maps. In this order we consider a linear map $\theta: (E^* \otimes E)_0 \to (\mathrm{Hom}(E,E))_0$ by putting :

$$\theta(u_k \otimes v_i)(v') = (-1)^k u_k(v') \cdot v_i, \quad \text{for } u_k \in E_k^*,\ v_i \in E_i,\ v' \in E_k,\quad i = k.$$

Important properties of the linear map θ are summarized in the following proposition (see [2] or [8]).

(2.1) *PROPOSITION*

Assume that E is a graded vector space of the finite type over Q and $f: E \to E$ is a linear map of degree zero. Then:

(i) θ is an isomorphism,

(ii) $\lambda(f) = e(\theta^{-1}(f))$, where e denotes the evaluation map (compare with section 1).

Consider a diagram

$$U \xleftarrow{p} \Gamma \xrightarrow{q} U,$$

in which U is an open subset of R^n. Assume that there is a finite polyhedron K such that $q(\Gamma) \subset K \subset U$. Then we can consider the following diagram:

$$K \xleftarrow{p'} p^{-1}(K) \xrightarrow{q'} K,$$

in which p' and q' are the respective contractions of p and q. We define *the Lefschetz number* $\lambda(p',q')$ of the pair (p',q') as follows:

$$\lambda(p',q') = \lambda(q'_* \circ (p')_*^{-1}).$$

By using Proposition (2.1) we can prove the following (see [8]).

(2.2) *PROPOSITION*

$\lambda(p',q') = e(a)$, (compare with (1.8), (iii)).

Recall, the generalized Lefschetz number as given by J. Leray. Let $f : E \to E$ be an arbitrary endomorphism of an arbitrary vector space E over Q. Put $N(f) = \{x \in E : f^{(n)}(x) = 0$, for some $n\}$. Since $f(N(f)) \subset N(f)$, we have the induced endomorphism $\tilde{f} : \tilde{E} \to \tilde{E}$, where $\tilde{E} = E/N(f)$. Call f admissible provided $\dim \tilde{E} < \infty$; for such f we define the *generalized trace* $\mathrm{Tr}(f)$ of f by putting $\mathrm{Tr}(f) = \mathrm{tr}(\tilde{f})$. Let $f : E \to E$ be an endomorphism of degree zero of a graded vector space E over Q. Call f *the Leray endomorphism* if the space $\tilde{E} = \{\tilde{E}_n\}$ is of finite type. For such f we define the (*generalized*) Lefschetz number $\Lambda(f)$ by putting

$$\Lambda(f) = \sum_n (-1)^n \, \mathrm{Tr}(f_n).$$

The following property of the Leray endomorphisms is a consequence of the well-known formula $\mathrm{tr}(uv) = \mathrm{tr}(vu)$ for the ordinary trace (compare with [8]).

(2.3) Assume that in the category of graded vector spaces the following diagram commutes

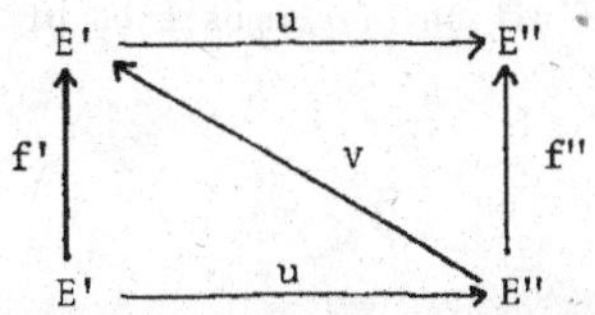

If one of the vertical arrows is a Leray endomorphism, then so is the other and in that case $\Lambda(f') = \Lambda(f'')$.

3. Morphisms

Given two spaces X and Y let $\mathcal{D}(X,Y)$ be the set of all diagrams of the form $X \xleftarrow{p} \Gamma \xrightarrow{q} Y$ where $q(\Gamma)$ is not necessarily contained in a compact polyhedron in Y. Every such a diagram we denote briefly by (p,q). Given two diagrams (p,q), $(p',q') \in \mathcal{D}(X,Y)$, we write $(p,q) \sim (p',q')$ if there are two maps $f : \Gamma \to \Gamma'$ and $g : \Gamma' \to \Gamma$ for which the following two diagrams commute

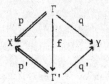

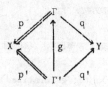

It is easy to see that $\sim$ is an equivalence relation in $\mathcal{D}(X,Y)$.

(3.1) *DEFINITION*

The equivalence class of a diagram $(p,q) \in \mathcal{D}(X,Y)$ with respect to $\sim$ is denoted by

$$\varphi = \{X \xLeftarrow{p} \Gamma \xrightarrow{q} Y\} : X \to Y$$

and is called a *morphism* from X to Y; we let $M_{X,Y}$ be the set of all such morphisms.

The above notion was first introduced in [10].

In what follows we denote morphisms by Greek letters and the ordinary maps by Latin letters; we identify $f : X \to Y$ with the morphism $f = \{X \xLeftarrow{Id} X \xrightarrow{f} Y\} : X \to Y$. Following [10] we give the definition of composition of morphisms.

(3.2) *DEFINITION*

To compose two morphisms $\varphi = \{X \xLeftarrow{p} \Gamma \xrightarrow{q} Y\} : X \to Y$ and $\psi = \{Y \xLeftarrow{p'} \Gamma \xrightarrow{q'} Z\} : X \to Y$ and we write a commutative diagram

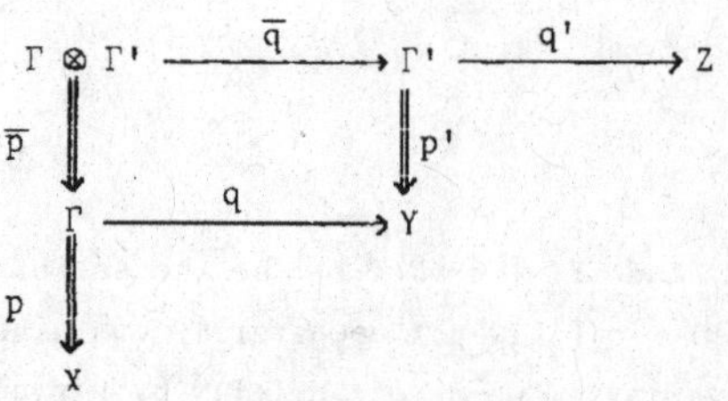

in which $\Gamma \boxtimes \Gamma'$ is the fibre product of q and p' and $\overline{p}, \overline{q}$ is the pull-back of p', q (compare with section 1); we define the composite $\psi \circ \varphi$ of φ and ψ by putting

$$\psi \circ \varphi = \{X \xleftarrow{p \circ \overline{p}} \Gamma \boxtimes \Gamma' \xrightarrow{q' \circ \overline{q}} Z\} : X \to Z.$$

Denote by $\mathcal{M}$ the category of all topological spaces and morphisms (it is not difficult to see that $\mathcal{M}$ is a category).

(3.3) *PROPOSITION (compare with* [10]*).*

The Čech homology functor with compact carriers and coefficients in Q, $H:\mathrm{Top} \to \mathcal{E}$ extends over $\mathcal{M}$ to a functor $\hat{H} : \mathcal{M} \to \mathcal{E}$ (where $\mathcal{E}$ denotes the category of graded vector spaces over Q and linear maps).

Indeed, for a morphism $\varphi = \{X \xleftarrow{p} \Gamma \xrightarrow{q} Y\} : X \to Y$ we let

$$\hat{H}(\varphi) = \varphi_* = q_* \circ p_*^{-1} ;$$

it is easy to see that the definition of φ_* does not depend on the choice of a representative of φ and also that if $\varphi = f$ then $\varphi_* = f_*$. For details see [10].

Using the functor $\hat{H}$ we can generalize the notion of Lefschetz maps to the case of morphisms.

(3.4) *DEFINITION*

A morphism $\varphi : X \to X$ is called a *Lefschetz morphism* provided $\varphi_* : H(X) \to H(X)$ is a Leray endomorphism; for such φ we define the generalized Lefschetz number by putting $\Lambda(\varphi) = \Lambda(\varphi_*)$.

From (2.3) we deduce:

(3.5) Assume that in the category $\mathcal{M}$ the following diagram commutes

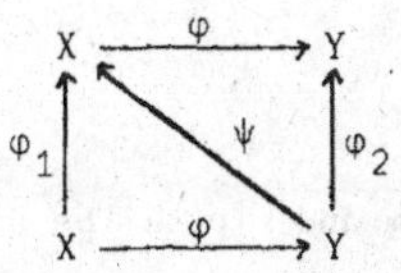

Then, if one of the vertical arrows is a Lefschetz morphism, then so is the other and in that case $\Lambda(\varphi_1) = \Lambda(\varphi_2)$.

(3.6) *DEFINITION*

We say that $\varphi : X \to X$ has a *coincidence* provided the set $\varkappa(\varphi) = p(\varkappa_{p,q})$ is nonempty. Clearly φ has a coincidence if and only if for any representative (p,q) of φ the set $\varkappa_{p,q}$ is nonempty.

(3.7) *LEMMA* ([10])

Let $\varphi : X \to Y$, $\psi : Y \to X$ be two morphisms. Then $\psi \circ \varphi$ has a coincidence if and only if $\varphi \circ \psi$ does.

Let $\varphi = \{X \xleftarrow{p} \Gamma \xrightarrow{q} Y\} : X \to Y$ be a morphism. For each $x \in X$ we define the image $\varphi(x)$ of x by φ as follows:

$$\varphi(x) = q \circ p^{-1}(x).$$

It is evident that the above definition is correct. A morphism $\varphi : X \to Y$ is called *continuous*, if for each open U in Y the set $\{x \in X : \varphi(x) \subset U\}$ is open.

Let $\varphi : X \to X$ be a morphism and let $\alpha \in \mathrm{Cov}(X)$. A point $x \in X$ is called an α-*coincidence* for φ if and only if there is a member $U \in \alpha$ such that $x \in U$ and $\varphi(x) \cap U \neq \phi$.

We have the following:

(3.8) *LEMMA (compare with* [10] *or* [4]*)*

Let $\varphi : X \to X$ be a continuous morphism and assume that for each $\alpha \in \mathrm{Cov}(X)$ there is an α-coincidence point for φ. Then φ has a coincidence.

4. Coincidence spaces

Let $\varphi : X \to Y$ be a morphism and (p,q) be a representative of φ. We define

$\varphi(X) \subset Y$ by $\varphi(X) = q(p^{-1}(X))$. From the definition of morphism immediately follows that $\varphi(X)$ does not depend on the choice of the representative of φ; φ is called *compact* provided φ is continuous and the set $\varphi(X)$ is relatively compact in Y.

Note the following observation:

(4.1) Let $\varphi : X \to Y$ and $\psi : Y \to Z$ be two morphisms. If φ or ψ is compact, then so is the composite $\psi \circ \varphi : X \to Z$.

(4.2) *DEFINITION*

A space X is said to be a *coincidence space* for compact morphisms, provided (i) any compact morphism $\varphi : X \to X$ is a Lefschetz morphism and (ii) $\Lambda(\varphi) \neq 0$ implies that φ has a coincidence.

From (3.5) and (3.7) we deduce.

(4.2) *PROPOSITION*

A retract of a coincidence space is coincidence space.

We establish now important special cases of the main result.

(4.3) *THEOREM*

Every open subset U of R^n is a coincidence space.

<u>*Proof*</u>

For the proof it is sufficient to show that for a representative (p,q) of φ the Lefschetz number $\Lambda(p,q)$ is defined and that $\Lambda(p,q) \neq 0$ implies $\varkappa_{p,q} \neq \phi$. Let (p,q) be a representative of φ. Because φ is compact we can find a finite polyhedron K such that $\varphi(U) \subset K \subset U$. So we have the following commutative diagram:

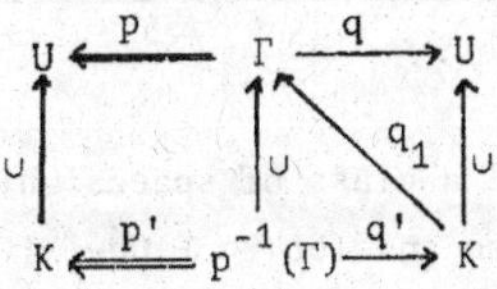

in which p', q' and q_1 are contractions of p and q respectively. Applying to the above diagram the functor H, from (2.3) we obtain:

$$\Lambda(p,q) = \Lambda(p',q') = \lambda(q'_*(p')_*^{-1}).$$

To prove the second part of our theorem assume that $\Lambda(p,q) \neq 0$; so $\lambda(q_*(p')_*^{-1}) \neq 0$. By (1.8) we have $I(p,q) = e(a)$. Using (2.2) we obtain $\lambda(q'_*(p')_*^{-1}) = e(a)$; so $I(p,q) \neq 0$ and by applying (1.8) again we have $\varkappa_{p,q} \neq \emptyset$. The proof is completed.■

(4.4) *THEOREM*

Every open subset U of a Tychonoff cube T is a coincidence space.

In the proof of (4.4) we will use the fact that any open subset of a locally convex space is a coincidence space. This fact is proved in [10] (compare with (7.6) in [10]).

Proof of (4.4)

We can assume that T is a retract of a locally convex space E. Let $r:E \to T$ denote the retraction map. Let U be an open set of T and $\varphi : U \to U$ a compact morphism. We have the following commutative diagram:

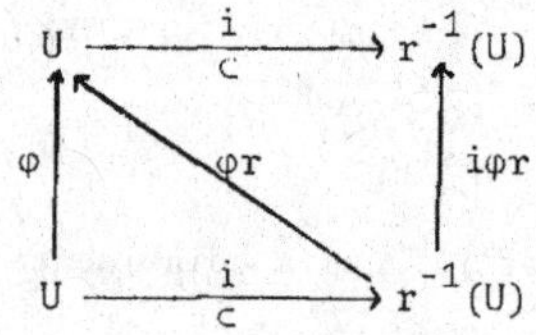

Because, as an open subset of a locally convex space E, $r^{-1}(U)$ is a coincidence space, our assertation immediately follows from (3.5) and (3.7) and the proof is completed. ■

5. THE LEFSCHETZ COINCIDENCE THEOREM FOR MORPHISMS OF ANES(COMPACT)-SPACES.

First, we shall introduce a class of spaces which will be important in our considerations. Let X be a space and let $\alpha \in Cov(X)$. For two given maps $f,g:Y \to X$

we will write $f \underset{\alpha}{\equiv} g$ if and only if (i) f and g are α-close and (ii) $f_* = g_*$.

(5.1) *DEFINITION*

A space X is called *an approximative neighbourhood extension space for the class of compact spaces* ($X \in$ ANES(compact)) provided for each $\alpha \in \mathrm{Cov}(X)$, for each pair (Y,A) of compact spaces and for each map $f : A \to X$ there exist an open neighbourhood U_α of A in Y and a map $f_\alpha : U \to X$ such that $f_\alpha|A \underset{\alpha}{\equiv} f$, where $f_\alpha|A$ denotes the restriction of f_α to A.

The class of ANES(compact)-spaces is quite general. For details see [6] and [13]. Note that it contains a class of NES(compact)-spaces and, in particular, metric ANR-s (see: [6], [13] and [14]). Moreover, any approximative ANR in the sense of H. Noguchi [19] belongs to ANES(compact)-spaces. By using theorem (1.1) it is not difficult to see that any approximative ANR, of finite type, in the sense of M.H. Clapp [1] belongs to ANES(compact)-spaces, too.

Now we are able to formulate the main result of this paper.

(5.2) *THEOREM*

Any $X \in$ ANES(compact) is a coincidence space.

<u>*Proof*</u>

Let X be ANES(compact) and $\varphi : X \to X$ be a compact morphism. We are going to show that (i) $\Lambda(\varphi)$ is defined and (ii) $\Lambda(\varphi) \neq 0$ implies that φ has a coincidence. Let (p,q) be a representative of φ and let K be a compact set containing $q(X)$. We embed K into Tychonoff cube T and denote by $s : K \to \tilde{K}$ the homeomorphism of K onto $\tilde{K} \subset T$. We may write the following commutative diagram of morphisms:

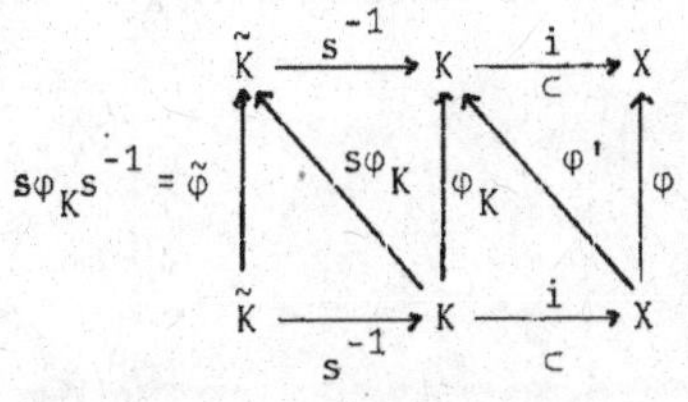

So, by (3.5) if one of the Lefschetz numbers $\Lambda(\varphi)$ or $\Lambda(\tilde{\varphi})$ is defined then so is the other and

$$\Lambda(\varphi) = \Lambda(\tilde{\varphi}). \tag{1}$$

Let $\alpha \in \mathrm{Cov}(X)$. Consider the map $is^{-1}\colon \tilde{K} \to X$. Since $X \in \mathrm{ANES(compact)}$ it follows that there is an open set U_α in T containing $\tilde{K}$ and a map $h_\alpha\colon U_\alpha \to X$ such that $h_\alpha j_\alpha \underset{\alpha}{\equiv} is^{-1}$, where $j_\alpha : \tilde{K} \to U_\alpha$ is the inclusion map. So the following diagram commutes:

$$\begin{array}{ccc}
H(U_\alpha) & \xleftarrow{\quad (j_\alpha)_* \quad} & H(\tilde{K}) \\
(\tilde{\tilde{\varphi}}_\alpha)_* = (j_\alpha)_* s_* \varphi_*^! (h_\alpha)_* \Big\uparrow & \nearrow s_*\varphi_*^!(h_L)_* & \Big\uparrow \tilde{\varphi}_* = s_*\varphi_*(S^{-1})_* \\
H(U_\alpha) & \xleftarrow[\quad (j_\alpha)_* \quad]{} & H(\tilde{K})
\end{array}$$

Now, from (4.4) follow that $\Lambda(\tilde{\tilde{\varphi}}_\alpha)$ is defined and, in view of (2.3) and (1) we have

$$\Lambda(\varphi) = \Lambda(\tilde{\varphi}) = \Lambda(\tilde{\tilde{\varphi}}_\alpha) ,$$

so the Lefschetz number of φ is well defined.

If $\Lambda(\varphi) \neq 0$ then for each $\alpha \in \mathrm{Cov}(X)$ we have $\Lambda(\tilde{\tilde{\varphi}}_\alpha) \neq 0$. By (4.4), we obtain that $\tilde{\tilde{\varphi}}_\alpha$ has a coincidence. Because the maps $h_\alpha j_\alpha$ and is^{-1} are α -close we deduce that φ has an α -coincidence for every $\alpha \in \mathrm{Cov}(X)$. Therefore our result follows from (3.8) and the proof is completed.■

(5.3) *COROLLARY*

Any $X \in \mathrm{NES(compact)}$ or, in particular, $X \in \mathrm{ANR(metric)}$ is a coincidence space.

In the last three sections of this paper we will give several applications of the above results.

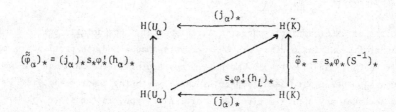

6. FIXED POINTS OF MULTIVALUED MAPS.

We recall first some terminology.

Let X and Y be two spaces and assume that for every point $x \in X$ a nonempty subset $\varphi(x)$ of Y is given; in this case we say that φ is *a multivalued map* from X to Y and we write $\varphi : X \to Y$. We associate with φ the diagram

$$X \xleftarrow{p_\varphi} \Gamma_\varphi \xrightarrow{q_\varphi} Y$$

in which $\Gamma_\varphi = \{(x,y) : y \in \varphi(x)\}$ is the graph of φ and the natural projections p_φ, q_φ are given by $p_\varphi(x,y) = x$ and $q_\varphi(x,y) = y$. If $\varphi : X \to Y$ and $\psi : Y \to Z$ are two multivalued maps, then their *composition* is the map $\psi \circ \varphi : X \to Z$ defined by $(\psi \circ \varphi)(x) = \bigcup_{y \in \varphi(x)} \psi(y)$. Let $\varphi : X \to X$ be a multivalued map; a point $x \in X$ is called *a fixed point* for φ, provided $x \in \varphi(x)$. A multivalued map is said to be *upper semi continuous* (u.s.c.) provided (i) $\varphi(x)$ is a compact set for each $x \in X$ and (ii) for each open set $U \subset Y$ the set $\{x \in X : \varphi(x) \subset U\}$ is open. A multivalued map $\varphi : X \to Y$ is called *compact* provided the image $\varphi(X) = \bigcup_{x \in X} \varphi(x)$ is a relatively compact subset of Y. An u.s.c. map is called *acyclic*, if for each $x \in X$ the set $\varphi(x)$ is acyclic.

From the definition clearly follows:

(6.1) If $\varphi : X \to Y$ is an acyclic map, then the projection $p_\varphi : \Gamma_\varphi \Rightarrow X$ is a Vietoris map.

Let $p : X \to Y$ be a (singlevalued) map from X onto Y. We associate with such a p the multivalued map $\varphi_p : Y \to X$ given by $\varphi_p(y) = p^{-1}(y)$.

It is not difficult to prove the following (see [8]):

(6.2) *PROPOSITION*

Let $p : X \to Y$ be a proper map from X onto the metric space Y. Then the map $\varphi_p : Y \to X$ is u.s.c..

(6.3) *REMARK (see [8]).*

For an arbitrary Hausdorff space Y Proposition (6.2) is not true.

A multivalued map $\varphi : X \to Y$ is called *admissible* (see [8]) if there exists a diagram $X \xleftarrow{p} \Gamma_{\varphi} \xrightarrow{q} Y$ such that $q(p^{-1}(x)) \subset \varphi(x)$, for each $x \in X$; in this case the pair (p,q) is called a *selected pair* for φ and we write $(p,q) \subset \varphi$.

From (6.4) immediately follows that any acyclic map is admissible. It is not difficult to verify (see [8]) that any composition of acyclic maps and moreover any composition of admissible maps is admissible.

An admissible, compact map $\varphi : X \to X$ is called a *Lefschetz map* provided for each selected pair $(p,q) \subset \varphi$ the endomorphism $q_* p_*^{-1} : H(X) \to H(X)$ is a Leray endomorphism. For a Lefschetz map $\varphi : X \to X$ we define a *Lefschetz set* $\Lambda(\varphi)$ by putting

$$\Lambda(\varphi) = \{\Lambda(q_* p_*^{-1}) : (p,q) \subset \varphi\} .$$

For details concerning admissible maps we recommended [8]. Recall that the Lefschetz fixed point theorem for multivalued maps were given by several authors (see [3], [4], [7], [8], [10], [11], [12], [17], [20], [21], [22]). We are interested to present consequences from our main result. In this order we shall discuss now the relationship between the multivalued maps and the morphisms.

A multivalued map $\varphi : X \to Y$ is said to be *determined* by a morphism $\{X \xleftarrow{p} \Gamma \xrightarrow{q} Y\}$ provided $\varphi(x) = q(p^{-1}(x))$ for each $x \in X$; the morphism which determines φ is also denoted by φ. Clearly every morphism determines a multivalued map, but not conversly. It is easy to see that any admissible map has a selector which is determined by some morphism. In view of (6.2) and (6.3), we deduced that morphisms of non metric spaces determines multivalued maps which are not necessarily u.s.c. and hence are not necessarily admissible.

In [10] we have proved the following:

(6.4) *THEOREM*

Let X be an NES(compact)-space and let $\varphi : X \to X$ be a multivalued map determined by some compact morphism, then φ is a Lefschetz map and $\Lambda(\varphi) \neq \{0\}$ implies that φ has a fixed point.

We remark that all Lefschetz-type fixed point results signalized above are special cases of Theorem (6.4). But as an immediate consequence of Theorem (5.2) we obtain the following generalization of Theorem (6.4).

(6.5) *THEOREM*

Let X be an ANES(compact)-space and let $\varphi : X \to X$ be a multivalued map determined by some compact morphism. Then φ is a Lefschetz map and $\Lambda(\varphi) \neq \{0\}$ implies that φ has a fixed point.

7. Families of multivalued maps.

In this section by K we will denote a compact ANES(compact)-space. We define the *Euler characteristic* $\chi(K)$ of such K by putting

$$\chi(K) = \Lambda(Id_K).$$

A *multivalued semi-flow* on K is an u.s.c. map $\varphi : K \times R^+ \to K$ such that:

(i) for each $t \in R^+$ the map $\varphi_t : K \to K$ given by formula $\varphi_t(x) = \varphi(x,t)$ is admissible,

(ii) the map φ_o is acyclic and for each $x \in K$ we have $x \in \varphi_o(x)$,

(iii) $\varphi(\varphi(x,t),\tau) = \bigcup_{y \in \varphi(x,t)} \varphi(y,\tau) \subset \varphi(x,t+\tau)$, for each $x \in K$, $t,\tau \in R^+$,

where R^+ denote the set of all nonnegative real numbers.

A *fixed point* for the multivalued semi-flow is a point x_o such that $x_o \in \varphi(x_o,t)$ for all $t \in R^+$.

(7.1) *THEOREM*

If K is a compact ANES(compact)-space with $\chi(K) \neq 0$, then any multivalued semi-flow on K must have a fixed point.

Proof

Consider the homotopy $(x,\rho) \to \varphi(x,(1-\rho)t_o)$, we see that φ_o is homotopic to φ_{t_o} for every $t_o \in R^+$. But φ_o is an acyclic map so the Lefschetz set $\Lambda(\varphi_o)$ of φ_o is a singleton (see [8]). From (ii) and $\varkappa(K) \neq 0$ it follows that $\Lambda(\varphi_o) \neq 0$. Now, by the homotopy argument (see [8]), we obtain that $\Lambda(\varphi_t) \neq \{0\}$ for every $t \in R^+$. Therefore from Theorem (6.5) it follows that φ_t has a fixed point for each $t \in R^+$. Let $A_n = \{x \in K : x \in \varphi(x, 2^{-n})\}$; each A_n is nonempty, closed and therefore compact. Moreover $A_n \supset A_{n+1} \dots$ since $x \in \varphi(x, 2^{-(n+1)})$ implies that $\varphi(x, 2^{-(n+1)}) \subset \varphi(\varphi(x, 2^{-(n+1)}), 2^{-(n+1)}) \subset \varphi(x, 2^{-(n+1)})$. By the finite intersection property, there is some $x_o \in \bigcap_n A_n$; and since $x_o \in \varphi(x_o, 2^{-n})$ for each natural n we have $x_o \in \varphi(x_o, m2^{-n})$ for all m and n. Because the set of dyadic rationals $\{m2^{-n}\}$ is dense in R^+, upper semicontinuity of φ assures $x_o \in \varphi(x_o,t)$ for all $t \in R^+$, and the proof is complete. ■

REMARK

It is easy to see that the above theorem remains true if in the definition of semi-flow we replace (ii) by the following assumption:

(ii) φ_o is admissible with $0 \notin \Lambda(\varphi_o)$ and $x \in \varphi_o(x)$ for each $x \in K$.

Now, by using Theorem (6.5), we shall give a generalization of the Holsztynski's result [15] from the case of singlevalued maps to the case of admissible maps. By I^n we will denote the n-th cartesian product of intervals $[0,1]$.

(7.2) *THEOREM*

Let K be a compact ANES(compact)-space and let $\varphi : K \times I^n \to K$ be an admissible map. Assume that $0 \notin \Lambda(\varphi_o)$, where $\varphi_o : K \to K$ is given by $\varphi_o(x) = \varphi(x,0)$. Then there exists $x \in K$ such that

$$\dim\{t \in I^n : x \in \varphi(x,t)\} \geq n - \dim K.$$

Proof

Let $A = \{(x,t) \in K \times I^n : x \in \varphi(x,t)\}$. By Theorem (6.5), in view of $\Lambda(\varphi_o) \neq \{0\}$, the set A is nonempty. Moreover from the upper semicontinuity of φ follows that F is a closed subset of $K \times I^n$ and therefore compact. Let $f : A \to K$ and $g : A \to I^n$ be two maps given by $f(x,t) = x$, $g(x,t) = t$. We claim that the map $g : A \to I^n$ is universal (see [16]) that is, for an arbitrary $h : A \to I^n$ there is $(x,t) \in A$ such that $h(x,t) = g(x,t)$. Indeed, let $h : A \to I^n$ be a map. Because A is a closed subset of $K \times I^n$ there is a continuous extension $h' : K \times I^n \to I^n$ of h. The map $\psi : K \times I^n \to K \times I^n$ defined by putting

$$\psi(x,t) = \{(y,s) \in K \times I^n : y \in \varphi(x,t), \quad s = h'(x,t)\}$$

is clearly admissible. So by the homotopy argument (compare with the proof of (7.1)) we obtain that $\Lambda(\psi) \neq \{0\}$. By using the definition it is easy to see that $(K \times I^n) \in \text{ANES(compact)}$. Therefore from Theorem (6.5) follows that ψ has a fixed point. If $(x,t) \in \psi(x,t)$, then we have $x \in \varphi(x,t)$ and $t = h'(x,t) = h(x,t) = g(x,t)$. Since $g : A \to I^n$ is universal we have that $\dim A \geq n$ (see [16]). Thus by the generalized Hurewicz theorem (see [15], [16]) relating maps and dimension we obtain

$$\dim f^{-1}(x) \geq n - \dim K, \quad \text{for some } x \in K.$$

Such an x satisfies our theorem and the proof is completed. ■

8. SPHERIC MAPS.

In this section we shall give some applications of the main theorem to the euclidean space. The presented results are strictly connected with the paper [9]. We shall start from some example.

Example. Let K^2 be the unit ball in the euclidean 2-space R^2 and let S^1 be the unit sphere in R^2. Consider a multivalued map $\varphi : K^2 \to K^2$ given as follows

$$\varphi(x) = \{y \in K^2 : \|y - x\| = \rho(x)\} \cup \{y \in S^1 : \|y - x\| \geq \rho(x)\},$$

where $\rho(x) = 1 - \|x\| + \|x\|^2$.

It is evident that φ is an u.s.c. map with images which have the same homology as S^1 but φ has no fixed points.

The above example was given in 1957 by B. O'Neill. We want to define a class of u.s.c. maps with images which have the same homology as the unit sphere in the euclidean space R^{n+1} for which, in particular, the Brouwer fixed point theorem holds.

Let A be a compact subset of R^{n+1} which has the same homology as the unit sphere S^n in R^{n+1}. Then from the Alexander duality theorem follows that $R^{n+1} \setminus A = B(A) \cup D(A)$, where $B(A)$ and $D(A)$ are nonempty components of $R^{n+1} \setminus A$. In what follows for such A we shall denote by $B(A)$ the bounded component of $R^{n+1} \setminus A$ and by $D(A)$ the unbounded component of $R^{n+1} \setminus A$.

Let X be a compact subset of R^{n+1}. An u.s.c. map $\varphi : X \to X$ is called *a spheric map* (in R^{n+1}) provided (i) for each $x \in X$ the set $\varphi(x)$ has the same homology as the unit sphere S^n in R^{n+1} and (ii) for each $x \in X$, if $x \in B(\varphi(x))$, then there is an open neighbourhood V_x of x in X such that $y \in B(\varphi(y))$, for every $y \in V_x$.

Let $\varphi : X \to X$ be a spheric map; we associate with such a φ the map $\tilde{\varphi} : X \to R^{n+1}$ given by

$$\tilde{\varphi}(x) = R^{n+1} \setminus D(\varphi(x)), \quad \text{for every} \quad x \in X.$$

(8.1) *REMARK*

Observe that if $R^{n+1} \setminus X$ is connected then $\tilde{\varphi}$ maps X into X. In particular, in view of the Alexander duality theorem, if X is acyclic, then $\tilde{\varphi}$ maps X into itself.

Recall the folowing proposition (see [9]).

(8.2) *PROPOSITION*

Let X be a compact, acyclic subset of R^{n+1}. If $\varphi : X \to X$ is a spheric map, then the associated map $\tilde{\varphi} : X \to X$ is acyclic.

Now, we are able to prove the following.

(8.3) *THEOREM*

Let X be a compact, acyclic subset of R^{n+1}. If $X \in$ ANES(compact), then any spheric map $\varphi : X \to X$ has a fixed point.

Proof

Consider the map $\tilde{\varphi} : X \to X$ associated with φ. From Proposition (8.2) and Theorem (6.5) it follows that $\tilde{\varphi}$ has a fixed point. Suppose that φ does not have a fixed point. Let $U = \{x \in X : x \in \tilde{\varphi}(x) \setminus \varphi(x)\}$. Then U, as the set of fixed points of an u.s.c. map, is a closed subset of X. On the other hand the assumption (ii) of the definition of spheric maps, implies that U is an open subset of X. We claim that $X \setminus U$ is nonempty. Indeed, let $x \in \partial X$ (where ∂X denotes the boundary of X in R^n). If $x \in \tilde{\varphi}(x)$, then by definition of $\tilde{\varphi}$ we would have that $x \in \varphi(x)$, but because by assumption φ has no fixed points we obtain that $x \in X \setminus U$. So $\partial X \neq \phi$ implies that $X \setminus U$ is nonempty. Thus we have proved that X is nonconnected, but it is a contradiction (X is acyclic). The proof of (8.3) is completed. ■

Note, that several consequences of (8.3) can be given.

References

[1] CLAPP, M.H.: On a generalization of absolute neighbourhood retracts, Fund. Math., 70 (1971), 17-30.

[2] DOLD, A.: Fixed point index and fixed point theorems for euclidean neighbourhood retracts, Topology., 4 (1965), 1-8.

[3] EILENBERG, S. and MONTGOMERY, D.: Fixed point theorems for multivalued transformations, Amer. J. Math., 58 (1946), 214-222.

[4] FOURNIER, G. and GÓRNIEWICZ, L.: The Lefschetz fixed point theorem for multivalued maps of non-metrizable spaces, Fund, Math., 92 (1976), 213-222.

[5] FOURNIER, G. and GRANAS, A.: The Lefschetz fixed point theorem for some classes of non-metrizable spaces, J. Math. Pures Appl., 52 (1973), 271-284.

[6] GAUTHIER, G. and GRANAS, A.: Note sur le théorème de Lefschetz pour les ANR approximatifs, Coll. Math., (to appear).

[7] GÓRNIEWICZ, L.: Fixed point theorems for multivalued mappings of approximative ANR-s, Bull. Acad. Polon. Sci., 8 (1970), 431-436.

[8] GÓRNIEWICZ, L.: Homological methods in fixed point theory of multivalued maps, Dissertationes Math., 29 (1976), 1-71.

[9] GÓRNIEWICZ, L.: Fixed point theorems for multivalued maps of subsets of euclidean spaces, Bull. Acad. Polon. Sci., 1 (1979), 111-116.

[10] GÓRNIEWICZ, L. and GRANAS, A.: Some general theorems in coincidence theory I, J. Math. Pures et Appl., (to appear).

[11] GÓRNIEWICZ, L. and GRANAS, A.: Fixed point theorems for multivalued mappings of ANR-s, J. Math. Pures et Appl., 49 (1970), 381-395.

[12] GÓRNIEWICZ, L. and SKORDEV, G.: Some coincidence theorems, Fund. Math. 101 (1978). 172-180, (in russian).

[13] GRANAS, A.: Points fixes pour les applications compactes en topologie et analyse fonctionnelle, Sem. de Math. Supérieures, Montréal (1973).

[14] HANNER, O.: Retraction and extension of mappings of metric and non-metric spaces, Ark. Mat., 2 (1952), 315-360.

[15] HOLSZTYNSKI, W.: Common fixed points of compact ANR-s, Bull. Acad. Polon. Sci., (to appear).

[16] HOLSZTYNSKI, W.: Une généralisation du théorème de Brouwer sur les points invariants, Bull. Acad. Polon. Sci., 10 (1964), 603-606.

[17] JAWOROWSKI, J.W.: Set-valued fixed point theorems for approximative retracts, Lec. Not. Math., 171 (1969), 34-39.

[18] KLEE, V.: Shrinkable neighbourhoods in Hausdorff linear spaces, Math. Ann., 4 (1960), 281-285.

[19] NOGUCHI, H.: A generalization of absolute neighbourhood retracts. Kodai Math., Sem. Rep., 1 (1953), 20-22.

[20] POWERS, M.: Fixed point theorems for non-compact approximative ANR-s, Fund. Math., 75 (1972), 61-68.

[21] POWERS, M.: Lefschetz fixed point theorem for a new class of multivalued maps, Pacific J. of Math., 42 (1972), 211-220.

[23] SKORDEV, G.: Fixed points of maps of AANR-s, Bull. Acad. Polon. Sci., 2 (1973), 73-80.

COBORDISMS OF MAPS

By

S.Y. Husseini

Department of Mathematics
University of Wisconsin
Madison, WI 53706
U.S.A.

0. Introduction.

Suppose that $f_0 : M_0 \to M_0$ and $f_1 : M_1 \to M_1$ are two maps of smooth and closed manifolds. The natural concept of cobordism between (M_0, f_0) and (M_1, f_1) is to require that there be a map $f : W \to W$ such that $f|\partial W = f_0 \coprod f_1$. Without further restrictions, it might happen that $f_0 : M_0 \to M_0$ has essential fixed points whereas $f_1 : M_1 \to M_1$ does not (see §3 below.) Thus this concept of cobordism needs further refinement. It one thinks of the usual fixed-point invariants of a map to be some sort of equivalence classes of neighborhoods of the fixed point set, then one is led to a notion of a cobordism that takes into account the local "action" of the maps. The precise formulation is given in §1 below, and such cobordisms are said to be *level-preserving*. In §1 we prove that level-preserving cobordisms preserve the usual Lefschetz number. But this does not take care of the case when the Lefschetz number is 0 yet the map has essential fixed points. To accomodate this case we introduce the concept of cobordisms for regular geometric settings, and prove that the generalized Lefschetz numbers defined for settings as in [6], are also preserved by level-preserving cobordisms. In this framework of settings, we obtain a natural correspondence between the Nielsen classes of $f_0 : M_0 \to M_0$ on the one hand, and those of $f_1 : M_1 \to M_1$ on the other. We prove that there are continua, one for each Nielsen class, going from M_0 to M_1. This generalizes the well-known result of F. Browder, [2], on the existence of continua for homotopies. Finally, in §3 we discuss some of the simple aspects of cobordisms which are not necessarily level-preserving, and give an example of a level-preserving cobordism which is not a homotopy.

I would like to acknowledge many useful conversations with E. Fadell pertaining to the existence of continua for homotopies as well as the other topics discussed herein.

1. Level-Preserving Cobordisms.

Suppose that $f_0 : M_0 \to M_0$ and $f_1 : M_1 \to M_1$ are smooth maps of the smooth closed manifolds M_0 and M_1. Then (M_0, f_0) and (M_1, f_1) are said to be *cobordant* if, and only if, there is a compact manifold W such that $\partial W = M_0 \coprod M_1$ and a map $f : (W, \partial W) \to (W, \partial W)$ such that $f|\partial W = f_0 \coprod f_1$.

(1.1) *DEFINITION*

A cobordism (W,f) *between* (M_0,f_0) *and* (M_1,f_1) *is said to be level-preserving if, and only if, there is an open neighborhood* U *of* $\mathrm{Fix}\, f$, *the fixed-point set of* f, *and a map*

$$p : U \to [0,1]$$

such that

(i) $p^{-1}(0) = U \cap M_0$,

(ii) $p^{-1}(1) = U \cap M_1$,

(iii) p *is a trivial local-fibration, and*

(iv) *the diagram*

$$\begin{array}{ccc} U \cap f^{-1}U & \xrightarrow{\quad f \quad} & U \\ & \searrow \quad \swarrow & \\ & [0,1] & \end{array}$$

commutes.

(1.2) *REMARK*

A. Dold pointed out to me that the condition on $f : U \to W$ that it be level-preserving makes it also an ENR_B, where $B = [0,1]$. See [3].

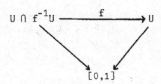

Recall that a *local-fibration* $p : E \to B$ is a map such that every point e in E has a neighborhood V_e which is fibered by p. A *trivial local-fibration* is a local fibration where the fibered neighborhoods V_e are trivial fibrations.

Homotopies connecting neighborhoods of the fixed-point sets at both ends are certainly examples of level-preserving cobordisms. We shall see in §3 below that there are level-preserving cobordisms which do not arise from homotopies. However, level-preserving cobordisms are in a certain sense sequences of local homotopies.

The following simple proposition provides a very useful tool in the study of level-preserving cobordisms. It allows us to erase, by homotopy, certain portions of the fixed-point set without introducing new fixed points.

(1.3) *PROPOSITION*

Suppose that (W,f) *is a level-preserving cobordism from* (M_0,f_0) *to* (M_1,f_1), *and let* C *be a closed and open subset of* $\mathrm{Fix}\, f$ *such that* $C \cap M_1 \neq \phi$. *Then there is a neighborhood* T *of* C, *and a deformation retraction*

$$r_t : W \to W$$

such that

$$r_1(W) \subset (W - \mathrm{int}\, T) \cup M_0$$

and

$$\mathrm{Fix}\,(fr_1) = (\mathrm{Fix}\, f - C) \cup \mathrm{Fix}\,(f_0).$$

<u>*Proof*</u>

Let U be a neighborhood of $\mathrm{Fix}\, f$, and $p : U \to [0,1]$ a trivial local fibration as in Definition (1.1) above. As

$$\mathrm{Fix}\, f \subset U \cap f^{-1}U$$

we can find a finite family $\{V_\alpha, I_\alpha\}$ such that

(i) for each α, V_α is a compact subset of U, I_α is a subinterval of

[0,1], and

$$(p|V_\alpha) : V_\alpha \to I_\alpha$$

is a trivial fibration;

(ii) if we put $\delta V_\alpha = \partial V_\alpha - p^{-1}(\partial I_\alpha)$, then

$$C \cap \delta V_\alpha = \phi ;$$

and

(iii) $V = \cup_\alpha V_\alpha$ covers C, and $V \cap (\text{Fix } f - C) = \phi$.

Next, using the partition of [0,1] defined by the end-points of the subintervals I_α, we can find a family $\{T'_\beta\}$ of compact subsets of U such that

(i) $T' = \cup_\beta T'_\beta$ covers C and $\{I_\beta\}$, where $I_\beta = p(T'_\beta)$ have disjoint interiors;

(ii) $(p|T_\beta) : T'_\beta \to I_\beta$ is a trivial fibration; and

(iii) $T' \cap (\text{Fix } f - C) = \phi$.

Now shrink $T' = \{T'_\beta\}$, if necessary, to a cover $T = \{T_\beta\}$ so that (i), (ii) and (iii) remain valid, and, in addition, the fibers F_β of the fibrations

$$(p|T_\beta) : T_\beta \to I_\beta$$

are compact submanifold with boundary of W of dimension m, where $m = \dim M_0 = \dim M_1$.

The deformation retraction

$$r_t : W \to (W - \text{int } T) \cup M_0$$

is constructed by induction on the number of sets in the families $\{I_\beta\}$. The induction step is similar to the case when $\{I_\beta\}$ consists of just a single set. So let T be a compact subset of U which covers C and is trivially fibered by

$$p : T \to I$$

where I is the closed interval $[a,1]$. Hence we have a commutative diagram

$$\begin{array}{ccc} I \times F & \xrightarrow{\varphi} & T \\ & {\scriptstyle p}\searrow \quad \swarrow{\scriptstyle p} & \\ & I & \end{array}$$

where the fiber is a compact manifold F with boundary. Next let $K \subset F$ be a compact submanifold such that $(F - \text{int } K)$ is homeomorphic to $\partial K \times I$ and

$$(\text{int } K) \times I \supset C.$$

Using the collar of ∂K, we can find a deformation retraction $r'_t : T \to T$ such that

(i) $r'_1(T) \subset T - \varphi(I \times \text{int } K) \cup M_0$

(ii) $r'_t | T - \varphi(I \times \text{int } K)$ is constant, and

(iii) r'_t is level-decreasing, that is

$$p(x) > p(r_t(x))$$

for x in $\varphi(I \times \text{int } K) - M_0$.

Clearly, r'_t gives rise to a deformation retraction $r_t : W \to W$ subject to the required conditions. The general case when the family $\{T_\beta\}$ covering C consists of many subsets can be easily dealt with inductively, with the inductive step being similar to the special case just discussed. ■

We shall call the family $T = \{T_\beta\}$ a *micro-tubular neighborhood of* C.

(1.4) *REMARK*

If (W,f) is a level-preserving cobordism and C is an open and closed subset of $\text{Fix } f \cap \text{int } W$, then one can show in a similar fashion that (W,f) is homotopically equivalent to another level-preserving cobordism (W,f') such that

$$\text{Fix } f' = \text{Fix } f - C.$$

But the deformation taking (W,f) to (W,f') need not be through level-preserving maps.

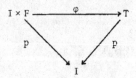

(1.5) *THEOREM*

Suppose that (W,f) *is a level-preserving cobordism from* (M_0,f_0) *to* (M_1,f_1). *Then*

$$L(f_0) = L(f) = L(f_1)$$

where $L(\cdot)$ *is the ordinary Lefschetz number.*

Proof

Take the set C of Proposition (1.2) to be all of Fix f, and let $r_t : W \to W$ be the deformation retraction of W onto $W_0 = (W - \text{int } T) \cup M_1$, where T is a micro-tubular neighborhood of C = Fix f, such that

$$\text{Fix}(fr_1) = \text{Fix}(f_1).$$

Then certainly $fr_1 : W \to W$ has the same Lefschetz number as $fr_1 : W_0 \to W$, since W_0 is a deformation retract of W. Also if one triangulates W_0 so that the subcomplex consisting of the simplices which meet the fixed-point set lies in $T \cap M_1$, then by direct calculation one sees that

$$L(f_1) = L(fr_1).$$

But, the maps

$$f, fr_1 : W \to W$$

are homotopic, and therefore,

$$L(f) = L(fr_1).$$

Similarly one proves that

$$L(f_0) = L(f). \blacksquare$$

(1.6) *REMARK*

The last part of the argument suggests the notion of a "local" trace. In fact one can develop such a theory not only for the ordinary Lefschetz number, but also for the generalized Lefschetz numbers with the appropriate homotopy invariance and commutativity. These details as well as the relation of the local trace to the local obstruction of [4] will be given in a forthcoming article by E. Fadell and the author. A simplicial treatment of the local trace which is appropriate for the ordinary theory is given in [5].

(1.7) *REMARK*

Using the fact that the Lefschetz number $L(\cdot)$ is equal to the Hopf-index, one can prove Theorem (1.5), by appealing to the homotopy invariance of the Hopf-index and interpreting $p : U \to [0,1]$ as a homotopy of ENR's.

2. Cobordisms of Geometric Settings.

According to §1, if $f_0 : M_0 \to M_0$ is connected to $f_1 : M_1 \to M_1$ by a level-preserving cobordism, then $L(f_0) \neq 0$ if, and only if, $L(f_1) \neq 0$. Thus if either f_0 or f_1 has a non-zero Lefschetz number then both have essential fixed-points. But if the Lefschetz numbers are zero, we cannot conclude that if one of the maps has an essential fixed point then so does the other. In order to handle this problem, we need to modify the notion of cobordisms to accomodate the role of the fundamental groups in fixed-point theory.

Suppose therefore that $\varphi'' : \pi'' \to \pi''$ is a homorphism and recall that ([6]) a (π'', φ'')-setting is a commutative diagram

$$S_{(\pi'', \varphi'')} : \qquad \begin{array}{ccc} \tilde{M}'' & \xrightarrow{\tilde{f}''} & \tilde{M}'' \\ {\scriptstyle p''}\downarrow & & \downarrow{\scriptstyle p''} \\ M & \xrightarrow{f} & M \end{array}$$

where $\tilde{p}'' : \tilde{M}'' \to M$ is a regular cover whose group of covering transformations is π''. Two (π'', φ'')-settings $S_{(\pi'', \varphi'')}[M_0, f_0]$ are said to be *cobordant* if, and only if, there is a (π'', φ'')-setting

$$S_{(\pi'',\varphi'')}[W,f] : \quad \begin{array}{ccc} \widetilde{W}'' & \xrightarrow{\widetilde{f}''} & \widetilde{W}'' \\ {\scriptstyle \widetilde{p}''}\downarrow & & \downarrow{\scriptstyle \widetilde{p}''} \\ W & \xrightarrow{f} & W \end{array}$$

such that (W,f) is a cobordism from (M_0,f_0) to (M_1,f_1) and $S_{(\pi'',\varphi'')}[W,f]$ restricted to ∂W is $S_{(\pi'',\varphi'')}[M_0,f_0] \coprod S_{(\pi'',\varphi'')}[M_1,f_1]$. The cobordism is said to be level-preserving if, and only if, the cobordism (W,f) is level-preserving in the sense of §1. Note that the notion of cobordism of (π'',φ'')-settings implies that the groups of covering transformations of the universal cover are related by the commutative diagram

$$\begin{array}{ccccc} \pi(M_0) & \xrightarrow{i_0} & \pi(W) & \xleftarrow{i_1} & \pi(M_1) \\ & \searrow & \downarrow & \swarrow & \\ & & \pi'' & & \end{array}$$

where the horizontal lines are induced by the natural injections, while the other homomorphisms are the natural surjections.

(2.1) *THEOREM*

Suppose that $S_{(\pi'',\varphi'')}[W,f]$ *is a level-preserving cobordism from* $S_{(\pi'',\varphi'')}[M_0,f_0]$ *to* $S_{(\pi'',\varphi'')}[M_1,f_1]$. *Then*

$$L_{(\pi'',\varphi'')}S[M_0,f_0] = L_{(\pi'',\varphi'')}S[W,f] = L_{(\pi'',\varphi'')}S[M_1,f_1]$$

where $L_{(\pi'',\varphi'')}S[\cdot,\cdot]$ *is the generalized Lefschetz number of the setting* $S_{(\pi'',\varphi'')}[\cdot,\cdot]$.

To prove the theorem let us recal ([6]) how one defines the generalized Lefschetz number $L_{(\pi'',\varphi'')}S[X,f]$ for a given regular (π'',φ'')-setting

$$S_{(\pi'',\varphi'')}[X,f] : \quad \begin{array}{ccc} \widetilde{X}'' & \xrightarrow{\widetilde{f}''} & \widetilde{X}'' \\ {\scriptstyle \widetilde{p}''}\downarrow & & \downarrow{\scriptstyle \widetilde{p}''} \\ X & \xrightarrow{f} & X \end{array}$$

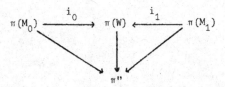

where X is a finite CW-complex. First the given setting is deformed to one where the maps are all cellular, and passing to the cellular chain level one then defines the generalized Lefschetz number $L_{(\pi'',\varphi'')}S[X,f]$ as the alternating (π'',φ'')-Reidemeister trace of the cellular chain map

$$C_*(\tilde{f}''): C_*(\tilde{X}'') \to C_*(\tilde{X}'') .$$

It turns out that $L_{(\pi'',\varphi'')}S[X,f]$ is well-defined, independent of the cellular structure and invariant under homotopy. (For details see [6].) Thus to prove Theorem (2.1) one proceeds as in the proof of Theorem (1.5).

Now it is possible to describe how the essential fixed-points of two cobordant maps are related.

(2.2) *THEOREM*

Suppose that $S_{(\pi'',\varphi'')}[M_0,f_0]$ *and* $S_{(\pi'',\varphi'')}[M_1,f_1]$ *are two settings related by a level-preserving cobordism. Then* $f_0 : M \to M_0$ *has an essential* π''*-Nielsen class if, and only if,* $f_1 : M_1 \to M_1$ *does.*

To prove the Theorem, recall that Theorem (1.13) of [6] states that

$$L_{(\pi'',\varphi'')}S[M_0,f_0] = \sum_{[\alpha]} \lambda_{[\alpha]}[\alpha] = L_{(\pi'',\varphi'')}S[M_1,f_1]$$

where $\lambda_{[\alpha]}$ is the index of the π''-Nielsen class $[\alpha]$ of $\tilde{p}''(\mathrm{Fix}\,\tilde{f}''_\alpha)$, with

$$f''_\alpha(\cdot) = f''(\cdot)\alpha^{-1}.$$

Now one sees immediately that $f_0 : M_0 \to M_0$ has an essential π''-Nielsen class if, and only if, $f_1 : M_1 \to M_1$ does, as required.

Note that in our formulation the Nielsen classes of $f_0 : M_0 \to M_0$ correspond naturally to those of $f_1 : M_1 \to M_1$. Thus arises the question of whether there is a continuum of fixed-points of the cobordism $f : W \to W$ which connects a given Nielsen class $[\alpha]$ of the map $f_0 : M_0 \to M_0$ to the map $f_1 : M_1 \to M_1$. (Cf. [2].)

(2.3) *THEOREM*

Suppose that $S_{(\pi'',\varphi'')}[M_0,f_0]$ *and* $S_{(\pi'',\varphi'')}[M_1,f_1]$ *are settings connected by level-preserving cobordism* $S_{(\pi'',\varphi'')}[W,f]$ *and let* $[\alpha]$ *denote an essential* π''*-Nielsen class of* $f_0 : M_0 \to M_0$, *as well as the corresponding class of* $f_1 : M_1 \to M_1$. *Then there is a continuum of essential* π''*-Nielsen classes of* $f : W \to W$ *connecting the class* $[\alpha]$ *of* f_0 *to the class* $[\alpha]$ *of* f_1.

Proof

The proof proceeds along lines similar to those of Theorem (1.5). Denote the π''-Nielsen class of $f : W \to W$ by C, and assume by way of contradiction that there is no continuum in C going from $M_0 \cap C$ to $M_1 \cap C$ and decompose C as the disjoint union $C_0 \coprod C_1$ of closed subsets C_0, C_1 where C_0 meets M_0 but not M_1 while C_1 meets M_1 but not M_0. Next surround C_0 and C_1 by two disjoint micro-tubular neighborhoods T_0 and T_1 respectively. Now arguing as in Proposition (1.3), we can find a deformation retraction

$$r : W \times I \to W \times I$$

of W onto $W-(\text{int } T \cup \text{int } T')$ such that

$$\text{Fix}\,(fr_1) = \text{Fix}\, f - C.$$

Consider now the (π'',φ'')-setting

$$S_{(\pi'',\varphi'')}[W,fr_1] : \quad \begin{array}{ccc} \widetilde{W}'' & \xrightarrow{\ \widetilde{f}''\widetilde{r}''_1\ } & \widetilde{W}'' \\ {\scriptstyle p''}\downarrow & & \downarrow{\scriptstyle p''} \\ W & \xrightarrow{\ fr_1\ } & W \end{array}$$

obtained by composing the given setting $S_{(\pi'',\varphi'')}[W,f]$ with the restriction of

$$S_{(\pi'',\varphi'')}[W,r] : \quad \begin{array}{ccc} \widetilde{W}''\times I & \xrightarrow{\ \widetilde{r}''\ } & \widetilde{W}''\times I \\ \downarrow & & \downarrow \\ W\times I & \xrightarrow{\ r\ } & W\times I \end{array}$$

to $W\times\{1\}$, where $\widetilde{r}''$ is the lift of r which begins with the identity. It is quite easy to see that $(\widetilde{f}''\widetilde{r}''_1)_\alpha$ where $\widetilde{r}''_1 = \widetilde{r}''|\widetilde{W}''\times\{1\}$ has no fixed points and hence the coefficient of $[\alpha]$ in $L_{(\pi'', \varphi'')}S[W, fr_1]$ is zero. But

$$L_{(\pi'', \varphi'')}S[W, fr_1] = L_{(\pi'', \varphi'')}S[W, f]$$

and hence $[\alpha]$ appears with a non-trivial coefficient, since it is assumed to be an essential π''-Nielsen class. Thus the theorem is proved.

3. Examples and Remarks.

It is quite easy to give examples to show that cobordisms which are not level-preserving need not preserve the Lefschetz number. In fact, let W be $S^1 \times D^2 - \text{int}\,\tilde{D}^3$ where $\tilde{D}^3$ is a 3-disk in the interior of the solid torus $S^1 \times D^2$, and take $f : W \to W$ to be the identity. Then at one end we have the map $f_0 = \text{id} : S^2 \to S^2$ with Lefschetz number 2 while at the other we have $f\ : \text{id} : S^1 \times S^1 \to S^1 \times S^1$ with Lefschetz number 0. However, note that $L(f_1) - L(f_0) = 0 \mod 2$.

To put the last relation in a general context, suppose that (W, f) is a cobordism from (M_0, f_0) to (M_1, f_1), and let $S_{(\pi'', \varphi'')}[W, f]$ be a regular (π'', φ'')-setting over $f : W \to W$. Then just as in the case of the ordinary Lefschetz number, one can show that

$$\begin{aligned} L_{(\pi'', \varphi'')}S[W, f] &= L_{(\pi'', \varphi'')}S[M_0, f_0] + L_{(\pi'', \varphi'')}S[W, M_0, f] \\ &= L_{(\pi'', \varphi'')}S[M_1, f_1] + L_{(\pi'', \varphi'')}S[W, M_1, f]. \end{aligned}$$

Hence

$$L_{(\pi'', \varphi'')}S[M_0, f_0] = L_{(\pi'', \varphi'')}S[M_1, f_1]$$

if, and only if,

$$L_{(\pi'', \varphi'')}S[W, M_0, f] = L_{(\pi'', \varphi'')}S[W, M_1, f]. \blacksquare$$

A natural tool for studying the last equality is Poincaré duality. The problem, it turns out, is fairly complicated and will be considered on a different occasion. Here we shall content ourselves with the following simple proposition.

(3.1) *PROPOSITION*

Suppose that $f : (W ; M_0, M_1) \to (W ; M_0, M_1)$ *is homotopic to the identity and that* $|\pi''| < \infty$. *Assume also that* W *is orientable. Then*

$$\text{(i)}\quad L_{(\pi'', \varphi'')} S[M_0, f_0] = L_{(\pi'', \varphi'')} S[M_1, f_1]$$

if dim W *is even, and*

$$\text{(ii)}\quad L_{(\pi'', \varphi'')} S[M_0, f_0] = L_{(\pi'', \varphi'')} S[M_1, f_1] + 2L_{(\pi'', \varphi'')} S[W, M_0, f]$$

if dim W *is odd.*

The proof is very easy: just note that the generalized Lefschetz numbers can be computed at the homology level, and that the only non-trivial class is [1] with the usual Euler characteristic as a coefficient.

Note that cobordisms such that

$$L_{(\pi'', \varphi'')} S[W, M_0, f] = 0 = L_{(\pi'', \varphi'')} S[W, M_1, f] \qquad (*)$$

preserve the generalized Lefschetz number $L_{(\pi'', \varphi'')} S[\cdot, \cdot]$. According to §2, level-preserving cobordisms fit in this case. It would be interesting to characterize those cobordisms for which (*) holds. For example, are they more general than those which are level-preserving? In the special case when f is homotopic to the identity, it is possible to answer this question.

(3.2) *PROPOSITION*

Suppose that $S_{(\pi'', \varphi'')}[W, f]$ *is a cobordism of settings from* $S_{(\pi'', \varphi'')}[M_0, f_0]$ *to* $S_{(\pi'', \varphi'')}[M_1, f_1]$, *and assume that* $f : (W ; M_0, M_1) \to (W ; M_0, M_1)$ *is homotopic to the identity,* dim W ≠ 3, *and*

$$L_{(\pi'', \varphi'')} S[W, M_0, f] = 0.$$

Then $S_{(\pi'', \varphi'')}[W, f]$ *is homotopic to a level-preserving setting.*

To prove the proposition, let us observe first of all that we can assume that the homomorphism $\varphi'' : \pi'' \to \pi''$ is the identity, after adjusting the given setting by a conjugation if necessary. Hence the Euler characteristic $\chi(W, M_0) = 0$, since $\chi(W, M_0)$ is the coefficient of the class [1] in $L_{(\pi'', \varphi'')} S[W, M_0]$. Since dim W ≠ 3 by assumption, it follows from [1] that W is obtained from M_0 by attaching round handles. The proposition is proved by induction on the dimension of the handle core. Therefore it suffices to consider the special case when

$$W = (M \times I) \cup_h S^1 \times D^p \times D^q$$

where

$$h : S^1 \times \partial D^p \times D^q \to M \times \{1\}$$

is an imbedding. Now define

$$g : S^1 \times D^p \times D^q \to S^1 \times D^p \times D^q$$

to be a small rotation in the first factor and the identity on the other two. Clearly g has no fixed points and is diffeotopic to the identity. Hence hgh^{-1} can be extended to a level-preserving diffeomorphism

$$(M \times I) - \text{image}(h \times 1) \to (M \times I) - \text{image}(h \times 1)$$

taking the boundary $h(S^1 \times \partial D^p \times \partial D^q) \times I$ to itself. Denote the map

$$W = M \times I \cup_h S^1 \times D^p \times D^q \to W = (M \times I) \cup_h S^1 \times D^p \times D^q$$

obtained by putting together the various maps defined above also by g, and note that g is actually a map of triples

$$g : (W, M_0, M_1) \to (W ; M_0, M_1)$$

where

$$M_0 = M \times \{0\},$$

and

$$M_1 = ((M \times \{1\}) - \text{image } h) \cup_h (S^1 \times \partial D^p \times \partial D^q).$$

Moreover,

$$\text{Fix } g \subset U$$

and $g : U \to U$ is level-preserving, where

$$U = W - [(S^1 \times D^p \times D^q) \cup (\text{image } h) \times I]$$

$$= (M - \text{image } h) \times I.$$

Therefore $g : W \to W$ is a level-preserving cobordism from $g_0 : M_0 \to M_0$ to $g : M_1 \times M_1$, and $g(W ; M_0, M_1) \to (W, M_0, M_1)$ is homotopic to f. Now it is an easy matter to deform the given setting $S_{(\pi'', \varphi'')}[W,f]$ to a setting $S_{(\pi'', \varphi'')}[W,g]$ over $g : W \to W$. This proves the proposition. ■

REFERENCES

[1] ASIMOV, D.: Round handles and nonsingular Morse-Smale flows, Ann. of Math. 104 (1975), 41-54.

[2] BROWDER, F.E.: On continuity of fixed points under deformations of continuous mappings, Summa Brasil. Math. 4 (1960), 183-191.

[3] DOLD, A.: The Fixed Point Index of Fibre-Preserving Maps, Inventiones math. 25 (1974), 281-297.

[4] FADELL, E and HUSSEINI, S.Y.: Local fixed-point theory for non-simply connected manifolds, to appear in the Illinois J. Math.

[5] FOURNIER, G.: These proceedings.

[6] HUSSEINI, S.Y.: Generalized Lefschetz Numbers, to appear.

FIBRE PRESERVING MAPS OF SPHERE BUNDLES INTO VECTOR SPACE BUNDLES

By

JAN JAWOROWSKI*
Forschungsinstitut für Mathematik
ETH Zürich
and
Department of Mathematics
Indiana University
Bloomington, Indiana.

0. Introduction

Let X be a (paracompact) space with a free involution $t : X \to X$. Conner and Floyd [2] and Yang [6] defined the cohomology index of (X,t), usually denoted by $\operatorname{coind}(X,t)$; we will denote it briefly by $c(t)$. It is an integer having, among others, the following properties:

(i) If X and Y are spaces with involutions $s : X \to X$ and $t : Y \to Y$, respectively, and if $f : X \to Y$ is an equivariant map then $c(s) \leq c(t)$.

(ii) For the antipodal involution $a : S^n \to S^n$, $c(a) = n$.

A theorem of Yang [6] and Bourgin [1] says that if X is a space with a free involution $t : X \to X$ such that $c(t) \geq n$ and if $f : X \to R^k$ is any map then the index of the set $A_f = \{x \in X : fx = ftx\}$ with the involution $t|A_f$ is at least $n-k$. This note presents a continuous or parametrized version of the Bourgin-Yang theorem in which a single X and $\mathbb{R}^n$ are replaced by fibre bundles and f by a fibre preserving map. The following theorem specializes the main result of this note to maps of sphere bundles. It says that the points which are carried, together with their antipodal points, to the same point are spread over the base in a certain systematic way.

---

*The author is grateful to the Forschungsinstitut für Mathematik, ETH, Zürich for the support during the writing of this paper.

THEOREM 0

Let $p : S \to B$ be an n-sphere bundle, $q : V \to B$ be an $\mathbb{R}^k$-bundle and $f : S \to V$ be a fibre preserving map. Let $\overline{A}_f$ be the set of pairs $\{x, -x\}$ such that $x \in S$ and $fx = f(-x)$. For any integers i,m there is a homomorphism $e_{i,m} : H^i(B) \to H^{i+m}(\overline{A}_f)$ where H^* is a continuous cohomology theory with coefficients in $\mathbb{Z}/2$. If the cohomology dimension d of B is finite then $e_{d,n-k} : H^d(B) \to H^{d+n-k}(\overline{A}_f)$ is injective. Moreover, if the Stiefel-Whitney classes $w_j(q)$ of q vanish for $1 \leq j \leq r$ then $e_{i,n-k}$ is injective for $i \geq d-r$. If all the Stiefel-Whitney classes of q are zero then $e_{i,n-k}$ is injective in every dimension i.

Also just as the Bourgin-Yang theorem generalizes the Borsuk-Ulam theorem, the result of this note is an extension of a "continuous" version of the Borsuk-Ulam theorem by the author in [5].

1. Preliminaries

Let X be a space with an involution $t : X \to X$. We denote by $\overline{X} = X/t$ the orbit space of t. If $p : X \to B$ is a map of X into a space B satisfying $ptx = px$ for each $x \in X$, we denote by $\overline{p} : \overline{X} \to B$ the induced map of the orbit space. If $f : X \to Y$ is a map of X into some space Y, we write

$$A_f = \{x \in X : fx = ftx\} .$$

Then A_f is an invariant subset of X, the restriction of t to A_f is an involution denoted by $t_{A_f} : A_f \to A_f$ in A_f and $\overline{A}_f = A_f/t$.

We will use the Alexander-Spanier cohomology theory H^* mod 2 throughout the paper so that the coefficient group $\mathbb{Z}/2$ will be suppressed from the notation. If Z is any space, A is a subset of Z and $i : A \to Z$ is the inclusion map, then the image of an element $z \in H^*(Z)$ under the induced homomorphism $i^* : H^*(Z) \to H^*(A)$ will sometimes be denoted by $z|A$ and called the restriction of z to A.

Let $\dim Z$ be the covering dimension of Z and let $d(Z) = \operatorname{Sup}\{m : H^m(Z) \neq 0\}$ be its cohomology dimension. Then $d(Z) \leq \dim Z$ if Z is a paracompact space. If $q : V \to B$ is a vector space bundle over B, we denote by $w_j(q)$ its j-th Stiefel-Whitney class.

We will assume throughout the paper that B is a paracompact space.

2. Maps of Sphere Bundles

THEOREM 1

Let $p : S \to B$ be an n-sphere bundle with the antipodal involution, let $q : V \to B$ be an $\mathbb{R}^k$-bundle and let $f : S \to V$ be a fibre preserving map. If $d(B) \leq d$ and $w_j(q) = 0$ for $1 \leq j \leq r$ then there is a monomorphism

$$e_{i,n-k} : H^i(B) \to H^{i+n-k}(\overline{A}_f)$$

for $i \geq d - r$.

Theorem 1 is a special case of Theorem 2 which will be proved in the last section.

COROLLARY 1

If $f : S \to V$ is a fibre preserving map of an n-sphere bundle $p : S \to B$ with the antipodal involution into an $\mathbb{R}^k$-bundle $q : V \to B$ and if $d = d(B) < \infty$, then there is a monomorphism $H^d(B) \to H^{d+n-k}(\overline{A}_f)$.

COROLLARY 2

If $f : S \to V$ is a fibre preserving map of an n-sphere bundle $p : S \to B$ with the antipodal involution into an $\mathbb{R}^k$-bundle $q : V \to B$ and if all the Stiefel-Whitney classes of q vanish then there is a monomorphism $H^*(B) \to H^*(\overline{A}_f)$ of degree $n - k$ in all dimensions.

COROLLARY 3

If B is a closed manifold and $f : S \to V$ is a fibre preserving map of an n-sphere bundle $p : S \to B$ with the antipodal involution into an $\mathbb{R}^k$-bundle $q : V \to B$ then $\dim A_f = \dim \overline{A}_f \geq \dim B + n - k$.

In Corollary 3, we have $d = d(B) = \dim B$ and $H^d(B) \neq 0$. It follows that $H^{d+n-k}(\overline{A}_f) \neq 0$, hence $d(\overline{A}_f) \geq d + n - k$. On the other and, $\dim \overline{A}_f = \dim A_f$ since the orbit map $A_f \to \bar{A}_f$ is a double covering.

3. Index of a Fibre Preserving Involution

Let X be a space with a free involution $t : X \to X$. It has its characteristic class which we will denote by $u(t)$. The characteristic class $u(t)$ is an element of $H^1(\bar{X})$ and is the Stiefel-Whitney class of the double covering $X \to \bar{X}$. This class is natural, that is, if X and Y are spaces with free involutions $s : X \to X$ and $t : Y \to Y$, respectively, and $f : X \to Y$ is an equivariant map then $f^*u(t) = u(s)$. The antipodal involution on S^n will be denoted by $a : S^n \to S^n$; the class $u(a)$ generates the polynomial ring $H^*(\mathbb{P}^n)$ of height n of the real projective space $\mathbb{P}^n = \bar{S}^n$. The cohomology index of a space X with an involution t is defined by $c(t) = \mathrm{Sup}\,\{n : u^n(t) \neq 0\}$.

Suppose now that X is a space over B, that is, together with a map $p : X \to B$, and suppose that $t : X \to X$ is a fibre preserving free involution, that is, a free involution satisfying $pt = p$. If $b \in B$ then t restricts to a free involution t_b on the fibre $p^{-1}b$ and the fibre inclusion $p^{-1}b \to X$ is an equivariant map. By the naturality of u, in the orbit bundle $\bar{p} : \bar{X} \to B$ the class $u(t) \in H^1(\bar{X})$ restricts to the characteristic class $u(t) \mid (\bar{p}^{-1}b) = u(t_b) \in H^1(\bar{p}^{-1}b)$ of the fibre.

DEFINITION

Let $p : X \to B$ be a map with a fibre preserving free involution $t : X \to X$. For any pair i, m of integers we define *the characteristic homomorphism of* (p,t)

$$e_{i,m}(p,t) : H^i(B) \to H^{i+m}(\bar{X})$$

by

$$x \mapsto (\bar{p}^* x) \cup u^m(t).$$

We define *the index of* (p,t) *in dimension* i, written $c_i(p,t)$, as follows: $c_i(p,t) \geq m$ if the map $e_{i,m}(p,t)$ is injective (if $H^i(X)=0$, we set $c_i(p,t)=-1$).

If B is a point then the index $c_i(p,t)$ in every dimension i is clearly equal to the Conner-Floyd-Yang cohomology index $c(t)$ of (X,t).

BASIC EXAMPLE

Let $p : S \to B$ be an n-sphere bundle with the antipodal involution $a : S \to S$. Then:

(i) $c_i(p,a) \geq n$ in every dimension i;

(ii) if $d(B) \leq d$ and the j-th Stiefel-Whitney class $w_j(p) = 0$ for $1 \leq j \leq r$ then $e_{i,m}(p,a) = 0$ for all $i \geq d - r$ and $m > n$. In particular, $c_i(p,a) = n$ in dimensions $i \geq d - r$;

(iii) if $d = d(B) < \infty$ then $e_{d,m}(p,a) = 0$ for $m > n$ and $c_d(p,a) = n$. This is the case, for instance, if B is a closed d-dimensional manifold;

(iv) if all the Stiefel-Whitney classes of p are zero then $e_{i,m}(p,a) = 0$ for all i and $m > n$. In particular, $c_i(p,a) = n$ in every dimension i.

<u>*Proof*</u>

(i) If $b \in B$ then the fibre $\bar{p}^{-1}b$ of $\bar{p} : \bar{S} \to B$ is the real projective n-space $\mathbb{P}^n$. Its cohomology ring is a polynomial ring of height n generated by the characteristic class $u(a_b)$, where a_b is the antipodal involution on the fibre, and $u(a_b)$ is the restriction of $u(a)$ to the fibre. By the Leray-Dold-Hirsch theorem ([4], p. 229), $H^*(\bar{S})$ is an $H^*(B)$-module freely generated by the powers $1, u(a), \ldots, u^n(a)$, where the action of $H^*(B)$ on $H^*(\bar{S})$ is given by $x \cdot y = (\bar{p}^* x) \cup y$ for $x \in H^*(B)$ and $y \in H^*(\bar{S})$. In particular, for every integer i, the map

$$e_{i,n} : H^i(B) \to H^{i+n}(\bar{S})$$

$$x \mapsto (\bar{p}^* x) \cup u^n(a)$$

is injective. This proves part (i).

(ii) Suppose that $d(B) \leq d$ and $w_j(p) = 0$ for $1 \leq j \leq r$. It suffices to show that $e_{i,n+1}(p,a)$ is zero whenever $i \geq d - r$. Let $x \in H^i(B)$ where $i \geq d - r$. Then $u^{n+1}(a)$ is a polynomial in $u^h(a)$, $h \leq n$, whose coefficients are the Stiefel-Whitney classes $w_j = w_j(p)$ ([4], p. 232):

$$u^{n+1}(a) = \sum_{j=1}^{n+1} (\bar{p}^* w_j) \cup u^{n+1-j}(a),$$

hence

$$e_{i,n+1}(p,a)\, x = (\bar{p}^* x) \cup u^{n+1}(a)$$

$$= (\bar{p}^* x) \cup \sum_{j=1}^{n+1} (\bar{p}^* w_j) \cup u^{n+1-j}(a)$$

$$= \sum_{j=1}^{n+1} \bar{p}^*(x \cup w_j) \cup u^{n+1-j}(a)\,.$$

If $j \le r$ then $w_j = 0$ by the assumption. If $j > r$ then

$$\deg(x \cup w_j) = i+j > i+r \ge d \ge d(B)$$

since $i \ge d-r$, hence $x \cup w_j = 0$. Therefore all the coefficients in this polynomial are zero and hence $e_{i,n+1}(p,a) = 0$.

(iii) Follows from (ii) when $r = 0$.

(iv) Follows from (ii) when $d = \infty$ and $r = n+1$. ■

4. A General Theorem

THEOREM 2

Let $p : X \to B$ be a map with a fibre preserving free involution $t : X \to X$, let $q : V \to B$ be an $\mathbb{R}^k$-bundle and let $f : X \to V$ be a fibre preserving map. If $w_j(q) = 0$ for $1 \le j \le r$ then $c_i(p|A_f, t_{A_f}) \ge c_i(p,t) - k$ in all dimensions $i \ge d-r$.

In particular, if $d = d(B) < \infty$ then for any $\mathbb{R}^k$-bundle $q : V \to B$, $c_d(p|A_f, t_{A_f}) \ge c_d(p,t) - k$.

If all the Stiefel-Whitney classes of q vanish then $c_i(p|A_f, t_{A_f}) \ge c_i(p,t) - k$ in every dimension i.

If B is a closed manifold then $\dim A_f = \dim \overline{A}_f \geq \dim B + c_d(p,t) - k$.

Proof

Suppose that $c_i(p,t) \geq n$. To prove that $c_i(p|A_f,t_{A_f}) \geq n-k$, we have to show that the map

$$e_{i,n-k}(p|A_f,t_{A_f}) : H^i(B) \to H^{i+n-k}(\overline{A}_f)$$

$$x \mapsto [(\overline{p}|\overline{A}_f)^* x] \cup u^{n-k}(t_{A_f})$$

is injective. Suppose then that $x \in H^i(B)$ and $e_{i,n-k}(p|A_f,t_{A_f})\, x = 0$, that is $[(\overline{p}^* x) \cup u^{n-k}(t)] \mid \overline{A}_f = 0$ as $u(t_{A_f}) = u(t) \mid \overline{A}_f$. By the continuity of H^*, there exists a neighborhood U of A_f in X such that $[(\overline{p}^* x) \cup u^{n-k}(t)] \mid \overline{U} = 0$ (recall that $\overline{U}$ denotes the image of U in the orbit space $\overline{X}$). By the exactness of the cohomology sequence of the pair $(\overline{X},\overline{U})$, there is a class $y \in H^{i+n-k}(\overline{X},\overline{U})$ such that

$$(\overline{p}^* x) \cup u^{n-k}(t) = \alpha^* y,$$

where $\alpha : X \to (\overline{X},\overline{U})$ is the inclusion map.

Let 0 be the zero section in V and $V_o = V - 0$. Then the antipodal map $a : V_o \to V_o$ is a free involution in V_o and the fibre of the bundle $q_o = q|V_o : V_o \to B$ is $\mathbb{R}^k_o = \mathbb{R}^k - (0)$. The bundle q_o, as a bundle with involution, is equivalent to its S^{k-1}-bundle, therefore

$$u^k(a) = \sum_{j=1}^{k} (\overline{q}_o^* w_j) \cup u^{k-j}(a),$$

where $w_j = w_j(q)$ are the Stiefel-Whitney classes of q.

Let $X_o = X - A_f$, let $t_o : X_o \to X_o$ be the involution defined by t and let $p_o = p|X_o : X_o \to B$. Then the map $g : X \to V$ defined by $gx = fx - f(-x)$ is equivariant, $A_g = A_f = g^{-1}0$ and the restriction of g to X_o defines a fibre preserving equivariant map $g_o : X_o \to V_o$. Therefore $u(t_o) = \overline{g}_o^* u(a)$ and hence

$$u^k(t_o) = \overline{g}_o^* [\sum_{j=1}^{k} (\overline{q}_o^* w_j) \cup u^{k-j}(a)]$$

$$= \sum_{j=1}^{k} (\overrightarrow{p}_o^* w_j) \cup u^{k-j}(t_o).$$

Consider

$$v = u^k(t) - \sum_{j=1}^{k} (\overrightarrow{p}^* w_j) \cup u^{k-j}(t).$$

Then

$$v|X_o = [u^k(t)]|X_o - [\sum_{j=1}^{k} (\overrightarrow{p}^* w_j) \cup u^{k-j}(t)]|X_o$$

$$= u^k(t_o) - \sum_{j=1}^{k} [(\overrightarrow{p}^* w_j)|X_o] \cup [u^{k-j}(t)|X_o]$$

$$= u^k(t_o) - \sum_{j=1}^{k} (\overrightarrow{p}^* w_j) \cup u^{k-j}(t_o) = 0.$$

By the exactness of the cohomology sequence of the pair $(\overline{X},\overline{X}_o)$, $v = \beta^* z$, for some $z \in H^k(\overline{X},\overline{X}_o)$, where $\beta : \overline{X} \to (\overline{X},\overline{X}_o)$ is the inclusion map. Since $(\overline{X};\overline{U},\overline{X}_o)$ is an excisive triad, $\alpha^* y \cup \beta^* z = y \cup z = 0$, hence

$$0 = (\overrightarrow{p}^* x) \cup u^{n-k}(t) \cup [u^k(t) - \sum_{j=1}^{k} (\overrightarrow{p}^* w_j) \cup u^{k-j}(t)]$$

$$= (\overrightarrow{p}^* x) \cup u^n(t) - \sum_{j=1}^{k} \overrightarrow{p}^*(x \cup w_j) \cup u^{n-j}(t).$$

Thus

$$(\overrightarrow{p}^* x) \cup u^n(t) = \sum_{j=1}^{k} \overrightarrow{p}^*(x \cup w_j) \cup u^{n-j}(t).$$

Now if $j \le r$ then $w_j = 0$ by the assumption. If $j > r$ then

$$\deg(x \cup w_j) = i+j > i+r \ge d \ge d(B)$$

since $i \ge d - r$ and hence $x \cup w_j = 0$. Thus all the coefficients in this polynomial are zero and hence $(\overrightarrow{p}^* x) \cup u^n(t) = 0$. But $(\overrightarrow{p}^* x) \cup u^n(t) = e_{i,n}(p,t)$ and $e_{i,n}(p,t)$ is injective since $c_i(p,t) \ge n$. It follows that x must be zero which proves that $e_{i,n-k}(p|A_f,t_{A_f})$ is injective. ∎

Theorem 1 can now be obtained by applying Theorem 2 to the Basic Example.

REMARK

A relative version of Theorems 1 and 2, for pairs of spaces over B, can also be proved in a similar way.

REFERENCES

[1] BOURGIN, D.G.: On some separation and mapping theorems, Comment. Math. Helv. 29 (1955), 199-214.

[2] CONNER, P.E. and FLOYD, E.E.: Fixed point free involutions and equivariant maps, Bull. Amer. Math. Soc. 64 (1960), 416-441.

[3] CONNETT, J.E.: On the cohomology of the fixed-point sets and coincidence-point sets, Indiana Univ. Math. J. 24 (1974-75), 627-634.

[4] HUSEMOLLER, D.: Fibre Bundles, McGraw-Hill, New York, (1966).

[5] JAWOROWSKI, J.: A Continuous Version of Borsuk-Ulam Theorem, Forschungsinstitut für Mathematik, ETH Zürich, June (1980) (preprint).

[6] YANG, C.T.: On Theorems of Borsuk-Ulam, Kakutani-Yamabe-Yujobo and Dyson, I, Ann. of Math. 60 (1954), 262-282.

FIXED POINT CLASSES FROM A DIFFERENTIAL VIEWPOINT

By

BOJU JIANG (PO-CHU CHIANG)

University of California,
Berkeley, California, USA
and
Beijing University
Beijing, China

1. Introduction

For a compact manifold M of dimension ≥ 3, the Nielsen number $N(f)$ of a continuous map $f:M \to M$ is exactly the least number of fixed points in the homotopy class of f. (See [2] for the definition of Nielsen number. General references for Nielsen's theory of fixed point classes are [1] and [4].) This fact was first proved by Wecken [7] for triangulated manifolds, then by Weier [8] for topological manifolds. Here we will sketch a proof of the corresponding statement in the smooth category:

THEOREM

If M is a smooth manifold (with or without boundary) of dimension $m \geq 3$, and $f:M \to M$ is a smooth map, then f can always be smoothly deformed to a map g with only $N(f)$ fixed points.

The key to the proof is a simple relationship between Nielsen's notion of fixed point class and the well known Whitney trick in differential topology.

We will also discuss what is known in dimension 2, where Whithey's trick fails but we have more knowledge about diffeomorphisms.

Unless otherwise stated, we work in the smooth category: Manifolds, maps and homotopies are meant to be smooth.

2. The approach

Let M^m be a compact manifold of dimension m, with or without boundary. Let $f:M \to M$ be a map. Then in the product manifold $M \times M$, we have two submanifolds, the diagonal $\Delta = \{(x,x) \mid x \in M\}$ and the graph $\Gamma f = \{(x,f(x)) \mid x \in M\}$. Fixed points of f correspond to intersection points of these two submanifolds, i.e. $\mathrm{Fix}\, f \cong \Gamma f \cap \Delta$. Roughly speaking, deforming the map f to minimize the number of fixed points corresponds to deforming the graph Γf to minimize the number of intersections with Δ.

Whitney's trick is a standard tool in differential topology to reduce the number of intersections of two submanifolds of complementary dimension. But at a closer look, there are some technical difficulties. Let us discuss them first.

3. The cancelling procedure

Whitney's trick deals only with transverse intersections. The corresponding notion in fixed point context is the following:

A fixed point x_0 of f is said to be a transverse fixed point if $x_0 \in \mathrm{Int}\, M$ and 1 is not eigenvalue of $df(x_0)$. It is equivalent to say that (x_0,x_0) is a transverse intersection of Γf and Δ.

A local orientation at x_0 of M determines local orientations of Γf, Δ, and $M \times M$, at (x_0,x_0). We can talk about the (local) intersection coefficient of Γf and Δ at (x_0,x_0), which is $+1$ if the local orientations of Γf and Δ give the local orientation of $M \times M$, and -1 otherwise. This coefficient does not depend on the choice of local orientation at x_0 of M. A simple computation shows that it equals the sign of $\det(I - A)$, where I is the identity matrix and A is the Jacobian of f at x_0. So it is nothing but the fixed point index of $x_0 \in \mathrm{Fix}\, f$.

Whitney's trick deals with the following situation. Suppose N^n is a smooth manifold, P^p and Q^q are two submanifolds, with $p+q = n$. Suppose $x,y \in P \cap Q$ are two transverse intersection points, α and β are two smooth arcs in P^p and Q^q respectively, from x to y and both free of other points of $P \cap Q$, such that $\alpha \simeq \beta$ in N^n. Now we can compare local orientations of P at x and y by continuation along α, similarly on Q along β, and on N along either α or β (because $\alpha \simeq \beta$ in N). Suppose under this comparison the intersection coefficients at x and y are opposite.

WHITNEY LEMMA

In the above situation, if $p,q \geq 3$, then there exists a smooth isotopy of $P \hookrightarrow N$ with support in any prescribed neighborhood of α, carrying P_1 to P', such that $P' \cap Q = P \cap Q-\{x,y\}$.

(This statement is adapted from [3]. See [5] Section 6 for a proof. By the support of an isotopy we mean the closure of the set of points where the isotopy is nontrivial, i.e. different from the original map.)

Now let x_1, x_2 be two transverse fixed points of $f:M \to M$, they correspond to two transverse intersections of Γf and Δ. Under what conditions can they be cancelled out by Whitney's trick?

(1) We need an arc α in Γf and an arc β in Δ from (x_1,x_1) to (x_2,x_2) such that $\alpha \simeq \beta$ in $M \times M$. This is clearly equivalent to the existence of an arc γ in Int M from x_1 to x_2 such that $\gamma \simeq f(\gamma)$ in M. In other words, x_1 and x_2 have to be in the same fixed point class of f.

(2) The intersection coefficients at (x_1,x_1) and (x_2,x_2) must be opposite. This amounts to saying that the fixed point indices of x_1 and x_2 have to be opposite.

Suppose these two conditions are satisfied, we propose to remove this pair of fixed points by homotoping f. Direct use of the Whitney Lemma to Γf and Δ in $M \times M$ can not do the job, in that the distorted Γf may no longer be a graph of a map. We must take care.

Pick an Euclidean neighborhood U missing all other fixed points of f, of the arc γ in M. By use of the Whitney Lemma on $U \times M$, we get an isotopy

$$\{h_t \times g_t\}_{t \in I} : U \to U \times M,$$

with compact support S, such that $h_0 \times g_0 = id \times f$ and $(h_1 \times g_1)(U) \cap \Delta = \phi$. By an abuse of language, we may identify U with $\mathbb{R}^m$, and by suitable scaling may assume that $S \cup h_1(S)$ is contained in the unit disc D^m of $\mathbb{R}^m$. It is not hard to construct a diffeomorphism $\varphi:\mathbb{R}^m \times \mathbb{R}^m \to \mathbb{R}^m \times \mathbb{R}^m$ of the form $\varphi(u,v) = (u,\psi_u(v))$, such that $\varphi(u,v) = (u,v-u)$ on $D^m \times D^m$ and $\varphi = id$ outside of a compact set. Extend φ (by the identity map) to a diffeomorphism $U \times M \to U \times M$ (and also ψ_u to $M \to M$). Consider the smooth homotopy $\{\bar{g}_t\}_{t\in I} : U \to M$ defined by

$$\bar{g}_t(u) = \psi_u^{-1} \circ \psi_{h_t(u)}(g_t(u)).$$

We see that $\bar{g}_0(u) = f(u)$, $\bar{g}_1(u) \neq u$ for all $u \in U$, and $\bar{g}_t(u) = f(u)$ outside of S, so that $\{\bar{g}_t\}$ can be extended over M to be a homotopy of $f:M \to M$. Thus we can indeed cancel this pair x_1, x_2 of fixed points by a smooth homotopy.

4. THE CREATING PROCEDURE

Our goal is to reduce each fixed point class to a single fixed point. If the index of the fixed point class is k, the resulting single fixed point will also have index k, hence cannot be a transverse one if $k \neq \pm 1$. In other words, non-transverse fixed points are unavoidable. This motivates the creating procedure.

Suppose $m \geq 2$. Given an isolated fixed point $x_0 \in \text{Fix } f$ and an integer k, we can always smoothly homotope f to create a new fixed point of index k along with $|k|$ new transverse fixed points, all of them lying in a neighborhood of x_0 and belonging to the same fixed point class as x_0. The construction follows.

Take an Euclidean neighborhood U of x_0 free of other fixed points. By abuse of language, we may identify U with $\mathbb{R}^m$, and by suitable scaling, we may assume $3 < \|x_0\| < 4$ and f maps $\bar{B}_4$ into $\mathbb{R}^m$; here B_r denotes the open ball in $\mathbb{R}^m$ with center 0 and radius r. Define $\theta:\bar{B}_4 \to \mathbb{R}^m$ by $\theta(x) = x - f(x)$, then $\theta(\bar{B}_3) \subset \mathbb{R}^m - \{0\}$, so that $\theta|_{\partial B_3} : \partial B_3 \to \mathbb{R}^m - \{0\}$ is inessential (i.e. of degree 0). As $m \geq 2$, we may decompose $\mathbb{R}^m$ as $\mathbb{C} \times \mathbb{R}^{m-2}$, and define a map $\zeta:\mathbb{R}^m \to \mathbb{R}^m$ by

$$\zeta(z,y) = \begin{cases} (z^k(\bar{z}^k - 1), y) & , \text{ if } k > 0, \\ (z\bar{z}, y) & , \text{ if } k = 0, \\ (\bar{z}^{|k|}(z^{|k|} - 1), y) & , \text{ if } k < 0. \end{cases}$$

By continuously changing the constant in the formula from 1 to 0, we see that

$$\zeta|_{\partial B_2} : \partial B_2 \to R^m - \{0\}$$

is also inessential. So we can extend $\theta|_{(\overline{B}_4 - B_3)}$ and $\zeta|_{\overline{B}_2}$ to a smooth map $\overline{\theta} : B_4 \to \mathbb{R}^m$ with $\overline{\theta}(\overline{B}_3 - B_2) \subset \mathbb{R}^m - \{0\}$. Define $\overline{f} : \overline{B}_4 \to \mathbb{R}^m$ to be $\overline{f}(x) = x - \overline{\theta}(x)$ and extend it over M in the obvious manner. It is easy to check that $\overline{f}$ is smoothly homotopic to f and has $|k| + 1$ new fixed points with all the desired properties.

5. Proof of the Theorem

By a standard transversality argument, we may assume the generic case: f maps M into $\text{Int}\, M$, and the graph Γf intersects the diagonal Δ transversely in $M \times M$. To every fixed point class of index $k \neq 0$, apply the creating procedure to get a fixed point of index k and $|k|$ transverse fixed points. Put the one of index k aside. The remaining fixed points in this fixed point class are all transverse ones and their indices pair off, so that by repeated use of the cancelling procedure they can all be cancelled out. Inessential fixed point classes can be swept away in the same way. What remains is a single fixed point for each essential fixed point class.

(1) *REMARK*

The two procedures of Section 3 and 4 are local in nature, so that the theorem (and proof) is also true for a map $U \to M$ (U open in M), which is discussed in Fadell [2].

(2) *REMARK*

The same theorem (with a similar proof) holds in the PL category, because there is a PL version of Whitney's Lemma.

6. The case of dimension 2

Little is known about the least number of fixed points in a homotopy class for a surface M^2. Our proof for $m \geq 3$ no longer works for $m = 2$, because the Whitney Lemma does not apply. (Note that the creating procedure still works.)

But we do know this least number equals the Nielsen number if the Euler characteristic $\chi(M^2) \geq 0$. This is also true in the smooth category, as can be checked case by case.

7. Fixed point classes of diffeomorphisms

We may also propose the fixed-point-minimizing problem for diffeomorphisms. Given a diffeomorphism $\varphi: M \to M$, what is the least number of fixed points among all the diffeomorphisms isotopic to φ? When is it equal to $N(\varphi)$? This problem seems accessible in the case $m = 2$.

Thurston [6] has proved the following theorems.

CLASSIFICATION THEOREM

For any diffeomorphism φ of a surface M^2 with $\chi(M^2) < 0$, φ is isotopic to a diffeomorphism φ' such that either

(i) φ' is an isometry with respect to some hyperbolic metric on M (in this case φ is said to be elliptic);

or

(ii) there is a number $\lambda > 1$ and a pair of transverse measured foliations F^s and F^u such that $\varphi'(F^s) = \frac{1}{\lambda}F^s$ and $\varphi'(F^u) = \lambda F^u$ (such a φ' is called a pseudo-Anosov diffeomorphism, in this case φ is said to be hyperbolic);

or

(iii) φ' is reducible by a system of disjoint simple closed curves $\Gamma = \{C_1,\ldots,C_k\}$, meaning that Γ is invariant by φ' (but the C_i may be permuted), and Γ has a φ'-invariant tubular neighborhood $\eta(\Gamma)$ such that on each (not necessarily connected) φ'-component of $M^2-\eta(\Gamma)$, φ' satisfies (i) or (ii); in this case φ is also said to be reducible.

MINIMIZING THEOREM

If φ is a pseudo-Anosov diffeomorphism, then φ has $N(\varphi)$ fixed points, so that φ has the minimum number of fixed points in its isotopy class.

This second theorem has settled (but not completely) the fixed-point-minimizing problem for hyperbolic diffeomorphism. A pseudo-Anosov diffeomorphism may have singularities (isolated points where the map is not differentiable in the usual sense). It is still an open question whether this minimum can be attained by an honest diffeomorphism.

For elliptic diffeomorphisms, we have the following.

PROPOSITION

Let φ be an orientation-preserving isometry of a hyperbolic surface (of constant Gaussian curvature -1 and with totally geodesic boundary). If φ is not the identity, then no two fixed points of φ are in the same fixed point class, so that φ has $N(\varphi)$ fixed points and it has the minimum number of fixed points in its isotopy class.

Proof

Suppose x_1, x_2 are two fixed points in the same fixed point class of φ. By definition there exists a path γ from x_1 to x_2 such that $\gamma \simeq \varphi(\gamma)$. From hyperbolic geometry we know that, in the path class of γ (homotopy class with end-points fixed) there is a unique geodesic curve $\bar{\gamma} \simeq \gamma$. So $\bar{\gamma} \simeq \varphi(\bar{\gamma})$. But φ is an isometry, hence $\varphi(\bar{\gamma})$ is also a geodesic curve, so by uniqueness we have $\bar{\gamma} = \varphi(\bar{\gamma})$, and $\bar{\gamma}$ is pointwise fixed under φ. Moreover, φ being orientation-preserving, does not switch the two sides of $\bar{\gamma}$. Then φ must be the identity on a neighborhood of $\bar{\gamma}$, hence on the whole surface. ■

The restriction "orientation-preserving" in the Proposition cannot be removed. A simple example: M is a disc with two holes, φ is a reflection with respect to an axis. It has Nielsen number $N(\varphi) = 1$,

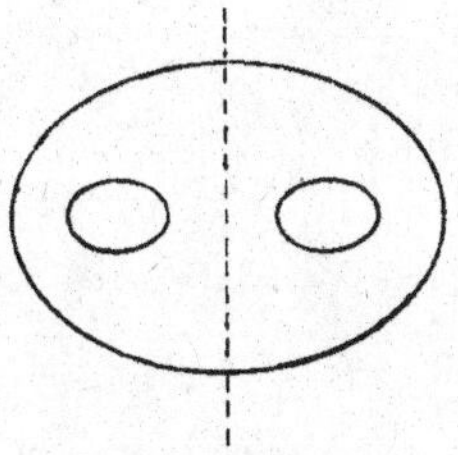

but any homeomorphism isotopic to φ will map the outer boundary onto itself, in an orientation-reversing manner, thereby producing at least two fixed points.

Little is known to date about fixed point classes of reducible diffeomorphisms.

References

[1] BROWN, R.F.: The Lefschetz Fixed Point Theorem, Scott, Foresman and Co. 1971.

[2] FADELL, E. and HUSSEINI, S.Y.: these Proceedings

[3] KERVAIRE, M.A.: Geometric and algebraic intersection numbers, Comm. Math. Helv. 39 (1965), 271-280.

[4] KIANG, T.H.: The Theory of Fixed Point Classes, Scientific Press, Peking, 1979.

[5] MILNOR, J.: Lectures on the h-cobordism Theorem, Princeton University Press, 1965.

[6] THURSTON, W.P.: On the geometry and dynamics of diffeomorphisms of surfaces,I (preprint).

[7] WECKEN, F.: Fixpunktklassen III, Math. Ann. 118 (1942), 544-577.

[8] WEIER, J.: Fixpunkttheorie in topologischen Mannigfaltigkeiten, Math. Z. 59 (1953), 171-190.

FIXED POINT SETS OF CONTINUOUS SELFMAPS ON POLYHEDRA

BY

BOJU JIANG AND HELGA SCHIRMER*

Boju Jiang (Po-chu Chiang)
University of California
Berkeley, California, U.S.A.
and
Beijing University
Beijing, China.

Helga Schirmer
Department of Mathematics
Carleton University
Ottawa, Ontario, Canada.

1. THE RESULT

It is known that every closed and nonempty subset of a 2-dimensionally connected finite polyhedron X is the fixed point set of a continuous selfmap of X, that is, that such a polyhedron has the complete invariance property [4], [5]. It is the purpose of this paper to extend this fact to a much wider class of polyhedra: we do not assume 2-dimensional connectedness, and we replace finiteness by local finiteness. It is doubtful that the assumption of local finiteness can be relaxed further. (See [5], Question 1.) Here is our result.

THEOREM

Let $|K|$ be a locally finite simplicial complex with the weak topology. Then every closed and nonempty subset of $|K|$ is the fixed point set of a selfmap of $|K|$.

Our proof needs the tools from the proof of the corresponding theorem for 2-dimensionally connected finite polyhedra, namely the definitions of a 2-dimensionally connected polyhedron, a proximity map, a path field, and the fact that a proximity map f on a polyhedron determines a path field whose singularities are exactly the fixed points of f. (See [5], Definition 2 and §3(iii).) In addition we use several ideas which have been developed by Shi in his study of fixed points of mappings of the identity class [7]. Shi defines a 2-*dimensionally connected component* of a simpli-

*The second author was partially supported by NSERC Grant A 7579.

cial complex K as a maximal 2-dimensionally connected subcomplex. A *part* P of K is either a 2-dimensionally connected component or a 1-dimensional maximal simplex. The elements of the set $|P| \cap |\overline{K - P}|$ (where the bar denotes the closure) are called the *welding vertices* of P, and the union of all welding vertices of all parts of K is the *welding set* of K. It is a subset of the set of all vertices of $|K|$.

Finally we need the following Lemma, in which $\mathrm{Fix}\, g$ denotes the fixed point set of g.

LEMMA

Let C be a 2-dimensionally connected component of a locally finite simplicial complex with the weak topology. Then there exists a proximity map $g: |C| \to |C|$ so that

(i) if C is finite, then $\mathrm{Fix}\, g$ is either empty, or it consists of a single point which is contained in a maximal simplex of $|C|$,

(ii) if C is infinite, then $\mathrm{Fix}\, g$ is empty.

Proof

If C is finite, this is a special case of Satz 1, p. 568, in [10]. It also follows easily from the techniques used in the proof of Theorem (1.4) in [6] or of Theorem 1, p. 143, in [2]. If C is infinite, this is a direct consequence of Lemma 4 and Theorem 2 of a recent paper by Shi [8], in which he extends results from [7] to infinite complexes. ■

The proof of the Theorem will be done in two steps. In the first, we order the parts P_i of K so that $\bigcup_{i=1}^{n} P_i$ is connected for every $n \geq 1$, and in the second, we use this ordering to construct inductively on each P_n a map f_n with $\mathrm{Fix}\, f_n = |P_n| \cap A$. This is simple if P_n is 1-dimensional or if $|P_n| \cap A = \phi$ (Cases 1 and 3). If P_n is a 2-dimensionally connected component with $|P_n| \cap A \neq \phi$ (Case 2), then the Lemma allows us a construction similar to the one previously used for 2-dimensionally connected finite polyhedra [4], [5] §3 (iii). But care has to be taken to adjust the map on the welding vertices to ensure that the final map is continuous. For this adjustment we use a modification of Shi's idea of "putting a tail on a map". (See [7], proof of Lemma 1.)

2. THE PROOF

Now let $|K|$ be a locally finite polyhedron with the weak topology, and A a closed and nonempty subset of $|K|$. We have to find a map $f : |K| \to |K|$ with $\mathrm{Fix}\, f = A$, and it follows from [5], Theorem 3 that we can assume that K is connected.

STEP 1

We will show that the parts of K can be arranged into a sequence $P_1, P_2, P_3, \ldots$ (which terminates if K is finite) so that

(i) $|P_1| \cap A \neq \phi$

(ii) $\bigcup_{i=1}^{n} P_i$ is connected for all $n \geq 1$.

For this purpose, let $v_0, v_1, \ldots, v_r$ be vertices of K, call r the length of the edge-path $v_0 v_1 \ldots v_r$ in K, and define the simplicial distance $sd(v,v')$ between any two vertices v and v' of K as the minimal length of all edge-paths from v to v'. As K is connected, $sd(v,v')$ is defined for all $v,v' \in K$.

Now select P_1 as any part of K with $|P_1| \cap A \neq \phi$, choose a vertex $v_0 \in P_1$, and assign to each part P of K the label

$$\ell = \min \{sd(v_0, v) : v \in P\} .$$

The local finiteness of K implies that there exist, for every integer ℓ, only a finite number of parts with label ℓ. Therefore we can arrange the parts of K into a sequence, starting with P_1, according to increasing order of their labels. As the label ℓ of a part $P \neq P_1$ is always attained at a welding vertex of P, this sequence satisfies $P_n \cap \bigcup_{i=1}^{n-1} P_i \neq \phi$ for all $n \geq 2$, and hence (ii).

STEP 2

Denote by K_n, where $n \geq 1$, the connected subcomplex $\bigcup_{i=1}^{n} P_i$, and by W_n, where $n \geq 2$, the nonempty subset $P_n \cap \bigcup_{i=1}^{n-1} P_i$ of the welding set. Also define $W_1 = \phi$.

We shall establish, by induction, the following statement:

(I_n) There exists a map $f_n : |P_n| \to |K_n|$ so that

(i) $\mathrm{Fix}\, f_n = |P_n| \cap A$,

(ii) $f_n|W_n = \varphi_n$, where $\varphi_n : W_n \to K_{n-1}$ is determined by the maps $f_1, f_2, \ldots, f_{n-1}$.

The argument that (I_{n-1}) implies (I_n) splits into three cases. As

$$|P_1| \cap A \neq \phi,$$

the statement (I_1) follows from Case 1 or 2.

CASE 1

P_n is a maximal 1-simplex and $|P_n - W_n| \cap A \neq \phi$. This case is easy. Details are similar to those given in [4], p. 225.

CASE 2

P_n is a 2-dimensionally connected component and $|P_n - W_n| \cap A \neq \phi$. According to the Lemma there exists a proximity map $g_n : |P_n| \to |P_n|$ so that $\mathrm{Fix}\, g_n$ is either empty or consists of a single point b_n which is contained in a maximal simplex of $|P_n|$. We first change g_n to a map $g_n' : |P_n| \to |K_n|$ with $g_n'(v) = \varphi_n(v)$ for all $v \in W_n$. To do so, select a point $a_n \in |P_n - W_n| \cap A$. Denote by d the metric of $|P_n|$ defined by barycentric coordinates, by $U(v,\delta)$ the open neighbourhood of v in $|P_n|$ with radius δ, and by $st_K(v)$ the open star of v. As $g_n(v) \neq v$ for all $v \in W_n$, we can choose for every $v \in W_n$ a $\delta_v > 0$ sufficiently small so that

$$a_n \notin U(v,\delta_v) \subset |st_K(v)|,$$

$$U(v,\delta_v) \cap U(v',\delta_{v'}) = \phi \quad \text{if} \quad v' \in W_n \quad \text{but} \quad v \neq v',$$

$$\overline{U}(v,\delta_v) \cap g_n(\overline{U}(v,\delta_v)) = \phi.$$

Select a welding vertex $v_1 \in W_n$. If $g_n(v_1)$ is not contained in a maximal simplex of $|K|$, let $z \in |st_K(v_1)| - \overline{U}(v_1,\delta_{v_1})$ be a point whose carrier $\varkappa(z)$ is a maximal simplex with $\varkappa(g_n(v_1))$ as its face. Otherwise put $z = g_n(v_1)$. Let p_{v_1} be the (possibly broken) line segment $[g_n(v_1),z] \cup [z,v_1]$ in $|st_K(v_1)|$ from $g_n(v_1)$ to v_1, and let q_{v_1} be a piecewise linear path in $|K_{n-1}|$ from v_1 to $\varphi_n(v_1)$. If $v \in W_n - \{v_1\}$, let p_v be a piecewise linear path in $|P_n| - U(v,\delta_v)$ from $g_n(v)$ to v_1, and q_v be a piecewise linear path in $|K_{n-1}|$ from v_1 to $\varphi_n(v)$. Parametrize all paths by $t \in [0,1]$ so that $p_v(0) = g_n(v)$, $p_v(1) = v_1 = q_v(0)$ and $q_v(1) = \varphi_n(v)$. Every point $x \in \overline{U}(v,\delta_v) - \{v\}$ can be written uniquely as $x = (1-t)w + tv$, where $d(w,v) = \delta_v$ and $0 \leq t \leq 1$. Therefore we can define g_n' on each $\overline{U}(v,\delta_v)$ by $g_n'(v) = \varphi_n(v)$ and

$$g_n'((1-t)w+tv) = \begin{cases} g_n((1-3t)w+3tv) & \text{if} \quad 0 \leq t \leq 1/3, \\ p_v(3t-1) & \text{if} \quad 1/3 \leq t \leq 2/3, \\ q_v(3t-2) & \text{if} \quad 2/3 \leq t < 1, \end{cases}$$

and extend it to $g_n' : |P_n| \to |K_n|$ by $g_n'(x) = g_n(x)$ for $x \in |P_n| - \cup\{U(v,\delta_v) : v \in W_n\}$. Write

$$V_n = \cup\{U(v,2\delta_v/3) : v \in W_n, \quad v \neq v_1\} \cup U(v_1,\delta_{v_1}/3).$$

Then g_n' is a proximity map on $|P_n| - V_n$ with $\text{Fix}\, g_n' = \text{Fix}\, g_n \cup \{c_n\}$, where $c_n \in U(v_1, 2\delta_{v_1}/3) - \overline{U}(v_1, \delta_{v_1}/3)$ lies in a maximal simplex of $|P_n|$, and $g_n'(v) = \varphi_n(v)$ for all $v \in W_n$.

Next we change g_n' to a map $g_n'' : |P_n| \to |K_n|$ which is still a proximity map on $|P_n| - V_n$, equals g_n' on V_n, and has a_n as its only fixed point. If P_n is finite and $\text{Fix}\, g_n' = \{b_n, c_n\}$, we use Lemmas (1.2) and (1.3) in [6] (also stated as Lemma 2, p. 126, and Lemma 3, p. 128, in [2]) to unite the fixed points b_n and c_n within a subset of $|P_n| - V_n$, and construct thus for arbitrary P_n a map which is a proximity map on $|P_n| - V_n$, equals g_n' on V_n, and has c_n as its only fixed point. If we use the technique from the proof of Lemma (2.4) in [4] to push c_n to the point $a_n \in |P_n| - V_n$, we obtain g''_n. g''_n determines a path field $\beta_n : |P_n| \to |K_n|^I$ which has on $|P_n| - W_n$ the point a_n as its only singularity. To see this, let $\beta_n(x)$, for $x \in |P_n| - V_n$, be the broken line segment determined by the proximity map g''_n (see [6], Lemma 1, [2], pp. 124-125, or [3], Lemma (2.1)). If $x \in \overline{U}(v, 2\delta_v/3)$ for $v \in W_n - \{v_1\}$, let ℓ_x be the broken line segment determined by the proximity map $\overline{U}(v, 2\delta_v/3) \to g_n(v)$, let $(p_v \circ q_v)_x$ be the part of the path $p_v \circ q_v$ from $g_n(v)$ to $g'_n(x)$, and define $\beta_n(x)$ as the composite path $\ell_x \circ (p_v \circ q_v)_x$. If finally $x \in \overline{U}(v_1, \delta_{v_1}/3)$, let $\beta_n(x) = m_x \circ (q_{v_1})_x$, where $m_x = [x, v_1]$ and $(q_{v_1})_x$ is the part of q_{v_1} from v_1 to $g'_n(x)$. Parametrize all $\beta_n(x)$ by $t \in [0,1]$ so that

$$\rho(\beta_n(x)(0), \beta_n(x)(t)) = t\,\rho(\beta_n(x)(0), \beta_n(x)(1)),$$

where $\rho(h,k)$ denotes the arclength from h to k measured along the piecewise linear path $\beta_n(x)$.

Now change d to a bounded metric $\overline{d} \le 1$, and define $f_n : |P_n| \to |K_n|$ by $f_n(x) = \beta_n(x)(\overline{d}(x,A))$. Then f_n satisfies (I_n).

CASE 3

$|P_n - W_n| \cap A = \phi$. Let T_n be a maximal tree of P_n [9], p. 139. As $|K_{n-1}|$ is path-connected, $\varphi_n : W_n \to K_{n-1}$ can be extended to a map $g_n : |T_n| \to |K_{n-1}|$. It follows from (6.2), p. 91, and (9.1), p. 96 in [1] that there exists a retraction $r_n : |P_n| \to |T_n|$, hence the map $f_n = g_n \circ r_n : |P_n| \to |K_n|$ satisfies (I_n).

This completes the induction of Step 2. If we now define $f : |K| \to |K|$ by $f \,\big|\, |P_n| = f_n$, then f is continuous and Fix $f = A$. ■

References

[1] BORSUK, K.: Theory of Retracts, Monografie Matematyczne, vol. 44, Polish Scientific Publishers, Warszawa, Poland, (1967).

[2] BROWN, R.F.: The Lefschetz Fixed Point Theorem, Scott, Foresman and Co., Glenview, Ill., (1971).

[3] FADELL, E.: A remark on simple path fields in polyhedra of characteristic zero, Rocky Mount. J. Math. 4 (1974), 65-68.

[4] SCHIRMER, H.: Fixed point sets of polyhedra, Pac. J. Math. 52 (1974), 221-226.

[5] SCHIRMER, H.: Fixed point sets of continuous selfmaps, these proceedings.

[6] SHI GEN HUA : On the least number of fixed points and Nielsen numbers, Chinese Math. 8 (1966), 234-243.

[7] SHI GEN HUA : The least number of fixed points of the identity mapping class, Acta Math. Sinica 18 (1975), 192-202.

[8] SHI GEN HUA : On the least number of fixed points for infinite complexes, preprint.

[9] SPANIER, E.H.: Algebraic Topology, McGraw Hill, New York, (1966).

[10] WECKEN, F.: Fixpunktklassen III, Math. Ann. 118 (1942), 544-577.

LOCALLY NONEXPANSIVE MAPPINGS IN BANACH SPACES

By

W.A. KIRK *

Department of Mathematics
The University of Iowa
Iowa City, Iowa 52242

1. Introduction

In this report we describe a fixed point theory for the *locally* nonexpansive nonlinear mappings in Banach spaces. While many aspects of this theory have been presented in detail elsewhere, we shall attempt to provide here, not only a more unified and readable treatment, but one which also illuminates for the first time the difference between what is known about the locally nonexpansive mappings and what is known about the more general continuous local pseudo-contractions.

Throughout the paper X will denote a (real) Banach space, $B(x;\varepsilon)$ will denote the closed ball centered at $x \in X$ with radius $\varepsilon > 0$, and $seg[x,y]$ will denote the algebraic segment joining points $x,y, \in X$ (with the usual convention for denoting open and half-open segments). A mapping $T: D \to X$, $D \subset X$, is said to be *locally* k-*lipschitzian* if to each point $x \in D$ there corresponds $\varepsilon > 0$ such that $\|T(u) - T(v)\| \le k\|u - v\|$ for all $u,v \in B(x;\varepsilon) \cap D$. For $k < 1$ (respectively, $k = 1$) T is said to be a *local contraction* (respectively, *locally nonexpansive*). More generally, $T: D \to X$ is said to be a *local* k-*pseudo-contraction* if corresponding to each $x \in D$ there exists $\varepsilon > 0$ such that for each $\lambda > k$,

$$(\lambda - k)\|u - v\| \le \|(\lambda I - T)(u) - (\lambda I - T)(v)\|, \quad u,v \in B(x;\varepsilon) \cap D. \tag{1}$$

For $k < 1$ (respectively, $k = 1$) these mappings are called *local strong pseudo-contractions* (respectively, *local pseudo-contractions*).

We call attention below to connections between the classes of mappings just described and the so-called accretive operators. But first, since we shall be inte-

*Research supported in part by National Science Foundation grant MCS 76-03945-A01.

rested primarily in the behavior of mappings which have local Lipschitz constant 1, we recall a fact which should serve to motivate this aspect of our study. Here we follow [9].

DIFFERENTIABILITY

Let G be an open subset of X and $T:G\to X$. A linear operator $T'_{x_0}\in L(X,X)$ is said to be the *Gateaux derivative* of T at $x_0\in D$ if for each $x\in X$:

$$t^{-1}[T(x_0+tx)-T(x_0)]-T'_{x_0}(x)\to 0 \quad \text{as} \quad t\to 0.$$

Suppose $T:G\to X$ has a Gateaux derivative at each point of G and let x_0 and x_1 be points of G for which $\mathrm{seg}[x_0,x_1]\subset G$. Select $x^*\in X^*$ such that

$$x^*(T(x_1)-T(x_0)) = \|x^*\|\|T(x_1)-T(x_0)\|,$$

let $x_t = x_0+t(x_1-x_0)$, $t\in[0,1]$, and define the mapping $\varphi:[0,1]\to\mathbb{R}$ by

$$\varphi(t) = x^*(T(x_t)).$$

It is a straightforward matter to see that φ is differentiable on $[0,1]$ with

$$\varphi'(t) = x^*(T'_{x_t}(x_1-x_0)),$$

so by the mean value theorem there exists $\xi\in(0,1)$ such that $\varphi'(\xi)=\varphi(1)-\varphi(0)$, that is

$$x^*(T'_{x_\xi}(x_1-x_0)) = x^*(T(x_1))-x^*(T(x_0)).$$

Thus

$$\|x^*\|\|T(x_1)-T(x_0)\| = \|x^*(T'_{x_\xi}(x_1-x_0))\| \le \|x^*\|\|T'_{x_\xi}\|\|x_1-x_0\|,$$

and in particular,

$$\|T(x_1)-T(x_0)\| \le \sup_{\xi\in[0,1]} \|T'_{x_\xi}\|\|x_1-x_0\|.$$

It follows that if T has a Gateaux derivative T'_x at each $x\in G$ with $\|T'_x\|\le k$, then the restriction of T to any convex neighborhood of a point in G is k-lipschitzian; thus T is locally k-lipschitzian. (This observation is well-known; see for example, [9, p. 661].)

While it is a trivial fact that mappings which are locally nonexpansive on convex sets must in fact be globally nonexpansive, nonexpansive mappings which arise in connection with the study of accretive operators need not have convex domains.

ACCRETIVITY

A mapping $f: D \to X$, $D \subset X$, is said to be *accretive* ([1],[10]) if for each $u,v \in D$ there exists $j \in J(u-v)$ such that

$$<f(u) - f(v), j> \geq 0 \tag{2}$$

where $<\cdot,\cdot>$ denotes the usual pairing between elements of X and elements of its dual X^*, and $J: X \to 2^{X^*}$ denotes the duality mapping defined by

$$J(x) = \{j \in X^* : \|j\|^2 = \|x\|^2 = <x,j>\}.$$

The notion of accretivity arose as an extension of the concept of monotonicity, a theory which coincides with accretivity in a Hilbert space setting and which has its origins in the theory of the direct method of the calculus of variations (see Browder [3]). The monotone operators are defined as follows. If X is a Banach space with dual X^*, and if D is a subset of X, then an operator $f: D \to X^*$ is said to be *monotone* if for each $u,v \in D$,

$$<f(u) - f(v), u - v> \geq 0.$$

(As a very special instance, if $f: X \to \mathbb{R}^1$ is of class C^1 and if f' is the Fréchet derivative of f (thus $f' \in L(X,\mathbb{R}^1) = X^*$), then f' is monotone if and only if f is a convex function.).

It has been known from the outset that there are close connections between the accretive operators and the nonexpansive mappings. In particular (see [1],[3]), if $f: D \to X$ is continuous and accretive for $D \subset X$ closed, and if the initial value problem

$$\begin{cases} u'(t) = -f(u(t)), & t \geq 0 \\ u(0) = u \in D \end{cases} \tag{3}$$

has a unique continuously differentiable solution for each $u \in D$, then the mappings $S(t) : D \to X$, $t \geq 0$, defined by $S(t)(u) = u(t)$ where $u(t)$ is the solution to (3) for which $u(0) = u$, form a semigroup of nonexpansive mappings. General conditions which assure existence of solutions to (3) are well-known. (For example, see Browder [3], **Deimling** [5], Martin [16], and works cited therein.). We are more interested here, however, in another connection between nonexpansiveness and accretivity. It is known ([10]) that the condition (2) has the following equivalent geometric formulation:

$$\|u - v\| \leq \|u - v + \lambda(f(u) - f(v))\|, \quad u,v \in D, \quad \lambda > 0. \tag{4}$$

Thus if $f_\lambda = I + \lambda f$, $\lambda > 0$, then the mapping f is accretive if and only if the resolvent $J_\lambda = (f_\lambda)^{-1}$ is a nonexpansive mapping defined on the range $R(f_\lambda)$ of f_λ. Moreover, a point $x_0 \in D$ is a zero of f (corresponding to a constant solution of (3)) if and only if $J_\lambda(x_0) = x_0$. However in order to reduce the problem of finding zeros of f to an application of fixed point theory for nonexpansive mappings, one must deal with nonexpansive mappings whose domains in general are not convex. This observation motivated our paper [11].

Finally we remark that from (4), it readily follows that f is accretive in the sense of (2) if and only if $T = I - f$ is pseudo-contractive (globally) in the sense of (1). Thus the fixed point theorem stated in Section 5 can be also formulated as a theorem which guarantees existence of zeros of continuous locally accretive operators.

2. Principal Results

We formulate our main result in sufficiently general terms so as to include its global counterpart (Corollary 1). In this theorem, we use $\partial_K D$ to denote the relative boundary of D in K. (We use ∂B to denote the boundary of a set $B \subset X$.)

THEOREM 1

Let X be a reflexive Banach space, let K be a closed convex subset of X, and let D be a bounded open subset of X with $D \cap K \neq \phi$. Suppose $T : \overline{D \cap K} \to K$ is a closed mapping which is locally nonexpansive on $G = D \cap K$, and suppose the following conditions hold:

(a) there exists $z \in G$ such that $\|z - T(z)\| < \|x - T(x)\|$ for all $x \in \sigma = \partial_K D$;

(b) for each $\varepsilon > 0$ sufficiently small, there exists a $\delta(\varepsilon) \in (0,\varepsilon]$ such that if $u,v \in G$ and if $\mathrm{seg}[u,v] \subset G$, then the conditions $\|u - T(u)\| \le \delta(\varepsilon)$ and $\|v - T(v)\| \le \delta(\varepsilon)$ imply $\|m - T(m)\| \le \varepsilon$ for all $m \in \mathrm{seg}[u,v]$.

Then T has a fixed point in G.

If $K \subset D$ in Theorem 1, then $\sigma = \phi$, (a) is satisfied vacuously, and $T : K \to K$. Since K is convex and T locally nonexpansive on K, T must be globally nonexpansive on K and Theorem 1 reduces to the following known fact (Browder [2]; cf. also Bruck [4]).

COROLLARY 1

Let X be a reflexive Banach space, let K be a nonempty bounded closed and convex subset of X, and suppose $T : K \to K$ is a nonexpansive mapping which satisfies condition (b) of Theorem 1. Then T has a fixed point in K.

Taking $K = X$ in Theorem 1 yields a fact known to hold in uniformly convex spaces for the continuous local pseudo-contractions. (See Section 5.)

COROLLARY 2

Let X be a reflexive Banach space, let D be a nonempty bounded open subset of X, and let $T : \overline{D} \to X$ be a closed mapping which is locally nonexpansive on D and which satisfies condition (b) of Theorem 1. Suppose there exists $z \in D$ such that $\|z - T(z)\| < \|x - T(x)\|$ for all $x \in \partial D$. Then T has a fixed point in D.

In each of the above, reflexivity of X and the assumption that T satisfies condition (b) may be replaced with the single assumption that X is uniformly convex. This follows from the fact that the mapping T is nonexpansive on $\mathrm{seg}[u,v]$ when $\mathrm{seg}[u,v] \subset D \cap K$, and from the following known geometric property of uniformly convex spaces.

LEMMA 1 *(Browder* [2]; *Göhde* [7])

Let X be uniformly convex and let B be a bounded subset of X. Then for each $\varepsilon > 0$ there exists a number $\xi = \xi(\varepsilon) \in (0,\varepsilon]$ such that if $u,v \in B$ and if $m \in \text{seg}[u,v]$, then the conditions

$$\|x - u\| \le \|m - u\| + \xi \quad \text{and} \quad \|x - v\| \le \|m - v\| + \xi$$

imply $\|x - m\| \le \varepsilon$.

There are also other settings in which condition (b) always holds.

CONTRACTIONS OF TYPE (γ) *(Bruck* [4])

Let C be a bounded convex subset of X and let Γ denote the set of strictly increasing, continuous, convex functions $\gamma : \mathbb{R}^+ \to \mathbb{R}^+$ with $\gamma(0) = 0$. A mapping $T : C \to X$ is said to be of type (γ) if $\gamma \in \Gamma$ and for all $x,y \in C$, $c \in [0,1]$,

$$\gamma(\|cT(x) + (1 - c)T(y) - T(cx + (1 - c)y)\|) \le \|x - y\| - \|T(x) - T(y)\|.$$

Clearly every contraction of type (γ) is nonexpansive (since $\gamma \ge 0$), and also every affine nonexpansive mapping is of type (γ). It is shown in [4] that if X is uniformly convex then there exists $\gamma \in \Gamma$ (definable in terms of the modulus of convexity of X) such that every nonexpansive mapping $T : C \to X$ is of type (γ), and it is also shown that in general, every mapping $T : C \to X$ of type (γ) (with C convex) satisfies condition (b) on C.

Finally, we remark that in Theorem 1 (and Corollary 2) we have assumed only that T is a *closed* mapping. This assumption, weaker than continuity, means that if $\{x_n\} \subset \overline{D \cap K}$, then the conditions $x_n \to x$ and $T(x_n) \to y$ as $n \to \infty$ imply $T(x) = y$. Of course, because of local nonexpansiveness, T will actually be continuous on $D \cap K$.

3. Preliminaries

Our proof of Theorem 1 rests upon some preliminary facts, the first of which is quite trivial.

PROPOSITION 1

Let X be a Banach space, let K be a closed convex subset of X, and let D be an open subset of X with $D \cap K \neq \phi$. Suppose $T : \overline{D \cap K} \to K$ is a closed mapping which is a local contraction on $G = D \cap K$, and suppose for $z \in G$,

$$\|z - T(z)\| \leq \|x - T(x)\| \quad \text{for all} \quad x \in \sigma = \partial_K D. \tag{a'}$$

Then the sequence $\{T^n(z)\}$ lies in G and converges to a fixed point of T.

Proof

Suppose $\mathrm{seg}[T^i(z), T^{i+1}(z)] \subset G$ for $i = 0,1,\ldots,n-1$, and suppose that there exists $m \in \mathrm{seg}[T^n(z), T^{n+1}(z)] \cap \sigma$ for which $\mathrm{seg}[T^n(z), m] \subset G$. Since the restriction of a local contraction to a convex set is a global contraction, $\|T^n(z) - T^{n+1}(z)\| < \|z - T(z)\|$, and hence

$$\|m - T(m)\| \leq \|m - T^{n+1}(z)\| + \|T^{n+1}(z) - T(m)\| \leq \|m - T^{n+1}(z)\| + \|T^n(z) - m\|$$

$$= \|T^n(z) - T^{n+1}(z)\| < \|z - T(z)\|.$$

Since $m \in \sigma$, this contradicts (a'). The argument can now be completed in the standard way by showing that $\{T^n(z)\}$ is Cauchy. ■

PROPOSITION 2

Let X, K and D be as in Proposition 1 and let $T : \overline{D \cap K} \to K$ be a closed mapping which is locally nonexpansive on $G = D \cap K$. Suppose $T(x_0) = t_0 x_0$ for some $x_0 \in G$ and $t_0 > 1$. Then for each $t > 1$ sufficiently near t_0, there exists a unique point $x_t \in G$ such that

(i) $$T(x_t) = t x_t$$

and for which

(ii) $$\|x_0 - x_t\| \leq [\,|t_0 - t\,|\,/(t-1)]\|x_0\|.$$

Proof

Because $\sigma = \partial_K D$ is closed and $x_0 \notin \sigma$, it is possible to select $\varepsilon > 0$ so that $B(x_0;\varepsilon) \cap K$ is a closed convex subset of G. Since $t_0^{-1}T$ is a contraction mapping on $B(x_0;\varepsilon) \cap K$, x_0 is the unique fixed point of $t_0^{-1}T$ in this set, and in particular, if $\sigma' = (\partial B(x_0;\varepsilon)) \cap K$,

$$\rho = \inf\{\|x - t_0^{-1}T(x)\| : x \in \sigma'\} > 0.$$

Hence for $t > 1$ sufficiently near t_0,

$$\|x - t_0^{-1}T(x)\| \geq \rho/2 \quad \text{for all} \quad x \in \sigma';$$

and

$$\|x_0 - t^{-1}T(x_0)\| < \rho/2.$$

It is now possible to apply Proposition 1 to the mapping $t^{-1}T$, replacing D with the interior of $B(x_0;\varepsilon)$, to obtain a (unique) fixed point x_t of $t^{-1}T$ in $B(x_0;\varepsilon)$. To obtain the estimate (ii), notice that

$$\|x_0 - x_t\| \geq \|T(x_0) - T(x_t)\| = \|t_0x_0 - tx_0 + tx_0 - tx_t\|$$

$$\geq t\|x_0 - x_t\| - |t_0 - t|\,\|x_0\|. \blacksquare$$

PROPOSITION 3

Let X, K and D be as in Proposition 1, and in addition suppose $0 \in G = D \cap K$. Let $T : \overline{D \cap K} \to K$ be a closed mapping which is locally nonexpansive on G, and suppose

$$\|T(0)\| < \|x - T(x)\| \quad \text{for all} \quad x \in \sigma = \partial_K D. \tag{a}$$

Then in G there exists a unique continuous path $t \mapsto x_t$, $t > 1$, for which $T(x_t) = tx_t$. Moreover, the function ρ defined by $\rho(t) = \|x_t - T(x_t)\|$ decreases as t decreases, and if D is bounded, $\rho(t) \to 0$ as $t \to 1^+$.

Proof

Select $\varepsilon > 0$ so that $B = B(0;\varepsilon) \cap D$ is a closed convex subset of G.

Since T is nonexpansive on B, $T(B)$ is a bounded subset of K and hence $t^{-1}T(B) \subset B$ for t sufficiently large. For such t there exist unique points x_t for which $T(x_t) = tx_t$, and it follows from Proposition 2 that there exists a smallest number $r \geq 1$ for which the path $t \mapsto x_t$ can be uniquely defined for all $t > r$.

The remainder of the proof combines ideas of Kirk-Morales [14] and Morales [18]. Suppose $r > 1$. Let $s,t \in (r,\infty)$ with $t > s$. Then by Proposition 2(ii) the interval $[s,t]$ may be covered by a finite number of overlapping (consecutive) subintervals $\{I_i\}_{i=0}^{n}$ such that if $\lambda,\mu \in I_i$, then

$$(\mu - 1)\|x_\lambda - x_\mu\| \leq |\lambda - \mu| \, \|x_\lambda\|.$$

Select $t_i \in I_i \cap I_{i+1}$, $i = 0,\ldots,n-1$, with $t_0 = t$, $t_n = s$. Since $t_1 > r$,

$$\begin{aligned}(r - 1)\|x_{t_1}\| &\leq (t_1 - 1)\|x_{t_1}\| \leq (t_1 - 1)[\|x_{t_0} - x_{t_1}\| + \|x_{t_0}\|] \\ &\leq (t_0 - t_1)\|x_{t_0}\| + (t_1 - 1)\|x_{t_0}\| = (t_0 - 1)\|x_{t_0}\|.\end{aligned} \tag{5}$$

Similarly,

$$\begin{aligned}(r - 1)\|x_{t_2}\| &\leq (t_2 - 1)\|x_{t_2}\| \leq (t_1 - 1)\|x_{t_1}\| && \text{(as above)} \\ &\leq (t_0 - 1)\|x_{t_0}\| && \text{(by (5))}.\end{aligned} \tag{6}$$

By continuing, we obtain

$$(r - 1)\|x_s\| \leq (t_0 - 1)\|x_{t_0}\|.$$

Since $s > r$ is arbitrary, this in particular proves that

$$M = \sup\{\|x_t\| : t > r\} < \infty .$$

Moreover,

$$\|x_s - x_t\| \leq \sum_{i=0}^{n-1} \|x_{t_i} - x_{t_{i+1}}\| \leq (M \sum_{i=0}^{n-1} (t_i - t_{i+1}))/(r - 1) = M(t - s)/(r - 1).$$

It follows that $x_t \to \bar{x} \in \bar{D}$ as $t \to r^+$, and in addition $T(x_t) \to r\bar{x}$ as $t \to r^+$. Since T is closed, $T(\bar{x}) = r\bar{x}$.

The fact that $\rho(t)$ is (locally, hence globally) decreasing as t decreases follows from (6):

$$\|x_{t_1} - T(x_{t_1})\| = (t_1 - 1)\|x_{t_1}\| \leq (t_0 - 1)\|x_{t_0}\| = \|x_{t_0} - T(x_{t_0})\|.$$

Since $x_t \in B$ for t sufficiently large and $T(B)$ is bounded,

$$\|x_t\| = \|x_t - Tx_t\| / (t - 1) \to 0 \quad \text{as} \quad t \to \infty.$$

Therefore,

$$\|\bar{x} - T(\bar{x})\| \leq \|x_t - T(x_t)\| \leq \|T(0)\|$$

proving (via (a)) that $\bar{x} \in G$. It is now possible to define $x_r = \bar{x}$ and apply Proposition 2 to contradict minimality of r. The final assertion in Proposition 3 is trivial. ■

4. Proof of Theorem 1

(The argument given here is implicit in [12]). Suppose $z \in G$ satisfies (a), with $z \neq T(z)$. Then

$$\rho = \inf\{\|x - T(x)\| : x \in \sigma\} > 0.$$

Fix $\varepsilon_0 > 0$, $\varepsilon_0 < \rho$, and, using (b), define $\{\varepsilon_j\} \subset \mathbb{R}^+$ by taking $\varepsilon_j = \delta(\varepsilon_{j-1})$, $j = 1,2,\dots$. Proposition 3, applied to the mapping $T_z : \overline{D \cap K} - z \to K - z$ defined by

$$T_z(x) = T(x + z) - z,$$

yields a path $t \mapsto x_t$, $t > 1$, in G for which

$$T(x_t) = tx_t + (1 - t)z,$$

and for which

$$\|x_t - T(x_t)\| \downarrow 0 \quad \text{as} \quad t \downarrow 1.$$

Since X is reflexive, it is possible to select points $x_i = x_{t_i}$ such that $t_i \downarrow 1$ as $i \to \infty$ and so that

$$x_i \to x \in X \quad \text{weakly as} \quad i \to \infty .$$

Moreover, by passing to a suitable subsequence, we may suppose

$$\|x_i - T(x_i)\| \le \varepsilon_i, \quad i = 1,2,\ldots . \tag{7}$$

We shall show that $x \in G$ and $T(x) = x$.

ASSERTION (I)

If $s,t \in (1, t_j]$, then $\mathrm{seg}[x_t,x_s] \subset G$ and $\|m - T(m)\| \le \varepsilon_{j-1}$ for all $m \in \mathrm{seg}[x_t,x_s]$.

Proof

Fix $t \in (1, t_j]$ and let

$$H = \{s \in (1, t] : \mathrm{seg}[x_t,x_s] \subset G\}.$$

First note that if $s \in H$, then $\|x_s - T(x_s)\| \le \varepsilon_j$. Hence by condition (b), $\|m - T(m)\| \le \varepsilon_{j-1} \le \varepsilon_0$ for all $m \in \mathrm{seg}[x_t,x_s]$. Since $\varepsilon_0 < \rho$, $\mathrm{seg}[x_t,x_s]$ is bounded away from σ for all $s \in H$. It follows that, since $t \mapsto x_t$ is continuous, H is closed in $(1,t]$, and it is obvious that H is open in $(1,t]$. Thus $H = (1,t]$, and the remainder of (I) follows from (b). ■

Now let $S_j = \{x_j,x_{j+1},\ldots\}$, $j = 1,2,\ldots$.

ASSERTION (II)

If $w \in \mathrm{conv}\, S_j$, then $w \in G$ and $\|w - T(w)\| \le \varepsilon_{j-1}$.

Proof

If $w \in \mathrm{seg}[x_j,x_i]$ for some $i > j$, (II) reduces to (I). Make the inductive assumption that (II) holds if w is a linear combination of n elements of any S_j, and suppose w is a linear combination of $n+1$ elements of some fixed S_j. Then w is on a segment joining some $x_i \in S_j$ with a point u which is itself a linear combination of n elements of S_{j+1}. By the inductive assumption, $u \in G$ and $\|u - T(u)\| \le \varepsilon_j$. Since $\|x_i - T(x_i)\| \le \varepsilon_j$, (II) will follow from (b) upon showing that $w \in G$. Fix $m > i$. Then $\text{seg}[x_i, x_m] \subset G$ by (I). Let $m_r = rx_m + (1-r)u$, and let $V = \{r \in [0,1] : \text{seg}[x_i, m_r] \subset G\}$. Then $V \neq \phi$ $(1 \in V)$, and as in the proof of (I) it is easily verified that V is both open and closed in $[0,1]$. Thus $0 \in V$, yielding $\text{seg}[x_i, u] \subset G$, whence $w \in G$. ■

Now, since $x_i \to x$ weakly, $x \in \overline{\text{conv}}\{x_1, x_2, \ldots\}$. But (II) implies that if $y \in \text{conv}\{x_1, x_2, \ldots\}$, then $y \in G$ and $\|y - T(y)\| \le \varepsilon_0$. In particular, such points y are bounded away from σ. It follows that $x \in G$ and by continuity, $\|x - T(x)\| \le \varepsilon_0$. Since $\varepsilon_0 > 0$ was arbitrary, $x = T(x)$. ■

REMARKS

(1) We should point out that *under the assumptions of* Corollary 1, Browder's proof of [2], which provided the germ of the above argument, yields the even stronger conclusion of demiclosedness of $I - T$ on K; thus for $\{x_n\} \subset K$, the conditions $x_n \to x$ weakly while $x_n - T(x_n) \to 0$ strongly imply $x = T(x)$. The weaker assumptions of our Theorem 1 do not seem to yield demiclosedness.

(2) It appears that if a suitable analogue of Proposition 1 could be obtained for continuous local pseudo-contractions, an analogue not requiring openness of the domain, then very likely Theorem 1 could be easily shown to hold for this wider class of mappings. We take up that is known for this class in the next section.

5. Local Pseudo-contractions

As we mentioned earlier, the following theorem is known.

THEOREM 2 *(Kirk* [12]*)*

Let X be a uniformly convex Banach space, let D be a bounded open subset of X, and let $T : \overline{D} \to X$ be a continuous mapping which is a local pseudo-contrac-

tion on D. Suppose there exists $z \in D$ such that
(a): $\|z - T(z)\| < \|x - T(x)\|$ for all $x \in \partial D$. Then T has a fixed point in D.

The proof of Theorem 2, in large measure, parallels the one given for Theorem 1, although some complications arise from topological considerations. Analogues of Propositions 1-3 under the assumption $K = X$ hold for continuous local pseudo-contractions. In particular, the existence part of Proposition 1 (so modified) is an immediate consequence of Theorem 3 of Kirk-Schöneberg [15]. The appropriate analogues of Propositions 2 and 3 follow, respectively, from results of Kirk-Morales, [13] and Morales [18], and the details are almost the same as given above. The essential differences between the nonexpansive and pseudo-contractive cases (aside from Proposition 1) occur in the actual proof of the theorem.

Proof of Theorem 2 (outline).

It follows from (1) (with $k = 1$) that the mapping $F = 2I - T$ satisfies, locally on D, the condition

$$\|u - v\| \leq \|F(u) - F(v)\|.$$

Furthermore, by Deimling's domain invariance theorem [5, Theorem 3], $F(D)$ is open in X. (Since $u - T(u) = F(u) - u$, $u \in D$, the boundary condition (a) carries over suitably to $\partial F(D)$, so if F were one-to-one, one could apply Theorem 1, or a generalized version of Theorem 1, directly to the locally nonexpansive mapping F^{-1} defined on $F(G)$. However since F need not be (globally) invertible, it seems that a separate argument is needed.).

From the analogue of Proposition 3, it is possible to obtain a path $t \mapsto x_t$, $t > 1$, for which (assuming $z = 0$),

$$F(x_t) = (2 - t)x_t.$$

As in the nonexpansive case it is possible to show that $\|x_t - F(x_t)\| \downarrow 0$ as $t \downarrow 1$ (see [18]). Moreover, if $z \neq T(z)$,

$$\rho = \inf\{\|x - F(x)\| : x \in \partial D\} > 0.$$

Let $\varepsilon_0 \in (0, \rho/4)$. It then follows that if $\|x - F(x)\| < \varepsilon_0$, $B(x;\rho/4) \subset F(D)$. (This is needed in the proof of (I') below.). Now select $x_i = x_{t_i}$ so that

$t_i + 1$ and $F(x_i) \to h_0 \in X$ weakly as $i \to \infty$, and pass to a subsequence so that $\|x_i - F(x_i)\| \le \varepsilon_i = \xi(\varepsilon_{i-1})$ (Lemma 1). Using the fact that F is a local homeomorphism along with a continuation argument, it is possible to prove:

(I') Let $s,t \in (1,t_j]$; then $seg[F(x_t), F(x_s)] \subset F(D)$ and there exists a unique path $\Gamma(x_t,x_s)$ which begins at x_t, terminates at x_s, and is mapped by F onto $seg[F(x_t), F(x_s)]$. Moreover, if $w \in \Gamma(x_t,x_s)$, $\|w - F(w)\| \le \varepsilon_{j-1}$.

Similarly, if $S_j = \{F(x_j), F(x_{j+1}),\ldots\}$, $j = 1,2,\ldots$, then the following holds.

(II') Suppose $w \in \operatorname{conv} S_j$, $i \ge j$. Then $seg[F(x_j), w] \subset F(D)$ and there is a unique path Γ beginning at x_j, terminating at a point $u \in F^{-1}(w)$, and such that $\|m - F(m)\| \le \varepsilon_{j-1}$ for each $m \in \Gamma$. In particular, $\|u - w\| \le \varepsilon_{j-1}$.

The argument concludes as follows. Since $F(x_i) \to h_0$ weakly as $i \to \infty$, there exists a sequence $\{y_i\} \subset \operatorname{conv}\{F(x_1), F(x_2),\ldots\}$ such that $y_i \to h_0$ strongly as $i \to \infty$, and by (II') there exists for each i, $v_i \in F^{-1}(y_i)$ such that $\|v_i - y_i\| \le \varepsilon_0$. It is thus possible to fix i_0 so that

$$\|v_{i_0} - h_0\| \le \varepsilon_0 + 1.$$

Now let $\bar{\varepsilon}_0 = \varepsilon_0/2$ and pass to a subsequence of $\{x_i\}$ for which

$$\|x_{i_n} - F(x_{i_n})\| \le \bar{\varepsilon}_n = \xi(\bar{\varepsilon}_{n-1}), \quad n = 1,2,\ldots .$$

By repeating the argument given for ε_0 it is possible to obtain i_1 and $v_{i_1} \in D$ such that both

$$\|v_{i_1} - F(v_{i_1})\| \le \bar{\varepsilon}_0 = \varepsilon_0/2;$$

$$\|v_{i_1} - h_0\| \le \bar{\varepsilon}_0 + 1/2 = (\varepsilon_0 + 1)/2.$$

By induction, it is possible to obtain points $v_{i_n} \in D$ such that

$$\|v_{i_n} - F(v_{i_n})\| \le \varepsilon_0/n+1;$$

$$\|v_{i_n} - h_0\| \le (\varepsilon_0 + 1)/n+1.$$

It follows that $v_{i_n} \to h_0 = F(h_0)$ as $n \to \infty$, and thus $h_0 = 2h_0 - T(h_0)$ yielding $T(h_0) = h_0$. ■

6. ITERATION

The following is a simple consequence of Proposition 3.

THEOREM 3

Let X be a Banach space, let K be a closed and convex subset of X, and let D be an open subset of X with $D \cap K \neq \phi$. Suppose $T : \overline{D \cap K} \to K$ is a closed mapping which is locally nonexpansive on $G = D \cap K$ and which has precompact range, and suppose for $z \in G$,

$$\|z - T(z)\| < \|x - T(x)\| \quad \text{for all} \quad x \in \sigma = \partial_K D.$$

Then there exists a continuous path $t \mapsto x_t \in G$, $t > 1$, satisfying

$$T(x_t) = tx_t + (1-t)z.$$

Moreover, $\lim_{t \to 1+} x_t = x \in G$, where x is a fixed point of T.

Proof

The existence of the asserted path follows directly from Proposition 3 applied to the mapping T_z as defined in the proof of Theorem 1. Since the range of T is precompact, there exists a sequence $\{t_i\} \subset (1,\infty)$ with $t_i \downarrow 1$ such that $x_{t_i} \to x \in G$ as $i \to \infty$, and since $\|x_{t_i} - T(x_{t_i})\| \to 0$, $x = T(x)$. Thus the points x_t eventually all lie in a convex subset of G which contains x. Since $\|x_t - T(x_t)\| \downarrow 0$ as $t \downarrow 1$ (by Proposition 3), it follows that $x_t \to x$ as $t \to 1^+$. ■

We shall now discuss two results which provide iteration techniques for approximating fixed points whose existence is assured by Theorem 3. The first, a localized version of Ishikawa's fundamental result of [8], basically is not new since it has already been noted in [14] that Ishikawa's proof carries over essentially without change to establish the local result.

THEOREM 4 (*cf.* [14]).

Suppose X, K, D and T satisfy all the assumptions of Theorem 3, and suppose z_1 is a point of $G = D \cap K$ for which

$$\|z_1 - T(z_1)\| < \|x - T(x)\| \quad \text{for all} \quad x \in \partial_K D.$$

Let $\{t_n\} \subset \mathbb{R}$ satisfy

$$\sum_{n=1}^{\infty} t_n = \infty \quad \text{and} \quad 0 \le t_n \le b < 1, \quad n = 1,2,\ldots .$$

Then the sequence $\{z_n\}$ defined by

$$z_{n+1} = (1 - t_n)z_n + t_n T(z_n)$$

lies in G and converges to a fixed point of T.

Convergence of Picard iterates of T results if the assumptions on T in Theorem 4 are slightly strengthened. A mapping $T : D \to X$, $D \subset X$, is said to be *locally contractive* on D if each point of D has a neighborhood N in D such that

$$\|T(u) - T(v)\| < \|u - v\| \quad \text{for all} \quad u,v \in N, \quad u \neq v.$$

THEOREM 5

Suppose X, K and D satisfy all the assumptions of Theorem 3, and suppose $T : \overline{D \cap K} \to K$ is a continuous mapping which is locally contractive on $G = D \cap K$ and which has precompact range. Then if $z \in G$ satisfies

$$\|z - T(z)\| < \|x - T(x)\| \quad \text{for all} \quad x \in \partial_K D,$$

the sequence $\{T^n(z)\}$ lies in G and converges to a fixed point of T.

We derive Theorem 5 from a much more abstract result, one which can be formulated in a metric space setting.

Let M be a metric space, let D be a subset of M, and let $\{F_n\}_{n=1}^{\infty}$ be a sequence of mappings of D into M. For $k \in \mathbb{N}$, define the iteration process P_k as follows: $P_k(x) = \{x_n^k\}_{n=0}^{\infty}$, where

$$x_0^k = x \in D; \; x_{n+1}^k = \begin{cases} F_{n+k}(x_n^k) & \text{if } x_n^k \in D \\ x_n^k & \text{if } x_n^k \notin D \end{cases} \qquad (n \in \mathbb{N}).$$

The process P_k is called *stable* on a subset S of M if for each $x \in S$ and $\varepsilon > 0$ there exists $\delta = \delta(x,\varepsilon,k) > 0$ such that if $u,v \in B(x;\delta)$, then $d(u_n^k, v_n^k) < \varepsilon$ for all $n \in \mathbb{N}$.

We use P to denote the set of points $x \in D$ for which the sequence $P_k(x)$ remains bounded away from the complement of D for each $k \in \mathbb{N}$. Specifically:

$$P = \{x \in D : \inf_n \operatorname{dist}(x_n^k, M \setminus D) > 0 \quad \text{for all } k \in \mathbb{N}\}.$$

Recall that if a complete metric space is *convex*, then each two of its points can be joined by an isometric image of a real-line interval whose length is equal to the distance between the points (cf. [17]).

THEOREM 6

Let G be an open subset of a complete and convex metric space M and let $\{F_n\}$ be a sequence of mappings of G into M for which the iteration processes P_k are stable on P $(k \in \mathbb{N})$. Let A be a given nonempty closed subset of G, and let

$$S = \{x \in P : P_k(x) \text{ converges to a point of } A \text{ for each } k \in \mathbb{N}\}.$$

Suppose also that the following condition holds:

$$\text{If } p \in A, \text{ then } B(p;\varepsilon) \subset S \text{ for some } \varepsilon > 0. \qquad \text{(c)}$$

Then:

(i) S is open in G, and

(ii) $\overline{S} \cap P \subset S$.

In particular, if C is any connected subset of P which contains a point of S, then $C \subset S$.

Proof

(i) Fix $x \in S$, $k \in \mathbb{N}$. By assumption, $P_k(x) = \{x_n^k\}$ with $x_n^k \to p \in A$ as $n \to \infty$. Since $A \subset G$ with G open, there exists $\varepsilon > 0$ such that $B(p;\varepsilon) \subset G$, and we may suppose ε to be chosen in conformance with (c). Select $\overline{n} > k$ so that $x_{\overline{n}}^k \in B(p;\varepsilon/2)$ and let $\rho_{\overline{n}} = \varepsilon/2$. Since each of the mappings $\{F_i\}_{i=1}^{\overline{n}-1}$ is continuous it is possible to choose $\rho_i \in (0,\varepsilon/2)$ so that $B(x_i^k;\rho_i) \subset G$ and so that

$$F_{i+k}(B(x_i^k;\rho_i)) \subset B(x_{i+1}^k;\rho_{i+1}), \qquad 0 \le i \le \overline{n}-1.$$

Let $u_0 \in B(x;\rho_0)$ and let $u_{\overline{n}} = F_{\overline{n}-1} \circ \dots \circ F_k(u_0)$. Then $u_{\overline{n}} \in B(p;\varepsilon)$. By (c), $P_{\overline{n}}(u_{\overline{n}})$ converges to a point of A. But $P_{\overline{n}}(u_{\overline{n}}) = P_k(u_0)$, proving $u_0 \in S$.

(ii) Now suppose $\{u_n\} \subset S$ with $u_n \to u \in P$ as $n \to \infty$, and fix $k \in \mathbb{N}$. Since $u \in P$ there exists $\rho > 0$ such that for all $n \in \mathbb{N}$, $\operatorname{dist}(u_n^k, M \setminus G) > \rho$. Also, since P_k is stable at u, there exists $\delta = \delta(u,\rho/2,k) > 0$ such that if $d(x,u) \le \delta$, then $d(x_n^k, u_n^k) < \rho/2$. It follows that $x_n^k \in G$ for all $n \in \mathbb{N}$, and moreover:

$$\inf_n \operatorname{dist}(x_n^k, M \setminus G) \ge \rho/2.$$

Thus $x \in P$. Now, select $n \in \mathbb{N}$ so that $d(u_n,u) < \delta$, and let $\operatorname{seg}[u_n,u]$ be any metric segment joining u_n and u. Let $\overline{u}$ be any point of $\operatorname{seg}[u_n,u]$ for which $\operatorname{seg}[u_n,\overline{u}) \subset S$, where we take $\operatorname{seg}[u_n,\overline{u})$ to be the points of $\operatorname{seg}[u_n,u]$ between u_n and $\overline{u}$, excluding $\overline{u}$. By (i), at least one such point $\overline{u}$ exists. We shall show that $\overline{u} \in S$ and conclude from this, via (i), that $\operatorname{seg}[u_n,u] \subset S$.

Suppose $\{z_i\}$ is a sequence of points of $\operatorname{seg}[u_n,\overline{u})$ such that $z_i \to \overline{u}$ as $i \to \infty$. Since each of the points z_i is in S, there exist points $p_i \in A$ such that $P_k(z_i) \to p_i$ as $n \to \infty$. Now let $\varepsilon > 0$ be arbitrary and select n_0 so that for $i \ge n_0$, $d(z_i,\overline{u}) < \delta(\overline{u},\varepsilon/2,k)$. (Note that $\overline{u} \in P$ because $d(\overline{u},u) < \delta = \delta(u,\rho/2,k)$.) Then if $i,j \ge n_0$, $d(p_i,p_j) \le d(p_i,(z_i)_n^k) + d((z_i)_n^k,(z_j)_n^k) + d((z_j)_n^k,p_j)$ from which, letting $n \to \infty$, $d(p_i,p_j) < \varepsilon$. It follows that there exists $p \in A$ such that $p_i \to p$ as $i \to \infty$. Moreover, since $\overline{u} \in P$,

$$d(\bar{u}_n^k,p) \le d(\bar{u}_n^k,(z_i)_n^k) + d((z_i)_n^k,p_i) + d(p_i,p).$$

Again, let $\varepsilon > 0$ be arbitrary. Select i so that $d(p_i,p) < \varepsilon/3$ and so that $d(\bar{u}_n^k,(z_i)_n^k) < \varepsilon/3$ for all n. With i thus fixed, select n so that $d((z_i)_n^k,p_i) < \varepsilon/3$, from which $d(\bar{u}_n^k,p) < \varepsilon$. Therefore, $\bar{u}_n^k \to p$ as $n \to \infty$ proving $\bar{u} \in S$. ■

Proof of Theorem 5

We apply Theorem 6 taking $K = M$, $D \cap K = G$, $F_n = T$ $(n = 0,1,2,\ldots)$, C to be the image of the path $t \mapsto x_t$, $t > 1$, of Theorem 3, and A to be the fixed point set of T in G. The fact that the iteration processes P_k associated with $\{F_n\}$ are stable on P follows from local nonexpansiveness of the mappings F_n. Thus it need only be shown that condition (c) holds, and that $C \subset P$, that is, that if $u \in C$, $T^n(u) \in G$ for all n and that

$$\inf_n \operatorname{dist}(T^n(u), \partial_K D) > 0.$$

By Proposition 3, $\|u - T(u)\| \le \|z - T(z)\|$ for all $u \in C$; thus

$$\|u - T(u)\| < \|x - T(x)\| \quad \text{for all} \quad x \in \partial_K D.$$

A repetition of the proof of Proposition 1 shows that $T^n(u) \in G$ for all n, and in fact that $\{T^n(u)\}$ is bounded away from $\partial_K D$. To see that (c) holds, let $p \in A$ and choose $\varepsilon > 0$ so that $B(p;\varepsilon) \cap K$ is a convex subset of G. Then T is a strictly contractive self-mapping of $B(p;\varepsilon) \cap K$. By an observation of Edelstein [6, (3.2)], $\{T^n(x)\}$ converges to p, the unique fixed point of T in $B(p;\varepsilon) \cap K$. ■

REMARK

Theorem 6 also can be used, in essentially the same way as above, to reduce Theorem 4 to a local application of Ishikawa's result [8].

REFERENCES

[1] BROWDER, F.E.: Nonlinear mappings of nonexpansive and accretive type in Ba-

nach spaces, Bull. Amer. Math. Soc. 73 (1967), 875-881.

[2] BROWDER, F.E.: Semicontractive and semiaccretive nonlinear mappings in Banach spaces, Bull. Amer. Math. Soc. 74 (1968), 660-665.

[3] BROWDER, F.E.: Nonlinear operators and nonlinear equations of evolution in Banach spaces, Proc. Symp. Pure Math. 18, pt. 2, Amer. Math. Soc., Providence, R.I., (1976).

[4] BRUCK, R.E.: A simple proof of the mean ergodic theorem for nonlinear contractions in Banach spaces, Israel J. Math. 32 (1979), 107-116.

[5] DEIMLING, K.: Zeros of accretive operators, Manuscripta Math. 13 (1974), 365-374.

[6] EDELSTEIN, M.: On fixed and periodic points under contractive mappings, J. London Math. Soc. 37 (1962), 74-79.

[7] GÖHDE, D.: Zum prinzip der kontraktiven Abbildung, Math. Nachr. 30 (1965), 251-258.

[8] ISHIKAWA, S.: Fixed points and iteration of nonexpansive mapping in a Banach space, Proc. Amer. Math. Soc. 59 (1976), 65-71.

[9] KANTOROVICH, L.V. and AKILOV, G.P.: Functional Analysis in Normed Spaces, Macmillan Co., N.Y., (1964).

[10] KATO, T.: Nonlinear semigroups and evolution equations, J. Math. Soc. Japan 19 (1967), 508-520.

[11] KIRK, W.A.: On zeros of accretive operators in uniformly convex spaces, Bollettino Un. Mat. Ital. (5) 17-A (1980), 249-253.

[12] KIRK, W.A.: A fixed point theorem for local pseudo-contractions in uniformly convex spaces, Manuscripta Math. 30 (1979), 89-102.

[13] KIRK, W.A. and MORALES, C.: Fixed point theorems for local strong pseudo-contractions, Nonlinear Analysis: Theory, Methods & Applications 4 (1980), 363-368.

[14] KIRK, W.A. and MORALES, C.: On the approximation of fixed points of locally

nonexpansive mappings, Canad. Math. Bull. (to appear).

[15] KIRK, W.A. and SCHÖNEBERG, R.: Mapping theorems for local expansions in metric and Banach spaces, J. Math. Anal. Appl. 71 (1979), 114-121.

[16] MARTIN, R.H., Jr.: Differential equations on closed subsets of a Banach space, Trans. Amer. Math. Soc. 179 (1973), 399-414.

[17] MENGER, K.: Untersuchungen über allgemeine Metrik, Math. Ann. 100 (1928), 75-163.

[18] MORALES, C.: On the fixed point theory for local k-pseudo-contractions, Proc. Amer. Math. Soc. 81 (1981), 71-74.

ASYMPTOTIC CENTERS IN PARTICULAR SPACES

By

E. LAMI DOZO

Département de Mathématique,
Université Libre de Bruxelles,
Belgium.

0. Introduction

We give a natural generalization of a fixed point theorem for nonexpansive maps with nonempty and compact values in ℓ^p spaces $(1 < p < \infty)$ to Orlicz sequence spaces ℓ^φ with φ convex or concave, in particular for ℓ^p $(0 < p \leq 1)$. We use an asymptotic center method. We point out that a similar result in L^φ does not occur.

1. Preliminaries

Let (X,d) be a metric space. For a given sequence $(x_n)_{n\in\mathbb{N}}$ in X we may associate two functionals

$$\overline{F}(x) = \limsup_n d(x_n,x)$$

$$\underline{F}(x) = \liminf_n d(x_n,x).$$

For C a subset of X, we call $\overline{c} \in C$ an *upper* (respectively $\underline{c}$ a *lower*) *asymptotic center in* C of $(x_n)_{n\in\mathbb{N}}$ if

$$\overline{F}(\overline{c}) = \inf_{x\in C} \overline{F}(x) \quad \text{(respectively } \underline{F}(\underline{c}) = \inf_{x\in C} \underline{F}(x)\text{)}.$$

The interest of this notion, due to M. Edelstein, in relation with fixed points for nonexpansive mappings comes from the following simple and essentially known fact.

PROPOSITION 1

Let $T : C \to C$ be a nonexpansive map (that is $d(Tx,Ty) \leq (d(x,y))$ and suppose that $(x_n)_{n\in\mathbb{N}}$ is a sequence of approximating fixed points (that is $\lim_n d(x_n,Tx_n) = 0$). If $(x_n)_{n\in\mathbb{N}}$ has an unique upper (respectively lower) asymptotic center $\overline{c}$ (respectively $\underline{c}$) in C, then $\overline{c}$ (respectively $\underline{c}$) is a fixed point of T. More generally the set of upper (respectively lower) asymptotic centers of $(x_{n_k})_{k\in\mathbb{N}}$ is T-invariant.

Proof

From $\lim_n d(x_n,Tx_n) = 0$ we deduce that $\overline{F}(x) = \limsup_n d(x,Tx_n)$ for all x in C, so $\overline{F}(\overline{c}) \leq \overline{F}(x)$, $\forall\, x \in C$, gives

$$\overline{F}(T\overline{c}) = \limsup_n d(T\overline{c},Tx_n) \leq \limsup_n d(x_n,\overline{c}) = \overline{F}(\overline{c}).$$

If $\overline{c}$ is unique, then $\overline{c} = T\overline{c}$, if not $T\overline{c}$ is also an upper asymptotic center. We proceed in the same way for lower asymptotic centers. ∎

Both functional $\overline{F}$ and $\underline{F}$ are 1-lipschitzian. When (X,d) is a Banach space, $(x_n)_{n\in\mathbb{N}}$ a bounded sequence, $\overline{F}$ is convex, so for C convex and weakly compact, $\overline{F}$ admits a minimum and the set of minimum points is convex and closed, that is the upper asymptotic centers form a nonempty closed convex subset of C. It is well known that upper asymptotic centers are unique in C a closed convex bounded subset of X a uniformly convex Banach space, such as ℓ^p or $L^p(0,1)$, $1 < p < \infty$. To identify these asymptotic centers we suppose the sequence $(x_n)_{n\in\mathbb{N}}$ *almost convergent in* C *to* c (cf. [2]) that is

$$\limsup_k d(x_{n_k},c) = \inf_{x\in C} \limsup_k d(x_{n_k},x)$$

for each subsequence $(x_{n_k})_{k\in\mathbb{N}}$ of $(x_n)_{n\in\mathbb{N}}$. In this case any upper asymptotic center is also a lower asymptotic center. One also has the following "compactness" result:

PROPOSITION 2 ([2]).

Let $(X,\|\cdot\|)$ be an uniformly convex Banach space and C a nonempty closed convex bounded subset of C. Then every sequence in C has a subsequence which is almost convergent in C to a unique c in C.

We recall that a Banach space $(X,\|\cdot\|)$ satisfies *Opial's condition* ([4]) if for each weakly convergent sequence $(x_n)_{n\in\mathbb{N}}$ to c,

$$\liminf_n \|x_n - c\| < \liminf_n \|x_n - x\|, \quad \forall\, x \in X \setminus \{c\};$$

in particular $\underline{F}(c) = \inf_{x\in X} \underline{F}(x)$.

PROPOSITION 3. ([5]).

$(X,\|\cdot\|)$ satisfies Opial's condition if and only if every weakly convergent sequence is almost convergent.

These are the ideas which we used in [5] to study asymptotic centers in some linear metric spaces. We apply them now to Orlicz sequence spaces.

2. Orlicz sequence spaces

An *Orlicz function* $\varphi : [0,\infty) \to [0,\infty]$ is an increasing function such that $\varphi(0) = 0$ with φ continuous at 0 and φ non identically 0. The associated *Orlicz sequence space* is defined by

$$\ell^\varphi = \left\{x : \mathbb{N} \to \mathbb{R};\ \sum_{n\in\mathbb{N}} \varphi\left(\frac{|x(n)|}{a}\right) < \infty \text{ for some } a > 0\right\}.$$

For simplicity we will only consider here two types of Orlicz functions

(CV) $\begin{cases} \varphi \text{ is continuous, convex and for all } R > 0, \text{ the map } k_R:(0,1) \to \mathbb{R}^+ \\ \text{defined by } k_R(\gamma) = \inf\limits_{0<t<R} \dfrac{\varphi(\gamma t)}{\varphi(t)} \text{ satisfies } 0 < k_R(\gamma) < 1 \text{ and } \lim\limits_{\gamma\to 1} k_R(\gamma)=1 \end{cases}$

(CC) φ is continuous, concave and unbounded.

The best known examples of Orlicz functions are $\varphi_p(t) = t^p$, $0 < p < \infty$ which satisfy one of these conditions. Under (CV) or (CC), $\ell^\varphi = \{x : \mathbb{N} \to \mathbb{R};\ \Sigma\, \varphi(|x(n)|) < \infty\}$ because in both cases φ fulfills the Δ_2-condition at $0 : \varphi(2t) \le M\varphi(t)$ for some $M > 0$ and all t near 0. In fact, (CV) implies the Δ_2-condition at 0:

$$\sup_{0<t<\varepsilon} \frac{\varphi(2t)}{\varphi(t)} = \frac{1}{\inf\limits_{0<s<2\varepsilon} \frac{\varphi(s/2)}{\varphi(s)}} = \frac{1}{k_R(1/2)} < \infty;$$

(CC) implies the Δ_2-condition at 0:

$$\varphi(2t) \leq 2\varphi(t) \text{ by the concavity of } \varphi.$$

We provide ℓ^φ with the quasi-norm

$$\|x\|_\varphi = \begin{cases} \inf\left\{a > 0;\ \sum\limits_{n\in\mathbb{N}} \varphi\left(\frac{|x(n)|}{a}\right) \leq 1\right\} & \text{if } \varphi \text{ satisfies (CV)} \\ \sum\limits_{n\in\mathbb{N}} \varphi(|x(n)|) & \text{if } \varphi \text{ satisfies (CC)} \end{cases}$$

We have that $(\ell^\varphi, \|\cdot\|_\varphi)$ is a complete linear metric space (that is an F-space) which is a Banach space when φ satisfies (CV). In other cases $(\ell^\varphi, \|\cdot\|_\varphi)$ may have a non-locally convex topological vector space topology (cf. [3], [8], [9]).

THEOREM 1

Let φ satisfy (CV) or (CC). Every bounded sequence in $(\ell^\varphi, \|\cdot\|_\varphi)$ has a subsequence which is pointwise and almost convergent in ℓ^φ to a unique c of ℓ^φ.

Proof

Let $(x_k)_{k\in\mathbb{N}}$ be a bounded sequence, in particular $\sup\limits_{k,n} |x_k(n)| < \infty$ because φ is increasing and unbounded. Using the Cantor diagonal method we construct a subsequence, still denoted $(x_k)_{x\in\mathbb{N}}$ such that $\lim\limits_k x_k(n) = c(n)$ exists and $c \in \ell^\varphi$.

The natural projections $P_N : \ell^\varphi \to \ell^\varphi : x \to P_N x$ defined by $(P_N x)(n) = x(n)$ if $n \leq N$, 0 if not, satisfy the estimate

$$(*) \quad \begin{cases} \forall\ 0 < \varepsilon < 1/2,\ \exists\ r > 0 \text{ such that } \forall\ N \geq 1 \\ \|x - P_N x\|_\varphi = 1,\ \|P_N x\|_\varphi > \varepsilon \Rightarrow \|x\|_\varphi \geq 1 + r \end{cases}$$

when φ satisfies (CV):

$$\|P_N x\|_\varphi > \varepsilon \quad \text{gives} \quad \sum_1^N \varphi\left(\frac{|x(n)|}{\varepsilon}\right) > 1,$$

$$\|x - P_N x\|_\varphi = 1 \quad \text{gives} \quad \sum_{N+1}^\infty \varphi\left(\frac{|x(n)|}{1-\delta}\right) > 1, \quad \forall\ 0 < \delta < 1/2.$$

$$\sum_1^\infty \varphi\left(\frac{|x(n)|}{1+r}\right) = \sum_1^N \varphi\left(\frac{|x(n)|}{\varepsilon}\right) \frac{\varphi(|x(n)|/1+r)}{\varphi(|x(n)|/\varepsilon)} + \sum_{N+1}^\infty \varphi\left(\frac{|x(n)|}{1-\delta}\right) \frac{\varphi(|x(n)|/1+r)}{\varphi(|x(n)|/1-\delta)}$$

$$\geq \left[\sum_1^N \varphi\left(\frac{|x(n)|}{\varepsilon}\right)\right] \inf_{0<t<R} \frac{\varphi(t/1+r)}{\varphi(t/\varepsilon)} + \left[\sum_{N+1}^\infty \varphi\left(\frac{|x(n)|}{1-\delta}\right)\right] \inf_{0<t<R} \frac{\varphi(t/1+r)}{\varphi(t/1-\delta)}$$

$$\geq \inf_{0<t<R} \frac{\varphi(t/1+r)}{\varphi(t/\varepsilon)} + \inf_{0<t<R} \frac{\varphi(t/1+r)}{(t/1-\delta)}, \quad \forall\ 0 < \delta < 1/2$$

$$= \inf_{0<s<R/\varepsilon} \frac{\varphi(s\varepsilon/1+r)}{\varphi(s)} + \inf_{0<s<R/1-\delta} \frac{\varphi(s(1-\delta)/1+r)}{\varphi(s)}, \quad \forall\ 0 < \delta < 1/2.$$

Taking $\delta = r$, we have $0 < \frac{1-r}{1+r} < 1$. As r tends to 0,

$$0 < k_R\left(\frac{1-r}{1+r}\right) = \inf_{0<t<R} \frac{\varphi((1-r)t/1+r)}{\varphi(t)} < 1$$

tends to 1 by (CV). Also $k_R(\varepsilon/1+r) \geq k_R(\varepsilon/2)$ for r small enough. We can now find $0 < r = r(\varepsilon) < 1/2$ such that, $k_R(\varepsilon/2) + k_R(\frac{1-r}{1+r}) > 1$, thus $\sum_1^\infty \varphi(|x(n)|/(1+r)) >$ which gives $\|x\|_\varphi \geq 1+r$, that is (*) is proved.

From (*) we deduce that

$$(^{**}) \qquad \liminf_k \|x_k - c\|_\varphi < \liminf_k \|x_k - x\|_\varphi \qquad \forall\ x \neq c$$

as in the proof of theorem (2.1) in [4] and the same inequality holds for any subsequence, so x_k is almost convergent only to c. When φ satisfies (CC) a direct proof of the almost convergence to c is as follows:

$$\|x_k - x\|_\varphi = \sum_1^\infty \varphi(|x_k(n) - x(n)|) = \sum_1^N \varphi(|x_k(n) - x(n)|) + \sum_{N+1}^\infty \varphi(|x_k(n) - x(n)|)$$

$$\geq \sum_1^N \varphi(|x_k(n) - x(n)|) + \sum_{N+1}^\infty \varphi(|x_k(n) - c(n)|) - \sum_{N+1}^\infty \varphi(|x(n) - c(n)|)$$

$$= \sum_1^N \varphi(|x_k(n) - x(n)|) + \sum_1^\infty \varphi(|x_k(n) - c(n)|) - \sum_1^N \varphi(|x_k(n) - c(n)|)$$

$$- \sum_{N+1}^{\infty} \varphi(|x(n) - c(n)|)$$

because $\varphi(a+b) \le \varphi(a) + \varphi(b)$. Then

$$\limsup_k \|x_k - x\|_\varphi \ge \sum_1^N \varphi(|c(n) - x(n)|) + \limsup_k \|x_k - c\|_\varphi - \sum_{N+1}^{\infty} \varphi(|x(n) - c(n)|)$$

because φ is continuous. Finally, for N tending to ∞, we obtain

$$\limsup_k \|x_k - x\|_\varphi \ge \|c - x\|_\varphi + \limsup_k \|x_k - c\|_\varphi .$$

The same inequality holds for a subsequence of $(x_k)_{k\in\mathbb{N}}$ because $\lim_k x_k(n) = c(n)$ also holds. ■

3. The main Result

Let us denote by D the Hausdorff metric on the nonempty subsets of a metric space (X,d). It is known that a map $T: C \to 2^C$ which is nonexpansive with respect to the Hausdorff metric (that is $D(Tx,Ty) \le d(x,y)$), nonempty and compact valued, has a fixed point when C is a bounded closed convex subset of an uniformly convex Banach space ([7]). A proof can be given using proposition 1, 2 ([2]). In particular this holds in $C \subset \ell^p$, $1 < p < \infty$. In this space C is bounded starshaped and compact for the topology of pointwise convergence. We generalize this result to ℓ^φ.

THEOREM 2

Let C be a bounded and starshaped subset of $(\ell^\varphi, \|\cdot\|_\varphi)$ which is compact for the topology of pointwise convergence. Let φ satisfy either (CC) or (CV) and $\sup_{0<t\le M} \varphi(\lambda t)/\varphi(t) < 1$ for each $M > 0$. Then each nonexpansive mapping $T: C \to 2^C$ with nonempty and compact values has a fixed point $c \in Tc$.

Proof

We shall construct a sequence $(x_n)_{n\in\mathbb{N}}$ of approximating fixed points (that is $\lim_n D_\varphi(x_n, Tx_n) = 0$) which we may suppose almost convergent to $c \in C$ in ℓ^φ by theorem 1 and also that $\lim_n \|x_n - c\|_\varphi$ exists. As $D_\varphi(Tc,Tx_n) \le \|c - x_n\|_\varphi$, we can find $c_n \in Tc$ such that $\lim\sup_n \|x_n - c_n\|_\varphi \le \lim_n \|c - x_n\|_\varphi$. Since Tc is compact we may suppose, by passing to a subsequence, that $\exists\, c' \in Tc$ such that

$$\lim_n \|c_n - c'\|_\varphi = 0.$$

Finally

$$\inf_{y\in\ell^\varphi} \lim\sup_n \|x_n - y\|_\varphi \le \lim\sup_n \|x_n - c'\|_\varphi = \lim\sup_n \|x_n - c_n\|_\varphi \le \lim_n \|c - x_n\|_\varphi$$

$$= \inf_{y\in\ell^\varphi} \lim\sup_n \|x_n - y\|_\varphi,$$

so $\lim\sup_n \|x_n - c'\|_\varphi = \lim_n \|x_n - c\|_\varphi$ with $c' \in Tc$. From the fact that c is unique, we deduce that $c' = c \in Tc$. The sequence of approximating fixed points is constructed as follows: Let $y_o \in C$ satisfy $(1 - \lambda)y_o + \lambda x \in C$, $\forall\, x \in C$, $\forall\, 0 \le \lambda < 1$, then the map

$$T_\lambda : C \to 2^C : x \to (1 - \lambda)y_o + \lambda Tx$$

is a strict contraction: $D_\varphi(T_\lambda x, T_\lambda y) \le h(\lambda)\ \|x - y\|_\varphi$, $0 \le h(\lambda) < 1$. In fact, because $\|\cdot\|_\varphi$ is a norm when φ satisfies (CV), we have that

$$D_\varphi(T_\lambda x, T_\lambda y) \le \lambda D_\varphi(Tx,Ty) \le \lambda\|x - y\|_\varphi;$$

when φ satisfies (CC), we get

$$D_\varphi(T_\lambda x,\ T_\lambda y) = \max\left\{\sup_{v\in T_\lambda y}\ \inf_{u\in T_\lambda x} \|u - v\|_\varphi,\ \sup_{u\in T_\lambda x}\ \inf_{v\in T_\lambda y} \|u - v\|_\varphi\right\}$$

$$= \max\left\{\sup_{v\in Ty}\ \inf_{u\in Tx} \|\lambda(u - v)\|_\varphi,\ \sup_{u\in Tx}\ \inf_{v\in Ty} \|\lambda(u - v)\|_\varphi\right\}$$

$$\le h(\lambda)D_\varphi(Tx,Ty) \le h(\lambda)\|x - y\|_\varphi, \quad 0 \le h(\lambda) < 1$$

because

$$\|\lambda(u - v)\|_\varphi = \sum_{n\in\mathbb{N}} \varphi(\lambda|u(n) - v(n)|) \le \left(\sup_{0<t\le M} \frac{\varphi(\lambda t)}{\varphi(t)}\right)\left(\sum_{n\in\mathbb{N}} \varphi(|u(n) - v(n)|) = h(\lambda)\|u - v\|_\varphi\right)$$

with $0 \le h(\lambda) = \sup_{0<t\le M} \dfrac{\varphi(\lambda t)}{\varphi(t)} < 1$. Since

C is closed in ℓ^{φ}, it is complete, so T_λ has a fixed point $x_\lambda \in T_\lambda x_\lambda$, that is $x_\lambda = (1-\lambda)y_o + \lambda w_\lambda$ with $w_\lambda \in Tx_\lambda$. From

$$D_\varphi(x_\lambda, Tx_\lambda) = \inf_{u \in Tx_\lambda} \|x_\lambda - u\|_\varphi \leq \|(1-\lambda)(y_o - w_\lambda)\|_\varphi$$

we obtain, for $\lambda_n = 1 - \frac{1}{n}$ and $x_n = x_{\lambda_n}$: $\lim_n D(x_n, Tx_n) = 0$ that is the sequence $(x_n)_{n\in\mathbb{N}}$ is of approximating fixed points. ■

REMARK 1

Theorem 2 can be also proved as a consequence of theorem 4 in [5] but we prefer to isolate it from the general framework of F-spaces because the corresponding situation in

$$L^{\varphi}(0,1) = \left\{x : [0,1] \to \mathbb{R};\ x \text{ Lebesgue measurable and } \int_0^1 \varphi\left(\frac{|x(t)|}{a}\right) dt < \infty, \text{ for some } a > 0\right\}$$

is not similar to that one in ℓ^{φ}. We have learnt at this meeting that there exists a nonexpansive map from a convex weakly compact subset of $L^1(0,1)$ into itself without fixed points ([1]). This implies that our generalization does not work in L^{φ} as it does in ℓ^{φ}.

REMARK 2

The other types of Orlicz functions, especially when the Δ_2-condition is not satisfied will be studied elsewhere [6].

ACKNOWLEDGEMENT

We wish to thank Ph. Turpin for pointing out a missing hypothesis in theorem 2 necessary for our proof.

REFERENCES

[1] ALSPACH, D.: To appear.

[2] GOEBEL, K.: On a fixed point theorem for multivalued nonexpansive mappings, Ann. Univ. Marie Curie. Lublin, Poland.

[3] KRASNOSEL'SKII, M.A. and RUTICKII, Ya.B.: Convex functions and Orlicz spaces, P. Noordhoff. Groningen, The Netherlands, (1961).

[4] LAMI DOZO, E.: Multivalued nonexpansive mappings and Opial's condition. Proc. Am. Math. Soc. 38 (1973), 286-292.

[5] LAMI DOZO, E.: Centres asymptotiques dans certains F-espaces. Boll. Un. Mat. Ital. (5), 17B (1980), 740-747.

[6] LAMI DOZO, E.: In preparation.

[7] LIM, T.C.: A fixed point theorem for multivalued nonexpansive mappings in an uniformly convex Banach space. Bull. Am. Math. Soc. 80 (1974), 1123-1126.

[8] ROLEWICZ, S.: Metric linear spaces. Monografic Matematyczne 56. Warszawa, (1972).

[9] TURPIN, P.: Convexités dans les espaces vectoriels topologiques généraux. Dissertationes Mathematicae CXXXI, Polska Akad. Nauk. Warszawa, (1976).

WU-LIKE CLASSES AND GENERALIZED PETERSON-STEIN CLASSES

By

T.Y. LIN

Louisiana State University

and

University of South Carolina at Aiken

1. INTRODUCTION

In 1968, G.E. Bredon introduced a new cohomology operation for studying involutions [2]. In 1973, M. Nakaoka used this operation to study the Lefschetz "fixed" point theorem for the equivariant points of the map of manifolds with free or trivial involutions [15], [16]. He named the operation Bredon operation. In 1974, A. Hattori extended the study to general involutions and discovered some characteristic classes which reflect some phenomena of fixed point set [9], [10]. Totally unaware of Nakaoka's and Hattori's works, P.E. Conner and E.Y. Miller set forth for a systematic study of the Bredon operation on free involutions [7]. (However, there is some philosophical difference between the viewpoint of [7] and that of the mere restriction of the Bredon operation to the free involution). To reflect the geometry, they called the operation equivariant self-intersection. This viewpoint provides a link with Browder-Livesay invariant [3], [12]. To study the effect of the Bredon operation on a manifold with free involution, Conner-Miller introduced a new type of characteristic classes θ_k. θ_k are defined in the same spirit as the tangential Wu classes via Steenrod squares. We call them Conner-Miller classes (abb. CM-classes). Using CM-classes θ_k, they studied the (not necessarily free) involution and introduced the Bredon classes B_k for any involution. These Bredon classes turned out to be the characteristic classes of Hattori. We will call them Bredon-Hattori classes (abb. BH-classes). One of our central efforts is to establish various vanishing theorems for CM-classes θ_k and BH-classes B_k. Our study reflects that θ_k and B_k are quite sensitive invariants of the involution. In fact, they reflect some phenomena of the G-structure on a manifold [5], [20] (see (3.9)-(3.11)).

This paper is a response to some questions raised in the introduction of [7].

In [7], they established a remarkable fact that the Bredon operation of a bundle involution is determined by the Peterson-Stein secondary classes when the bundle ψ has vanishing top Stiefel-Whitney classes. In this note we find the similar formalism for general bundles (without any assumption on the Stiefel-Whitney classes). That is, the Bredon operation of the bundle involution of any sphere bundle $S(\psi)$ is determined by certain classes. We name them generalized Peterson-Stein classes (abb. GPS-classes). These classes are determined, up to certain ambiguity, by the Steenrod operation on the cohomology of the sphere bundle. If the bundle has vanishing top Stiefel-Whitney class, then the generalized Peterson-Stein classes reduce to Peterson-Stein secondary classes and the ambiguity disappears (see (5.3)).

The author would like to express his warmest gratitude to Professor P.E. Conner for his kind advice, guidance and generous sharing of his time, insight and knowledge. His ideas and observations are almost everywhere in the paper. The author would like to thank the Department of Mathematics for opportunity for visiting Louisiana State University.

We shall quote some sample results:

(3.9) *THEOREM*

Let M^m be an orientable manifold with free involution T. Then,

(i) if T preserves the orientation, then $\theta_k = 0$, k is even and $k > \frac{m+1}{2}$

(ii) if T reverses the orientation, then $\theta_k = 0$, k is odd and $k > \frac{m+1}{2}$

Using this vanishing theorem we have

(4.3) *THEOREM*

Let V^n be an orientable manifold with involution T (not necessarily free).

(i) If T preserves the orientation, then $B_{2k+1} = 0$, for all k.

(ii) If T reverses the orientation, then $B_{2\ell} = 0$, for all ℓ.

In particular, if T is the identity, $Wu_{2k+1} = 0$, for all k.

Next is a restatement of theorem (5.3).

(5.3) *THEOREM*

Let $\psi \to X$ be an $(m+1)$-plane bundle. Let $t \in H^{m+\ell}(S(\psi))$. Then there is a unique class $\Phi_k(t) \in H^{m+2\ell+k}(X)$, for each k such that

$$Q(t) = \sum_{k=0}^{m} \pi^*(\Phi_k(t))c^{m-k}.$$

Moreover,

$$p^*\Phi_k(t) = Sq^{\ell+k}t + \sum_{j=0}^{\ell-1}\sum_{s=0}^{\ell-j-1} p^*w_{k+\ell-j-s}\overline{w}_s Sq^j t + tp^*(p_!(t)w_k)$$

where $\Phi_k(t)$ is called the generalized Peterson-Stein class and w_k is the Stiefel-Whitney class.

This theorem implies that Q can be described by Sq, up to some ambiguity.

$\Phi_j(t)$ is quite mysterious, we only have some partial explanation (see (5.11)).

As applications, we would like to quote a Borsuk-Ulam type theorem.

(4.12) *COROLLARY*

Let S^m be a mod 2 homology sphere with a free involution T. Let $f : S^m \to V^n$ be a map $(m > n)$, then $\dim A(f) \geq m - n$, where $A(f) = \{x : f(x) = f(Tx)\}$.

2. Bredon operations

Let (T,X) be a fixed point free involution. Let $c \in H^1(X/T)$ be the Stiefel-Whitney class of 0-spere bundle $\nu : X \to X/T$. We shall call c the fundamental cohomology class of the involution. Let us recall the quadratic construction: let (A,S^∞) be the antipodal map of infinite dimensional sphere. Its orbit space S^∞/A is the infinite dimensional real projective space RP^∞. Let τ be the free involution acting on $X \times X \times S^\infty$ by

$$\tau(x_1,x_2,s) = (x_2,x_1,As).$$

The orbit space, denoted by $X^2 \times_\tau S^\infty$, is the quadratic construction. This construction is functorial: let $f : X \to Y$ be a map. Then

$$f^2 \times_\tau 1 : X^2 \times_\tau S^\infty \to Y^2 \times_\tau S^\infty$$

is a map of quadratic constructions.

Since S^∞ is contractible, the projection

$$\nu : X^2 \times S^\infty \to X^2 \times_\tau S^\infty$$

gives rise to $\nu^* : H^*(X^2 \times_\tau S^\infty) \to H^*(X^2)$ and the transfer $tr : H^*(X^2) \to H^*(X^2 \times_\tau S^\infty)$.

According to Steenrod [19, ch VII] or [10], there is an external cohomology operation

$$P : H^k(X) \to H^{2k}(X^2 \times_\tau S^\infty)$$

satisfying the following properties

(a) $P(f^*\alpha) = (f^2 \times_\tau 1)^* P(\alpha)$

(b) $\nu^* P(\alpha) = \alpha \times \alpha$ (cross product)

(c) $P(\alpha + \beta) = P(\alpha) + P(\beta) + tr\,(\alpha \times \beta)$

(d) $P(\alpha\beta) = P(\alpha)P(\beta)$

(e) $(d \times_\tau 1)^* P(\alpha) = \sum_{i=0}^{k} Sq^i\alpha \times W^{k-i}, \quad \alpha \in H^k(X).$

Suppose now that (T,X) has a fixed point free involution. The diagonal involution, denoted by T again, $(T, X \times S^\infty)$ has an equivariant embedding

$$e : X \times S^\infty \to X \times X \times S^\infty$$

given by $e(x,s) = (x,Tx,s)$. This induces an embedding of the orbit spaces

$$E : X \times_T S^\infty \to X \times X \times_\tau S^\infty.$$

In [2], Bredon defines an operation

$$Q : H^r(X) \to H^r(X \times_T S^\infty)$$

by

$$Q(\alpha) = E^* (P(\alpha)) .$$

Since T is a free involution, the equivariant projection

$$X \times S^\infty \to X$$

induces a fibration

$$p : X \times_T S^\infty \to X/T$$

with fibre S^∞, that is, p is a homotopy equivalence. By regarding p^* as an identification, we will regard the Bredon operation as an operation

$$Q : H^k(X) \to H^{2k}(X/T).$$

Corresponding to (a) - (e) above, we have

(2.1) *THEOREM*

Let (T,X) be a fixed point free involution. Then there is an operation

$$Q : H^k(X) \to H^{2k}(X/T)$$

with the following properties

1) Q commutes with homomorphisms induced by equivariant maps;

2) $\nu^*Q(\alpha) = \alpha T^*\alpha$;

3) $Q(\alpha + \beta) = Q(\alpha) + Q(\beta) + tr(\alpha T^*\beta)$;

4) $Q(\alpha\beta) = Q(\alpha)Q(\beta)$;

5) For $\alpha \in H^k(X/T)$, $Q(\nu^* (\alpha)) = \sum_{j=0} c^{k-j} Sq^j\alpha$.

Next, we will recall the relations between the Bredon operation and Steenrod squares. The following theorem is extracted from [7]; its proof follows immediately from the computation of Sq^i on $H^*(X^2 \times_T S^\infty)$ (see, for example, [17], [13]).

(2.2) *THEOREM*

Let (T,X) be a fixed point free involution. Then for $\alpha \in H^k(X)$,

$$Sq^j Q(\alpha) = \sum_{0 \le i \le j/2} \begin{Bmatrix} k-i \\ j-2i \end{Bmatrix} c^{j-2i} Q(Sq^i \alpha) + \sum_{0 \le i \le j/2} [Sq^{j-i}\alpha, Sq^i \alpha]$$

where $[x,y] = tr(xT^*y)$.

3. WU-LIKE CLASSES FOR FREE INVOLUTIONS

In [7], P.E. Conner and E.Y. Miller introduced some Wu-like θ_k for each free involution. We will call the θ_k Conner-Miller class. These classes reflect some effect of the Bredon operation on manifolds. In this section we will study the right action of Sq^i on θ_k, and use the relations to establish some vanishing theorems. For convenience of expressing the right actions of Sq^i, we introduce the generalized Conner-Miller (abb. GCM-classes). However, GCM-classes are not an essential generalization. We can see from the propositions below that all the GCM-classes are all arised from ordinary CM-classes of some other manifolds.

Let (T,M) be a fixed point free involution. Let $\nu : M \to M/T$ be the projection to the orbit space. Then there is the Smith-Gysin exact sequence [14, (12.3)]

$$\xrightarrow{\nu^*} H^j(M) \xrightarrow{tr} H^j(M/T) \xrightarrow{\cup c} H^{j+1}(M/T) \xrightarrow{\nu^*} H^{j+1}(M) \longrightarrow$$

where $tr : H^j(M) \to H^j(M/T)$ is the transfer homomorphism and $\cup c$ is the cup product with the Stiefel-Whitney class of the 0-sphere bundle $\nu : M \to M/T$. (c is the fundamental cohomology of the involution.)

From the remarks above, it is clear that $\cup c \cdot tr = 0$ and hence the correspondence

$$\alpha \mapsto c^r Q(\alpha)$$

is linear if $r > 0$. In fact, we can have more, let $z \in H^s(X/T)$, then the correspondence

$$\alpha \mapsto c^r zQ(\alpha)$$

is linear if $r > 0$. By this observation, we can define the generalized θ_k.

(3.1) *DEFINITION*

Let (T,M^n) be a fixed point free involution. If $\frac{n}{2} < k \leq n$, then, for a fixed $z \in H^s(M/T)$, the class $\theta_k(z)$ is the unique cohomology class for which

$$<\theta_k(z)\ \alpha,\ \sigma(M^n)> = <c^{2k-n-s} zQ(\alpha),\ \sigma(M^n/T)>$$

for all $\alpha \in H^{n-k}(M^n)$.

It is clear that $\theta_k(0) = 0$, $\theta_k(c^j) = \theta_k$, the Conner-Miller class. We shall call $\theta_k(z)$ generalized Conner-Miller class (abb. GCM-class). Before we use the GCM-classes to express the right action of Sq^i, we shall see that $\theta_k(z)$ comes from some θ_k. First we need to recall the notion of Umkehr homomorphism. Let $f : V \to M$ be a map between two manifolds, then the Umkehr homomorphisms (or, Gysin homomorphism) $f_!$ is defined as follows: for $x \in H^*(V)$, we define $f_!(x)$ to be the unique cohomology class for which

$$<xf^*(y),\ \sigma(V)> = <f_!(x)y,\ \sigma(M)>$$

for all y in $H^*(M)$.

Now, we have the following

(3.2) *PROPOSITION*

Let $f : (T_1,M_1^n) \to (T,M^m)$ be an equivariant map between two fixed point free involutions on closed manifolds. Let $F : M_1^n/T_1 \to M^m/T$ be the induced map on orbit spaces. If $z \in H^s(M_1^n/T_1)$ then, for all $k > \frac{n+s}{2}$

$$\theta_{k+(n-m)}(F_!(z)) = f_!(\theta_k(z)) .$$

Proof

For $\alpha \in H^{n-k}(M^m)$, we consider

$$<\theta_k(z)f^*(\alpha), \sigma(M_1^n)> = <c_1^{2k-n-s} zQ_1(f^*(\alpha)), \sigma(M_1^n/T_1)>.$$

where Q_1 is the Bredon operation on (T_1, M_1^n) and c_1 is the fundamental cohomology class of involution (T_1, M_1^n). Thus,

$$\begin{aligned}<f_!\theta_k(z)\alpha, \sigma(M^n)> &= <\theta_k(z)f^*(\alpha), \sigma(M_1^n)> \\ &= <c_1^{2k-n-s} zQ(f^*(\alpha)), \sigma(M_1^n/T_1)> \\ &= <F^*(c^{2k-n-s}Q(\alpha))z, \sigma(M_1/T_1)> \\ &= <c^{2k-n-s}Q(\alpha)F_!(z), \sigma(M/T)> \\ &= <\theta_{k+(m-n)}(F_!(z))\alpha, \sigma(M)>.\end{aligned}$$

This gives the desired formula. ■

(3.3) *PROPOSITION*

For any $z \in H^s(M/T)$, there is a fixed point free involution (T_1, M_1^{m-s}) and an equivariant map $f : (T_1, M_1^{m-s}) \to (T, M^m)$ with $F_!(1) = z$, and

$$f_!(\theta_k) = \theta_{k+s}(z).$$

Proof

From [6], it is clear that for every $z \in H^s(M/T)$ there is a manifold (T_1, M_1) such that $F_!(1) = z$. This proves the proposition. ■

REMARK

If $z \in H^1(M/T)$, then we can say there is, in fact, a closed T-invariant submanifold $M_1^{m-1} \subset M^m$ with $I_!(1) = z$. If $z \in H^s(M/T)$ and is the reduction mod 2 of an integral class, then there is a closed T-invariant submanifold $M_1^{m-2} \subset M^m$ with

$I_!(1) = z$. This proposition indicates that the generalized characteristic class $\theta_k(z)$ is arised from θ_k. However, special choices of z may have some interests, we use them to study the Steenrod actions on θ_k.

Let us recall the right action of the Steenrod squares: let $u \in H^i(M^m)$. Following Adams [1], we define uSq^j to be the unique element determined by

$$<(uSq^j)v, \sigma(M^m)> = <u(Sq^j v), \sigma(M^n)>$$

for all $v \in H^{m-i-j}(M^m)$. According to [4]

$$uSq^j = \sum_{j=0}^{j} Wu_t(M^m)\,\chi(Sq^{j-t})u$$

where χ is the canonical antiautomorphism of Steenrod algebra.

(3.4) *PROPOSITION*

Let $Wu_t = Wu_t(M/T)$ be the Wu class of the quotient. Then

$$\sum_{0\leq i\leq j/2} \begin{Bmatrix} m-k-i \\ j-2i \end{Bmatrix} \theta_{k-i}Sq^i = \sum_{t=0}^{j} \begin{Bmatrix} 2k-m-j \\ j-t \end{Bmatrix} \theta_k(Wu_t)$$

where $2k-m-j > 0$.

<u>*Proof*</u>

From definition of θ_k, we have

$$<\theta_{k-i} \cdot Sq^i\alpha, \ \sigma(M^m)> = <c^{2(k-i)-m}Q(Sq^i\alpha), \ \sigma(M^m/T)>.$$

Multiply $\begin{Bmatrix} m-k-i \\ j-2i \end{Bmatrix}$ to this equation and sum up, we have

$$<\sum_{0\leq i\leq j/2} \begin{Bmatrix} m-k-i \\ j-2i \end{Bmatrix} \theta_{k-i}Sq^i\alpha, \ \sigma(M^m)> = <\sum_{0\leq i\leq j/2} \begin{Bmatrix} m-k-i \\ j-2i \end{Bmatrix} c^{2(k-i)-m}Q(Sq^i\alpha), \ \sigma(M^m/T)>.$$

By (2.2) this becomes

$$<c^{2k-m-j}Sq^jQ(\alpha), \ \sigma(M^m/T)>.$$

By the formula of right action of Sq^i on c^{2k-m-j}, this reduces to

$$< \sum_{t=0}^{j} Wu_t \chi(Sq^{j-t}) c^{2k-m-j} Q(\alpha),\ \sigma(M^m/T)> = < \sum_{t=0}^{j} Wu_t \left\{ {2k-m-j \atop j-t} \right\} c^{2k-m-t} Q(\alpha),\ \sigma(M^m/T)>$$

$$= < \sum_{t=0}^{j} \left\{ {2k-m-j \atop j-t} \right\} \theta_k(Wu_t)\alpha,\ \sigma(M^m)>$$

for all $\alpha \in H^{n-k}(M^m)$. So we have proved the proposition. ∎
We quote two special cases as corollary.

(3.5) *COROLLARY*

Let $Wu_t = Wu_t(M/T)$ again. Then

(i) $(k-1)\theta_k = \theta_k(Wu_1), \quad k > \frac{m+1}{2}$

(ii) $\theta_{k-1}Sq^1 = \left[\left\{ {m-k \atop 2} \right\} + \left\{ {2k-m-2 \atop 2} \right\}\right]\theta_k + (2k-m-2)\theta_k(Wu_1) + \theta_k(Wu_2), \quad k > \frac{m+2}{2}$

REMARK

$$\left\{ {m-k \atop 2} \right\} + \left\{ {2k-m-2 \atop 2} \right\} = \begin{cases} 1 \pmod 2, & \text{if } k \equiv 1 \pmod 4 \\ 0 \pmod 2, & \text{if } k \equiv -1 \pmod 4 \end{cases}.$$

This can be seen directly or we use Lemma (3.8) below.

With suitable assumption on $Wu_t = Wu_t(M/T)$, we can get various vanishing theorems. For lower t, the conditions: $Wu_t = 0$ or $Wu_t = c^t$, arise quite naturally from geometry. We will elaborate them later. For now, we will state the general vanishing theorem.

(3.6) *THEOREM*

Let M be an m-dimensional manifold. Suppose $Wu_t(M/T) = 0$ for $t \leq 2^a$. Then, for some $b = 0,1,2,\ldots, a$

$$\theta_k(M) = 0, \not\equiv k \equiv -1 \pmod{2^{b+1}}, \not\equiv k > \frac{m}{2} + 2^{b-1}.$$

Proof

We will prove this by induction on a: for a = 0. By (3.5), we have $(k-1)\theta_k(M) = 0$. That is,

$$\theta_k(M) = 0, \quad k \not\equiv -1 \pmod{2^{0+1}}, \quad k > \frac{m}{2} + 2^{0-1}.$$

Therefore, for $b = 0,1,2,\ldots,a+1$, we only need to show that (when $b = a+1$)

$$\theta_k(M) = 0, \quad k \not\equiv -1 \pmod{2^{a+2}}, \quad \text{and} \quad k > \frac{m}{2} + 2^a.$$

However, from the case $b = a$, we know the only possibly non-zero classes are

1) $\theta_k(M)$, $k \equiv 2^{a+1}-1 \pmod{2^{a+2}}$, $k > \frac{m}{2} + 2^a$

2) $\theta_k(M)$, $k \equiv -1 \pmod{2^{a+2}}$, $k > \frac{m}{2} + 2^a$.

To complete the induction we only need to show that, for case 1), $k \equiv 2^{a+1}-1 \pmod{2^{a+2}}$

$$\theta_k(M) = 0, \quad k > \frac{m}{2} + 2^a.$$

From (3.4), we have

$$\left(\begin{Bmatrix} m-k \\ 2^{a+1} \end{Bmatrix} + \begin{Bmatrix} 2k-m-2^{a+1} \\ 2^{a+1} \end{Bmatrix}\right) \theta_k + \sum_{1\le t\le 2^a} \begin{Bmatrix} m-k-t \\ 2^{a+1}-2t \end{Bmatrix} \theta_{k-t} Sq^t = 0.$$

By Lemma (3.8) below, we have

3) $$\theta_k + \sum_{1\le t\le 2^a} \begin{Bmatrix} m-k-t \\ 2^{a+1}-2t \end{Bmatrix} \theta_{k-t} Sq^t = 0, \quad k \equiv 2^{a+1}-1 \pmod{2^{a+2}}, \quad k > \frac{m}{2} + 2^a.$$

Next, observe that $k - t = 2^{a+1}-1-t \pmod{2^{a+2}}$ and hence

$$k - t \equiv 2^{a+1}-1-t \pmod{2^{a+1}}.$$

Thus for $1 \le t \le 2^a$

$$k - t \equiv 2^{a+1} - 1 - t \equiv -1 - t \not\equiv -1 \pmod{2^{a+1}}.$$

By induction hypothesis,

$$\theta_{k-1} = 0, \quad 1 \leq t \leq 2^a.$$

Therefore the relation 3) reduces to

$$\theta_k(M) = 0, \quad k \equiv 2^{a+1} - 1 \pmod{2^{a+2}}, \ k > \frac{m}{2} + 2^a.$$

Thus, from 1) and 2), we conclude that

$$\theta_k(M) = 0, \quad k \not\equiv -1 \pmod{2^{a+2}}, \quad k > \frac{m}{2} + 2^a.$$

This completes the induction. ■

(3.7) *COROLLARY*

Suppose $Wu_t(M/T) = 0$, $t > 0$. Then the Conner-Miller class θ_k are zero if k does not satisfy the following conditions: for $b = 0,1,2...$

$$k \equiv -1 \pmod{2^b}$$

$$\frac{m}{2} + 2^b \geq k > \frac{m}{2} + 2^{b-2} .$$

(When $b = 0$, $k \equiv -1 \pmod 1$ means k is an integer.) In other words, the only possibly non zero θ_k are indexed by those k satisfying the condition above.

REMARK

1) For sphere of dimension $2^i - 1$ with antipodal map, it satisfies the conditions of (3.6) and has non-zero Conner-Miller class at dimension $2^i - 1$. So (3.6) is the best possible for some dimension m.

2) The vanishing theorem really indicates that Conner-Miller classes are rather sensitive. They reflect how T^* moves G-structures of manifolds [8]. (See (3.11)).

(3.8) *LEMMA*

For $k > \frac{m}{2} + 2^{a+1}$ and $k \equiv -1 \pmod{2^{a+1}}$, $k \not\equiv -1 \pmod{2^{a+2}}$,

$$\left\{\begin{matrix} m-k \\ \\ 2^{a+1} \end{matrix}\right\} + \left\{\begin{matrix} 2k-m-2^{a+1} \\ \\ 2^{a+1} \end{matrix}\right\} \equiv 1 \pmod{2}.$$

Proof

Let $\ell = m-k$. Then $2k-m-2^{a+1} = k - \ell - 2^{a+1}$. Write $\ell = s2^{a+2} + \sum_{h=0}^{a+1} n_h 2^h$, where $n_h = 1$ or 0. Then

$$k - 2^{a+1} = (t+1)2^{a+2} - 2^0 = t2^{a+2} + 2^{a+1} + 2^a + \ldots + 2^0.$$

Thus $k - \ell - 2^{a+1} = (t-s)2^{a+2} + \sum_{h=0}^{a+1} (1-n_h)2^h$. By [19,I,(2.6)]

$$\left\{\begin{matrix} m-k \\ \\ 2^{a+1} \end{matrix}\right\} \equiv 1 \pmod{2} \text{ if and only if } n_{a+1} = 1$$

and

$$\left\{\begin{matrix} 2k-m-2^{a+1} \\ \\ 2^{a+1} \end{matrix}\right\} \equiv 1 \pmod{2} \text{ if and only if } n_{a+1} = 0.$$

Thus we have

$$\left\{\begin{matrix} m-k \\ \\ 2^{a+1} \end{matrix}\right\} + \left\{\begin{matrix} 2k-m-2^{a+1} \\ \\ 2^{a+1} \end{matrix}\right\} \equiv 1 \pmod{2}.$$

This proves the Lemma. ■

The next few theorems, we will examine the cases $Wu_1(M/T) = 0$ or c, and $Wu_2(M/T) = 0$ or c^2. The first one is the case $Wu_1(M/T) = 0$ and c.

(3.9) *THEOREM*

Let (T,M^n) be a fixed point free involution on orientable manifold. Then,

(i) if T preserves the orientation, then $\theta_k = 0$, k is even and $k > \frac{n+1}{2}$

(ii) if T reverses the orientation, then $\theta_k = 0$, k is odd and $k > \frac{n+1}{2}$.

REMARK

If $\frac{n+1}{2}$ is an integer, then $\theta_{\frac{n+2}{2}}$ is defined and is not covered by this theorem.

Proof of theorem (3.9)

If T preserves the orientation, then $Wu_1(M/T) = 0$. By (3.5), $(k-1)\theta_k = 0$ Thus we have (i).

If T reverses the orientation, then $Wu_1(M/T) = c$. By (3.5), $(k-1)\theta_k = \theta_k$. Thus we have (ii). ∎

Next results are on the case $Wu_1(M/T) = Wu_2(M/T) = 0$ and its generalization. Recall that a manifold is called spin-manifold if $W_1 = W_2 = 0$ (that is,

$$Wu_1(M) = Wu_2(M) = 0).$$

Now, if T preserves the spin-structure, M/T is still a spin-manifold, that is, $Wu_1(M/T) = Wu_2(M/T) = 0$. Thus we have

(3.10) *COROLLARY*

Let (T,M^m) be a fixed point free involution of a spin-manifold. Suppose T preserves the spin-structure. Then,

$$\theta_k(M^m) = 0, \quad k \not\equiv 3 \pmod 4, \quad k > \frac{n+2}{2}$$

and

$$\theta_{\frac{n+2}{2}}(M^m) = 0, \quad \text{if} \quad \frac{n+2}{2} \quad \text{is even integer.}$$

Let M be a manifold with (B,f)-structure in the sense of Stong [20; ch II]. M will be called an Wu_a-manifold if $f^*(Wu_i) = 0 \quad 0 < i \le 2^a$, where Wu_i are the universal Wu-classes. In particular, Wu_0-manifold is an orientable manifold, and Wu_1-manifold is a spin-manifold. Now, we have the following generalization of (3.10).

(3.11) *COROLLARY*

Let M be a Wu_a-manifold and T be a free involution preserving Wu_a-structure. Then,

$$\theta_k(M) = 0, \quad k \not\equiv -1 \pmod{2^{b+1}}, \quad k > \frac{m}{2} + 2^{b-1}$$

for some $b = 0, 1, \ldots, a$.

Next is a slightly odd case.

(3.12) *COROLLARY*

Let (T,M^m) be a fixed point free involution and assume that $Wu_1(M/T) = c$ and $Wu_2(M/T) = 0$. Then,

(i) if m is odd

$$\theta_k(M) = 0, \quad k \not\equiv 0 \pmod 4, \quad k > \frac{m+2}{2}$$

$$\theta_{\frac{m+2}{2}}(M) = 0, \quad \text{if} \quad \frac{m+2}{2} \quad \text{is odd.}$$

(ii) if m is even

$$\theta_k(M) = 0, \quad k \not\equiv 2 \pmod 4, \quad k > \frac{m+2}{2}$$

$$\theta_{\frac{m+2}{2}} = 0 \quad \text{if} \quad \frac{m+2}{2} \quad \text{is odd.}$$

Proof

This follows immediately from (3.5) and the remarks for θ_k. For $\theta_{\frac{m+2}{2}} = 0$, it follows from (3.9). ■

4. Bredon-Hattori classes and Borsuk-Ulam type theorems

This section is a study of θ_k for some manifolds which are product spaces. By considering the θ_k for $V \times S^m$, m large, we, following [7], introduce the Bredon-Hattori classes (it is called Bredon class in [7]; see §1) and establish some vanishing theorems. We also obtain some formula for θ_k for some other type of product spaces. We use them to establish some Borsuk-Ulam type theorems. Following [7], we define

(4.1) *DEFINITION*

Let (T,V^n) be an involution on a closed manifold (not necessarily free). With $m > n$ we define the Bredon-Hattori characteristic classes to be the unique cohomology class for which

$$\langle B_k \alpha, \sigma(V^n)\rangle = \langle c^{2k+m-n} Q(\alpha \otimes 1), \sigma(V^n \times_\tau S^m)\rangle$$

for all $\alpha \in H^{n-k}(V)$, where τ is the diagonal action of T and the antipodal map A.

These classes B_k can be expressed by the data of fixed point set. Let (T,V^n) be an involution. For each integer r, let F^r denote the union of all r-dimensional components of the fixed point set. (For any particular value r, the F^r may be empty). Let $\eta_r \to F^r$ be the normal bundle. Then we quote from [7].

(4.2) *THEOREM*

(1) $B_k = \sum_{r \geq n-k} i_!^r \, (Wu_{r+k-n} \, (\eta_r - F^r))$

(2) $B_* = Sq^{-1} i_! \, (W_*(F))$

(3) $B_k = 0$ if $k < \dim V - \dim F$ or $k > \frac{1}{2} \dim V$

(4) Suppose $B_{k_0} = 0$ for some k_0, then if $H^{n-k_0}(V) \to H^{n-k_0}(F)$ is onto it follows that $F^{n-k_0} = \phi$ and $Wu_{r+k_0-n}(\eta_r - F_r) = 0$, $r > n-k_0$.

(1) and (2) are used by A. Hattori as definition for B_k. (3) are vanishing theorems obtained by both [7] and [10]. (4) gives some geometric significante to the vanishing of B_k. We will give more vanishing theorems below.

(4.3) *THEOREM*

Let V^n be a manifold with involution T (not necessarily free). Then, if V^n is orientable and

(i) T^* preserves the orientation, then $B_k = 0$ for k odd

(ii) T^* reserves the orientation, then $B_k = 0$ for k even.

REMARK

When T is the identity, then (i) gives us the usual result that $Wu_k = 0$ for k odd if M is orientable.

Proof

By the definition of B_k, we see that

$$\theta_{k+m} = B_k \otimes \mu_m$$

where μ_m is the cofundamental class of odd dimensional sphere S^m (that is $m = 2\ell + 1$). Then we have (i) and (ii) by (3.9). ■

This theorem can be strengthened by using the arguments on bundle involutions.

(4.4) *PROPOSITION*

Let $\psi \to X^r$ be an $(m+1)$-plane bundle. Let $(T, S(\psi))$ be the bundle involution on the associated sphere bundle. Suppose that $w_1(\psi) = w_1(x)$.

(i) If m is odd, then $Wu_j(\psi - \tau_X) = 0$ for j odd.

(ii) If m is even, then $Wu_j(\psi - \tau_X) = 0$ for j odd.

Proof

By adding some trivial line bundles to ψ, we can assume that ψ has high fibre dimensions. Hence by the formula of θ_k in [7; 7.2]

$$\theta_k = p^*Wu_{k-m}(\psi - \tau_X)a$$

we can conclude (i), (ii) from (3.9). ■

REMARKS

1) We can also prove (4.4) by direct arguments without using (3.9): it is clear that

$$\tau_{S(-\psi)} \oplus \varepsilon = p^! (-\psi) \oplus (\tau_X)$$

where ε is a trivial line bundle. Hence

$$Wu(S(-\psi)) = p^*(Wu(\psi))\ p^*(Wu(-\tau_X))$$
$$= p^*(Wu(\psi - \tau_X))$$

By assumption $w_1(\psi) = w_1(-\psi) = w_1(X)$, so $S(-\psi)$ is an orientable manifold. Hence

$$Wu_j(\psi - \tau_X) = 0 \quad \text{for } j \text{ odd}.$$

2) We can use (4.4) to strengthen (4.3). Note that if T preserves the orientation, then η_r must be an odd dimensional plane bundle. Then, by (4.4) (i), we have

$$\sum_{r \geq n-k} Wu_{r+k-n}(\eta_r - \tau_{F^r}) = 0, \quad k \text{ odd}.$$

If we apply Umkehr homomorphism, we have (4.3) (i). Next, note if T reverses the orientation η_r must be an even dimensional plane bundle, by (4.4) (ii), we have

$$\sum_{r\geq n-k} Wu_{r+k-n}(\eta_r - \tau_F r) = 0, \quad k \text{ even.}$$

As above, we have (4.3) (ii).

Next, we will investigate vanishing theorems for manifolds with special structures.

(4.5) *THEOREM*

Let V^n be an Wu_a-manifold (see (3.11)). Suppose T preserves Wu_a-structure. Then

$$B_k(V) = 0, \quad k \not\equiv 0 \pmod{2^{a+1}}.$$

In particular $Wu_k(V) = 0$, $k \not\equiv 0 \pmod{2^{a+1}}$.

<u>*Proof*</u>

Let S^m be a sphere ($m = s \cdot 2^{a+1}-1$ and s large). Let $T_1 = T \times A$ be the diagonal involution on $V^n \times S^m$. It is clearly that T_1 preserves the Wu_a-structure. (Note that the antipodal map preserves the Wu_a-structure, because of the m we choose.) Then by (3.11), we have

$$\theta_\ell(V \times S^m) = B_{\ell-m}(V) \otimes \mu_{S^m} = 0, \quad \ell \not\equiv -1 \pmod{2^{a+1}}, \quad \ell > \frac{n+m}{2} + 2^{a-1}.$$

That is, by writing $k = \ell - m$, we have

$$B_k(V) = 0, \quad k \not\equiv 0 \pmod{2^{a+1}}.$$

For $a = 0$, this gives (4.3). For $a = 1$, this gives a vanishing theorem for spin-manifold, namely, for spin-manifold with T preserving spin-structure we have

$$B_k(V) = 0, \quad k \not\equiv 0 \pmod 4.$$

In particular, when $T = I$, we have

$$Wu_k(V) = 0, \quad k \not\equiv 0 \pmod 4).$$

This concludes our vanishing theorem. ■

In the rest of this section we will prove some product formulas.

(4.6) *PROPOSITION*

Let (I, V^n) be the identity involution and (T, M^m) be a fixed point free involution. Let $(I \times T, V \times M)$ be the diagonal involution. Then, for $k > \frac{n+m}{2}$,

$$\theta_k(V^n \times M^m) = \sum_{k \geq \ell > \frac{n}{2}} B_{k-\ell}(V^n) \otimes \theta_\ell(M^m)$$

$$= \sum_{k \geq \ell > \frac{n}{2}} Wu_{k-\ell}(V^n) \otimes \theta_\ell(M^m).$$

Proof

Note that $V \times_{I \times T} M = V \times M/T$. So the fundamental cohomology class of $H^*(V \times_{I \times T} M)$ is $1 \otimes c$, where c is the fundamental cohomology class of $H^*(M/T)$.

Let

$$\delta = \sum_{i+j=n+m-k} \alpha_i \otimes \beta_j \in H^{n+m-k}(V \times M) = \sum_{i+j=n+m-k} H^i(V) \otimes H^j(M).$$

Consider: $<\theta_k \delta, \sigma(V \times M)> = <(1 \otimes c)^{2k-n-m} Q(\delta), \sigma(V \times M/T)>$

$$=<1 \otimes c^{2k-n-m} Q(\Sigma\, \alpha_i \otimes B_j), \sigma(V) \otimes \sigma(M/T)> .$$

Since $\alpha_i \otimes 1_M = \nu^*(\alpha_i \otimes 1_{M/T})$, this becomes

$$\sum_{i+j=n+m-k} <1 \otimes c^{2k-n-m} \left(\sum_{t=0}^{i} Sq^t \alpha_i \otimes c^{i-t}\right) Q(1 \otimes B_j), \sigma(V) \otimes \sigma M/T)>$$

$$= \sum_{i,j} \sum_t <Sq^t \alpha_i, \sigma(V)> <c^{2k-n-m+i-t} Q(B_j), \sigma(M/T)> .$$

In summing up t, the only non-trivial contribution comes from $t = n - i$. Also note that $2k - m - n > 0$ and $(i - t) > 0$, so $2k - n - m + i - t > 0$. Thus, by definition of

Wu_k and θ_ℓ, this reduces to

$$\sum_{i+j=n+m-k} \langle Wu_{n-i}\alpha_i, \sigma(V)\rangle \langle \theta_{m-j} B_j, \sigma(M)\rangle$$

$$= \sum_{i+j=n+m-k} \langle (Wu_{n-i} \otimes \theta_{m-j})(\alpha_i \otimes B_j), \sigma(V) \otimes \sigma(M)\rangle$$

$$= \left\langle \left(\sum_{i+j=n+m-k} Wu_{n-i} \otimes \theta_{m-j}\right)\left(\sum_{i+j=n+m-k} \alpha_i \otimes B_j\right), \sigma(V) \otimes \sigma(M)\right\rangle$$

$$= \left\langle \left(\sum_{\ell>\frac{n}{2}}^{k} Wu_{k-\ell} \otimes \theta_\ell\right)\delta, \sigma(V) \otimes \sigma(M)\right\rangle.$$

This proves $\theta_k(V \times M) = \sum_{\ell > \frac{n}{2}} Wu_{k-\ell}(V) \otimes \theta_\ell(M)$. ■

(4.7) *PROPOSITION*

Let (T,V^n) be an involution (not necessarily free). Let (A,S^m) be the antipodal map. Let $(T \times A, V^n \times S^m)$ be the diagonal action. Then, for $k > \frac{n+m}{2}$

$$\theta_k(V^n \times S^m) = B_{k-m}(V^n) \otimes \theta_m(S^m)$$

where $B_i(V^n) = 0$ if $i < 0$.

Proof

It is clear that $S^r \subset S^{r+1}$, as embedded in the equator, is a characteristic submanifold of S^{r+1}. So, by [7]

$$i_!(\theta_k(V \times S^r)) = \theta_{k+1}(V \times S^{r+1})).$$

On the other hand

$$i_!(B_{k-r}(V) \otimes \theta_r(S^r)) = B_{k-r}(V) \otimes \theta_{r+1}(S^{r+1}).$$

Next, from the definition of $B_k(V)$, for $m >> 0$

$$B_{k-m}(V) \otimes \theta_m(S^m) = \theta_k(V \times S^m).$$

Let us assume that we have chosen such m. Then, by argument above we have

$$i_!(B_{k-m+1}(V) \otimes \theta_{m-1}(S^{m-1})) = i_!(\theta_k(V \times S^{m-1})).$$

Now, let us note that $(1_m \in H^0(S^m))$

$$<(B_{k-m+1}(V) \otimes \theta_{m-1}(S^{m-1}))\ \alpha \otimes 1_{m-1},\ \sigma(V \times S^{m-1})>$$

$$= <i_!((B_{k-m+1}(V) \otimes \theta_{m-1}(S^{m-1}))\ \alpha \otimes 1_m,\ \sigma(V \times S^m)>$$

$$= <i_!(\theta_k(V \times S^{m-1}))\ \alpha \otimes 1_m,\ \sigma(V \times S^m)>$$

$$= <\theta_k(V \times S^{m-1})\ \alpha \otimes 1_{m-1},\ \sigma(V \times S^{m-1})>,$$

for all $\alpha \in H^{n-k}(V)$.

Thus, $\theta_k(V \times S^{m-1}) = B_{k-m+1}(V) \otimes \theta_{m-1}(S^{m-1})$. By induction, we have shown the proposition. ∎

Next we shall try to use these to prove some Borsuk-Ulam type theorems.

Let (T_1, M^m) be a fixed point free involution and (T, V^n) be a (not necessarily free) involution. Let $f : M^m \to V^n$ be any map and define

$$A(f) = \{x : f(T_1 x) = Tf(x)\} .$$

The set $A(f)$ is closed and T-invariant (the point in $A(f)$ is called equivariant point by M. Nakaoka).

Consider the diagonal involution $T_2 = T_1 \times T$ on $M^m \times V^n$ (dim $M = m$, dim $V = n$). Let

$$g : M^m \to M^m \times V^n$$

be a map defined by $g(x) = (x, f(x))$. Clearly, the image G of g is the graph of f. Note that $g|A(f)$ is an equivariant homeomorphism of $A(f)$ onto the "equivariant self-intersection" $G \cap T_2(G)$. By appealing to the geometric meaning of the Bredon operation (recall that it is called equivariant self-intersection in [7]) we introduce the following "self-intersection number": let $1 \in H^0(M^m)$. Then $g_!(1) \in H^n(M^m \times V^n)$ is the cohomology class dual to the homology class represented by G. Then the "self-intersection number" of $G \cap T_2(G)$ is defined to be

$$C(f) = \langle c_2^{m-n} Q_2(g_!(1)),\ \sigma(M \times_{T_2} V)\rangle$$

where C_2 and Q_2 are the fundamental cohomology class of involution and the Bredon operation on $M \times V$. This number is suggested by P. Conner.

(4.8) *PROPOSITION*

If $C(f) = 1$, then $c_1^{m-n} \in H^{m-n}(M/T_1)$ restricts to non-zero element in $H^{m-n}(A(f)/T_1)$ and hence $\dim A(f) \geq m - n$.

<u>*Proof*</u>

Note that $C(f) = \langle c_2^{m-n} Q(g_!(1)),\ \sigma(M \times_{T_2} V)\rangle$

$$= \langle c_2^{m-n}, Q(g_!(1)) \cap \sigma(M \times_{T_2} V)\rangle = 1.$$

Then, by the geometric meaning of the Bredon operation (equivariant self-intersection), we see that $Q(g_!(1)) \cap \sigma(M \times_{T_2} V) = g_1^*(\sigma(A(f)/T_1))$ where $g_1 = g|A(f)$. Thus $\langle c_2^{m-n}, g_1^*\sigma(A(f)/T_1)\rangle = \langle (g_1^* c_2)^{m-n}, \sigma(A(f)/T_1)\rangle = 1$. But $g_1^* c_2$ is the restriction of c_1 to $A(f)$. This proves the proposition. ∎

If $m > n$, then by the definition of θ_k, we have

$$C(f) = \langle \theta_m(M \times V) g_!(1),\ \sigma(M \times V)\rangle .$$

For some product formulas we establish above, we can re-express $C(f)$ in the data of M and V. Namely,

(4.9) *PROPOSITION*

If $\theta_m(M \times V) = \sum_{j=0}^{[\frac{n}{2}]} \theta_{m-j}(M) \otimes B_j(V)$, (verified by S. Kahn [21])

then,
$$C(f) = \sum_{j=0}^{[\frac{n}{2}]} \langle \theta_{m-j}(M) f^*(B_j(V)),\ \sigma(M)\rangle$$

$$= \sum_{j=0}^{m} \langle B_j(V) f_!(\theta_{m-j}(M)), \sigma(V)\rangle .$$

Proof

This follows immediately from noting that $g(x) = (x, f(x))$. ■

We do know that the product formula is true when $T = I$. Note that $B_j(V) = Wu_j(V)$ if $T = I$. Also note that cup by $Wu_j(V)$ to the top dimension can also be achieved by Sq^j. So, we have

(4.10) *COROLLARY*

$$C(f) = \sum_{j=0}^{[\frac{n}{2}]} \langle Sq^j f_!(\theta_{m-j}(M)), \sigma(V)\rangle .$$

This corollary is useful, for example, it can be applied to the situation when $Wu_j(V) = 0$, $j > 0$: note that

$$Sq^j f_!(\theta_{m-j}(M)) = Wu_j(V) f_!(\theta_{m-j}(M)) = 0, \quad j > 0 .$$

Then,

$$C(f) = \langle f_!(\theta_m(M)), \sigma(V)\rangle .$$

$$= \langle \theta_m(M), \sigma(M)\rangle = \langle c_1^m, \sigma(M/T)\rangle .$$

The last two expressions are very often computable. For example, if V is a Lie group and c_1^m is non-zero. Then we have $C(f) = 1$, and we have the conclusion of (4.8).

One can generalize the product formula (4.7) to a free involution T on the homology sphere S^m.

(4.11) *COROLLARY*

$$C(f) = \langle \theta_m(S^m) f^*(B_0(V)), \sigma(S^m)\rangle .$$

If $(T,V) = (I,V)$, we have, $1 = Wu_0(V) = B_0(V)$, so

$$C(f) = \langle \mu_m, \sigma(S^m) \rangle = 1.$$

This gives the following Borsuk-Ulam type theorem:

(4.12) *COROLLARY*

Let $f : S^m \to V^n$ be a map $(m > n)$, then $\dim A(f) \geq m - n$, where

$$A(f) = \{x : f(x) = f(Tx)\} .$$

Another situation is interesting too. Let $S(\psi)$ be a m-sphere bundle with bundle involution. Let $f : S(\psi) \to X$ be a map homotopic to the projection $p : S(\psi) \to X^n$ of the bundle. Note that $\dim (S(\psi)) = m + n$. Then,

$$C(f) = C(p) = \sum_{j=0}^{[\frac{n}{2}]} \langle Sq^j f_!(\theta_{m+n-j}(S(\psi)), \sigma(X^n) \rangle.$$

Recall from [7] that

$$f_!(\theta_{m+n-j}(S(\psi)) = p_!(\theta_{m+n-j}(S(\psi))$$
$$= Wu_{n-j}(\psi - \tau_X) .$$

Thus,

$$C(f) = \sum_{j=0}^{[\frac{n}{2}]} \langle Sq^j(X) Wu_{n-j}(\psi - \tau_X), \sigma(X^n) \rangle$$
$$= \sum_{j=0}^{[\frac{n}{2}]} \langle Wu_j(-\tau_X) Wu_{n-j}(\psi - \tau), \sigma(X^n) \rangle$$
$$= \langle Wu_n(\psi - 2\tau_X), \sigma(X^n) \rangle .$$

Or another expression

(4.13) $$C(f) = \langle w_n(\psi - \tau_X), \sigma(X^n) \rangle$$

(Recall **that** $w_n(\psi - \tau_X) = \sum_{j=0}^{[\frac{n}{2}]} Sq^j Wu_{n-j}(\psi - \tau_X)$).

5. GENERALIZED PETERSON-STEIN CLASSES

In studying an $(m+1)$-plane bundle ψ with $w_{m+1}(\psi) = 0$, Peterson and Stein [18] introduced the secondary characteristic classes Φ_k. In studying the bundle involutions, Conner and Miller [7] related these secondary characteristic classes to the Bredon operation. The main results in this section are generalizations of these results to general $(m+1)$-plane bundles without any assumption on $w_{m+1}(\psi)$.

Let $\psi \to X$ be an $(m+1)$-plane bundle over a space X. Let $S(\psi)$ be the total space of the associated sphere bundle, $p : S(\psi) \to X$. On $S(\psi)$ there is the bundle involution T, that is, the fibre preserving fixed point free involution which on each fibre agrees with the antipodal map. The orbit space $RP(\psi) = S(\psi)/T$ fibre over X with fibre $RP(m)$ — m-dimensional real projective space. There is a commutative diagram

$$\begin{array}{ccc} S(\psi) & \xrightarrow{\ \nu\ } & RP(\psi) \\ {\scriptstyle p}\downarrow & \swarrow {\scriptstyle \pi} & \\ X & & \end{array}$$

Let c be the cohomology class of the bundle involution. Via π^*, $H^*(RP(\psi))$ can be regarded as a graded module over $H^*(X)$. It is free with basis $1, c, c^2, \ldots, c^m$. The fundamental relation is

$$c^{m+1} = \sum_{p=1}^{m+1} \pi^*(w_p) c^{m+1-p}$$

where w_p denote the Stiefel-Whitney class of ψ. In general, we have

$$c^{m+1+t} = \sum_{p=1}^{m+1} \pi^*(w_{p,t}) c^{m+1-p} \qquad (5.1)$$

where $w_{p,t} \in H^{p+t}(X)$, $1 \le p \le m+1$, $0 \le t$. These cohomology classes were defined in [8] (see [7] too).

For our future purpose, we would like to rearrange the indices of $w_{i,j}$ and combine the fundamental relations and (5.1) into a single family of relations. Let us write

$$c^r = \sum_{j=0}^{m} \pi^*(w_j^r)\, c^{m-j} \qquad r = 0,1,2,\ldots . \qquad (5.2)$$

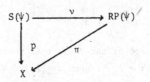

(1) For $0 \le r \le m$, $w_j^r = \delta^r_{m-j}$ (Kronecker delta)

(2) $w_j^{m+1} = w_{j+1}$; $w_j^r = w_{j+1,r-m-1}$, $w_0^r = \overline{w}_{r-m}$ for $r \ge m$

(3) $w_j^r = \sum_{s=0}^{r-m-1} w_{r+j-m-s}\overline{w}_s$

(4) $w_k^{r+s} = \sum_{d=0}^{2m} \left(\sum_{i+j=d} w_i^r w_j^s \right) w_k^{2m-d}$.

We shall call these w_j^r the generalized Stiefel-Whitney classes. (Abb GSW-classes).

Let us first recall the following two relations from [7], [18].

$$p^*(\Phi_k) = Sq^k a + p^*(w_k)a \tag{5.3.1}$$

$$Q(a) = \sum_{k=0}^{m} \pi^*(\Phi_k)c^{m-k} \tag{5.3.2}$$

where $a \in H^m(S(\psi))$ is a choice of the class that restricts to the co-fundamental class of the fibre of $p : S(\psi) \to X$. (When $w_{m+1}(\psi) = 0$, such class exists.) For any choice of such a class a, we will call it fibre-sphererical class.

First, we will generalize the second expression to general situation: let $\psi \to X$ be an $(m+1)$-plane bundle (no assumption on Stiefel-Whitney classes). Let

$$y \in H^{m+\ell}(S(\psi)) .$$

Then the value of the Bredon operation $Q(y)$ is lie in $H^{2(m+\ell)}(RP(\psi))$. By the cohomological structure of $H^*(RP(\psi))$, there is a unique class $\Phi_j(y) \in H^{m+2\ell+j}(X)$, for each j such that

$$Q(y) = \sum_{j=0}^{m} \pi^*(\Phi_j(y))c^{m-j} . \tag{5.4}$$

We shall call these classes *generalized Peterson-Stein classes* (*abb. GPS-classes*). Though $\Phi_j(y)$ has no secondary effect here, however, its image $p^*(\Phi_j(y)) \in H^*(S(\psi))$ has secondary effect. (Note that $p^*(w_{m+1}(\psi)) = 0$).

Next, we will establish an expression similar to the first expression (5.3.1). (This is our main result, see theorem (5.3)). Let $p : S(\psi) \to X$ be, as usual, an m-sphere bundle. Let $\Psi = p^!(\psi)$ be the pull-back of $(m + 1)$-plane bundle ψ. Then

$$\tilde{p} : S(\Psi) \to S(\psi)$$

is a m-sphere bundle over $S(\psi)$, induced by ψ via $p : S(\psi) \to X$. Thus we have the following commutative diagram.

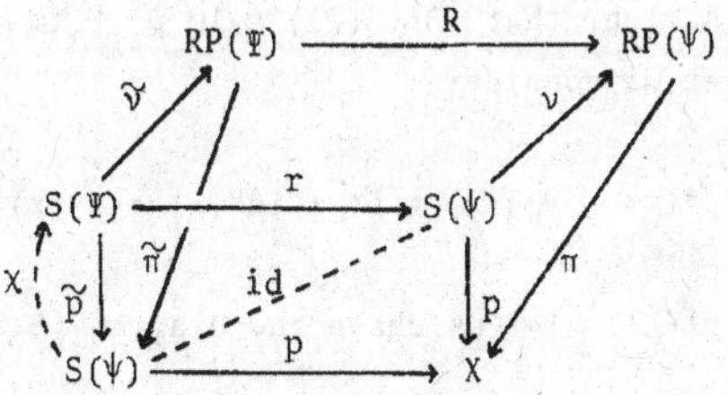

(5.5)

where

1) $\Psi = p^!(\psi)$ - the induced bundle over $S(\psi)$,

2) r is the bundle map over p, and hence equivariant with respect to the bundle involutions,

3) R is induced by the map r on the orbit spaces,

4) $\tilde{p}$, $\tilde{\pi}$ and $\tilde{\nu}$ are the corresponding maps of p, π and ν which are explained in the beginning of this section,

5) χ is the canonical cross section of $\tilde{p}$ and $r \cdot \chi = id$.

Next, let us examine the Gysin sequences of $\tilde{p}$. Since $\tilde{p}$ has cross section $\chi : S(\psi) \to S(\Psi)$, the Gysin sequence reduces to a splitting short exact sequence of modules over $\tilde{p}^*(H^*(S(\psi))$

$$0 \longrightarrow H^*(S(\psi)) \underset{\chi^*}{\overset{\tilde{p}^*}{\rightleftarrows}} H^*(S(\Psi)) \underset{\eta}{\overset{\tilde{p}_!}{\rightleftarrows}} H^{*-m}(S(\psi)) \longrightarrow 0 \qquad (5.6)$$

where $\chi^* \cdot \tilde{p}^* = id$ and $\tilde{p}_! \cdot \eta = id$.

(5.1) *PROPOSITION*

Let $a = \eta(1) \in H^m(S(\Psi))$. Then, for $x \in H^*(S(\psi))$

$$r^*(x) = \tilde{p}^*(x) + \tilde{p}^*(p^*p_!(x))a.$$

Proof

By the splitting, it is clear that $H^*(S(\Psi)) = \text{Im}\ \tilde{p}^* \oplus \text{Im}\ \eta$. To determine the first component of $r^*(x)$, let us consider

$$\tilde{p}^*\chi^*(r^*(x)) = \tilde{p}^*(\chi^* \cdot r^*)(x) = \tilde{p}^* \cdot id^*(x) = \tilde{p}^*(x).$$

For the second component of $r^*(X)$, let us chase the diagram (5.5) and the short exact sequence (5.6)

$$\eta(\tilde{p}_! r^*(x)) = \eta(p^*p_!(x)), \qquad \text{by commutativity of (5.5)}$$

$$= \tilde{p}^*(p^*p_!(x))\eta(1), \quad \text{since } \eta \text{ is a } \tilde{p}^*(H^*(S(\psi))) \text{ - module map.}$$

$$= \tilde{p}^*(p^*p_!(x))a, \qquad \text{since } \eta(1) = a.$$

Thus, by expressing $r^*(x)$ into its components, we have

$$r^*(x) = \tilde{p}^*(x) + \tilde{p}^*(p^*p_!(x))a.$$

This proves the proposition. ■

Next let us recall a result from [7].

(5.2) *PROPOSITION*

Let $\varphi \to Y$ be an $(m + 1)$-plane bundle. Assume that $p : S(\varphi) \to Y$ has cross section $\chi : Y \to S(\varphi)$. Let $a \in H^m(S(\varphi))$ such that $\chi^*(a) = 0$. Then $Q(a) = 0$.

(5.3) *THEOREM*

Let $\psi \to X$ be an $(m + 1)$-plane bundle. Let $t \in H^{m+\ell}(S(\psi))$. Then

$$p^*(\Phi_k(t)) = \sum_{j=0}^{m+\ell} p^*(w_k^{m+\ell-j})Sq^j t + tp^*(w_k(\psi)p_!(t))$$

$$= Sq^{\ell+k}t + \sum_{j=0}^{\ell-1} p^*(w_k^{m+\ell-j})Sq^j t + tp^*(w_k(\psi)p_!(t))$$

where $w_k(\psi)$ are Stiefel-Whitney classes and w_j^r are generalized Stiefel-Whitney classes as explained in (5.2).

Proof

Let $\Psi = p^!(\psi)$ as explained above, and consider the diagram (5.5). Let a be the class, as in proposition (5.1). Let Q_ψ and Q_Ψ be the Bredon operation of the bundle involutions on $S(\psi)$ and $S(\Psi)$ respectively. Consider

$$R^*(Q_\psi(t)) = Q_\Psi(r^*(t)), \qquad \text{by naturality of } Q$$

$$= Q_\Psi(\tilde{p}^*(t) + \tilde{p}^*(p^*p_!(t))a, \qquad \text{by (5.1)}$$

$$= Q_\Psi(\tilde{p}^*(t)) + Q_\Psi(\tilde{p}^*(p^*p_!(t))a + tr(T^*(\tilde{p}^*(t)) \cdot \tilde{p}^*(p^*p_!(t))a), \quad \text{by (2.1)}$$

$$= Q_\Psi(\tilde{p}^*(t)) + Q_\Psi(\tilde{p}^*(p^*p_!(t))Q_\Psi(a) + \pi^*(t \cdot p^*p_!(t))tr(a),$$

by (2.1) and properties of tr

$$= \sum_{j=0}^{m+\ell} c_\Psi^{m+\ell-j}\,\tilde{\pi}^*(Sq^j t) + 0 + \tilde{\pi}^*(t \cdot p^*p_!(t)) \cdot \left(\sum_{j=0}^{m} c_\Psi^{m-j}\,\tilde{\pi}^*(p^*w_j)\right),$$

by (5.2) and formula for tr(a) (see (5.6) below)

$$= \sum_{k=0}^{m}\left[\tilde{\pi}^*\left[\sum_{j=0}^{m+\ell} p^*(w_k^{m+\ell-j})Sq^j t\right]\right]c_\Psi^{m-k} + \sum_{k=0}^{m}\left[\tilde{\pi}^*(t \cdot p^*p_!(t))\,\tilde{\pi}^*(p^*w_k)\right] c_\Psi^{m-k},$$

by equation (5.2)

$$= \sum_{k=0}^{m} \tilde{\pi}^*\left[\sum_{j=0}^{m+\ell} p^*(w_k^{m+\ell-j})Sq^j t + t \cdot p^*(p_!(t)w_k)\right] c_\Psi^{m-k}$$

$$= \sum_{k=0}^{m} \tilde{\pi}^*\left[Sq^{\ell+k}t + \sum_{j=0}^{\ell-1} p^*(w_k^{m+\ell-j})Sq^j t + t \cdot p^*(w_k p_!(t))\right] c_\Psi^{m-k}.$$

On the other hand

$$R^*(Q_\psi(t)) = R^*\left(\sum_{k=0}^{m} \pi^*\Phi_k(t)c_\psi^{m-k}\right)$$

$$= \sum_{k=0}^{m} R^*\pi^*(\Phi_k(t))R^*(c_\psi^{m-k})$$

$$= \sum_{k=0}^{m} \tilde{\pi}^*p^*(\Phi_k(t))c_\psi^{m-k}, \text{ by diagram (5.5)}$$

and $R^*(c_\psi) = c_\Psi$.

Note that $\tilde{\pi}^*$ is a monomorphism, and compare the coefficients of both sides, we have

$$p^*\Phi_k(t) = Sq^{\ell+k}t + \left[\sum_{j=0}^{\ell-1} p^*(w_k^{m+\ell-j})\ Sq^j t\right] + tp^*(p_!(t)w_k).$$

This proves the theorem. ■

REMARKS

1) If we substitute the explicit formula ((5.2), (4)), then we have

$$p^*\Phi_k(t) = Sq^{\ell+k}t + \left[\sum_{j=0}^{\ell-1}\left(\sum_{s=0}^{\ell-j-1} p^*(w_{k+\ell-j-s}\overline{w}_s)\right) Sq^j t\right] + t\,p^*(w_k p_!(t)).$$

2) This theorem implies that $p^*(\Phi_k(t))$ is determined by the Steenrod square Sq on $H^*(S(\psi))$ and other known data: note that, in describing the cohomological structure of $H^*(S(\psi))$ via Gysin sequence (see §6), we will choose the generators of annihilator of $w_{m+1}(\psi)$; they are $p_!(t)$. Then we choose its preimage t. Therefore, the last term in the equation above causes no difficulty. Since $\operatorname{Ker} p^*$ is the ideal generated by $w_{m+1}(\psi)$, we can conclude from this theorem that, up to the ideal generated by $w_{m+1}(\psi)$, the GPS-classes $\Phi_k(t)$ and hence the Bredon operation Q, are determined by Sq on $H^*(S(\psi))$.

Some special cases are worth mentioning:

(5.4) *COROLLARY*

$$p^*\Phi_m(t) = t^2 + t\cdot p^*(p_!(t)w_m)$$

$$p^*\Phi_0(t) = \sum_{j=0}^{\ell} (p^*\overline{w}_{\ell-j})Sq^j t + t\, p^*(p_!(t)).$$

The key to understand the Bredon operation on $H^*(S(\psi))$ is to understand these generalized Peterson-Stein classes. So we shall prove few properties about them. In developing these properties the generalized Stiefel-Whitney classes (see (5.2)) play an important role.

(5.5) *PROPOSITION*

Let $s,t \in H^*(S(\psi))$. Then,

$$\Phi_d(st) = \sum_{\ell=0}^{2m}\left[\sum_{j+k=\ell} \Phi_j(s)\Phi_k(t)\right] w_d^{2m-\ell} .$$

Proof

By definition,

$$Q(st) = \sum_{d=0}^{m} \pi^*(\Phi_d(st))c^{m-d} .$$

By (2.1), we have

$$Q(st) = Q(s)Q(t) = \left[\sum_{j=0}^{m} \pi^*\Phi_j(s)c^{m-j}\right]\left[\sum_{k=0}^{m} \pi^*\Phi_k(t)c^{m-k}\right]$$

$$= \sum_{j,k=0}^{m} \pi^*(\Phi_j(s)\Phi_k(t))c^{2m-j-k}$$

$$= \sum_{\ell=0}^{2m} \pi^*\left[\sum_{j+k=\ell} \Phi_j(s)\Phi_k(t)\right] c^{2m-\ell}$$

$$= \sum_{d=0}^{m} \pi^*\left[\sum_{\ell=0}^{2m}\left[\sum_{j+k=\ell} \Phi_j(s)\Phi_k(t)\right] w_d^{2m-\ell}\right] c^{m-d}, \qquad \text{by (5.2).}$$

Comparing the coefficient (and note that π^* is a monomorphism), we have the proposition. ■

Let us recall from [7, supplement (2.2)].

(5.6) *LEMMA*

If $y \in H^{m\ j}(S(\psi))$, then

$$tr(y) = \sum_{i=0}^{m} \pi^*(w_i p_!(y))c^{m-i} .$$

The next proposition is immediate from (2.1) and this Lemma.

(5.7) *PROPOSITION*

Let $s,t \in H^*(S(\psi))$

$$\Phi_d(s+t) = \Phi_d(s) + \Phi_d(t) + w_d p_!(sT^*t) .$$

Next we shall investigate the behavior of GPS-classes when the bundle is a Whitney sum.

Let ψ_1 and ψ_2 be the $(m+1)$- and n-plane bundles over X. Let $\psi_1 \oplus \psi_2$ be the Whitney sum. Let

$$p_i : S(\psi_i) \to X, \quad i = 1,2$$

$$p : S(\psi_1 \oplus \psi_2) \to X$$

be their associate sphere bundles. Let

$$i_k : S(\psi_k) \to S(\psi_1 \oplus \psi_2), \quad k = 1,2$$

be the inclusions. Let $t_1 \in H^{m+\ell_1}S(\psi_1))$. Let $t = i_{1!}(t_1) \in H^{m+n+\ell_1}(S(\psi_1 \oplus \psi_2))$.

(5.8) *THEOREM*

Let ψ_1, ψ_2, t and t_1 be as above. Then

$$\Phi_k(t) = w_n(\psi_2)\left\{\sum_{r=0}^{k}\Phi_{k-r}(t_1)w_r(\psi_2)\right\}.$$

Proof

The proof of [7, Theorem (13.4)] holds for this situation by taking t_1 as a_1 and t as a in [7].■

The relations between Sq^i and $\Phi_k(t)$ are rather complicated. We summarize in

(5.9) *PROPOSITION*

Let $t \in H^{m+\ell}(S(\psi))$.

$$\sum_{0\le i\le j/2}\sum_{s=0}^{m}\begin{Bmatrix}m+\ell-i\\ j-2i\end{Bmatrix}\Phi_s(Sq^i t)w_k^{m+j-2i-s} + w_k p_!\left(\sum_{0\le i\le j/2} Sq^{j-i}t\cdot Sq^i T^*t\right)$$

$$= \sum_{i=0}^{m}\sum_{s=0}^{j}\begin{Bmatrix}m-i\\ s\end{Bmatrix}Sq^{j-s}\Phi_i(t)w_k^{m+s-i}.$$

Proof

By (2.2), we have

$$Sq^j\left(\sum_{i=0}^{m}\Phi_i(t)c^{m-i}\right) = \sum_{0\le i\le j/2}\begin{Bmatrix}m+\ell-i\\ j-2i\end{Bmatrix}c^{j-2i}\left(\sum_{s=0}^{m}\Phi_s(Sq^i t)c^{m-s}\right)$$

$$+ \operatorname{tr}\left\{\sum_{0\le i\le j/2} Sq^{j-i}t\cdot T^*Sq^i t\right\}.$$

The left hand side equals

$$\sum_{i=0}^{m}\sum_{s=0}^{j} Sq^{j-s}\Phi_i(t)Sq^s c^{m-i} = \sum_{i=0}^{m}\sum_{s=0}^{j}\begin{Bmatrix}m-i\\ s\end{Bmatrix}Sq^{j-s}\Phi_i(t)c^{m+s-i}$$

$$= \sum_{k=0}^{m} \left(\sum_{i=0}^{m} \sum_{s=0}^{j} \begin{Bmatrix} m-i \\ s \end{Bmatrix} Sq^{j-s} \Phi_i(t) w_k^{m+s-i} \right) c^{m-k}, \qquad \text{by (5.2).}$$

The first term of right hand side equals , by (5.2),

$$\sum_{k=0}^{m} \left(\sum_{0 \le i \le j/2} \sum_{s=0}^{m} \begin{Bmatrix} m+\ell-i \\ j-2i \end{Bmatrix} \Phi_s(Sq^i t) w_k^{m\ j-2i-s} \right) c^{m-k} .$$

The second term of right hand side equals , by (5.6),

$$\sum_{k=0}^{m} \pi^* \left[w_k p_! \left(\sum_{0 \le i \le j/2} Sq^{j-i} t \cdot Sq^i T^* t \right) \right] c^{m-k} .$$

Compare the coefficients, we have the proposition. ■

Let us examine the case $j = 1$.

(5.10) *COROLLARY*

$$Sq^1 \Phi_k(t) = (2m+\ell-1-k)\Phi_{k+1}(t) + \ell w_{k+1} \Phi_0(t) + w_k p_!(Sq^1 t \cdot T^* t).$$

(Note $\Phi_{m+1}(t) = 0$.)

The GPS-classes are quite mysterious. For some cases, we have some explanation in the spirit of Massey-Peterson [11].

(5.11) *PROPOSITION*

Suppose $H^*(S(\psi))$ is a free $p^*(H^*(X))$-module with basis $\{1 = t_0, t_1, \ldots, t_s\}$, then the generalized Peterson-Stein classes $p^*\Phi_k(t_i)$, $i = 0,1,\ldots,s$; $k = 0,1,\ldots,m$, determine the extension (as an $p^*(H^*(X)) \odot A$-module) of the following short exact sequence

$$0 \to p^*(H^*(X)) \to H^*(S(\psi)) \to \mathrm{Ann}(w_{m+1}(\psi)) \to 0$$

where $p^*(H^*(X)) \odot A$ is the semi-tensor product of Steenrod algebra A and $p^*(H^*(X))$.

Proof

We shall construct the following commutative diagram

$$\begin{array}{ccccccc} \xrightarrow{d_2} & C_1 & \xrightarrow{d_1} & C_0 & \xrightarrow{\varepsilon} & \mathrm{Ann}(w_{m+1}) & \longrightarrow 0 \\ & \Big\downarrow \varphi_1 & & \Big\downarrow \varphi_0 & & \Big\downarrow = & \\ 0 \longrightarrow & p^*H^*(X) & \longrightarrow & H^*(S(\psi)) & \longrightarrow & \mathrm{Ann}(w_{m+1}) & \longrightarrow 0 \end{array}$$

where the top horizontal line is a $S = R \otimes A$-free resolution $(R = p^*H^*(X))$.

Let C_0 be a free S-module with basis $\{y_0, y_1 \ldots y_s\}$ and $\varphi_0(y_i) = t_i$, $0 \le i \le s$. From (5.3), we see that

$$z_{ki} = \left[\sum_{j=0}^{m+\ell} (p^* w_k^{m+\ell-j}) \otimes Sq^j + p^*(w_k p_!(t)) \otimes Sq^0\right] y_i$$

is in the Ker ε (note $\deg y_i = \deg t_i = m+\ell$). So, we choose C_1 with basis containing, among others, c_{ki}, $0 \le i \le s$, $0 \le k \le m$ and define d_1, φ_1, so that $d_1(c_{ki}) = z_{ki}$, $\varphi_1(c_{ki}) = p^*(\Phi_k(t_i))$. Next, choose C_2 and d_2 so tha $\varphi_1 d_2 = 0$. It is well-known that the "cocyle" $\varphi_1 \varepsilon \mathrm{Hom}_S(C_1, R)$ determine the extension, and the $p^*(\Phi_k(t_i)$ determine φ_1 partially.

6. θ_k AND $\Phi_j(t)$.

In [7], the CM-classes were computed for two types of bundle involutions, namely, bundles with cross section and those with vanishing top Stiefel-Whitney classes. We generalized their computations.

(6.1) *THEOREM*

Let $p : S(\psi) \to X^r$ be an m-sphere bundle. Suppose that the annihilator ideal $\mathrm{Ann}(w_{m+1}(\psi))$ of the top Stiefel-Whitney class $w_{m+1}(\psi)$ is a principal ideal. Let t be an element of $H^{m+\ell}(S(\psi))$ such that $p_!(t)$ is a generator of $\mathrm{Ann}(w_{m+1}(\psi))$. Then,

$$p^*(p_!(t))\theta_k = p^*[Wu_*(\psi - \tau_X)Sq^{-1}\Phi_*(t)]_k + p^*[Wu_{k-m-\ell}(\psi - \tau_X)]t.$$

Moreover $Wu_*(\psi - \tau_X)Sq^{-1}\Phi_*(t)$ and $Wu_{k-m-\ell}(\psi - \tau_X)$ are divisible by $p_!(t)$.

REMARK

If $w_{m+1}(\psi) = 0$, then the fibre spherical class a exist (see §5, (5.3.1)). Then the theorem reduces to

$$\theta_k = p^*[Wu_*(\psi - \tau_X)Sq^{-1}\Phi_*]_k + p^*[Wu_{k-m}(\psi - \tau_X)]\, a.$$

This is [7, (13.6)]. Moreover if ψ has cross section then $\Phi_* = 0$ and the theorem reduces to

$$\theta_k = p^*[Wu_{k-m}(\psi - \tau_X)]\, a.$$

This is [7, (7.2)].

<u>*Proof*</u>

From Gysin sequence, it is clear that $H^*(S(\psi))$ is a module over $p^*(H^*(x))$ with two generators 1, t (or if we like, 1, T^*t). So, we can write

$$\theta_k = p^*(e) + p^*(d)t \qquad (6.1)$$

for some classes e and d in $H^*(X)$; our goal is to determine e and d.

It is shown in [7] that $p_!(\theta_k) = Wu_{k-m}(\psi - \tau_X)$. So we have, by (6.1)

$$Wu_{k-m}(\psi - \tau_X) = d \cdot p_!(t) .$$

Next, let us consider the Kronecker product $\langle p^*(e)y, \sigma(S(\psi))\rangle$, for all

$y \in H^{r+m-k}(S(\psi))$. By the dimension reason, it is clear that the product is zero if y comes from the base. So, we only need to take $y = p^*(\beta)T^*t$, where $\beta \in H^*(X)$.

Thus we have

$$<p^*(e)y, \sigma(S(\psi))> = <(\theta_k - p^*(d)t)(p^*(\beta)T^*t), \sigma(S(\psi))> .$$

Note that $tT^*t = \nu^*(Q(t)) = p^*(\Phi_m(t))$ is a class coming from the base, so the equality becomes

$$<\theta_k p^*(\beta)T^*t, \sigma(S(\psi))> = <c^{2k-m-r}Q(p^*(\beta))Q(T^*t), \sigma(S(\psi))>.$$

Note that $Q(t) = Q(T^*t)$, so the equality becomes

$$<\sum_{i=0}^{r-k-\ell}\sum_{j=0}^{m} c^{k-\ell-i-j}\pi^*(\Phi_j(t)Sq^i\beta), \sigma(RP(\psi))> = <\sum_{i=0}^{r-k-\ell}\sum_{j=0}^{m}\pi_!(c^{k-\ell-i-j})\Phi_j(t)Sq^i\beta, \sigma(X^r)>.$$

It is shown in [7] that $\pi_!(c^j) = \bar{w}_{j-m}$ if $j \geq m$ and $\pi_!(c^j) = 0$ if $0 \leq j < m$. Thus the equality becomes

$$\sum_{i=0}^{r-k-\ell}\sum_{j=0}^{m} <\bar{w}_{k-\ell-i-j-m}\Phi_j(t)Sq^i\beta, \sigma(X^r)> = <\bar{w}_*(\psi)\Phi_*(t)Sq^*\beta, \sigma(X^r)> ;$$

where $\bar{w}_*(\psi)$, $\Phi_*(t)$ are the total classes, and Sq^* is the total Steenrod square. On the other hand

$$<p^*(e)y, \sigma(S(\psi))> = <e\, p_!(y), \sigma(X^r)> = <e\ \beta\ p_!(t), \sigma(X^r)> .$$

By equating the two computations of the Kronecker product, we have

$$\begin{aligned} e\, p_!(t) &= [(\bar{w}_*(\psi)\Phi_*)Sq^*]_k \\ &= [Wu_*(\psi - \tau_X)Sq^{-1}\Phi_*]_k . \end{aligned}$$

Multiplying (6.1) by $p^*(p_!(t))$, we have our conclusions. ∎

This theorem can easily be generalized:
Let $t_1, t_2, \ldots, t_s \in H^*(S(\psi))$ such that $p_!(t_i)$ are the generators of $Ann(w_{m+1}(\psi))$. Then, we have the following generalization.

(6.2) *THEOREM*

Assume that $t_i T^* t_j \in p^*(H^*(X))$ for all i, j. Then

$$\theta_k = p^*(e) + \sum_{j=1}^{s} p^*(d_j)t_j$$

where e and d_j are classes in $H^*(X)$ satisfying the following relations

(1) $e\, p_!(t_j) = [Wu_*(\psi - \tau_X)Sq^{-1}\Phi_*(t_j)]_{k+\ell_j}$

where $\deg e = k$, $\deg t_j = m + \ell_j$.

(2) $\sum_{j=1}^{s} d_j p_!(t_j) = [Wu_*(\psi - \tau_X)]_{k-m}$.

The proof of this theorem can be carried out in the same spirit as in (6.1). We shall omit the details.

REMARK

1) If $s = 1$, then (6.2) reduces to (6.1).

2) The choice of t_j, $j = 1,\ldots,s$ is not unique, further, even with fixed choice of t_j, $j = 1,\ldots,s$, the expression of θ_k is not unique (unless $H^*(S(\psi))$ is a free module over $p^*(H^*(X))$ with basis t_j, $j = 1,\ldots s$). Therefore, e and d_j, $j = 1,\ldots,s$, are not unique; and this is reflected in the statement too (that is, the relations (1) and (2) do not characterize e and d_j, $j = 1,\ldots,s$, uniquely).

REFERENCES

[1] ADAMS, J.F: On formula of Thom and Wu, Proc. Lond. Math. Soc. 11 (1961), 741-752.

[2] BREDON, G.E.: Cohomological aspects of transformation groups, Proc. Conf. Transformation Groups (New Orleans, La. 1967) Springer-Verlag, (1968), 245-280.

[3] BROWDER, W. and LIVESAY, R.: Free involutions on homotopy spheres, Tohoku Math. J. 25 (1972), 69-88.

[4] BROWDER, W.: The Kervaire invariant of framed manifolds and its generalization. Ann. of Math. 90 (1969), 157-186.

[5] BROWN, E.H. Jr. and PETERSON, F.P.: Relations among characteristic classes I, Topology 3 (1969), Suppl. 1, 39-52.

[6] CONNER, P.E.: Differentiable periodic maps, 2nd ed., Lecture Notes in Math. No. 783 Springer-Verlag (1979).

[7] CONNER, P.E. and MILLER, E.Y.: Equivariant self-intersection, Preprint, (1979).

[8] CONNER, P.E.: Diffeomorphisms of period two. Mich. Math. Jour. 16 (1963), 341-352.

[9] HATTORI, A.: The fixed point set of an involution and theorems of the Borsuk-Ulam type, Proc. Japan Acad. 50 (1974), 537-541.

[10] HATTORI, A.: The fixed point set of a smooth periodic transformation I., J. Faculty of Science, Univ. Tokyo Ser IA 24 (1977), 137-165.

[11] MASSEY, W. and PETERSON, F.: The mod 2 cohomology structure of certain fibre spaces, Memoirs AMS, No. 74 (1967).

[12] MILGRAM , R.J. and HAMBELTON, I.: An obstruction to Poincaré transversality, Proc. Sym. Pure Math. Vol. XXXII, Part I, Amer. Math. Soc. (1978),161-166.

[13] MILGRAM , R.J.: Unstable homotopy from the stable point of view, Lecture Notes in Math. No. 368, Springer-Verlag, (1974) (Sec. 3).

[14] MILNOR, J.W.: Characteristic classes, Ann. Math. Studies, No. 76, Princeton Univ. Press. (1974).

[15] NAKAOKA, M.: Continuous maps of manifolds with involutions I, Osaka J. Math. 11 (1974),129-145.

[16] NAKAOKA, M.: Continuous maps of manifolds with involutions II, Osaka J. Math. 11 (1974),147-162.

[17] NISHIDA, G.: Cohomology operations in iterated loop spaces, Proc. Japan Acad., 44 (1968), 104-109.

[18] PETERSON, F. and STEIN, N.: Secondary characteristic classes, Ann. Math. (2) 16 (1962), 510-523.

[19] STEENROD, N. and EPSTEIN, D.B.A.: Cohomology operations, Ann. Math. studies No. 50, Princeton Univ. Press, (1962).

[20] STONG, R.: Notes on cobordism theory, Princeton Univ. Press, (1968).

[21] KAHN, S.: The Conner-Miller classes of a product involution and Borsuk-Ulam type theorem. Preprint (1980).

SEMI-FREDHOLM OPERATORS AND HYPERBOLIC PROBLEMS

BY

MARIO MARTELLI

Mathematics Department
Bryn Mawr College
Bryn Mawr, PA 19010 U.S.A.
and
Instituto Matematico U. Dini
Viale Morgagni 67/A
50134 Firenze, Italy

§1.

Operator equations of the form

$$Lx = N(x) \tag{1.1}$$

where L is linear and N is a nonlinear map have been extensively studied in recent years mainly in connection with boundary value problems arising both in ordinary and partial differential equations (see, for example, J. Mawhin [10], M. Furi, M. Martelli, A. Vignoli [7], L. Cesari [4]). The most common assumptions under which the solvability of (1.1) is studied are the following:

i) $L : E \to F$ is a linear Fredholm operator of non-negative index, acting between a normed space E and a Banach space F and having a compact right inverse $K : \mathrm{Im}L \to E$;

ii) $N : E \to F$ is demicontinuous, that is N is continuous from the norm topology of E to the weak topology of F and sends bounded sets into bounded sets.

Assumptions i), ii) are very natural in a large variety of boundary value problems but, as it is easily seen, are not sufficient for the solvability of (1.1). Existence results depend therefore on further hypotheses on L and N and their mutual behavior. These hypotheses, which in general imply some "a priori" bounds on the norm of the solutions of (1.1), lead frequently to the applications of Theorem (1.1) below or of some of its equivalent formulations or of some of its Corollaries.

(1.1) *THEOREM*

Let $A : E \to F$ be a linear compact operator such that $L + A$ is onto. Assume that there exists an open, bounded neighborhood Ω of $0 \in E$ such that

$$Lx + Ax \neq \lambda(N(x) + Ax) \tag{1.2}$$

for every $x \in \partial\Omega$ and $\lambda \in (0,1)$. Then (1.1) has a solution.

The condition ind $L \geq 0$ is needed in order to ensure the existence of A such that $L + A$ is onto. The proof of Theorem (1.1) is based on the fact that under the existing assumptions the operator $L + A$ has a compact right inverse B and the map $B \circ N$ turns out to be continuous and compact. One then considers the compact vector field $\Phi : \overline{\Omega} \to E$ defined by

$$\Phi(x) = x - B(N(x) + A(x)) \tag{1.3}$$

Condition (1.2) ensures that if $\Phi(x) \neq 0$ for every $x \in \partial\Omega$ (if $\Phi(x) = 0$ for some $x \in \partial\Omega$ the problem is solved) then Φ is essential in the sense of A. Granas [8]. This implies that $\Phi(x) = 0$ for at least one $x_0 \in \overline{\Omega}$, that is

$$x_0 = B(N(x_0) + Ax_0). \tag{1.4}$$

By applying $L + A$ on both sides of (1.4) we obtain

$$Lx_0 = N(x_0). \tag{1.5}$$

It may happen that Φ is essential even though condition (1.2) is not satisfied, thus leading to results more general than Theorem (1.1) (see, for example, L. Nirenberg [12]).

Unfortunately there are problems, whose solvability can be again reduced to the solvability of an equation of the form (1.1), but with the proviso that L is not Fredholm even though it has some of the features of Fredholm operators, namely

j) Ker L is closed and admits a direct summand, Im L is closed and there exists a compact (or, more generally, continuous) right inverse $K : \text{Im}\, L \to E$.

These problems, the so-called hyperbolic problems, have been studied by many authors (see, for example, L. Cesari [4], H. Brézis [1], H. Brézis, L. Nirenberg [2]) and the present paper is intended as a further contribution to this study. The general idea is to extend the results obtained before for Fredholm operators having a non-negative index to right semi-Fredholm operators and to approximate a problem involving an operator, L, which is not right semi-Fredholm via a sequence of problems involving right semi-Fredholm operators. The solution of $Lx = N(x)$ is then obtained as a limit of the sequence of solutions of the approximating problems.

§2.

Let E, F be normed spaces. A linear operator $L : E \to F$ is said to be right semi-Fredholm if

1) $\operatorname{Ker} L$ is closed and there exists a closed subspace $E_1 \subset E$ such that $E = \operatorname{Ker} L \oplus E_1$;

2) $\operatorname{Im} L$ is closed and $\dim F/\operatorname{Im} L < +\infty$.

The index of L is the extended integer

$$\operatorname{ind} L = \dim \operatorname{Ker} L - \dim F/\operatorname{Im}L. \tag{2.1}$$

Through this paper we always assume that $\operatorname{ind} L \geq 0$ unless otherwise stated. A linear operator $K : \operatorname{Im}L \to E$ is said to be a pseudo-right inverse of L if the following two diagrams hold

$$\begin{array}{ccc} \operatorname{Im} L & \xrightarrow{K} & E \\ & \overset{I}{\searrow} & \downarrow L \\ & & \operatorname{Im} L \end{array} \qquad\qquad \begin{array}{ccc} E_1 & \xrightarrow{L} & \operatorname{Im} L \\ & \overset{J}{\searrow} & \downarrow K \\ & & E \end{array}$$

where J is the inclusion of E_1 into E. We shall always assume that K is continuous.

Let $\Omega \subset E$ be a bounded open set. A continuous map $\Phi : \overline{\Omega} \to E$ is said to be a *compact vector field* if $\Phi - I$ is compact. A singular point $x_0 \in \overline{\Omega}$ of Φ is a point such that $\Phi(x_0) = 0$. If for some $V \subset \Omega$ we have $\Phi(x) \neq 0$ for every $x \in V$ then we say that Φ is *singularity free* on V. Observe that if $V \subset \overline{\Omega}$ is closed and Φ is singularity free on V then there exists $\varepsilon > 0$ such that $\|\phi(x)\| \geq \varepsilon$ for every $x \in V$.

Two compact vector fields $\Phi, \Psi : \overline{\Omega} \to E$ are said to be *homotopic* if there exists a continuous map $H : \overline{\Omega} \times [0,1] \to E$ such that $H = I - T$ and $T : [0,1] \times \overline{\Omega} \to E$ is compact.

Let $\Phi, \Psi : \overline{\Omega} \to E$ be two compact homotopic vector fields which are singularity free on $\partial\Omega$. The homotopy H is said to be *admissible* if H is singularity free on $\partial\Omega$ for every $t \in (0,1)$.

A compact vector field $\Phi : \overline{\Omega} \to E$ which is singularity free on $\partial\Omega$ is said to be *inessential* if there exists a compact vector field $\Psi : \overline{\Omega} \to E$ which is singularity free on Ω and coincides with Φ on $\partial\Omega$. If Φ is not *inessential* then Φ is said to be essential. Obviously, in this case, Φ has at least one singular point x_0.

§3.

Let $L : E \to F$ be a right semi-Fredholm operator. A pseudo-right inverse K of L is said to be associated to a continuous projection $P : E \to E$ and it is denoted by K_P if $\operatorname{Im} P = \operatorname{Ker} L$ and $\operatorname{Im} K_P = \operatorname{Ker} P$.

(3.1) *LEMMA*.

Let $L : E \to F$ be a right semi-Fredholm operator and let K be a continuous pseudo-right inverse of L. Assume that K is α-Lipschitz with constant k. Then any pseudo-right inverse associated to a projection P is continuous and α-Lipschitz with constant less than or equal to $\|P\|k$.

<u>*Proof*</u>

The commutativity of the diagram

$$\begin{array}{ccc} \mathrm{Im}\,L & \xrightarrow{\;K\;} & E \\ & \underset{K_P}{\searrow} & \Big\downarrow{\scriptstyle I-P} \\ & & \mathrm{Ker}\,P \end{array}$$

ensures the result. ■

We do not know if the continuity of K ensures the continuity of every pseudo-right inverse of L. It is obvious that for any pseudo-right inverse K_1 we have $\mathrm{Ker}\,L \oplus \mathrm{Im}\,K_1 = E$ algebraically but we do not know if this is true topologically. On the other hand it is obvious that a closed subspace of E, say E_0, which admits a topological direct summand, say E_1, that is $E = E_0 \oplus E_1$ it admits also direct summand which are not topological but only algebraic. In our situation we have one more piece of information at our disposal, namely the continuity of K; but we do not know if this further assumption and the existence of a pseudo-right inverse associated to a projection P, will ensure that every pseudo-right inverse is associated to a suitable continuous projection or, more generally, is continuous.

(3,2) *LEMMA*

Let $L : E \to F$ be a right semi-Fredholm operator having a continuous pseudo-right inverse K, which is α-Lipschitz with constant k. Let $A : E \to F$ be a compact linear map. Then

i) $\mathrm{Im}(L+A)$ is closed;

ii) $L+A$ admits a continuous pseudo-right inverse T;

iii) T is α-Lipschitz with constant not larger than $\|P\|k$ where P is the continuous projection such that $\mathrm{Ker}\,P = E_1$, $\mathrm{Im}\,P = \mathrm{Ker}\,L$.

Proof

i) By Lemma (3.1) K_P is continuous and α-Lipschitz with constant $\|P\|k$. Consider the bounded linear operator $I + AK_P : \mathrm{Im}\,L \to F$. We want to show that $\mathrm{Im}\,(I + AK_P)$ is closed, $\dim F/_{\mathrm{Im}\,(I+AK_P)} < +\infty$ and $\mathrm{Im}\,(I+AK_P) \subset \mathrm{Im}\,(L+A)$. From these facts will follow that $\mathrm{Im}\,(L+A)$ is closed (and $\dim F/_{\mathrm{Im}(L+A)} < +\infty$).

It is obvious that $\mathrm{Im}\,(I+AK_P) = \mathrm{Im}\,(Q+QAK_P)$ where $Q : F \to F$ is a continuous (linear) projection such that $\operatorname{Im} Q = \operatorname{Im} L$. Since $Q = I - R$, where R is compact, we obtain that $\operatorname{Im}(I + AK_P)$ is closed and $\dim F/_{\operatorname{Im}(I+AK_P)} < +\infty$. Now let $y = z + AK_P z$ for some $z \in \operatorname{Im} L$. Then $z = Lx_1$ for some $x_1 \in E_1$. Thus

$$y = Lx_1 + AK_P Lx_1 = Lx_1 + Ax_1 \tag{3.1}$$

and $y \in \operatorname{Im}(L + A)$.

ii) Write $\operatorname{Im} L = \operatorname{Ker}(I + AK_P) \oplus F_2$ and let $B : \operatorname{Im}(I + AK_P) \to F_2$ be a bounded linear operator such that $(I + AK_P)Bz = z$ for every $z \in \operatorname{Im}(I + AK_P)$. Then $(L + A)K_P Bz = z$ for every $z \in \operatorname{Im}(I + AK_P)$. Since $\dim \operatorname{Ker}(I + AK_P) < +\infty$ the result follows.

iii) $B = I + K_1$ where K_1 is compact. Thus $K_P B$ is α-Lipschitz with constant not larger than $\|P\|k$. ■

We are now ready to obtain a result which extends Theorem (1.1) to right semi-Fredholm operators.

(3.1) *THEOREM*

Let $L : E \to F$ be a right semi-Fredholm operator and let $A : E \to F$ be a compact linear map. Let $\Omega \subset E$ be an open bounded neighborhood of the origin and let $N : \overline{\Omega} \to F$ be a demicontinuous map which is α-Lipschitz with constant r. Assume that L admits a continuous pseudo-right inverse K which is α-Lipschitz with constant k. Assume moreover that

i) $\|P\|kr < 1$

ii) $L + A$ is onto and $Lx + Ax \neq t(N(x) + Ax)$ for every $x \in \partial\Omega$.

Then $0 \in F$ is an interior point of $\operatorname{Im}(L - N)$.

<u>*Proof*</u>

By Lemma (3.2) there exists a right inverse T of $L + A$ which is α-Lipschitz with constant not larger that $\|P\|k$. Therefore the map

$$T(N + A) : \overline{\Omega} \to E \tag{3.2}$$

is an α-contraction. Moreover it is easy to show that TN is continuous. The vector field $\Phi : \overline{\Omega} \to E$ defined by $\Phi(x) = x - T(N(x) + Ax)$ is homotopic to the identity because of assumption ii). Thus Φ is essential in the sense of Granas [8] (A. Granas in [8] develops his theory for compact vector fields, but the extension to vector fields of the form $I - T$ where T is an α-contraction is straightforward). Moreover there exists $\varepsilon > 0$ such that $\|\Phi(x)\| \geq \varepsilon$ for every $x \in \partial\Omega$. Hence the equation

$$Lx - N(x) = y \tag{3.3}$$

has a solution, provided that $\|y\|$ is sufficiently small. ■

In [13] P. Zezza proved the following:

(3.2) *THEOREM*

Let $L : E \to F$ be a linear map such that $\operatorname{Ker} L$ is finite dimensional and let $K : \operatorname{Im} L \to E$ be a linear right inverse of L. Let $N : E \to F$ be a (possibly) nonlinear map. Assume that KN is continuous and compact and set

$$C = \{x \in E : N(x) \in \operatorname{Im} L\} = N^{-1}(\operatorname{Im} L).$$

Suppose that there exists an open bounded set Ω, $\Omega \subset C$, such that $0 \in \Omega$ and

$$Lx \neq \lambda N(x) \tag{3.4}$$

for every $x \in \partial\Omega$ and $\lambda \in (0,1]$. Then $Lx = N(x)$ for some $x \in \Omega$.

P. Zezza pointed out that the advantage of this theorem over Mawhin results (see [10]) is the fact that no assumptions are made on the codimension of $\operatorname{Im} L$ which need not be closed. In the proof given by Zezza the assumption $\dim \operatorname{Ker} L < +\infty$ plays a crucial role. We want to show here that this assumption is unnecessary and in fact Theorem (3.2) can be extended to hyperbolic problems.

(3.3) *THEOREM*

Let $L : E \to F$ be linear and such that $E = E_1 \oplus \operatorname{Ker} L$. Let $N : E \to F$ be a (*possibly*) nonlinear map. Assume that KN is continuous and compact where $K : \operatorname{Im} L \to E$ is such that $LK : \operatorname{Im} L \to \operatorname{Im} L$ is the identity. Let $\Omega \subset N^{-1}(\operatorname{Im} L)$ be open and bounded, $0 \in \Omega$, and such that

$$Lx \neq \lambda N(x) \tag{3.5}$$

for every $\lambda \in (0,1]$ and $x \in \partial\Omega$. Then the equation

$$Lx = N(x) \tag{3.6}$$

has a solution $x \in \Omega$.

Proof

The compact vector field $\Phi : \bar{\Omega} \to E$ defined by $\Phi(x) = x - KN(x)$ is obviously essential. ■

It should be pointed out that the solvability of $Lx = N(x)$ is not equivalent to the solvability of $x = KN(x)$. If one wants to formulate the solvability of $Lx = N(x)$ in equivalent terms then the approach proposed by P. Zezza in [13] is one of the possible alternatives.

We would like to notice at this point that the class of maps N previously considered can be enlarged so as to include maps S such that $S(x)$ is an acyclic compact set for every $x \in \bar{\Omega}$. The demicontinuity can be replaced by the condition: $x_n \to x_0$ and $y_n \in S(x_n)$ implies that there exists $y_0 \in S(x_0)$ such that $y_n \rightharpoonup y_0$. The theoretical results which are needed in this case are obtained in [5], [6].

§4.

Our next goal is to apply Theorem (3.1) to hyperbolic problems. The setting is the following.

a) L is a linear operator acting between a normed space E and a Banach space F;

b) Ker L is closed, it admits a direct summand E_1 and $\dim \operatorname{Ker} L = +\infty$;

c) Im L is closed, it admits a direct summand F_0, $\dim F_0 = +\infty$ and there exists a continuous linear map $K : \operatorname{Im} L \to E$ such that the two diagrams commute

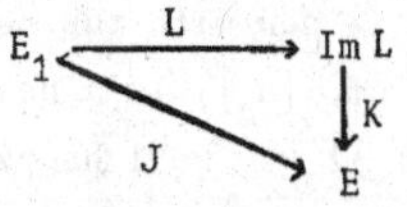

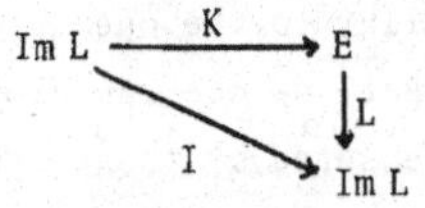

We assume, moreover, that

d) there exists a family $\{Q_n : n = 1,2,\dots\}$ of continuous projections, $Q_n : F \to F$, $Q_n F = F_n$, $Q_n F \subset Q_{n+1} F$, $\dim Q_n F = n$, $\overline{\bigcup_n Q_n F} = F_0$, $Q_n x \to Qx$ for every $x \in F$, where $Q : F \to F$ is a continuous projection such that $\operatorname{Im} Q = F_0$. Briefly we assume that F_0 admits a projectional scheme.

What we would like to obtain is a theorem analogous to Theorem (3.1) on the solvability of an operator equation of the form

$$Lx = N(x) \tag{4.1}$$

where N is demicontinuous and send bounded sets into bounded sets. The first idea is to consider the operator L as a map from E into $\operatorname{Im} L \oplus F_n$. Now for every $n \in \mathbb{N}$ we have that L is right semi-Fredholm having continuous pseudo-right inverse. We then modify N to the new setting by replacing it with the demicontinuous map $N_n(x) := (I - Q)\, N(x) + Q_n N(x)$, which sends bounded sets into bounded sets. Finally we consider the problem of solving

$$Lx = (I - Q)\, N(x) + Q_n N(x) \tag{4.2}$$

which is obviously equivalent to the system

$$\begin{cases} Lx = (I - Q)\, N(x) \\ Q_n N(x) = 0 \end{cases} \tag{4.3}$$

Suppose that (4.3) can be solved for every n and that $\{x_n\}$ is a sequence such that

$$\begin{cases} Lx_n = (I - Q)\, N(x_n) \\ Q_n N(x_n) = 0 \end{cases} .$$

We would like then to show that $x_n \to x_0$ which is a solution of (4.1).

The solvability of (4.2) (or, equivalently, (4.3)) can be proved under assumptions similar to the ones proposed in Theorem (3.1) and with the same technique used there. But, as one can expect, the convergence of $\{x_n\}$ is a new element, which is not ensured by those assumptions and in fact it way well happen that $\{x_n\}$ is only weakly convergent or not convergent at all. The following example wants to clarify this situation. The operator L involved is continuous and it has a continuous pseudo-right inverse K with $\|K\| = 1$. The non-linear operator N is continuous and α-Lipschitz with constant $r = \frac{1}{2}$.

(4.1) *EXAMPLE*

Let H be the Hilbert space of square summable sequences of real numbers and let $L : H \to H$ be defined by $Lx = (x_1, 0, x_3, 0, \ldots)$. Then $L + Q = I$ where $Qx = (0, x_2, 0, \ldots)$. Consider the map $N : D \to H$, $D = \{x \in H : \|x\| \leq 1\}$, $N(x) = (\sqrt{1 - \|x\|^2}, \frac{1}{2} x_1, \frac{1}{2} x_2, \ldots\ldots\ldots)$. N is obviously continuous and α - Lipschitz with constant $\frac{1}{2}$. Let $Q_n = (0, x_2, 0, \ldots, x_{2n}, 0, \ldots)$. The system

$$\begin{cases} Lx = (I - Q)N(x) \\ \\ Q_n N(x) = 0 \end{cases} \tag{4.4}$$

has a solution for every $n \in \mathbb{N}$. On the other hand the equation

$$Lx = N(x)$$

has no solutions.

The following result (see [3] for a similar one where L is Fredholm of index 0) gives sufficient conditions for the solvability of (4.3) for every $n \in \mathbb{N}$. We are working in a Hilbert space H and we are assuming that $\operatorname{Ker} L$ is closed and there exists a sequence of finite dimensional spaces $X_1 \subset X_2 \ldots$ such that $\overline{\bigcup_n X_n} = \operatorname{Ker} L$ and a family of orthogonal projections $P_n : H \to H$, $\operatorname{Im} P_n = X_n$ such that $P_n x \to Px$ for every $x \in H$, where P projects H orthogonally onto $\operatorname{Ker} L$.

(4.1) *THEOREM*

Let H be a Hilbert space and let $L : H \to H$ such that $\operatorname{Ker} L$ is closed and $\operatorname{Im} L = \operatorname{Ker} L^{\perp}$. Let $N : H \to H$ be demicontinuous and α-Lipschitz with constant r.

Assume that

i) there exists $a > 0$ such that $-a\|Lx\|^2 \leq (Lx,x)$ for every $x \in H$;

ii) there exists $b > a$, p,q positive numbers such that $((I - P)\ N(x),\ x - Px) \leq -b\ \|(I - P)\ N(x)\|^2 + p\|Px\| + q(\|x - Px\| + 1)$ for every $x \in H$, where $P : H \to H$ is an orthogonal projection such that $\operatorname{Im} P = \operatorname{Ker} L$.

iii) there exists $c > 0$ such that $a\left(1 + \frac{p}{c}\right) < b$ and $c < \liminf\limits_{\substack{t \to +\infty \\ u \to y}} ((I - P)N(tu),\ -u)$ for every $y \in \operatorname{Ker} L$, $\|y\| = 1$.

Then the system

$$\begin{cases} Lx = (I - P)\ N(x) \\ \\ P_n N(x) = 0 \end{cases} \tag{4.5}$$

has a solution provided that there exists a linear continuous map K such that $LKy = y$ for every $y \in \operatorname{Im} L$, and K is α-Lipschitz with constant k such that $rk < 1$.

Proof

Consider the operator $L + P_n : H \to \operatorname{Im} L \oplus X_n$ and let T be a right inverse of it, which is α-Lipschitz with constant not larger than k. If the set $S = \{x \in H : x = \lambda T((I - P)\ N(x) + P_n(x + N(x))), \lambda \in (0,1]\}$ is bounded then (4.5) has a solution (by Theorem (3.1)). Assume therefore that S is unbounded, that is there exist sequences $\{\lambda_m\}$, $\lambda_m \in (0,1)$, $\{x_m\}$, $\|x_m\| \to +\infty$ such that

$$x_m = \lambda_m T((I - P)\ N(x_m) + P_n(x_m + N(x_m))) \tag{4.6}$$

which implies

$$\text{a)}\ \ Lx_m = \lambda_m((I - P)\ N(x_m)) \qquad \text{b)}\ \ (1 - \lambda_m)\ P_n x_m = \lambda_m\ P_n N(x_m). \tag{4.7}$$

Hence

$$\|Lx_m\|^2 = \lambda_m^2 \|(I-P)\,N(x_m)\|^2 . \tag{4.8}$$

In (4.7) a) we take scalar product with $x_m - Px_m$ and we use i), ii) to obtain

$$(b-a)\ \|(I-P)\,N(x_m)\|^2 \le p\|Px_m\| + q\ (\|x_m - Px_m\| + 1). \tag{4.9}$$

Using

$$\|x_m - Px_m\| = \|KLx_m\| \le \|K\|\ \|(I-P)\,N(x_m)\|$$

we obtain that given $a < \beta < b$ there exists D depending on β, q, $\|K\|$ such that

$$(b-\beta)\ \|(I-P)\,N(x_m)\|^2 \le p\|Px_m\| + D .$$

This implies that $\|Px_m\| \to +\infty$ and $\dfrac{x_m - Px_m}{\|Px_m\|} \to 0$ as $m \to +\infty$.

Applying P to equation (4.6) and dividing by $\|Px_m\|$ we see that $\dfrac{Px_m}{\|Px_m\|} \to y$ as $m \to +\infty$. Thus $\dfrac{x_m}{\|Px_m\|} \to y$. On the other hand using i) and (4.9) we obtain

$$(b-\beta)\ ((I-P)\,N(x_m),\ -x_m) \le \beta(p\|Px_m\| + D)$$

and setting $\|Px_m\| = t_m$ and $x_m = t_m y_m$ we have

$$(b-\beta)((I-P)\,N(t_m y_m),\ -y_m) \le \beta\left[p + \frac{D}{\|Px_m\|}\right]. \tag{4.10}$$

Hence, by iii)

$$\frac{ap}{b-a} < c < \frac{\beta p}{b-\beta}$$

which is absurd if β is sufficiently close to a. ∎

Theorem (4.1) ensures that the system

$$\begin{cases} Lx = (I-P)\,N(x) \\ P_n N(x) = 0 \end{cases} \tag{4.5}$$

has a solution x_n for every $n \in N$. As the example (4.1) shows the solvability of (4.5) for every n is not enough to guarantee the solvability of $Lx = N(x)$.

We present now some possible additional assumptions, which will guarantee the solvability of $Lx = N(x)$ from the solvability of (4.5). The given list does not pretend to be complete and does not want to present the most general conditions either. We shall deal with the case when K, the pseudo-right inverse of L, is compact and N is a demicontinuous map sending bounded sets into bounded sets. We leave to the reader the natural extensions to the case when K and N are α-Lipschitz.

We remark, first, that under the assumptions of Theorem (4.1) the sequence $\{x_n\}$ is bounded. This condition is supposed to be verified in the sequel.

Case (4.1) Assume that $\{Px_n\}$ is a compact sequence. Then the equation $Lx = N(x)$ has a solution.

In fact, in this case, we obtain that $x_n \to x_0$ and $Lx_0 = (I - P)\,N(x_0)$. On the other hand it is not hard to show that $PN(x_n) \rightharpoonup 0$ and $PN(x_n) \rightharpoonup PN(x_0)$. Thus $PN(x_0) = 0$ and $Lx_0 = N(x_0)$.

Case (4.2) Assume that $\{PN(x_n)\}$ is compact. Then $Lx = N(x)$ has a solution provided that N is monotone.

In fact, in this case, we first obtain that $PN(x_n) \to 0$. To see this recall that $\{x_n\}$ is bounded and therefore $x_n \rightharpoonup x_0$. Moreover $\{N(x_n)\}$ is bounded. Hence $N(x_n) \rightharpoonup y$. Since $\mathrm{Ker}\, P_i \supset \mathrm{Ker}\, P_{i+1}$, $i = 1,2,\ldots$ we have $P_iN(x_n) = 0$ for $n \geq i$ and so $P_iy = 0$ for every $i = 1,2,\ldots$. This implies $Py = 0$ and $PN(x_n) \rightharpoonup 0$. The compactness of $\{PN(x_n)\}$ implies that $PN(x_n) \to z$ (by possibly using a suitable subsequence) and therefore $PN(x_n) \to 0$. Now observe that the sequence $\{x_{n1}\}$

$$x_{n1} = x_n - Px_n = K(I - P)\,N(x_n) \qquad n = 1,2,\ldots$$

is compact since K is compact. Thus $x_{n1} \to x_{01}$. Since $Lx_n = (I - P)\,N(x_n)$ we have that $\{Lx_n\}$ is bounded and therefore $Lx_n \rightharpoonup w$. So $KLx_n \rightharpoonup Kw$. But $KLx_n = x_{n1} \to x_{01}$.

Thus $Kw = x_{01}$ and $Lx_n \rightharpoonup LKw = Lx_{01} = Lx_0$. From

$$(N(x_n) - N(y), x_n - y) \geq 0$$

and

$$N(x_n) = Lx_n + PN(x_n)$$

we obtain

$$(Lx_n + PN(x_n) - N(y), \quad x_n - y) \geq 0 .$$

Since $Lx_n \rightharpoonup Lx_0$, $x_{n1} \to x_{01}$ and $(Lx_n, x_n) = (Lx_n, x_{n1})$ we obtain that

$$(Lx_n, x_n - y) \to (Lx_0, x_0 - y).$$

Since $PN(x_n) \to 0$ and $x_n \rightharpoonup x_0$ we obtain that $(PN(x_n), x_n - y) \to 0$. We therefore have

$$(Lx_0 - N(y), \quad x_0 - y) \geq 0$$

and, using Minty's trick [11] with $y = x_0 + tv$, we get

$$(Lx_0 - N(x_0), v) \leq 0$$

for every $v \in H$. Thus $Lx_0 = N(x_0)$.

A particular case of (4.2) is obtained when H is compactly embedded in a larger Hilbert space $\mathcal{H}$ and N is monotone with respect to the scalar product of $\mathcal{H}$ (see [4] for similar assumptions).

Both Theorem (4.1) and Case (4.2) are presented in Hilbert spaces. It is obvious that they can be obtained in suitable Banach spaces and for a larger class of maps which includes some multivalued ones. We leave these natural extensions to the reader.

REFERENCES

[1] BREZIS, H.: Nonlinear equations at resonance, preprint.

[2] BREZIS, H. and NIRENBERG, L.: Forced vibrations for nonlinear wave equation, Comm. Pure Appl. Math. 31 (1978), 225-326.

[3] BREZIS, H. and NIRENBERG, L.: Characterization of the range of some nonlinear operators and applications to BVP, Ann. Scuola Norm. Sup. Pisa, IV, Vol V, (1978), 225-326.

[4] CESARI, L.: Hyperbolic problems: existence and applications, EQUADIFF 78, Firenze, Italy.

[5] FENSKE, C.C. and PEITGEN, H.O.: Repulsive fixed points for multivalued transformations and the fixed point index, Math. Annalen 218, (1975), 9-18.

[6] FURI, M. and MARTELLI, M.: A degree for a class of acyclic valued vector fields in Banach spaces, Ann. Scuola Normale Superiore Pisa, 1 (1974), 301-310.

[7] FURI, M., MARTELLI, M. and VIGNOLI, A.: Boundary value problems, Fredholm operators and their index, preprint.

[8] GRANAS, A.: Topics in infinite dimensional topology, Sem. Collège de France, (1969-1970).

[9] LANDESMAN, E.M. and LAZER, A.C.: Nonlinear perturbations of linear elliptic boundary value problems at resonance, J. Math Mech 19, (1970), 609-623.

[10] MAWHIN, J.: Equivalence theorems for nonlinear operator equations and coincidence degree for some mappings in locally convex topological spaces, J. Diff. Eq. 12 (1972), 601-636.

[11] MINTY, G.J.: Monotone nonlinear operators in Hilbert spaces, Duke Math. Journal 29 (1962), 341-346.

[12] NIRENBERG, L.: An application of generalized degree to a class of non-linear problems, Coll. d'Analyse Math., Liège (1971), 54-74.

13] ZEZZA, P.: An equivalence theorem for nonlinear operator equations and an extension of Leray-Schauder continuatien theorem. Boll. Un. Mat. Ital. (5) 15-A (1978), 545-551.

MULTI-APPLICATIONS DU TYPE DE KANNAN

PAR

SYLVIO MASSA*

Instituto di Matematica
Università degli Studi di Milano
via Saldini, 50
20133 Milano (Italia)

1. LE THEOREME.

Soient (X,d) un espace métrique, $cb(X)$ la famille des sous-ensembles fermés et bornés de X, et H la métrique de Hausdorff induite par d sur $cb(X)$.

Nous considérons les applications $F : X \to cb(X)$ telles que, pour tout x,y dans X

$$H(F(x),F(y)) \leq a(x,y)\, d(x,F(x)) + a(y,x)\, d(y,F(y)) \tag{1.1}$$

où $a: X \times X \to [0,1)$ est telle que

$$a(x,y) + a(y,x) \leq 1 \tag{1.2}$$

$$a(x,y) \to 1 \Rightarrow \text{Max}\{d(x,F(x)),\ d(y,F(y))\} \to 0 \text{ ou } \infty. \tag{1.3}$$

Ces applications ont été introduites par R. Kannan [1] dans le cas: $F : X \to X$ et $a(x,y) \equiv \alpha < 1/2$. Elles ont été étudiées pendant les dernières douze années dans le cas général, soit sous la condition de non expansivité (1.2) soit sous la condition de contractivité

$$a(x,y) + a(y,x) < 1. \tag{1.2'}$$

(Pour la bibliographie voir aussi, p.e., [3].)

---

*Cette recherche a été conduite avec la contribution du GNAFA.

Il est bien connu que les applications de Kannan ne satisfont, de par leur nature, à aucune condition de continuité: il est donc intéressant d'obtenir des résultats de point fixe sans introduire d'hypothèses de ce genre. Pour le cas des applications à valeurs simples, on peut voir [3] et [4].

Le cas des multi-applications est plus complexe ; dans [2] on a assuré l'existence de points fixes, lorsque X est un sous-ensemble faiblement compact d'un espace normé, $F(x)$ est un sous-ensemble faiblement fermé de X et F est de Kannan contractive (c'est-à-dire que les conditions (1.1), (1.2') et (1.3) sont vérifiées).

Ici nous considérons le cas F de Kannan non expansive (c'est-à-dire que (1.1) et (1.3) sont vérifiées et (1.2') est remplacée par (1.2)) et nous prouvons le suivant.

THEOREME

Soient X un sous-ensemble fermé et convexe d'un espace de Banach uniformément convexe, F une application de X dans ses sous-ensembles non vides, fermés, bornés et convexes qui satisfait les conditions (1.1), (1.2) et (1.3). Alors il existe un sous-ensemble non vide K de X tel que $F(x) = K$ pour tout x dans K.

Ce théorème est l'extension aux multi-applications d'un résultat de [6] (Théorème 3).

2. LA DEMONSTRATION.

Pour demontrer le théorème, nous utiliserons le lemme suivant.

LEMME

Pour tout $r \geq 0$, si $A_r = \{x \in X : d(x, F(x)) \leq r\} \neq \phi$ on a

a) $\operatorname{diam}(A_r) < \infty$

b) $F(A_r) \subseteq A_r$

c) $\overline{co}\,(F(A_r)) \subseteq A_r$.

(Le lemme ne dépend pas de la condition de convexité uniforme.)

DEMONSTRATION DU LEMME

a) Si $\operatorname{diam}(A_r) = \infty$, alors on peut trouver $\{y_n\}$ dans A_r et, pour tout n, z_n dans $F(y_n)$ tels que

$$\lim_{n\to\infty} \|y_n\| = \lim_{n\to\infty} \|z_n\| = \infty .$$

Si l'on fixe y_0 dans A_r, on a

$$d(z_n, F(y_0)) \leq H(F(y_n), F(y_0)) \leq r$$

ce qui est absurde.

b) Si $y \in F(x)$ avec $x \in A_r$

$$d(y,F(y)) \leq H(F(x),F(y)) \leq a(x,y)\, r + a(y,x)\, d(y,F(y))$$

donc, puisque $a(y,x) < 1$, par (1.2) on obtient que $d(y,F(y)) \leq r$.

c) Soit $y \in \operatorname{co}(F(A_r))$, c'est-à-dire $y = \sum_1^n \lambda_i y_i$, $\lambda_i > 0$, $\sum_1^n \lambda_i = 1$ avec $y_i \in F(x_i)$ et $x_i \in A_r$.

Pour tout z dans X il existe $\bar{x} = \bar{x}(z,y)$ dans A_r tel que

$$d(z,F(z)) \leq r + \frac{d(z,y)}{1 - a(z,\bar{x})} .$$

En effet, on a

$$d(y,F(z)) \leq \sum_1^n \lambda_i\, H(F(x_i),F(z)) \leq \operatorname*{Max}_i \{H(F(x_i),F(z))\}$$

$$= H(F(\bar{x}), F(z)) \leq a(\bar{x},z)\, r + a(z,\bar{x})\, d(z , F(z)) .$$

Si $z \in \overline{\operatorname{co}}\,(F(A_r))$, il existe, pour tout entier k, y_k dans $\operatorname{co}(F(A_r))$ tel que $\|z - y_k\| < \frac{1}{k}$. Appelons x_k le point $\bar{x}(z,y_k)$. On a

$$d(z,F(z)) \leq r + \frac{1}{k[1 - a(z,x_k)]}$$

d'où c). ■

DEMONSTRATION DU THEOREME

Soit r tel que $A_r \neq \phi$. Considérons la restriction de F à $\overline{co}\,(F(A_r))$. Soit

$$r_0 = \text{Inf}\{r : A_r \neq \phi\}.$$

Le Théorème 2 de [2] assure que $A_{r_0} \neq \phi$. (Voir aussi [5], théorème 5.)

Soit B un ensemble minimal par rapport à la relation d'inclusion parmi les ensembles non vides A tels que

(i) A est fermé et convexe

(ii) $F(A) \subseteq A$

(iii) $d(x,F(x)) = r_0 \quad \forall\, x \in A.$

Par le lemme, $B = \overline{co}\,(F(B))$. Considérons $F|_B$. Il est facile de vérifier que

$$d(y,F(x)) \leq r_0 \quad \forall\, x,y \in B. \tag{2.1}$$

Considérons alors deux suites $\{x_n\}$, $\{y_n\}$ dans B telles que

$$\|x_n - y_n\| \to \delta = \text{diam } B.$$

On peut trouver (à cause de (2.1)) $\alpha_n, \beta_n \in F(\frac{1}{2}x_n + \frac{1}{2}y_n)$ tels que

$$\|x_n - \alpha_n\| \leq r_0\,, \quad \|y_n - \beta_n\| \leq r_0.$$

Nous avons (pour tout n)

$$r_0 \leq \|\tfrac{1}{2}(x_n + y_n) - \tfrac{1}{2}(\alpha_n + \beta_n)\| \leq \tfrac{1}{2}\left(\|x_n - \alpha_n\| + \|y_n - \beta_n\|\right) \leq r_0$$

donc $x_n - \alpha_n = y_n - \beta_n$, par la convexité uniforme, et par conséquent

$$\|x_n - \alpha_n\| = r_0 \tag{2.2}$$

$$x_n - y_n = \alpha_n - \beta_n. \tag{2.3}$$

Par (2.3), comme $\|x_n - \beta_n\| \leq \delta$ et $\|y_n - \alpha_n\| \leq \delta$, et que

$$\|x_n - y_n\| = \| \tfrac{1}{2}(x_n - y_n) + \tfrac{1}{2}(\alpha_n - \beta_n)\| = \| \tfrac{1}{2}(x_n - \beta_n) + \tfrac{1}{2}(\alpha_n - y_n)\| \to \delta,$$

nous obtenons par la convexité uniforme que

$$\|x_n - \beta_n + y_n - \alpha_n\| = 2\|x_n - \alpha_n\| \to 0.$$

De (2.2) on déduit $r_0 = 0$ et $K = B$. ■

References

[1] KANNAN, R.: Some results on fixed points, Bull, Calcutta Math. Soc., 60 (1968), 71-76.

[2] MASSA, S. et ROUX, D.: Multicontractive Kannan maps in normed spaces, Instit. Lombardo Accad. Sci. Lett. Rend. A (à paraître).

[3] ROUX, D. et MALUTA, E.: Contractive Kannan maps in compact spaces, Riv. Mat. Univ. Parma, (4) 5 (1979), 141-145.

[4] ROUX, D. et ZANCO, C.: Kannan maps in normed spaces, Atti. Accad. Naz. Lincei Rend. Cl. Sci. fis. mat. natur., (8) 65 (1978), 252-258.

[5] SHIAU, C., TAN, K.K. et WONG, C.S.: Quasi-nonexpansive multi-valued maps and selections, Fund. Math., 87 (1975), 109-119.

[6] SOARDI, P.: Su un problema di punto unito di S. Reich, Boll. Un. Mat. Ital., (4) 4 (1971), 841-845.

ON THE SOLVABILITY OF NONLINEAR EQUATIONS IN BANACH SPACES

By

I. MASSABO, P. NISTRI, J. PEJSACHOWICZ
Dipartimento di Matematica
Università della Calabria
C.P. Box 9
87030 Roges (Cosenza)
Italia

0. Introduction

In this note we are concerned with the solvability of nonlinear systems of the form

$$(1) \quad \begin{cases} 0 = f(x,y) \\ 0 = g(x,y) \end{cases}$$

with f and g continuous maps defined on the closure of an open subset U of the product of two Banach spaces X, Y of the form

$$f(x,y) = y - \overline{f}(x,y)$$

and

$$g(x,y) = x - \overline{g}(x,y)$$

where $\overline{f} : \overline{U} \to Y$ and $\overline{g} : \overline{U} \to X$ are compact maps.

In our approach, instead of considering sufficient conditions for the existence of fixed points of the compact map

$$(\overline{g},\overline{f}) : \overline{U} \to X \times Y,$$

we follow closely the "alternative method" developed by Cesari and his co-workers (see [4] for an extensive bibliography) in solving equations arising from nonlinear perturbations of Fredholm operators. Roughly speaking, we "solve" the first equation in term of y as a "function" of x and hence we introduce the solution $S(x)$ in the second one. The main goal of this paper consists in dropping the assumption on the *unique solvability* of the first equation. Thus we consider the multivalued *solution* map

$$x \multimap S(x) = \{y \in Y : \ f(x,y) = 0\}$$

whose graph, in the $X \times Y$ space is the *solution* set

$$S = \{(x,y) : \ f(x,y) = 0\}$$

of the first equation. Then we seek zeros of the multivalued map $x \multimap T(x) = g(x,S(x))$ which are obviously the only solutions of (1).

When $X = \mathbb{R}$ (the real line), the above approach has been used implicitly by several authors (see for example [15, 1, 10, 2] and it will be briefly described below.

Let us make the following assumption

(i) U is locally bounded over X

(2) (ii) the equation $0 = f(x,y) = y - \overline{f}(x,y)$ has no solutions on ∂U

(iii) for some $x \in X$, $\deg(f(x,\cdot), U(x), 0) \neq 0$, where $U(x) = \{y \in Y : (x,y) \in U\}$.

Under the assumption (2), given any continuous map $g : S \to \mathbb{R}$ the multivalued map $T(x) = g(x,S(x))$ has the intermediate value property. Namely:

(3) if for some $x_0, x_1 \in \mathbb{R}$ we have that $T(x_0) \subset \mathbb{R}_+$ and $T(x_1) \subset \mathbb{R}_-$, then $0 \in T(x)$ for some $x \in [x_0, x_1]$.

Thus (3) gives a sufficient condition for the solvability of system (1).

The proof of the above assertion follows directly from the existence of a connected subset C of S joining $\{x_0\} \times S(x_0)$ and $\{x_1\} \times S(x_1)$. The last has been proved in [3] under assumption (2). Notice that the converse is also true. Indeed, if for each continuous function $g : S \to \mathbb{R}$, the map $T(x) = g(x,S(x))$ has the intermediate value property (between x_0 and x_1) then we can ensure the existence of a connected set C as above.

The natural extension of the intermediate value property of a continuous function of a real variable to higher dimensional spaces is the well known Borsuk-Ulam theorem (see [9, p. 21]). Suppose for the moment that the solution map is a single-valued one. Then, by the Borsuk-Ulam theorem we can assert the existence of a solution of (1) provided that for the compact vector field

$$T(x) = g(x,S(x)) = x - \overline{g}(x,S(x))$$

the following holds

(4) there exists $r > 0$ such that for each x with $\|x\| = r$ $T(x) \neq tT(-x)$ for any $t \geq 0$.

In Theorem (1.1) we show that also in the case when the solution map is not single-valued then a Borsuk-Ulam type condition is sufficient to ensure the existence of a solution of system (1). More precisely:

Let f and g be as in (1). If f verifies assumption (2) then the multivalued map $T(x) = g(x,S(x))$ has a zero in a ball $B \subset X$ provided that

(5) for each $x \in \partial B$ there exists a continuous functional x^* on X such that $x^*(T(x)) \subset \mathbb{R}_+$ and $x^*(T(-x)) \subset \mathbb{R}_-$.

Notice that if T is a singlevalued map then (4) and (5) are equivalent.

Moreover we show, in Theorem (1.2), that under assumptions (2) if $\overline{g} : S \to X$ is any continuous map, then the multivalued map $T(x) = \overline{g}(x,S(x))$ has the fixed point property with respect to compact and convex sets. Namely, if $K \subset X$, is compact convex with $T(K) \subset K$, then there exists $x \in K$ such that $x \in T(x)$ and hence the equation

$$\begin{cases} y - \overline{f}(x,y) = 0 \\ x - \overline{g}(x,y) = 0 \end{cases}$$

has a solution (x,y) with $x \in K$.

Thus, Theorems (1.1) and (1.2) can be viewed from one side as an extension of Liapunov-Schmidt to the case of nonuniqueness of solutions of the equation $f(x,y) = 0$ and on the other side as an improvement to dimensions higher than one of the connectivity property described in [15]. This will be the object of a forthcoming paper.

Although several particular cases of Theorem (1.1) can be proved by direct degree arguments, our proof is of independent interest. Roughly speaking, we show that the solution map $x \multimap S(x)$ and hence also $x \multimap T(x)$ can be approximated *in graph* by a particular class of finite valued upper-semicontinuous maps to which the ordinary degree and the fixed point theory of continuous maps can be extended. This class of multivalued maps, called weighted maps (w-maps), was first introcuced by Darbo [5]. The existence of zeros for the multivalued map T will be obtained, using w-maps, in the same way as the zeros of convex valued vector fields can be obtained by approximation with singlevalued maps.

We would like to add, in passing, that the same method used in Theorem (1.1) could lead to a different approach to define some invariants of parametrized families of maps known in algebraic topology as *transfer homomorphism* and *generalized index* (see [7, 8, 12]). This was already observed in [8].

The paper is divided into two parts.

In the first part we state our main results (Theorem (1.1) and Theorem (1.4)) and some corollaries. Finally, we consider the particular case of equations arising from nonlinear perturbations of Fredholm operators.

The second part is devoted to the proof of Theorem (1.1). First of all, we establish some basic properties of the solution map $x \multimap S(x)$. Next, we show that the map f can be arbitrarily approximated by maps f' having, at each level x, a finite number of zeros. Then we discuss briefly the notion of w-maps and we show that the solution maps of our **approximation** are actually w-maps. In the last part we prove the Borsuk-Ulam theorem for w-maps and we use it in order to prove Theorem (1.1). We close this section by proving Theorem (1.2) and Corollary (1.3).

1. General results

Throughout this section X, Y will denote real Banach spaces and $B = B(0,r)$ the closed ball centered at zero with radius r. Furthermore, given any subset S of $X \times Y$ we denote by

$$S(x) = \{y \in Y : (x,y) \in S\}$$

$$S(A) = \{y \in Y : (x,y) \in S, \ x \in A\} \quad \text{for } A \subset X,$$

$$S_x = S \cap (\{x\} \times Y) \text{ and by } S_A = S \cap (A \times Y) \quad \text{for } A \subset X.$$

Let U be a subset of $X \times Y$. We shall say that U is *locally bounded over* X if for any $(x,y) \in U$ there exists a neighborhood N of x such that U_N is a bounded subset of $X \times Y$.

Let $U \subset X \times Y$ be open and locally bounded over X. We shall say that $f : \overline{U} \to Y$ is a *parametrized compact vector field* if $f(x,y) = y - \overline{f}(x,y)$ with $\overline{f}$ continuous and $\overline{f}(D)$ relatively compact in Y for any bounded subset D of U. Given any parametrized compact vector field $f : \overline{U} \to Y$ we shall denote by

$$S^f = \{(x,y) \in \overline{U} : \ f(x,y) = 0\}$$

$$\mathcal{D}^f = \{x \in X : \ S^f_x \cap \partial U = \phi\}.$$

When no confusion will arise, we shall omit the superscript f in the above notations.

REMARK

The set valued map $x \multimap S(x) = S^f(x)$ from $\mathcal{D}$ into Y is compact valued. Furthermore, $\mathcal{D}$ is an open subset of X and the map $x \multimap S(x)$ from $\mathcal{D}$ into Y is upper-semicontinuous (u.s.c.)(see Proposition (2.1)).

If f is defined on all $X \times Y$, taking

$$\mathcal{D} = \left\{ x \in X \;\middle|\; \begin{array}{l} x \text{ is not bifurcation point from infinity of the} \\ \text{equation } f(x,y) = 0 \end{array} \right\}$$

we have that $\mathcal{D}$ is an open subset of X and the map $x \multimap S(x)$ is u.s.c. from $\mathcal{D}$ into Y (see Proposition (2.2)).

Let $A_1, A_2 \subset X$. We shall say that A_1 and A_2 are *strictly separated by an hyperplane* (s.s.h.) if there exists a continuous functional $x^* \in X^*$ (the dual of X) such that $x^*(x) > 0$ for all $x \in A_1$ and $x^*(x) < 0$ for all $x \in A_2$. We are now able to state the main results. For this the following assumption will be taken once and for all.

(A) $\begin{cases} \text{The set } U \subset X \times Y \text{ is open in } X \times Y \text{ and locally bounded over } X. \text{ The map} \\ f : U \to Y \text{ is a parametrized compact vector field such that } 0 \in \mathcal{D} \text{ and the} \\ \text{Leray-Schauder degree } \deg(f(0,\cdot), U(0), 0) \neq 0. \end{cases}$

For any parametrized compact vector field $g : U \to X$ let us define $T : \mathcal{D} \multimap X$ by $T(x) = g(x, S(x))$.

(1.1) *THEOREM*

Under assumption (A) suppose that:

(i) there exists $r > 0$ such that $B = B(0,r) \subset \mathcal{D}$

(ii) for any $x \in \partial B$ such that $0 \notin T(x)$, $T(x)$ and $T(-x)$ are strictly separated by an hyperplane.

Then there exists $x \in B$ such that $0 \in T(x)$ and hence the system

$$\begin{cases} f(x,y) = 0 \\ \\ g(x,y) = 0 \end{cases} \tag{1.1}$$

has a solution in U.

REMARK

Let us note that Theorem (1.1) holds for g of the form $g(x,y) = x - \bar{g}(x,y)$, where $\bar{g}$ is a continuous map defined on S_B and such that $\bar{g}(S_B)$ is a relatively compact subset of X.

The following result states that if $\overline{g}$ is any continuous map defined on the solution set S of a map f verifying (A) and with values in X then the multivalued map $\overline{T}(x) = \overline{g}(x,S(x))$ has the fixed point property with respect to compact convex sets. Namely:

(1.2) *THEOREM*

Assume that (A) holds. Furthermore, let $K \subset X$ be a compact convex set such that $0 \in K \subset \mathcal{D}$. Let $\overline{g} : S \to X$ be any continuous map and let $\overline{T}(x) = \overline{g}(x,S(x))$, if $\overline{T}(x) \subset K$ for any $x \in K$, then $\overline{T}$ has a fixed point in K, hence the system.

$$\begin{cases} y = \overline{f}(x,y) \\ x = \overline{g}(x,y) \end{cases} \qquad (1.1)'$$

has a solution.

REMARK

The above theorem holds for maps $f : \overline{U} \to X$ of the form $f(x,y) = y - \overline{f}(x,y)$ where $\overline{f} : \overline{U} \to Y$ is a map such that

(i) $\overline{f}(x,\cdot) : \overline{U}(x) \to Y$ is compact for each x

(ii) $\overline{f}$ is continuous in x uniformly in y.

Let $g : \overline{U} \to X$ be a parametrized compact vector field. Under assumption (A) the following corollaries are consequences of Theorem (1.2) and they can be regarded, via the associated multivalued map $\overline{T}$, as an extension of Leray-Schauder and Krasnosel'skiĭ fixed point theorems for single-valued compact maps.

(1.3) *COROLLARY*

Suppose that (A) and (i) of Theorem (1.1) hold. Then (1.1)' has a solution in U_B provided that the following holds:

(L.S.) If $x \in \partial B$ and $tx \in \overline{T}(x) = \overline{g}(x,S(x))$ then $t \le 1$.

(1.4) *COROLLARY*

Let X be a Hilbert space (we denote by $\langle\cdot,\cdot\rangle$ its scalar product). Assume that (A) and (i) of Theorem (1.1) hold. Then system (1.1) has a solution in U_B provided that $x \in \partial B$ and $y \in T(x) = g(x,S(x))$ imply $\langle x,y\rangle \geq 0$.

SOLVABILITY OF NONLINEAR PERTURBATIONS OF FREDHOLM OPERATORS

Although Theorem (1.1) covers as particular cases several coincidence problems we would like to specialize it to the case of solvability of equations arising from nonlinear perturbations of Fredholm operators.

In the real Hilbert space $H = L^2(\Omega)$, where Ω is a bounded domain in a finite dimensional real Euclidean space, let $L : D(L) \subset H \to H$ be a closed linear operator with dense domain and closed range. Assume that dim Ker $L < +\infty$ and Im $L = (\text{Ker } L)^{\perp}$ (the orthogonal complement). Hence H has an orthogonal decomposition $H = \text{Ker } L \oplus \text{Im } L$. For $u \in H$, we set $u = v + w = Pu + (I - P)u$ with $v \in \text{Ker } L$ and $w \in \text{Im } L$. L is therefore a one to one map of $D(L) \cap \text{Im } L$ onto $\text{Im } L$. Assume furthermore that the "inverse" $L^{-1} : \text{Im } L \to \text{Im } L$ is compact. Let $h : H \to H$ be a continuous and bounded map (that is h maps bounded sets into bounded sets). We are interested in the solvability of the nonlinear equation

$$Lu = h(u) \quad \text{in } H. \tag{1.2}$$

Using the *alternative method*, equation (1.2) can be converted into the following system.

$$\begin{cases} w = L^{-1}(I - P)h(v + w) & \text{auxiliary equation} \\ 0 = Ph(v + w) & \text{bifurcation equation.} \end{cases} \tag{1.3}$$

The solvability of (1.3) can now be handled in two different ways. Roughly speaking, first solve the auxiliary equation considered as an equation in w with v as parameter. Introduce then the solution map $S(v)$ in the bifurcation equation. Solve the resulting equation

$$0 \in Ph(v + S(v)).$$

Conversely, we could solve first the bifurcation equation considered as an equation in v with w as parameter. Introduce then the solution map $S(w)$ in the first equation and then solve

$$w \in L^{-1}(I-P)\, h(S(w)+w).$$

We state now two results on the solvability of equation (1.2). Theorem (1.5) below is related to the first approach described above and this can be viewed as an abstract formulation of the so called Landesman-Lazer type conditions.

In Theorem (1.6) we will follow the second approach on the solvability of (1.2).

(1.5) *THEOREM*

Let $S(v) = \{w \in \mathrm{Im}\, L : Lw = (I-P)h(v+w)\}$. Assume that there exist $r > 0$ and $R > 0$ such that

(i)' $S(v) \subset B(0,R) \subset \mathrm{Im}\, L$ for all $v \in B(0,r) \subset \mathrm{Ker}\, L$

(ii)' $\deg(I - L^{-1}(I-P)h|_{\mathrm{Im}\, L}, B(0,R), 0) \neq 0$

(iii)' for $v \in \mathrm{Ker}\, L$ with $\|v\| = r$ either $0 \in Ph(v+S(v))$ or there exists $v' \in \mathrm{Ker}\, L$ such that

$$\int_\Omega h(v+w)v' < 0 \qquad \text{for } w \in S(v)$$

$$\int_\Omega h(-v+w)v' > 0 \qquad \text{for } w \in S(-v)$$

Then equation (1.2) has a solution in H.

REMARK

(i)' is equivalent to say that the auxiliary equation has no bifurcation points from infinity in $B(0,r)$.

(i)' and (ii)' are certainly verified if h is uniformly bounded.

Proof

Let $V = \text{Ker } L$ and $W = \text{Im } L$. Since equation (1.2) is equivalent to system (1.3), via the obvious isomorphism $V \times W \to H$ we can rewrite system (1.3) as

$$\begin{cases} f(v,w) = 0 \\ g(v,w) = 0 \end{cases}$$

where

$$f(v,w) = w - L^{-1}(I - P)h(v + w)$$

and

$$g(v,w) = Ph(v + w).$$

Assumption (i)' implies that $B(0,r) \subset \mathcal{D}^f$, while (ii)' implies (A).

The last assumption in Theorem (1.1) is satisfied by the functional v^* defined on Ker L as

$$v^*(v) = \int_\Omega v\, v' . \blacksquare$$

(1.6) *THEOREM*

Set $S(w) = \{v \in \text{Ker } L : 0 = Ph(v + w)\}$. Assume that

(i)" there exists a continuous function $\rho : \text{Im } L \to \mathbb{R}_+$ such that

$$Ph(v + w) = 0 \quad \text{implies} \quad \|v\| \le \rho(w)$$

(this is equivalent to assuming that Ph does not have any bifurcation points from infinity, see the proof of Proposition (2.2)).

(ii)" $\deg (Ph|_{\text{Ker } L \cap B(0,\rho(0))}, B(0, \rho(0)), 0) \ne 0$;

(iii)" there exists $r > 0$ such that if $w \in D(L)$ $\|Lw\| = r$ then either

$Lw = h(v+w)$ for some $v \in S(w)$ or there exists $w' \in \operatorname{Im} L$ such that

$$\int_\Omega Lw\, w' > \int_\Omega h(v+w)\, w' \qquad \text{whenever } v \in S(w) \tag{1.4}$$

and

$$-\int_\Omega Lw\, w' < \int_\Omega h(\bar{v}-w)\, w' \qquad \text{whenever } \bar{v} \in S(-w). \tag{1.5}$$

Then equation (1.2) has a solution in H.

Proof

Since L is a closed Fredholm operator, $D(L)$, endowed with the norm $|||u||| = \|Lu\| + \|Pu\|$, becomes a Banach space in which $L : D(L) \to H$ is bounded. Furthermore $D(L)$ splits in a topological direct sum of $W = D(L) \cap \operatorname{Im} L$ and $V = \operatorname{Ker} L$.

Equation (1.2) is equivalent to the following system

$$\begin{cases} f(v,w) = 0 \\ g(v,w) = 0 \end{cases}$$

where $f(v,w) = Ph(v,w)$ and $g(v,w) = w - L^{-1}(I-P)h(v+w)$. Under the assumptions (i)' and (ii)', $\mathcal{D}^f = W$ and both (A) and (i) in Theorem (1.1) are satisfied. Let $w \in W$ with $|||w||| = \|Lw\| = r$ such that $Lw \neq h(v,w)$ whenever $Ph(v+w) = 0$. Let w^* be defined by

$$w^*(z) = \int_\Omega Lz\, w' \qquad \text{for } z \in W.$$

By the boundedness of L it follows that $w^* \in W^*$. By (1.4),

$$\int_\Omega Lw\, w' > \int_\Omega (I-P)h(v+w)w'$$

since

$$Ph(v+w) = 0.$$

Hence

$$\int_\Omega [Lw - LL^{-1} (I-P)h(v+w)]\,w' > 0$$

and so

$$\int_\Omega Lg(v,w)w' = w^*(g(v,w)) > 0 \qquad \text{for all } v \in S(w).$$

Similarly

$$w^*(g(v,-w)) < 0 \qquad \text{for all } v \in S(-w).$$

This completes the proof. ■

2. BORSUK-ULAM THEOREM AND THE PROOFS

PROPERTIES OF THE "SOLUTION MAP" $x \multimap S(x)$

(2.1) *PROPOSITION*

Let U be an open subset of $X \times Y$, locally bounded over X. Let $f: \overline{U} \to Y$ be a parametrized compact vector field. Then $\mathcal{D}$ is an open subset of X and the map $x \multimap S(x)$ from $\mathcal{D}$ into Y is u.s.c. (that is, $S(x)$ is compact and if V is a neighborhood of $S(x)$ then $S(x') \subset V$ provided that x' is close enough to x.).

Proof

Let $x_0 \in \mathcal{D}$. We shall prove that if V is any open set with $S(x_0) \subset V$ then there exist neighborhoods N of x_0 in X and V' of $S(x_0)$ in V such that $N \times V' \subset U$ and $S(x) \subset V'$ for any $x \in N$. Indeed, given $y \in S(x_0)$ consider neighborhoods of the form $N_y \times V_y$ where N_y is a neighborhood of x_0 in $\mathcal{D}$ and V_y, a neighborhood of y in Y, such that

$$V_y \subset U(x_0) \cap V \qquad \text{and} \qquad N_y \times V_y \subset U.$$

By the compactness of S_{x_0}, we can choose a finite number of neighborhoods of the above type, say $N_1 \times V_1$, $N_2 \times V_2, \ldots, N_r \times V_r$ which cover S_{x_0}. Let

$$N_0 = \bigcap_{i=1}^{r} N_i \qquad \text{and} \qquad V' = \bigcup_{i=1}^{r} V_i .$$

Clearly for each neighborhood N of x_0, with $N \subset N_0$ we have that $N \times V' \subset U$. It remains to show that

(*) there exists a neighborhood N of x_0 such that $S(x) \subset V'$ for all $x \in N$.

Without loss of generality we can assume that U_{N_0} is a bounded set. Suppose now that there are no $N \subset N_0$ for which (*) holds. Then, we can construct a bounded sequence $\{x_n, y_n\}$ such that $\{x_n\}$ converges to x_0, $y_n \notin V'$ and $y_n = \overline{f}(x_n, y_n)$. By the compactness of $\overline{f}$ we can assume (by passing to an appropriate subsequence) that $\{y_n\}$ converges to some $y_0 \in S(x_0)$, contradicting $S(x_0) \subset V'$. ∎

(2.2) *PROPOSITION*

Let $f : X \times Y \to Y$ be a parametrized compact vector field. Let

$$\mathcal{D} = \left\{ x \in X \,\middle|\, \begin{array}{l} x \text{ is not a bifurcation point from infinity for the equation} \\ f(x,y) = 0 \end{array} \right\}.$$

Then $\mathcal{D}$ is an open subset of X and the map $x \multimap S(x)$ from $\mathcal{D}$ into Y is u.s.c.

<u>*Proof*</u>

Assume that x is not a bifurcation point from infinity for the equation $f(x,y) = 0$. Then there exist a positive number r_x and a neighborhood N_x of x such that $S(x') \subset B(0,r_x)$ for any $x' \in N_x$. In particular it follows that $\mathcal{D}$ is an open subset of X. Let W_i be a locally finite refinement of $\{N_x\}_{x \in \mathcal{D}}$. Hence, for any i, there exists x_i such that $W_i \subset N_{x_i}$. Let $r_i = r_{x_i}$, then for any $x' \in W_i$ we have that $S(x') \subset \mathring{B}(0,r_i)$. If s_i is the partition of unity subordinated to $\{W_i\}$, then the function

$$r(x) = \sum_j s_j(x) r_j$$

is continuous from $\mathcal{D}$ into $\mathbb{R}_+$ and so the following set

$$U = \{(x,y) \in X \times Y : x \in \mathcal{D} \text{ and } \|y\| < r(x)\}$$

is open in $X \times Y$ and locally bounded over X. The assertion follows from Proposition (1.1) applied to $f|_{\overline{U}}$ (that is, the restriction of f to the closure $\overline{U}$ of U). ■

APPROXIMATION LEMMAS

In the following. U will denote an open subset of $X \times Y$ locally bounded over X and $f : \overline{U} \to Y$ a parametrized compact vector field.

(2.3) *LEMMA*

Let $X = \mathbb{R}^n$ and let $B = B(0,r) \subset \mathcal{D}$. Then for any neighborhood W of S_B^f in U there exists $\varepsilon > 0$ such that if $f_1 : \overline{U} \to Y$ verifies $\|f_1(p) - f(p)\| < \varepsilon$ for all $p \in \overline{U}_B$, then $S_B^{f_1} \subset W$.

Proof

Set $A = \overline{U}_B \setminus W$. Then A is closed and bounded in $X \times Y$. Since B is compact, the projection $\pi : B \times Y \to Y$ is a proper map. Hence $f = \pi - \overline{f}$ is a closed map being a compact perturbation of a proper map. Therefore

$$\inf_{(x,y) \in A} \{\|f(x,y)\|\} = \varepsilon > 0.$$

If $f_1 : \overline{U} \to Y$ is an ε-approximation of f in $\overline{U}_B$ then clearly $S_B^{f_1} \subset W$. ■

For the proof of Lemma (2.4) we shall need the following result due to Kurland and Robbin [13, Theorem (6.1)].

THEOREM (K.R.)

Let P, M and N be manifolds with $\dim(M) = \dim(N)$. Then there is an open dense $G \subset C^\infty(P \times M, N)$, endowed with the fine C^∞-topology, such that each $f \in G$ has the property that the map $f(p,\cdot) : M \to N$ is locally finite to-one (that is, every point of M has a neighborhood U such that $f^{-1}(n) \cap U$ is finite for all $n \in N$). In particular the inverse image of each point is discrete.

REMARK

The proof of the above result involves an "Infinite codimensional Lemma" (see [13, p. 139] and [16, p. 150]) and transversality theory.

(2.4) *LEMMA*

Let $X = \mathbb{R}^n$, $Y = \mathbb{R}^m$ and $B = B(0,r) \subset \mathcal{D}$. For any $\varepsilon > 0$, there exists a $f_1 : \overline{U} \to Y$ such that $f_1|_{\overline{U}_B}$ is an ε-approximation of $f|_{\overline{U}_B}$ and

(i) $S^{f_1}(x)$ is a finite subset of $U(x)$ for all $x \in B$;

(ii) $\deg(f(0,\cdot), U(0), 0) = \deg(f_1(0,\cdot), U(0), 0)$.

Proof

Let B_1 be a closed ball such that $B \subset \mathring{B}_1 \subset \mathcal{D}$. Then $\overline{U}_{B_1}$ is a bounded subset of $X \times Y$ which is contained in $B_1 \times B_2$ for some ball $B_2 \subset Y$. Let us still denote by f any continuous extension of f to all $K = B_1 \times B_2$. Let $\varepsilon > 0$, and $\varepsilon' = \min\{\varepsilon, \rho\}$ where

$$\rho = \inf_{x \in B_1} \operatorname{dist}(S_x, \partial U).$$

Since K is compact, f is a uniformly continuous map on K and so there exists a $\delta > 0$ such that $\|f(p) - f(p')\| < \varepsilon'/2$ for any $p, p' \in K$ with $\|p - p'\| < \delta$. Let us consider a finite covering of K consisting of balls D_i centered at the points $p_i \in K$ of radius $\delta/2$. Let $\{s_i\}$ be a C^∞-partition of unity subordinated to $\{D_i\}$ and let

$$f'(p) = \sum_i s_i(p) f(p_i).$$

Clearly $f'(p)$ is a C^∞-map on K and for $p \in K$ we have that

$$\|f(p) - f'(p)\| < \varepsilon'/2.$$

By Theorem (K.R.) there exists a C^∞-map $f_0 : \mathring{K} \to Y$ such that $f_0(x,\cdot)$ has only isolated zeros for any $x \in \mathring{B}$ and

$$\|f'(p) - f_0(p)\| < \varepsilon'/2$$

for all $p \in \mathring{K}$. Let f_1 be any continuous extension of $f_0|_{\overline{U}_B}$ to all $\overline{U}$. Then clearly f_1 is an ε-approximation of f which verifies (i). The assertion (ii) follows directly from the invariance property of the degree for small perturbations.■

WEIGHTED MAPS

Now we introduce a particular class of multivalued maps which is of fundamental importance in the proof of Theorem (1.1) and we will show that the solution map $S'(x)$ of the ε-approximation given by Lemma (2.4) is actually a w-map. The reason will be briefly described below.

Let X, Y be finite dimensional spaces. As we mentioned in the introduction in solving system (1.1) we seek zeros of the multivalued map $T(x) = g(x, S(x))$. Under the hypotheses of Theorem (1.1), T is u.s.c. and verifies the Borsuk-Ulam condition on the boundary ∂B of a ball in X. (see definition below).

Furthermore using Lemmas (2.3), (2.4) we can modify f to f' in such a way that $S'(x) = S^{f'}(x)$ is arbitrarily near to S and it is a finite valued map. Hence also T can be approximated by the finite valued u.s.c. map $T'(x) = g(x, S'(x))$. Clearly the fact that S' is a finite valued map is not sufficient to ensure the existence of zeros for T'. But the map S', being a "solution map", has another nice characteristic: namely, each point $y \in S(x)$ has an assigned multiplicity as solution of the equation $0 = f(x,y)$ and roughly speaking the multiplicity changes "nicely" with respect to x.

Such a class of finite valued u.s.c. maps has been introduced by G. Darbo [5, 6], under the name of weighted maps. Weighted maps form a good category of multivalued maps that enlarges that of single valued ones, remaining adequate for the fixed point and degree theory. Now since T' is arbitrarily close to T, it follows that also T' verifies the Borsuk-Ulam condition on ∂B. Then, by the Borsuk-Ulam theorem for w-maps, T' has a zero in B and this ensures the existence of a zero for T.

Let us recall briefly Darbo's results: let X, Y be topological Hausdorff spaces.

(2.5) *DEFINITION*

A finite valued u.s.c. map $F : X \multimap Y$ will be called *weighted map* (shortly w-map) if to each x and $y \in F(x)$ a multiplicity or weight $m(y, F(x)) \in \mathbb{Z}$ is assigned in such a way that the following property holds

(a) if U is an open set in Y with $\partial U \cap F(x) = \phi$, then

$$\sum_{y \in F(x) \cap U} m(y, F(x)) = \sum_{y' \in f(x') \cap U} m(y', F(x'))$$

whenever x' is close enough to x.

(2.6) *REMARK*

The number $i(F(x), U) = \sum_{y \in F(x) \cap U} m(y, F(x))$ will be called the index or multiplicity of $F(x)$ in U. Notice that (a) in Definition (2.5) is a local invariance property of the index. It states that the index of $F(x')$ in U for x' close enough to x coincides with the index of $F(x)$ in U whenever $\partial U \cap F(x) = \phi$. In particular if X is connected the number $i(F(x), Y)$ does not depend on $x \in X$. In this case the number $i(F) = i(F(x), Y)$ will be called the index of the weighted map F.

Actually as defined by Darbo a wieighted map is an equivalence class of maps verifying (a). But Definition (2.5) is more adequate to our purposes and all the results proved in [5] hold also in this context.

Let X, Y. Z denote topological Hausdorff spaces. The following properties have been proved in [5] (see also [11, 14]).

1) The sum of two w-maps $F, G : X \multimap Y$ defined as the u.s.c. map $F + G(x) = F(x) \cup G(x)$ with multiplicities given by $m(y, F + G(x)) = m(y, F(x)) + m(y, G(x))$, is a w-map. In an analogous form is defined λF for $\lambda \in \mathbb{Z}$ (where we pose $m(y, F(x)) = 0$ whenever $y \notin F(x)$).

2) Given any two w-maps $F : X \multimap Z$ and $G : Y \multimap W$, the product $F \times G : X \times Y \multimap Z \times W$ defined as the u.s.c. map $F \times G(x,y) = \{(z,w) : z \in F(x) \text{ and } w \in G(y)\}$ with $m((z,w), (F \times G)(x,y)) = m(z, F(x)) \cdot m(w, G(y))$ is a w-map.

3) If $F : X \multimap Y$ and $G : Y \multimap Z$ are w-maps, then the composition $G \circ F : x \multimap Z$ becomes a w-map by assigning multiplicities

$$m(z,\ G \circ F(x)) = \sum_{\substack{y \in F(x) \\ z \in G(y)}} m(z,\ G(y)) \cdot m(y,\ F(x)).$$

Moreover, if X and Y are connected then

$$i(G \circ F) = i(G) \cdot i(F).$$

4) Any continuous singlevalued map $f : X \to Y$ can be considered as a w-map by assigning multiplicity 1 to each $f(x)$.

It follows from properties 1), 2), 3), 4) that the category having as objects Hausdorff spaces and as morphisms w-maps is an additive category containing as subcategory that of singlevalued continuous maps. The notion of homotopy in this category, so called σ-homotopy in [5, 14], is defined in the same way as for continuous maps, that is, two weighted maps $F, G : X \multimap Y$ are σ-homotopic if there exists a weighted map $H : X \times [0,1] \multimap Y$ such that $H(x,0) = F(x)$ and $H(x,1) = G(x)$. Notice that if X is connected then F is σ-homotopic to G implies that $i(F) = i(G)$. Furthermore, in [5] Darbo constructed a homology functor $H = \{H_n\}_{n \geq 0}$ defined in this category, compatible with the σ-homotopy, and such that it verifies the Eilenberg-MacLane axioms for a homology theory.

Hence the restriction of H to the category consisting of continuous maps between compact absolute-neighborhood retracts coincides with the ordinary singular homology functor. We would like to add in passing that by means of this functor, Darbo extended the Lefschetz fixed-point theorem to w-maps from a compact A.N.R. into itself (see [6]). Let X, Y be Banach spaces.

(2.7) *LEMMA*

Let $f : \overline{U} \subset X \times Y \to Y$ be a parametrized compact vector field defined in the clousure of an open and locally bounded set U. If $f(x,\cdot)$ has only isolated zeros, then the map $x \multimap S(x)$ is a w-map from $\mathcal{D}^f$ into Y with $i(S) = \deg(f(x,\cdot),\ U(x),\ 0)$.

<u>*Proof*</u>

By Proposition (2.1), the map $x \multimap S(x)$ is an u.s.c. finite valued map from $\mathcal{D}$ into Y. Hence it is enough to show that to every $y \in S(x)$ we can assign an integer $m(y, S(x))$ with the property described in Definition (2.5). For this, let $y \in S(x)$. Since y is an isolated zero of $f(x,\cdot): U(x) \to Y$, we define $m(y, S(x))$ to be the multiplicity of y as a zero of $f(x,\cdot)$ (that is $m(y, S(x)) = \deg(f(x,\cdot), \Omega, 0)$ where Ω is an open neighborhood of y such that $\Omega \cap S(x) = \{y\}$. By the excision property of the degree $m(y, S(x))$ does not depend on the particular choice of Ω). We will see that $m(\cdot, S(\cdot))$ verifies property (a) of Definition (2.5). Let W be an open subset of Y such that $S(x) \cap \partial W = \phi$. Then by the uppersemicontinuity of S, there exists a ball $B(x,r)$ such that for any $x' \in B(x,r)$ we have $S(x') \cap \partial W = \phi$. Without loss of generality, we can assume that $B(x,r) \times W \subset U$. For $x' \in B(x,r)$, let $H : W \times [0,1] \to Y$ be defined by $H(y,t) = f(tx + (1-t)x', y)$. Since $tx + (1-t)x' \in B(x,r)$ for all $t \in [0,1]$, H is an admissible homotopy between $f(x,\cdot)|_W$ and $f(x',\cdot)|_W$. This and the additivity of the degree imply that

$$\sum_{y \in S(x) \cap W} m(y, S(x)) = \deg(f(x,\cdot), W, 0) = \deg(f(x',\cdot), W, 0)$$

$$= \sum_{y \in S(x') \cap W} m(y, S(x')). \blacksquare$$

BORSUK-ULAM THEOREM FOR W-MAPS

In the following we shall extend, in a suitable form, the classical Borsuk-Ulam theorem for continuous map to the context of weighted maps.

Let $B \subset X$ be a closed ball centered at the origin. We shall say that an u.s.c. map $F : B \to Y$ verifies the *Borsuk-Ulam property on* ∂B if

(B.U.) for each $x \in \partial B$, $F(x)$ and $F(-x)$ are strictly separated by a hyperplane.

(2.8) *THEOREM*

Let B be the unit ball in $\mathbb{R}^n$. Let $F : B \multimap \mathbb{R}^n$ be a w-map with $i(F) \neq 0$. If F verifies (B.U.) on ∂B then there exists $x \in \mathring{B}$ such that $0 \in F(x)$.

Proof

Notice that it sufficies to show that if $\widetilde{F}$ denotes the restriction of F

to ∂B then the homomorphism induced by $\tilde{F}:\partial B \multimap \mathbb{R}^n\setminus\{0\}$ in the n-th homology group is not trivial (that is $\tilde{F}_*: H_{n-1}(\partial B)\to H_{n-1}(\mathbb{R}^n\setminus\{0\})$ is different from the zero map). In fact, if $0\notin F(B)$ we get that $H:\partial B\times I\to\mathbb{R}^n\setminus\{0\}$ defined by $H(x,t) = F(tx)$ is a σ-homotopy between $\tilde{F}$ and the "constant" w-map $G(x) = F(0)$. Hence $\tilde{F}_* = G_* = 0$. We shall show that $\tilde{F}_*$ is not trivial by constructing a σ-homotopy between $\tilde{F}$ and $i(F)f$ where $f:\partial B\to\mathbb{R}^n\setminus\{0\}$ is a singlevalued odd continuous map. This will prove the theorem since the oddness of f implies that $f_*: H_{n-1}(\partial B)\approx\mathbb{Z}\to H_{n-1}(\mathbb{R}^n\setminus\{0\})\approx\mathbb{Z}$ is a multiplication by an odd number and hence, if $i(F)\neq 0$, we have that $(i(F)f)_* = i(F)f_*\neq 0$. For this, let us observe that the (B.U.)-condition in $\mathbb{R}^n$ states that

$$(*)\qquad \begin{array}{l}\text{for each } x\in\partial B \text{ there exists } y\in\partial B \text{ such that}\\ \qquad <y,z> \,>0 \quad\text{for all}\quad z\in\tilde{F}(x)\\ \\ \qquad <y,z> \,<0 \quad\text{for all}\quad z\in\tilde{F}(-x).\end{array}$$

For $y\in\partial B$, let $V_y = \{x\in\partial B: (*)\text{ holds}\}$. Since $co(\tilde{F}(x))$ and $co(\tilde{F}(-x))$ are compact and since $\tilde{F}$ is u.s.c. it follows that V_y is an open subset of ∂B for each y. By the (B.U.)-condition we have that $\{V_y\}_{y\in\partial B}$ is covering of ∂B. Let $\{V_{y_i}\}$, $0\le i\le m$, be a subcovering of $\{V_y\}_{y\in\partial B}$ and let $s_i:\partial B\to[0,1]$, $0\le i\le m$ be the partition of the unity subordinated to $\{V_{y_i}\}$.

Set $f(x) = \sum_{i=0}^{m}(s_i(x) - s_i(-x))y_i$. Then $f:\partial B\to\mathbb{R}^n$ is an odd continuous map.
Consider $H:\partial B\times I\multimap\mathbb{R}^n$ defined as the composition

$$\partial B\times I \xrightarrow{\Delta\times Id} \partial B\times\partial B\times I \xrightarrow{\tilde{F}\times f\times Id} \mathbb{R}^n\times\mathbb{R}^n\times I \xrightarrow{g} \mathbb{R}^n$$

where $\Delta:\partial B\to\partial B\times\partial B$ is the diagonal map and

$$g(x,y,t) = tx + (1-t)y.$$

Clearly, H is a w-map (as composition of w-maps). Furthermore, $H(x,0) = i(F)f(x)$ and $H(x,1) = \tilde{F}(x)$ for all $x\in\partial B$. Actually we shall prove that the image of H is contained in $\mathbb{R}^n\setminus\{0\}$. In fact, if $0\in H(x,t)$ then there exist $z\in\tilde{F}(x)$ and $t\in[0,1]$ such that

$$tz = -(1-t)f(x) = -(1-t)\left(\sum_{i=0}^{m} s_i(x)y_i - \sum_{i=0}^{m} s_i(-x)y_i\right).$$

Clearly, $t\neq 1$. Then for $\mu = -\frac{t}{1-t} < 0$ we have that

$$\mu z = \sum_{i=0}^{m} s_i(x)y_i - \sum_{i=0}^{m} s_i(-x)y_i .$$

Hence

$$(*) \qquad \mu \|z\|^2 = \sum_{i=0}^{m} s_i(x) \langle y_i, z\rangle - \sum_{i=0}^{m} s_i(-x) \langle y_i, z\rangle.$$

But, if $\langle y_i, z\rangle < 0$ then $x \notin V_{y_i}$ and so $s_i(x) = 0$. On the other hand,

$$\text{if } \langle y_i, z\rangle > 0 \text{ then } -x \notin V_{y_i} \text{ and so } s_i(-x) = 0.$$

This implies that the right-hand side of equality (*) is always **positive**, which is impossible. Hence $H : \partial B \times I \multimap \mathbb{R}^n \setminus \{0\}$ is a σ-homptopy between $\tilde{F}$ and $i(F)f : \partial B \to \mathbb{R}^n \setminus \{0\}$. This achieves the proof. ■

REMARK

Notice that the degree of a w-map $F : \partial B \multimap \mathbb{R}^n \setminus \{0\}$ verifying the (B.U.)-condition is not necessarily odd but it depends on the parity of $i(F)$. By degree we mean the unique $d \in \mathbb{Z}$ such that

$$F_*(\text{generator } H_{n-1}(\partial B)) = d \cdot \text{generator } H_{n-1}(\mathbb{R}^n \setminus \{0\}).$$

MORE ON THE BORSUK-ULAM CONDITION

Now we shall give a more geometric formulation of the (B.U.)-condition. This will be used in proving that an u.s.c. map close enough to one that verifies (B.U.) also has the (B.U.)-property.

First of all notice that if A, B are compact subsets of X, then A, B are s.s.h. (strictly separated by a hyperplane) if and only if $\overline{co}A$, $\overline{co}B$ are s.s.h. where $\overline{co}A$ denotes the closed convex hull of A.

Let $A \subset X$ be bounded and convex. We shall denote by

$$K(A) = \{\lambda x : \lambda \in [0,1] \text{ and } x \in A\}$$

the bounded convex cone of A with vertex at the origin. Notice that $K(A)$ is compact whenever A is compact.

(2.9) *PROPOSITION*

Let $A, B \subset X$ be compact. A, B are s.s.h. if and only if

$$\begin{cases} K(\overline{coA}) \cap \overline{coB} = \phi \\ \\ K(\overline{coB}) \cap \overline{coA} = \phi . \end{cases} \tag{2.1}$$

Proof

The "only if" part is trivial. To show the converse, let us notice that by Hahn-Banach separation property from (2.1) it follows that there exist $\varphi, \psi \in X^*$ and $\alpha, \beta \in \mathbb{R}$ such that

$$\varphi(K(\overline{coA})) \subset (\alpha,+\infty), \qquad \varphi(\overline{coB}) \subset (-\infty,\alpha)$$

and

$$\psi(K(\overline{coB})) \subset (\beta,+\infty), \qquad \psi(\overline{coA}) \subset (-\infty,\beta).$$

Since $0 \in K(\overline{coA}) \cap K(\overline{coB})$ it follows that $\alpha, \beta < 0$. If we consider $\theta = \alpha\psi - \beta\varphi$, it is easy to show that A, B are strictly separated by the hyperplane defined by θ. ■

(2.10) *PROPOSITION*

$F : B \multimap X$ verifies the (B.U.)-property if and only if the multivalued map $\tilde{F} : B \multimap X$ defined by

$$\tilde{F}(x) = K(\overline{coF(x)}) - \overline{coF(-x)}$$

has no zeros on ∂B.

Proof

Let $x \in \partial B$. Clearly

$$0 \notin \tilde{F}(x) \quad \text{if and only if} \quad K(\overline{co}F(x)) \cap \overline{co}F(-x) = \phi.$$

Analogously,

$$0 \notin \tilde{F}(-x) \quad \text{if and only if} \quad K(\overline{co}F(-x)) \cap \overline{co}F(x) = \phi.$$

Now the assertion follows from Proposition (2.9).■

(2.11) *PROPOSITION*

If $F : B \multimap X$ is a multivalued compact vector field then $\tilde{F}$ is u.s.c. and proper.

Proof

The uppersemicontinuity of $\tilde{F}$ follows easily from the definition. In order to show that $\tilde{F}$ is a proper map, let us observe that if $F(x) = x - \overline{F}(x)$ with $\overline{F}$ u.s.c. and compact, then

$$\tilde{F}(x) = K(x - \overline{co}\overline{F}(x)) + x + \overline{co}\overline{F}(-x).$$

Let C be compact subset of X and let $\{x_n\}$ be a sequence in $\tilde{F}^{-1}(C)$. Let $\{y_n\}$, $y_n \in C \cap \tilde{F}(x_n)$. By passing to a subsequence, if necessary, we can assume that $y_n \to y \in C$. By the definition of $\tilde{F}$, we can write $y_n = \lambda_n(x_n - u_n) + x_n + v_n$ with $u_n \in \overline{co}\overline{F}(x_n)$ and $v_n \in \overline{co}\overline{F}(-x_n)$. Since $\overline{co}\overline{F}(B)$ is a precompact subset of X and since $\lambda_n \in [0,1]$, passing to subsequences if it is necessary, we can assume that $\lambda_n \to \lambda \in [0,1]$, $u_n \to u$, $v_n \to v$. Then

$$x_n = \frac{y_n + \lambda_n u_n - v_n}{\lambda_n + 1}$$

converges to some $x \in B$. By the uppersemicontinuity of $\tilde{F}$ and since $y_n \in \tilde{F}(x_n)$ and $y_n \to y$, we have that $y \in \tilde{F}(x)$ so $x \in \tilde{F}^{-1}(C)$. ■

(2.12) *REMARK*

Let $F, F' : X \multimap Y$ be u.s.c. maps and $\varepsilon > 0$. Denote by GrF the graph of F and εA the set of points of distance less than ε from A.

The equivalence of the following statements is easy to check

(i) $GrF' \subset \varepsilon\, GrF$

(ii) $\rho^+(GrF'. GrF) < \varepsilon$ where ρ^+ is the upper Hausdorff separation

(iii) for each $x \in X$, $F'(x) \subset \varepsilon F(\varepsilon x)$.

Here $X \times Y$ is considered with the norm $\|(x,y)\| = \sup(\|x\|, \|y\|)$.

DEFINITION

We shall say that an u.s.c. multivalued map $F': X \multimap Y$ is an *ε-approximation of* $F : X \multimap Y$ if one and hence any one of the above statements is satisfied.

(2.13) *PROPOSITION*

Let $F : B \multimap X$ be a compact vector field satisfying the (B.U.)-property. Then there exists $\varepsilon > 0$ such that any ε-approximation $F' : B \multimap X$ of F satisfies the (B.U.)-property.

Proof

By Proposition (2.10), we have that $\tilde{F}$ has no zeros on ∂B. Actually, using the properness of $\tilde{F}$, we shall show that there exists $\varepsilon' > 0$ such that

$$0 \notin \varepsilon' \tilde{F}(\varepsilon' x) \quad \text{for all } x \in \partial B. \qquad (2.2)$$

In fact, assuming the contrary, there exist sequences $\{\varepsilon_n\}$, $\{x_n\}$ with $\varepsilon_n \to 0$ $x_n \in \partial B$ and such that $0 \in \varepsilon_n \tilde{F}(\varepsilon_n x_n)$, that is, there exist $\{x'_n\}$ and $\{y_n\}$ with $x'_n \in \partial B$, $\|x'_n - x_n\| < \varepsilon_n$, $\|y_n\| < \varepsilon_n$, $y_n \in \tilde{F}(x'_n)$. Since $y_n \to 0$, by the properness of $\tilde{F}$, it follows that $\{x'_n\}$ has a convergent subsequence to some $x \in \partial B$, and so $0 \in \tilde{F}(x)$. Contradicting the fact that $\tilde{F}$ has no zeros of ∂B. Let $\varepsilon = \frac{1}{2}\varepsilon'$.

If $F' : B \multimap X$ is an ε-approximation of F we have that

$$F'(x) \subset \varepsilon F(\varepsilon x) \subset \varepsilon F(\varepsilon' x). \qquad (2.3)$$

Let us notice that if $A, B \subset X$ then the following relations hold

(i) $\overline{co}\varepsilon A \subset \varepsilon\overline{co}A$

(ii) $K(\varepsilon A) \subset \varepsilon K(A)$ (2.4)

(iii) $\varepsilon A - \varepsilon B \subset 2\varepsilon(A - B)$.

From (2.3) and (2.4) it follows easily that for each $x \in \partial B$ we have

$$\widetilde{F}'(x) = K(\overline{co}F'(x)) - \overline{co}F'(-x) \subset \varepsilon'\widetilde{F}(\varepsilon' x).$$

Hence $0 \notin \widetilde{F}'(x)$ for all $x \in \partial B$ and so the assertion follows from Proposition (2.10). ■

Proof of Theorem (1.1)

Let us consider the multivalued map $T : B \multimap X$ defined by $T(x) = g(x, S(x)) = x - \overline{g}(x, S(x))$. By Proposition (2.1), T is u.s.c.. Moreover, since $\overline{g}$ is compact, T is a multivalued compact vector field. Hence T is proper. Suppose now that the system (1.1) has no solutions (x,y) with $x \in B$. Then $0 \notin T(x)$ for all $x \in B$ and so, as in the proof of Proposition (2.13), there exists $\varepsilon_1 > 0$ such that

(*) $$0 \notin \varepsilon_1 T(\varepsilon_1 x) \quad \text{for all } x \in B.$$

On the other hand from the assumptions in Theorem (1.1), it follows that T verifies the (B.U.)-property on ∂B and so, by Proposition (2.13), there exists ε_2 such that any u.s.c. multivalued map $T' : B \multimap X$ with $GrT' \subset \varepsilon_2 GrT$ has the (B.U.)-property on ∂B. Let $\delta = \min\{\varepsilon_1, \varepsilon_2\}$ and let $V \subset U$ be defined by

$$V = \{(x,y) \in U : (x, g(x,y)) \in \delta GrT\}.$$

Clearly V is an open set being the inverse image of the open set δGrT by a continuous map. Furthermore, V is a neighborhood of S_B. We will divide the rest of the proof in three steps.

1.st step. $X = \mathbb{R}^n$, $Y = \mathbb{R}^m$.

Let ε' be given by Lemma (2.3), that is any ε'-approximation f_1 of f has the property that $S_B^{f_1} \subset V$. By Lemmas (2.4) and (2.7), there exists a continuous map $f_1 : V \to \mathbb{R}^n$ which ε'-approximates f on $\overline{V}_B$ and such that the multivalued map $x \mapsto S'(x) = S^{f_1}(x)$ from B into $\mathbb{R}^n$ is a w-map with

$$i(S'(x)) = \deg(f(0,\cdot), V(0), 0) = \deg(f(0,\cdot), U(0), 0) \neq 0$$

and such that $S'_B \subset V$. By properties 3) and 4) following Remark (2.6), the multivalued map

$$T'(x) = g(x, S'(x))$$

is a w-map with $i(T') = i(S') \neq 0$. Furthermore, since $S'_B \subset V$, we have that $GrT' \subset \delta GrT$. By our choice of δ it follows that T' verifies the (B.U.)-property on ∂B and hence by Theorem (2.8) there exists $x \in \mathring{B}$ such that $0 \in T'(x)$. Hence $0 \in \delta T(\delta x)$, contradicting (*).

2.nd step. $X = \mathbb{R}^n$ and Y any Banach space.

Since U_B is a bounded subset of $X \times Y$ and f is a parametrized compact vector field, there exists an ε'-approximation $\overline{f}_1$ of $\overline{f}$ on $\overline{U}_B$ with range contained in a finite dimensional subspace Y_1 of Y. Set $f_1 = Id - \overline{f}_1$. By Lemma (2.3) and the homotopy property of the degree we have that

$$S_B^{f_1} \subset V \cap X \times Y_1 = V_1,$$

and

$$\deg(f_1(0,\cdot), V(0), 0) = \deg(f(0,\cdot), V(0), 0) \neq 0.$$

Hence $S_B^{f_1} \neq \phi$ and so V_1 is a nonempty subset of $X \times Y_1$ which is locally bounded over X. Furthermore, by the reduction property of the degree for

$$f_2 = f_1\Big|_{V_1}$$

we have that

$$\deg(f_2(0,\cdot), V_1(0), 0) = \mathrm{def}(f_1(0,\cdot), V(0), 0) \neq 0.$$

Let $g_2 = g|_{V_1}$. It is clear that the pair (g_2, f_2) satisfies the assumptions of Theorem (1.1). Hence, by the 1st step, the multivalued map

$$T_2(x) = g_2(x, S^{f_2}(x))$$

has a zero in B. But, $S_B^{f_2}(x) \subset V$ and so $GrT_2 \subset \delta\, GrT$, contradicting (*).

3rd step. X, Y any Banach spaces.

Let $\overline{g}_1 : \overline{U}_B \to X$ be a finite dimensional ε-approximation of $\overline{g}$ on $\overline{U}_B$. Let X_1 denote any finite dimensional subspace of X containing the range of $\overline{g}_1$. Set

$$g_1 = Id - \overline{g}_1 \Big|_{X_1 \times Y \cap \overline{U}_B}$$

and

$$f_1 = f \Big|_{X_1 \times Y \cap \overline{U}_B} .$$

Let $T' : B' = B \cap X_1 \multimap X_1$ be defined by $T'(x) = g_1(x, S(x))$. Since the multivalued map T', considered as a map from B' into X, is an ε-approximation of the restriction of T to B', it follows, from Proposition (2.13), that T' has the (B.U.)-property on $\partial B'$. Therefore the pair (f_1, g_1) verifies the assumption (ii) of Theorem (1.1). Since (i) and (A) are direct consequences of the definition of f_1, by the 2nd step we have that T' has a zero in $B' \subset B$. But this contradicts (*). ∎

Proof of Theorem (1.2)

It is easy to see that under the assumptions (A) the map $\overline{f} : \overline{U}_K \to Y$ is compact. On the other hand since K is an absolute retract and S_K is closed in $X \times Y$ the map $\overline{g} : S_K \to K$ can be extended to a continuous map defined on all of $X \times Y$ with values in K that we still denote by $\overline{g}$.

Now let $r : X \to K$ be any retraction and let

$$U' = \{(x,y) \in X \times Y : (r(x), y) \in U\}.$$

Then U' is an open subset of $X \times Y$ which is locally bounded over X. Let us consider the maps

$$f : \overline{U}' \to Y,$$

$$g : \overline{U}' \to X$$

defined by

$$f(x,y) = y - \overline{f}(r(x), y),$$

$$g(x,y) = x - \overline{g}(x,y).$$

Since $\overline{f}(\overline{U}_K)$ is precompact and $\overline{g}(X \times Y) \subset K$ we have that f and g are parametrized compact vector fields. We shall see that f and g verify the hypotheses of Theorem (1.1). First of all notice that under our assumptions $\mathcal{D} = X$. Moreover, by the generalized homotopy invariance of degree

$$\deg(f(0,\cdot), U'(0), 0) = \deg(f(x,\cdot), U'(x), 0) = \deg(Id - \overline{f}(x,\cdot), U(x), 0)$$

for each $x \in K$. Therefore by (A) , $\deg(f(0,\cdot), U'(0), 0)$ is different from zero.

Let $B = B(0,r)$ be such that $K \subset \mathring{B}$. Let us show that the multivalued map $T : B \multimap X$ defined by

$$T(x) = g(x, S(x))$$

verifies the (B.U.)-property on ∂B. By Proposition (2.9) this is equivalent to prove that $K(\overline{co}\ T(x)) \cap \overline{co}\ T(-x) = \phi$ for all $x \in \partial B$. Indeed, if this is not true then there exist $\lambda \in [0,1]$ $y_1 \in \overline{co}\ \overline{g}(x, S(x)) \subset K$ and $y_2 \in \overline{co}\ \overline{g}(-x, S(-x)) \subset K$ such that

$$\lambda(x - y_1) = -x - y_2 .$$

Then

$$(\lambda + 1)\|x\| = \|\lambda y_1 - y_2\| < (\lambda + 1)r.$$

Contradicting $x \in \partial B$. Thus (ii) of Theorem (1.1) is verified and so Theorem (1.2) follows. ■

Proof of Corollary (1.3)

Let $p : X \to B$ be the canonical retraction of the space X into the unit ball B of X. The system

$$\begin{cases} y = \bar{f}(x,y) \\ x = p\bar{g}(x,y) \end{cases}$$

verifies the *hypothesis of Theorem* (1.2) with respect to the compact convex set $K = \overline{co}(p(\bar{T}(B))) \subset B$ and hence has a solution $(\bar{x},\bar{y})$ with $\bar{x} \in B$. Now because of the assumption (L.S) of Corollary (1.3) $\|g(\bar{x},\bar{y})\| \leq 1$. Therefore $p\bar{g}(\bar{x},\bar{y}) = \bar{g}(\bar{x},\bar{y})$ and hence $(\bar{x},\bar{y})$ is also a solution of (1.1)'. ■

References

[1] AMANN, H., AMBROSETTI, A. and MANCINI, G.: Elliptic equations with non invertible Fredholm part and bounded non-linearities, Math Zeit., 158 (1978), 179-194.

[2] AMBROSETTI, A. and HESS, P.: Pairs of solutions for some nonlinear elliptic equations. To appear in Boll. U.M.I.

[3] BROWDER, F.: On continuity of fixed points under deformations of continuous mappings, Summa Brasil. Math., 4 (1960), 183-190.

[4] CESARI, L.: Functional analysis, nonlinear differential equations, and the alternative method. Nonlinear Functional Analysis and Differential Equations, L. Cesari, R. Kannan and J.D. Schuur - Ed. M. Dekker Inc., New York (1976).

[5] DARBO, G.: Teoria dell'omologia in una categoria di mappe plurivalenti ponderate. Rend. Sem. Mat. Univ. Padova, 28 (1958), 188-224.

[6] DARBO, G.: Estensione alle mappe ponderate del teorema di Lefschetz sui punti fissi, Rend. Sem. Mat. Univ. Padova, 31 (1961), 46-57.

[7] DOLD, A.: A coincidence - fixed-point index. L'enseignement Mathématique, XXIV (1978), 41-53.

[8] DOLD, A.: The fixed point transfer of fibre-preserving maps, Math, Zeit., 148 (1976), 215-244.

[9] GRANAS, A.: The theory of compact vector fields and some applications to the topology of functional spaces, Rozprawy Matematyczne, Warszawa, 30 (1962).

[10] HESS, P. and RUF, B.: On a superlinear elliptic boundary value problem, Math Zeit., 164 (1978), 9-14.

[11] JERRARD, R.: Homology with multivalued functions applied to fixed points, Trans. Amer. Math. Soc., 213 (1975), 407-428.

[12] KNILL, R.J.: On the homology of fixed point set. Bull. Amer. Math. Soc. 77 (1971), 184-190.

[13] KURLAND, H. and ROBBIN, J.: Infinite codimension and transversality. Dynamical Systems, Warwick (1974). Lectures Notes in Mathematics N. 468, Springer Verlag.

[14] PEJSACHOWICZ, J.: The homotopy theory of weighted mappings, Boll. U.M.I. (5) 14-B (1977), 702-721.

[15] SHAW, H.: A nonlinear elliptic boundary value problem at resonance, J. Diff. Eq. 26 (1977), 335-346.

[16] TOUGERON, J.C.: Idéaux de fonctions différentiables. Springer-Verlag (1972).

CONTRACTION PRINCIPLE IN PSEUDO-UNIFORM SPACES

By

PEDRO MORALES
Département de mathématiques et d'informatique
Université de Sherbrooke,
Sherbrooke, Québec, Canada.

1. Introduction

The purpose of this paper is to establish a quite general fixed point theorem in a pseudo-uniform space X (uniform convergence space in the original terminology of Cook and Fischer [6]), using the notion of contraction of a mapping $f : X \to X$ introduced by Kneis in [16]. This contraction principle gives not only the existence of a unique fixed point of f, but also a iterative realization of it, like the classical Banach contraction principle [1, p. 160]. Besides to generalize the fixed point theorem for well-chained pseudo-uniform spaces of Kneis [16, p. 159], it contains the well-known results of Boyd and Wong [2], Browder [3], Davis [7], Edelstein ([8], [9]), Geraghty [12], Janos [13], Keeler and Meier [14], Knill [17], Naimpally [20], Rakotch [21], Tan [22], Tarafdar [23] and Taylor [24].

The reader not being familiar with the theory of convergence structures is referred to Gähler's book [11] or to the fundamental paper of Fischer [10].

2. Preliminaries

We begin with pertinent notions defining our setting. Let X be a nonempty set and let $\underline{F}(X)$ be the set of all proper filters on X. A natural partial order on $\underline{F}(X)$ is given by the inclusion. If $F_1, F_2 \in \underline{F}(X)$ and $F_1 \subset F_2$, we say that F_2 is finer that F_1. An $\wedge$-*ideal* in $\underline{F}(X)$ is a subset u of $\underline{F}(X)$ such that the following conditions are verified:

I.1) If $F_1, F_2 \in u$, then $F_1 \cap F_2 \in u$.

I.2) If $F_1 \in u$ and F_2 is an element of $\underline{F}(X)$ which is finer than F_1 then $F_2 \in \mu$.

We will denote by $\underline{I}(X)$ the set of all $\wedge$-ideals in $\underline{F}(X)$. A *pseudo-topology* on X is a mapping $\tau : X \to \underline{I}(X)$ such that, for every $x \in X$, the principal filter $[x]$ belongs to $\tau(x)$. For $F \in \tau(x)$ we say that the filter F *τ-converges* to x or that x is a *τ-limit* of F, and we write $F \xrightarrow{\tau} x$. If τ is a pseudo-topology on X, the pair (X,τ) is called a *pseudo-topological space*. An important example of a pseudo-topology which, in general, is not a topology is given by the continuous convergence on the set of all continuous mappings from one topological space to another (see [5]). A pseudo-topological space (X,τ) is said to be *separated* if $\tau(x) \cap \tau(y) = \phi$ for all $x,y \in X$ with $x \neq y$. In this case, if $F \xrightarrow{\tau} x$ and $F \xrightarrow{\tau} y$, then $x = y$. Let f be a mapping on X into itself. The mapping f is said to be *continuous at a point* $x \in X$ if $F \xrightarrow{\tau} x$ implies $f(F) \xrightarrow{\tau} f(x)$, and it is *continuous* if f is continuous at each $x \in X$.

For any sequence (x_n) in X, the set $B = \left\{\{x_n : n \geq k\} : k = 1, 2, \ldots\right\}$ is a filter base on X. The filter generated by B is called the *Fréchet filter* determined by (x_n), and is denoted by $F(x_n)$.

A *pseudo-uniformity* on X is a subset u of $\underline{I}(X \times X)$ satisfying the following conditions:

P.1) If $F \in u$, then $F^{-1} = \{F^{-1} : F \in F\}$ belongs to u.

P.2) The principal filter $[\Delta]$, where Δ denotes the diagonal of $X \times X$, belongs to u.

P.3) If $F_1, F_2 \in u$ and the filter $F_1 \circ F_2$ on $X \times X$ with the base $\{F_1 \circ F_2 : F_i \in F_i \text{ for } i = 1,2\}$ exists, then $F_1 \circ F_2 \in u$.

If u is a pseudo-uniformity on X, the pair (X,u) is called a *pseudo-uniform space*. There are two special and important examples of pseudo-uniform spaces: a) *a uniform space* (X,u) where u is the principal $\wedge$-ideal $[U]$ generated by a fixed filter U on $X \times X$. In this case, it can be shown that U is a uniformity on X in the usual sense; conversely, if U is a uniformity on X it can be shown that $[U]$ is a pseudo-uniformity on X [6, p. 293]; b) *a uniform Choquet space* (X,u) where a filter $F \in u$ if and only if every ultrafilter finer than F belongs to u (see [4]). Since every filter F on $X \times X$ is the intersection of all ultrafilters finer than F, every uniform space is a uniform Choquet space.

A pseudo-uniformity u on X induces a pseudo-topology $\lambda(u)$ on X as follows: for $x \in X$, let $\lambda(u)(x) = \{F \in \underline{F}(X) : F \times [x] \in u\}$. Then it can be shown that $\lambda(u)$ is a pseudo-topology on X [6, p. 291]. This generalization of the usual uniformity owes its importance to the fact that every separated pseudo-topological space is pseudo-uniformizable ([15], [11, p. 314]).

Let $X = (X,u)$ be a pseudo-uniform space. A filter F on X is said to be a *Cauchy filter* if $F \times F \in u$. The space X is called *sequentially complete* if every Fréchet filter, which is Cauchy, converges. A mapping $f : X \to X$ is said to be *pseudo-uniformly continuous* if $V \in u$ implies $(f \times f)(V) \in u$, where $f \times f$ is the mapping on $X \times X$ into itself defined by the formula: $(f \times f)(x,y) = (f(x),f(y))$. It can be shown that if f is pseudo-uniformly continuous, then f is continuous [11, p. 318]. We note finally that if we consider f as a binary relation on X, then $(f \times f)(A) = f \circ A \circ f^{-1}$ for all subset A of $X \times X$.

3. CONTRACTION THEOREM

Let $X = (X,u)$ be a pseudo-uniform space and let $f : X \to X$. We say that f is an *occasionally small mapping* if, for every pair of points x and y of X, there is a filter $V = V(x,y) \in u$ such that, for every $W \in V$, there exists a positive integer $k = k(W)$ with $(f^k(x), f^k(y)) \in W$. Following Kneis [16, p. 159] we say that f is a *contractive mapping* if, for every filter $V \in u$, there exists a base B of a filter $U \subset [\Delta]$ of u such that, for every $B \in B$, there is a set $V = V(B) \in B \cap V$ with $(f \times f)(B \circ V) \subset B$.

It will be shown that if f is a contractive mapping, then f is pseudo-uniformly continuous. In fact, let $V \in u$, and let B be the base of a filter $U \subset [\Delta]$ of u according to the contraction property of f. To show that $(f \times f)(V) \in u$ it suffices to prove that $(f \times f)(V) \supset U$. Let $U \in U$. Choose $B \in B$ such that $B \subset U$. Then there exists a set $V \in B \cap V$ with $(f \times f)(B \circ V) \subset B$. Since $B \supset \Delta$, $V \subset B \circ V$. So $(f \times f)(V) \subset U$, and therefore $U \in (f \times f)(V)$.

If x_0 is a fixed point of f, we say that x_0 is *iteratively realizable* if the Fréchet filter $F(f^n(x))$ converges to x_0, independent of the choice of the starting point x.

(3.1) *THEOREM*

Let $X = (X,\mathcal{U})$ be a separated sequentially complete pseudo-uniform space, and let f be a mapping on X into itself. If f is a contractive occasionally small mapping, then f has one and only one fixed point which is iteratively realizable.

Proof

We divide the proof in several steps:

1) It will be shown that f has at most one fixed point. Let $u,v \in X$ be fixed points of f. Since f is an occasionally small mapping, there is a filter $\mathcal{V} = \mathcal{V}(u,v) \in \mathcal{U}$ such that, for every $W \in \mathcal{V}$, there exists a positive integer $k = k(W)$ with $(u,v) = (f^k(u), f^k(v)) \in W$. This implies that $[u]\times[v] \supset \mathcal{V}$, so $[u]\times[v] \in \mathcal{U}$. Since X is separated, $u = v$ ([6, p. 292], [11, p. 314]).

2) Let $x \in X$ be arbitrary. It will be shown that the Fréchet filter $\mathcal{F} = \mathcal{F}(f^n(x))$ is a Cauchy filter. Since f is an occasionally small mapping, there is a filter $\mathcal{V} = \mathcal{V}(x) \in \mathcal{U}$ such that, for every $W \in \mathcal{V}$, there exists a positive integer $k = k(W)$ with $(f^k(x), f^{k+1}(x)) = (f^k(x), f^k(f(x))) \in W$. Let $\mathcal{B} = \mathcal{B}(x)$ be the base of a filter $\mathcal{U} = \mathcal{U}(x) \subset [\Delta]$ of $\mathcal{U}$ according to the contraction property of f. Let $U,V \in \mathcal{U}$. Since $U,V \supset \Delta$, $U^{-1} \circ V$ is nonempty. Hence the filter $\mathcal{U}^{-1} \circ \mathcal{U}$ exists and, since $\mathcal{U}^{-1} \in \mathcal{U}$, $\mathcal{U}^{-1} \circ \mathcal{U}$ belongs to $\mathcal{U}$. Then to prove that $\mathcal{F}\times\mathcal{F}$ is a Cauchy filter it is sufficient to prove that $\mathcal{U}^{-1} \circ \mathcal{U} \subset \mathcal{F}\times\mathcal{F}$. Let $Z \in \mathcal{U}^{-1} \circ \mathcal{U}$. Then there exist $U_1,U_2 \in \mathcal{U}$ such that $U_1^{-1} \circ U_2 \subset Z$. Choose $B_i \in \mathcal{B}_i$ such that $B_i \subset U_i$ for $i = 1,2$. Fix $i = 1,2$. There exists a set $V_i = V_i(B_i) \in \mathcal{B} \cap \mathcal{V}$ with $(f\times f)(B_i \circ V_i) \subset B_i$. Taking into account that $(f^{k(i)}(x), f^{k(i)+1}(x)) \in V_i$, where $k(i) = k(i)(V_i)$, and $V_i \subset B_i \circ V_i$ (because $B_i \supset \Delta$), it follows that $(f^n(x), f^{n+1}(x)) \in B_i$ for all $n \geq k(i)$. Since $V_i \in \mathcal{B}$ there exists a set $W_i = W_i(V_i) \in \mathcal{B} \cap \mathcal{V}$ with $(f\times f)(V_i \circ W_i) \subset V_i$. Because $\Delta \subset W_i$, $(f\times f)(V_i) \subset V_i$, and therefore $(f^{n-1}(x), f^n(x)) \in V_i$ for all $n \geq k(i)+1$. It will be shown that, if $n > k(i)$, $(f^n(x), f^{n+m}(x)) \in B_i$ for all $m = 1,2,3\dots$. This being true for $m = 1$, we suppose it for an arbitrary m. The induction hypothesis together with $(f^{n-1}(x), f^n(x)) \in V_i$ imply $(f^{n-1}(x), f^{n+m}(x)) \in B_i \circ V_i$. It follows that $(f^n(x), f^{n+m+1}(x)) \in B_i$, so the induction is complete. Let now $k = \max\{k(1), k(2)\}$. Take $m,n > k$. Then $(f^n(x), f^{n+m}(x)) \in B_1$ and $(f^m(x), f^{m+n}(x)) \in B_2$, and therefore $(f^m(x), f^n(x)) \in B_1^{-1} \circ B_2 \subset U_1^{-1} \circ U_2 \subset Z$. Since $F = \{f^j(x) : j > k\} \in \mathcal{F}$ and $F\times F \subset Z$, Z belongs to $\mathcal{F}\times\mathcal{F}$. So $\mathcal{U}^{-1} \circ \mathcal{U} \subset \mathcal{F}\times\mathcal{F}$.

3) Since X sequentially complete, there exists a point $u = u(x) \in X$ such that $\mathcal{F} \xrightarrow{\lambda(\mathcal{U})} u$. We will show that u is a fixed point of f. Being contractive, f is, in particular, continuous. So $f(\mathcal{F}) \xrightarrow{\lambda(\mathcal{U})} f(u)$. But $\mathcal{F} = f(\mathcal{F})$. So $\mathcal{F} \xrightarrow{\lambda(\mathcal{U})} f(u)$. Since X is separated, this implies that $f(u) = u$. Since f has a unique fixed point, u is independent of x. So u is iteratively realizable, and the proof is complete. ■

(3.2) *COROLLARY*

Let $X = (X,\mathcal{U})$ be a separated sequentially complete pseudo-uniform space, and let f be a mapping on X into itself. If at least one iterate f^k is a contractive occasionally small mapping, then f has one and only one fixed point u such that, for every $x \in X$, the Fréchet filter determined by the sequence $(f^{nk}(x))_{n=1}^{\infty}$ converges to u.

Proof

Let $h = f^k$. By the Theorem (3.1), h has a unique fixed point u such that, for every $x \in X$, the Fréchet filter $\mathcal{F} = \mathcal{F}(h^n(x))$ converges to u. So $\mathcal{F}$ is determined by the sequence $(f^{nk}(x))_{n=1}^{\infty}$ and $\mathcal{F} \xrightarrow{\lambda(\mathcal{U})} u$. Since

$$h(f(u)) = f(h(u)) = f(u),$$

$f(u)$ is a fixed point of h, and therefore $f(u) = u$. But a fixed point of f is also a fixed point of h, so u is the unique fixed point of f. ■

Following Kneis [16, p. 157] we say that X is *well-chained* if, for every pair of points x and y of X, there is a filter $\mathcal{V} = \mathcal{V}(x,y) \in \mathcal{U}$ such that, for every $W \in \mathcal{V}$, there exists a positive integer $n = n(W)$ with $(x,y) \in W^n$. He shows that every uniformly arcwise connected pseudo-uniform space is well-chained. It is clear that every connected uniform space is well-chained.

(3.3) *LEMMA*

Let $X = (X,\mathcal{U})$ be a well-chained pseudo-uniform space, and let f be a mapping on X into itself. If f is contractive, then f is an occasionally small mapping.

Proof

Let $x,y \in X$. Since X is well-chained, there exists a filter $\mathcal{V} = \mathcal{V}(x,y) \in \mathcal{U}$ such that, for every $W \in \mathcal{V}$, there is a positive integer $n = n(W)$ with $(x,y) \in W^n$. Let $\mathcal{B} = \mathcal{B}(x,y)$ be the base of a filter $\mathcal{U} = \mathcal{U}(x,y) \subset [\Delta]$ of $\mathcal{u}$ according to the contraction property of f. Let $U \in \mathcal{U}$. Choose $B = B(U)$ such that $B \subset U$. Then there exists a set $V = V(B) \in \mathcal{B} \cap \mathcal{V}$ with $(f\times f)(B\circ V) \subset B$ and also a set $V' = V'(V) \in \mathcal{B} \cap \mathcal{V}$ with $(f\times f)(V\circ V') \subset V$. Since $\Delta \subset V'$, $(f\times f)(V) \subset V$, and therefore $(f\times f)(V^n) \subset V^n$ for all $n = 1,2,3,\dots$. It will be shown inductively that $(f^n \times f^n)(B\circ V^n) \subset B$ for all $n = 1,2,3,\dots$. This being true for $n = 1$, we prove it for $n+1$, assuming it for n. Noting that, for

$$A,\ B \subset X\times X,\quad A\circ B \subset A\circ f^{-1}\circ f\circ B,$$

we have

$$(f^{n+1}\times f^{n+1})(B\circ V^{n+1}) = f^{n+1}\circ B\circ V^{n+1}\circ f^{-(n+1)} = f^n\circ f\circ B\circ V\circ V^n\circ f^{-1}\circ f^{-n}$$

$$\subset f^n\circ(f\circ B\circ V\circ f^{-1})\circ(f\circ V^n\circ f^{-1})\circ f^{-n} \subset f^n\circ B\circ V^n\circ f^{-n} \subset B.$$

This established, let $n = n(V) = n(V(B(U)))$. Then $(x,y) \in V^n$ and, because $B \supset \Delta$, $V^n \subset B\circ V^n$. So $(f^n(x), f^n(y)) \in B \subset U$, showing that f is an occasionally small mapping. ■

Using this Lemma, Corollary (3.2) yields the following generalization of the contraction principle of Kneis [16, p. 159], improving also the basic lemma of Taylor [24, p. 166], the uniform contraction principle of Knill [17, p. 451] and the theorem 2 of Davis [7, p. 984] for a T_1 uniform space (with this correction, because to assure the uniqueness of the sequential limit, his condition T_0 should be replaced by T_1):

(3.4) *THEOREM*

Let $X = (X,\mathcal{u})$ be a separated sequentially complete well-chained pseudo-uniform space, and let f be a mapping on X into itself. If at least one iterate f^k is a contractive mapping, then f has one and only one fixed point u such that, for every $x \in X$, the Fréchet filter determined by the sequence $(f^{nk}(x))_{n=1}^{\infty}$ converges to u.

(3.5) *REMARKS*

1) Since every (ε,λ)-uniformly local contraction on a ε-chainable metric space $X = (X,d)$ is a contractive occasionally small mapping relative to $\mathcal{U} = [\mathcal{U}_d]$, where $\mathcal{U}_d$ is the filter with base $\left\{\{(x,y) \in X \times X : d(x,y) < \varepsilon\} : \varepsilon > 0\right\}$, the fixed point theorem of Edelstein [8, p. 8] is a corollary of Theorem (3.1).

2) With trivial modifications in the proof, corollary (1.2) of [19] remains valid for a T_1 sequentially complete uniform space. Therefore, Theorem (3.1) contains the fixed point theorems of Janos [13, p. 69], Tan [22, p. 361] and Tarafdar [23, p. 212].

3) Using the results of Section 3 of [19] we can deduce that Theorem (3.1) contains the fixed point theorems of Keeler and Meier [14, p. 326] and Edelstein [9, p. 75], the theorem of Geraghty [12, p. 811] and its corollaries (3.1) (Rakotch [21]), (3.2), (3.3) (Boyd and Wong [2]) and (3.4) (Browder [3]), and the fixed point theorem of Naimpally [20, p. 479] for a λ-globally contraction on a uniform space generated by a generalized metric.

4) In a recent paper Kwapisz [18], using some earlier ideas of Ważewski, established a fixed point theorem on a space equipped with a special convergence structure called G-metric space, where G is a partially ordered semigroup with a notion of convergence for decreasing sequences in G. He showed some applications of this theorem to the theory of functional equations in Banach spaces.

REFERENCES

[1] BANACH, S.: Sur les opérations dans les ensembles abstraits et leur application aux équations intégrales, Fund. Math. 3 (1922), 138-181.

[2] BOYD, D.W. and WONG, J.S.W.: On nonlinear contractions, Proc. Amer. Math. Soc. 20 (1969), 458-464.

[3] BROWDER, F.E.: On the convergence of successive approximations for nonlinear functional equations, Indag. Math. 30 (1968), 27-35.

[4] CHOQUET, G.: Convergences, Ann. Univ. Grenoble Sect. Math. Phys. (N.S.) 23 (1948), 57-112.

[5] COOK, C.H. and FISCHER, H.R.: On equicontinuity and continuous convergence, Math. Ann. 159 (1965), 94-104.

[6] COOK, C.H. and FISCHER, H.R.: Uniform convergence structures, Math. Ann. 173 (1967), 290-306.

[7] DAVIS, A.S.: Fixpoint for contractions of a well-chained topological space, Proc. Amer. Math. Soc. 14 (1963), 981-985.

[8] EDELSTEIN, M.: An extension of Banach's contraction principle, Proc. Amer. Math. Soc. 12 (1961), 7-10.

[9] EDELSTEIN, M.: On fixed and periodic points under contractive mappings, J. London Math. Soc. 37 (1962), 74-79.

[10] FISCHER, H.R.: Limesräume, Math. Ann. 137 (1959), 269-303.

[11] GÄHLER, W.: Grundstrukturen der Analysis, Akademie-Verlag Berlin, Vol. I (1977).

[12] GERAGHTY, M.A.: An improved criterion for fixed points of contraction mappings, J. Math. Anal. Appl. 48 (1978), 811-817.

[13] JANOS, L.: A converse of the generalized Banach's contraction theorem, Arch. Math. 21 (1970), 69-71.

[14] KEELER, E. and MEIER, A.: A theorem of contraction mappings, J. Math. Anal. Appl. 28 (1969), 326-329.

[15] KELLER, H.H.: Die Limes-Uniformisierbarkeit der Limesräume, Math. Ann. 176 (1968), 334-341.

[16] KNEIS, G.: Contribution to the theory of pseudo-uniform spaces, Math. Nachr. 89 (1979), 149-163.

[17] KNILL, R.J.: Fixed points of uniform contractions, J. Math. Anal. Appl. 12 (1965), 449-455.

[18] KWAPISZ, M.: Some remarks on abstract form of iterative methods in functional equation theory, Preprint no. 22 (1979), Mathematics, University of Gdańsk, Poland.

[19] MORALES, P.: Topological contraction principle, Fund. Math. 110 (1981).

[20] NAIMPALLY, S.A.: Contractive mappings in uniform spaces, Indag. Math. 31 (1969), 474-481.

[21] RAKOTCH, E.: A note on contractive mappings, Proc. Amer. Math. Soc. 13 (1962), 459-465.

[22] TAN, K.K.: Fixed point theorems for non expansive mappings, Pacific J. Math. 41 (1972), 829-842.

[23] TARAFDAR, E.: An approach to fixed-point theorems on uniform spaces, Trans. Amer. Math. Soc. 191 (1974), 209-225.

[24] TAYLOR, W.W.: Fixed-point theorems for nonexpansive mappings in linear topological spaces, J. Math. Anal. Appl. 40 (1972), 164-173.

EIGENVECTORS OF NONLINEAR POSITIVE OPERATORS AND THE LINEAR KREIN-RUTMAN THEOREM

BY

ROGER D. NUSSBAUM*

Mathematics Department
Rutgers University
New Brunswick, New Jersey 08903

0. INTRODUCTION

In a recent paper [13] Massabo and Stuart prove an existence theorem for non-zero eigenvectors of a nonlinear operator which maps a normal cone into itself. They conjecture that normality of the cone is unnecessary; in Section I below we prove their conjecture. Our proof is quite different from that of Massabo and Stuart and involves some results from asymptotic fixed point theory. We hope that even the relatively simple case considered here will illustrate the usefulness of these ideas.

In the second section of this paper, which is essentially independent of the first, we prove a new fixed point theorem for nonlinear cone mappings. We then prove that our nonlinear theorem implies as a corollary the most general versions of the linear Krein-Rutman theorem. Finally, we discuss briefly an example of a linear operator which is best studied in non-normal cones.

Although the linear theorem we obtain is new, our central point is methodological. The linear Krein-Rutman theorem has played an important role in the study of nonlinear cone mappings, particularly in computing the so-called fixed point index of such mappings. Our results cone full circle and show that the linear Krein-Rutman theorem follows from a simple fixed point theorem. Partial results in this spirit have been obtained before [3, 8, 13, 18, 20, 21], but here we avoid unnecessary hypotheses like normality of cones (see Section 5 of [18]).

The approach to the linear Krein-Rutman theorem given here is suitable for a

---

*Partially supported by a National Science Foundation Grant.

course on nonlinear functional analysis, and in fact that was our original motivation for obtaining the results in Section 2. After development of the Leray-Schauder degree theory, the most general versions of the Krein-Rutman theorem for linear compact maps can be obtained in one lecture by our method.

1. Eigenvectors of nonlinear cone mappings

By a cone K in a Banach space X we mean a closed subset of X such that (1) if $x,y \in K$ and λ and μ are nonnegative reals, then $\lambda x+\mu y \in K$ and (2) if $x \in K-\{0\}$, then $-x \notin K$. If K only satisfies (1), K is a "wedge". Notice that K induces a partial ordering on X by $x \leq y$ if and only if $y-x \in K$. A cone K is "normal" if there exists a positive constant γ such that for all elements x and y of K, $\|x+y\| \geq \gamma\|x\|$. The cone of nonnegative functions in $C[0,1]$ or $L^p[0,1]$, $1 \leq p \leq \infty$, is normal; the same cone in $C^k[0,1]$, $k \geq 1$, or in a Sobolev space (other than L^p) is not.

We also need to recall Kuratowski's notion of measure of noncompactness [10]. If S is a bounded subset of a Banach space X (or, more generally, of a metric space) define $\alpha(S)$, the measure of noncompactness of S, by

$$\alpha(S) = \inf\left\{d>0 : S = \bigcup_{i=1}^{n} S_i,\ n < \infty \text{ and diameter } (S_i) \leq d \text{ for } 1 \leq i \leq n\right\}.$$

In general suppose that β is a map which assigns to each bounded subset S of X a nonnegative real number $\beta(S)$. We will call β a generalized measure of noncompactness if β satisfies the following properties:

(1) $\beta(S) = 0$ if and only if the closure of S is compact.

(2) $\beta(\overline{co}(S)) = \beta(S)$ for every bounded set S in X ($\overline{co}(S)$ denotes the convex closure of S, i.e., the smallest closed, convex set which contains S).

(3) $\beta(S+T) \leq \beta(S)+\beta(T)$ for all bounded sets S and T, where $S+T = \{s+t:\ s \in S,\ t \in T\}$.

(4) $\beta(S \cup T) = \max(\beta(S), \beta(T))$.

The measure of noncompactness $\alpha(S)$ is well-known to satisfy properties 1-4. Only property 2, first proved by Darbo [5], presents any difficulties.

If D is a subset of a Banach space X, β is a generalized measure of noncompactness, and $f:D \to X$ a continuous map, f is called a k-set-contraction with respect to β if

$$\beta(f(S)) \leq k\beta(S) \tag{1.1}$$

for every bounded set S in D. If $\beta = \alpha$ we shall simply say that f is a k-set-contraction. Now suppose that C is a closed, convex subset of X and that W is a bounded, relatively open subset of C (so $W = \mathcal{O} \cap C$ for some open subset $\mathcal{O}$ of X). Assume that $f:W \to C$ is a k-set-contraction with respect to β and that $k < 1$. If $\{x \in W: f(x) = x\}$ is compact or empty or (less generally) if f is a k-set-contraction on $\overline{W}$ and $f(x) \neq x$ for $x \in \overline{W}-W$, it is proved in [16] that there is defined an integer $i_C(f, W)$, the fixed point index of $f:W \to C$, which is roughly an algebraic count of the fixed points of f in W. We shall only need a few facts about the fixed point index. If $i_C(f, W) \neq 0$, then f has a fixed point in W. If $f_s(x) = sf(x)$ for $0 \leq s \leq 1$ and $f_s(x) \neq x$ for $x \in \overline{W}-W$, then $i_C(f_s,W)$ is constant for $0 \leq s \leq 1$ and $i_C(f,W) = i_C(f_o,W) = 1$ if $0 \in W$. If C is a cone (or a wedge), $x_o \in C$, $f_t(x) = f(x)+tx_o$ for $t \geq 0$ and $f_t(x) \neq x$ for $x \in \overline{W}-W$, then $i_C(f_t,W)$ is constant for $0 \leq t$.

In the situation described above, the fixed point index can be described in terms of Leray-Schauder degree. Define $C_1 = \overline{co}\, f(W)$, $C_n = \overline{co}\, f(W \cap C_{n-1})$ and let D denote a compact, convex set such that $\bigcap_{n\geq 1} C_n \subseteq D \subset C$ and $f(W \cap D) \subset D$ (such a D exists). Let $\mathcal{O}$ be any bounded open set in X such that $\mathcal{O} \cap D = W \cap D$ and let $g:\mathcal{O} \to D$ be a continuous map such that $g|W \cap D = f|W \cap D$ (g exists by virtue of a theorem of Dugundji [6]). One can <u>define</u> $i_C(f, W) = \deg(I-g, \mathcal{O}, 0)$ (observe that the fixed point set of g in $\mathcal{O}$ is compact, so the Leray-Schauder degree can be defined) and prove that the definition is independent of the particular D, g and $\mathcal{O}$ as above. Properties of the fixed point index now follow from properties of the Leray-Schauder degree, and this is especially easy to see if f is compact.

We want to generalize now following theorem of Massabo and Stuart [13].

THEOREM (Massabo-Stuart [13]*).*

Let C be a normal cone in a Banach space X, let Ω be a bounded, relatively open subset of C containing 0, and let $f:C \to C$ be a k-set-contraction. Suppose that

$$\delta > kd/\gamma \tag{1.2}$$

where $\delta = \inf\{\|f(z)\|: z \in \bar{\Omega}-\Omega\}$, $d = \max\{\|z\|: z \in \bar{\Omega}-\Omega\}$ and γ is the constant appearing in the definition of a normal cone. Then there exist $t > 0$ and $z \in \partial_K(\Omega)$ (the boundary of Ω as an open subset of K) such that $f(z) = tz$.

The chief tool we shall use is an "asymptotic fixed point theorem". Theorem 1.1 below is a special case of Theorem 3 in [15] or Propositions 2.4 and 3.1 in [17].

(1.1) *THEOREM*

Let C be a closed, convex subset of a Banach space X, U a bounded, relatively open subset of C and $f: U \to U$ a k-set-contraction with $k < 1$. Assume that $\overline{f^n(U)}$ is contained in U for some integer n. Then $i_C(f, U)$ is defined, $L_{gen}(f)$, the generalized Lefschetz number of $f: U \to U$ is defined and $L_{gen}(f) = i_C(f, U)$. In particular, if $L_{gen}(f) \neq 0$, f has a fixed point in U.

We have not defined here Leray's generalized Lefschetz number [11], but it suffices to know a few facts. The generalized Lefschetz number agrees with the ordinary Lefschetz number for $f: U \to U$ if the ordinary Lefschetz number is defined. If $f^m(U) \subset Y \subset U$, $f(Y) \subset Y$ and g denotes $f: Y \to Y$, then $L_{gen}(f) = L_{gen}(g)$. In particular, if Y is homotopic in itself to a point, $L_{gen}(f) = 1$.

We can now show that the assumption of normality in Theorem 1 is unnecessary, at least if the set Ω is "radial".

(1.2) *THEOREM*

Let C be a cone in a Banach space X and let Ω be a bounded, relatively open neighborhood of 0 in C. Assume that for each $x \in S \stackrel{\text{def}}{=} \{x \in C: \|x\| = 1\}$ there is a unique real number $t = t_x > 0$ such that $tx \in \partial_C(\Omega)$, where $\partial_c(\Omega)$ denotes the boundary of Ω as a subset of C. Let $f: \partial_C(\Omega) \to C$ be a k-set-contraction and suppose that

$$\delta > kd \tag{1.3}$$

where $\delta = \inf\{\|f(z)\|: z \in \partial_C(\Omega)\}$ and $d = \sup\{\|z\|: z \in \partial_C(\Omega)\}$. Then there exists $t \geq \delta d^{-1}$ and $z \in \partial_C(\Omega)$ such that $f(z) = tz$.

Proof

Define $\Sigma = \{x \in C: \|x\| = d\}$. For each $x \in \Sigma$, there exists a unique $s = s_x$ such that $sx \in \partial_C(\Omega)$, and by the definition of d one has $0 < s_x \le 1$. The assumption that s_x is unique implies easily that $R(x) \overset{\text{def}}{=} s_x x$ is a continuous map. Since the image of any set A in Σ lies in $\overline{co}(A \cup \{0\})$, we find that R is a 1-set-contraction. It follows that $f_1(x) \overset{\text{def}}{=} f(Rx)$ is a k-set-contraction and

$$\inf_{x\in\Sigma} \|f_1(x)\| = \delta$$

If we can prove that Theorem 1.2 is valid for the special case $\Omega = \{x \in C: \|x\| < d\}$, the discussion above shows that there exists $x \in \Sigma$ and $s \ge \delta d^{-1}$ such that $f_1(x)=sx$. If we write $z = R(x) \in \partial_C(\Omega)$, it is clear that z is an eigenvector of f with eigenvalue $t \ge \delta d^{-1}$.

The above discussions shows that it suffices to prove Theorem 1.2 in the case that $\Omega = \{x \in C: \|x\| < d\}$. In this case define $g:\Sigma \to \Sigma$ by $g(x) = \frac{df(x)}{\|f(x)\|}$. If A is any subset of Σ, it is clear that

$$\begin{aligned} g(A) &\subset d\delta^{-1}\{tf(x): 0 \le t \le 1, x \in A\} \\ &\subset d\delta^{-1}\overline{co}(f(A) \cup \{0\}) \end{aligned} \tag{1.4}$$

where for a set T, $\lambda T \overset{\text{def}}{=} \{\lambda x: x \in T\}$. Using (1.4) and the basic properties of the α measure of noncompactness, one finds that

$$\alpha(g(A)) \le kd\delta^{-1}\alpha(A) \tag{1.5}$$

so g is a c-set-contraction for $c = kd\delta^{-1} < 1$.

For $\varepsilon > 0$ let $U = \{x \in C: d-\varepsilon < \|x\| < d+\varepsilon\}$ and define a retraction $r: U \to \Sigma$ by $r(x) = \frac{dx}{\|x\|}$. By reasoning like that above one can see that r is a $d(d-\varepsilon)^{-1}$-set-contraction, so $h(x) = g(r(x))$ is a $cd(d-\varepsilon)^{-1}$-set-contraction. lect ε so small that $c_1 = cd(d-\varepsilon)^{-1} < 1$ and observe that fixed points of $h:U \to U$ are the same as the fixed points of $g:\Sigma \to \Sigma$. Since fixed points of g correspond to the desired eigenvectors, it suffices to find a fixed point of h. By Theorem 1.1 and the remark immediately following it, h will have a fixed point in U and $L_{gen}(h)=1$ if Σ is homotopic in itself to a point. To show Σ is homotopic to a point, let $\rho:K-\{0\} \to \Sigma$ be radial retraction, fix $x_0 \in \Sigma$, and define a deformation $\varphi: \Sigma\times[0,1]\to\Sigma$ by

$$\varphi(x,t) = \rho[(1-t)x+tx_o]. \tag{1.6}$$

The fact that K is a cone insures that $(1-t)x+tx_o \neq 0$ for $0 \leq t \leq 1$ and $x \in \Sigma$, so φ is well-defined and continuous. ■

(1.1) *REMARK*

If X is an infinite dimensional Banach space, $S = \{x \in X: \|x\| = d > 0\}$ and $f: S \to S$ is a k-set-contraction with $k < 1$, then it follows from a remark on p. 373 of [15] that f has a fixed point in S. (Note that S is a continuous retract of $B = \{x \in X: \|x\| \leq d\}$ when X is infinite demensional, so S can be deformed in itself to a point). The proof of Theorem 1.2 follows by essentially the same trick.

(1.2) *Remark*

We have not proved Theorem 1.2 for general neighborhoods Ω of the origin in K, although we conjecture it is true for such neighborhoods. In fact, Massabo and Stuart use their theoren for general Ω in proving Theorem 1.2 of [13]. However, as is observed in [13], Theorem 1.2 of [13] is a global bifurcation result which can be proved (for general cones) by using the fixed point index in the same way degree theory is used to prove the classical Banach space version of Theorem 1.2 in [13].

(1.3) *Remark*

The problem of proving Theorem 1.2 for general Ω may be related to an extension problem for certain functions. I am indebted to Heinrich Steinlein for a conversation which led to the following observation.

(1.3) *THEOREM*

Let Ω be a bounded, relatively open neighborhood of the origin in a cone K and $f: \partial_K(\Omega) \to K-\{0\}$ a continuous map. Define

$$S_1 = \{x \in K: \|x\| = 1\},\ d_1 = \inf\{\|x\|:\ x \in \partial_K(\Omega)\},$$

$d_2 = \sup\{\|x\| : x \in \partial_K(\Omega)\}$ and $A = \{(x,\lambda) \in S_1 \times [d_1, d_2] : \lambda x \in \partial_K(\Omega)\}$.

Define a map $\varphi : A \to S_1$ by

$$\varphi(x,\lambda) = \frac{f(\lambda x)}{\|f(\lambda x)\|} .$$

Assume that there exists a continuous map $\Phi : S_1 \times [d_1, d_2] \to S_1$ such that (i) $\Phi|A = \varphi$ and (ii) Φ is a c-set-contraction for some $c < 1$. Then there exists $x \in \partial_K(\Omega)$ and $t > 0$ such that $f(x) = tx$.

Outline of proof

Define $\Phi_\lambda(x) = \Phi(x,\lambda)$. One can associate a fixed point index to $\Phi_\lambda : S_1 \to S_1$ and prove that this fixed point index equals one. Using this fact one can prove that there exists a connected set $C \subset S_1 \times [d_1, d_2]$, $C \subset \{(x,\lambda) : \Phi(x,\lambda) = x\}$, such that C has nonempty intersection with $S_1 \times \{d_1\}$ and $S_1 \times \{d_2\}$. It follows that C must intersect $\partial_K(\Omega)$, which is the desired result. ∎

Under the hypotheses of Theorem 1.2, one can prove that there exists an extension Φ as in Theorem 1.3, so Theorem 1.2 is a consequence of Theorem 1.3. Furthermore one can derive the original Massabo-Stuart theorem from Theorem 1.3. Thus the known results reduce to extending φ to Φ in such a way that Φ is a c-set-contraction, $c < 1$.

2. LINEAR AND NONLINEAR KREIN-RUTMAN THEOREMS

The linear Krein-Rutman theorem has been used to calculate the fixed point index of certain nonlinear cone mappings and to obtain fixed point theorems for such mappings. We shall show here that the most general linear Krein-Rutman actually follows by elementary arguments using the fixed point index for cone mappings. We wish to emphasize the elementary nature of our proof. We do not need the apparatus of asymptotic fixed point theory, and (at least for compact linear operators) our proof is suitable for a course in nonlinear functional analysis as an application of the Leray-Schauder degree.

We begin by recalling a trivial but useful observation of Bonsall [1].

(2.1) *LEMMA (Bonsall* [1]*)*

Let $\{a_m: m \geq 1\}$ be an unbounded sequence of nonnegative reals. Then there exists a subsequence $\{a_{m_i} : i \geq 1\}$ such that

(1) $a_{m_i} \geq i$

(2) $a_{m_i} \geq a_j$ for $1 \leq j \leq m_i$.

If C is a cone in a Banach space and $f:C \to C$ is a map, we shall say that f is "order-preserving" if $0 \leq u \leq v$ implies $f(u) \leq f(v)$ for all u and v in C. We shall say that f is "positively homogeneous of degree 1" if $f(tu) = tf(u)$ for all real numbers $t \geq 0$ and all $u \in C$.

(2.1) *THEOREM*

Let C be a cone in a Banach space X and $f:C \to C$ an order-preserving map which is positively homogeneous of degree 1. Assume that there exists a generalized measure of noncompactness β such that f is a k-set-contraction with respect to β and $k < 1$. Assume that there exists $u \in C$ such that $\{\|f^m(u)\|: m \geq 1\}$ is unbounded. Then there exists $x \in C$ with $\|x\| = 1$ and $t \geq 1$ such that $f(x) = tx$, and if $f(y) \neq y$ for $\|y\| = 1$ and $U = \{y \in C: \|y\| < 1\}$, $i_C(f,U) = 0$.

(2.1) *REMARK*

If C is a normal cone, Theorem 2.1 is a very special case of Proposition 6 on p. 252 of [18], so the whole point of the following argument is that it applies to nonnormal cones.

(2.2) *REMARK*

Suppose that $g:V = \{x \in C: \|x\| < R\} \to C$ is a k-set-contraction w.r.t. β, $k < 1$, and that f is as in Theorem 2.1. If $tf(x)+(1-t)g(x) \neq x$ for $\|x\| = R$, and $0 \leq t \leq 1$, then there exists $\lambda \geq 1$ and $x \in C$ with $\|x\| = R$ such that $f(x) = \lambda x$. If not, the homotopy $g_s(x) = sg(x)$, $0 \leq s \leq 1$, shows that $i_C(g, V) = 1$, while the homotopy $tf(x)+(1-t)g(x)$ shows that $i_C(g, V) = i_C(f, V) = 0$, a contradiction.

Proof of theorem (2.1)

Let $\Omega = \{x \in C: \|x\| < 1\}$ and assume that there does not exist $t \geq 1$ and a point $x \in C$ with $\|x\| = 1$ such that $f(x) = tx$. If we define $f_s(x) = sf(x)$ for $0 \leq s \leq 1$, the above assumption shows that $f_s(x) \neq x$ for $\|x\| = 1$ and $0 \leq s \leq 1$. The homotopy property of the fixed point index implies that

$$i_C(f_1, \Omega) = i_C(f, \Omega) = i_C(f_o, \Omega) = 1. \tag{2.1}$$

If we can prove that $i_C(f,\Omega) = 0$, we will have a contradiction of our original assumption and the theorem will be proved. It is well-known that $I-f|\overline{\Omega}$ is a proper map (the inverse image of any compact set is compact), so there exists $\delta > 0$ such that $\|x-f(x)\| \geq \delta$ for $x \in C$, $\|x\| = 1$ (since we are assuming $x-f(x) \neq 0$ for $\|x\| = 1$). Let u be as in the statement of the theorem. Because f is homogeneous of degree 1, we can, by multiplying u by a small positive constant, assume that $0 < \|u\| < \delta$ and $\{\|f^m(u)\|: m \geq 1\}$ is unbounded. Define a function $g(x)$ by

$$g(x) = f(x)+u . \tag{2.2}$$

The homotopy $f(x)+su$ for $0 \leq s \leq 1$ has no fixed points x with $\|x\| = 1$ so

$$i_C(f,\Omega) = i_C(g,\Omega). \tag{2.3}$$

To complete the proof it suffices to show $i_C(g,\Omega) = 0$; and to prove $i_C(g,\Omega) = 0$, it suffices to prove that g has no fixed points in $\overline{\Omega}$. Thus we assume that $g(x) = x$ for $x \in \overline{\Omega}$ and try obtain a contradiction.

If $g(x) = x = f(x)+u$ we have

$$x \geq u . \tag{2.4}$$

In general, for purposes of induction, assume that

$$x \geq f^m(u). \tag{2.5}$$

It follows from the order-preserving property of f that

$$\begin{aligned} x = g(x) = f(x)+u \geq f(x) \geq f(f^m u) \\ = f^{m+1}(u) \end{aligned} \tag{2.6}$$

so equation (2.5) holds for all $m \geq 0$ by mathematical induction.

Now we apply Lemma 2.1. Define $a_m = \|f^m(u)\|$, an unbounded sequence of nonnegative reals, and let $\{a_{m_i} : i \geq 1\}$ satisfy the conclusions of Lemma 2.1. Define $v_m = \dfrac{f^m(u)}{\|f^m(u)\|}$ and let $\Sigma = \{v_{m_i} : i \geq 1\}$. We claim that $\beta(\Sigma) = 0$, so that Σ has compact closure. To see this observe that for $j \geq 1$ we can write (using homogeneity of f)

$$\Sigma = \bigcup_{i=1}^{j} \{v_{m_i}\} \cup f^j(T_j)$$

$$T_j \overset{\text{def}}{=} \left\{ \frac{f^{m_i - j}(u)}{\|f^{m_i}(u)\|} : i > j \right\} \tag{2.7}$$

Since $\|f^{m_i - j}(u)\| \leq \|f^{m_i}(u)\|$ for $i \geq j$, we have

$$T_j \subset B = \{x \in K: \|x\| \leq 1\} \tag{2.8}$$

Equations (2.7) and (2.8) imply that

$$\Sigma \subset \bigcup_{i=1}^{j} \{v_{m_i}\} \cup f^j(B) \tag{2.9}$$

Since f^j is a k^j-set-contraction with respect to β, equation (2.9) implies

$$\beta(\Sigma) \leq k^j \beta(B) \tag{2.10}$$

The right hand side of (2.10) approches zero as $j \to \infty$, so $\beta(\Sigma) = 0$, $\bar{\Sigma}$ is compact, and for some subsequence $v_{m_{i_j}} \overset{\text{def}}{=} w_j$ we can assume that w_j converges strongly to w. Of course $\|w\| = 1$ and $w \in K$.

If we now return to equation (2.5) and divide both sides by $a_{m_{i_j}}$ we obtain

$$(a_{m_{i_j}})^{-1} x - w_j \in K \tag{2.11}$$

Taking the limit as $j \to \infty$ yields that $-w \in K$, and this contradiction proves the theorem. ■

Our first corollary generalizes Theorem 4.2 in [13] by removing the assumption of normality of the cone. If the cone C is normal, however, Corollary 2.1 is an easy corollary of an earlier result, Proposition 6 on page 252 of [18]. If the operator g below is compact and $p = 1$, Corollary 2.1 below is a result of Krein and Rutman [9, Theorem 9.1].

(2.1) *COROLLARY*

(Compare [13] and [18]). Let C be a cone in a Banach space X, β a generalized measure of noncompactness such that $\beta(tS) = t\beta(S)$ for every bounded set S in C and every real $t \geq 0$, and $g:C \to C$ a k-set-contraction w.r.t. β. Assume that g is order-preserving and positively homogeneous of degree 1 and that there exists $u \in C-\{0\}$ and $c > k^p$ such that $g^p(u) \geq cu$. Then there exists $x_o \in C-\{0\}$ and and $\lambda \geq c^{(p-1)}$ such that $g(x_o) = \lambda x_o$.

Proof

Let b be any real such that

$$k^p < b < c.$$

If $q = q^{-1}$, define $f(x) = b^{-q}g(x)$. Using the homogeneity of f, it is clear that

$$f^p(u) \geq ru, \quad r = cb^{-1} > 1 \tag{2.12}$$

and that f is a k_1-set-contraction w.r.t. β, where $k_1 = kb^{-q} < 1$.

By Theorem 2.1, f will have an eigenvector in C of norm 1 if $\{\|f^m(u)\|: m \geq 1\}$ is unbounded. But equation (2.12) implies that

$$\frac{f^{jp}(u)}{r^j} - u \geq 0. \tag{2.13}$$

If $\|f^{jp}(u)\|$ were bounded, we would obtain by letting $j \to \infty$ in (2.13) that $-u \in C$, a contradiction.

Theorem 2.1 thus implies that there exists $x \in C$, $\|x\| = 1$, and $t \geq 1$ such that $f(x) = tx$. Writing in terms of g, there exists $\lambda \geq b^q$ such that

$$g(x) = \lambda x. \tag{2.14}$$

Select an integer N such that $c-N^{-1} > k^p$. If $n \geq N$ and if we define $c-n^{-1} = b$, the above remarks show that there exist $x_n \in K$, $\|x_n\| = 1$ and $\lambda_n \geq (c-n^{-1})^q$ such that

$$g(x_n) = \lambda_n x_n. \tag{2.15}$$

If we define $\Sigma = \{x_n : n \geq N\}$, $B = c-N^{-1}$ and $f(x) = B^{-q}g(x)$, previous remarks show that f k_1-set-contraction with respect to β for some $k_1 < 1$. Equation (2.15) shows that

$$\Sigma \subset \overline{co}(f(\Sigma) \cup \{0\}). \tag{2.16}$$

Equation (2.16) implies that $\beta(\Sigma) \leq k_1\beta(\Sigma)$, so $\overline{\Sigma}$ is compact. Therefore, by taking a subsequence we can assume $x_{n_i} \to x$ and $\lambda_{n_i} \to \lambda \geq c^q$ and

$$f(x) = \lambda x, \quad \lambda \geq c^q \tag{2.17}$$

which is the desired result. ■

We now wish to show how Theorem 2.1 can be used to obtain a general version of the linear Krein-Rutman theorem. First we need to recall some definitions. If X is a Banach space and $L:X \to X$ is a bounded linear operator, define $r(L)$, the spectral radius of L, by

$$r(L) \stackrel{def}{=} \lim_{n\to\infty} \|L^n\|^{\frac{1}{n}} = \inf_{n\geq1} \|L^n\|^{\frac{1}{n}}. \tag{2.18}$$

If X is real and $\tilde{X}$ denotes the complexification of X, $\tilde{X} = \{x+iy: x,y \in X\}$ with

$$\|x+iy\| = \sup_{0\leq\theta\leq2\pi} \|(\cos\theta)\, x+(\sin\theta)y\|$$

then L has an obvious linear extension $\tilde{L}$ and $\|\tilde{L}\| = \|L\|$. If $\sigma(\tilde{L})$ denotes the spectrum of $\tilde{L}$, then of course

$$r(L) = \sup\{|z|: z \in \sigma(\tilde{L}),\ z \text{ complex}\}.$$

If $\alpha(L)$ is defined by

$$\alpha(L) = \inf\{c \geq 0: L \text{ is a } c\text{-set-contraction}\} \tag{2.19}$$

it is clear that $\alpha(L) \leq \|L\|$. Because $\alpha(L_1L_2) \leq \alpha(L_1)\,\alpha(L_2)$, $\lim\ (\alpha(L^n))^{\frac{1}{n}}$ exists, and we can define $\rho(L)$ by

$$\sigma(L) = \lim_{n\to\infty} (\alpha(L^n))^{\frac{1}{n}} = \inf_{n\geq1} (\alpha(L^n))^{\frac{1}{n}}. \tag{2.20}$$

If $\tilde{L}$ is as above, one can prove that $\alpha(\tilde{L}) = \alpha(L)$. There is a subset $ess(\tilde{L})$ of $\sigma(\tilde{L})$, the essential spectrum of $\tilde{L}$, and it is proved in [14] that

$$\rho(L) = \sup\{\ |z|: z \in ess(L)\}. \tag{2.21}$$

Actually, there are several inequivalent definitions of the essential spectrum, but (2.21) is valid for all of them. It is proved in [14] that if $\varepsilon > 0$, there is at most a finite number of complex numbers z such that $z \in \sigma(\tilde{L})$ and $|z| \geq \rho(L)+\varepsilon$. If z is such a complex number and Γ a simple closed curve in $\mathbb{C}$ which contains no other point of $\sigma(\tilde{L})$ in its interior or on Γ, then the spectral projection P corresponding to z,

$$P = \left(\frac{1}{2\pi i}\right) \int_{\Gamma} (\zeta-\tilde{L})^{-1}\, d\zeta$$

is finite dimensional.

If K is a cone in X and $L:X \to X$ is a bounded linear operator such that $L(K) \subset K$, define numbers $\|L\|_K$ and $\alpha_K(L)$ by

$$\|L\|_K \overset{\text{def}}{=} \sup\{\|Lu\|: u \in K, \|u\| \leq 1\}$$
$$\alpha_K(L) = \inf\{c \geq 0: L|K \text{ is a } c\text{-set-contraction}\}. \tag{2.22}$$

In analogy $r(L)$ and $\rho(L)$, define $r_K(L)$, the cone spectral radius of L, and $\rho_K(L)$, the cone essential spectral radius of L, by

$$r_K(L) \overset{\text{def}}{=} \lim_{n\to\infty} (\|L^n\|_K)^{\frac{1}{n}} \tag{2.23}$$

$$\rho_K(L) \overset{\text{def}}{=} \lim_{n\to\infty} (\alpha_K(L^n))^{\frac{1}{n}}. \tag{2.23}$$

As was remarked in [18] it is easy to see that $\rho(L) \leq r(L)$, $r_K(L) \leq r(L)$ and $\rho_K(L) \leq \rho(L)$; if the cone K is "reproducing" (so $X = \{u-v: u,v \in K\}$) one can also prove that $\rho_K(L) \leq r_K(L)$. Note also that $\rho(L) = 0$ if L is compact and $\rho_K(L) = 0$ if $L|K$ is compact.

We need to recall one more definition. If K is a cone in a Banach space X with norm $\|\cdot\|$, define $Y = \{u-v: u,v \in K\}$. Define a norm $|\cdot|$ on Y by

$$|y| = \inf\{ \|u\|+\|v\|: u, v \in K, y = u-v\}. \tag{2.24}$$

It is remarked in [2, 22] (and is not hard to prove) that Y is a Banach space in this norm, $\|y\| \leq |y|$ for $y \in Y$, and $\|y\| = |y|$ for $y \in K$. If $L: X \to X$ is a bounded, linear operator such that $L(K) \subset K$, then $L(Y) \subset Y$. Furthermore, if $|L|_Y$ denotes the norm of L considered as a map from the Banach space Y to itself, it is easy to see that

$$|L|_Y = \|L\|_K \tag{2.25}$$

so that

$$\lim_{n\to\infty} |L^n|_Y^{\frac{1}{n}} \overset{\text{def}}{=} r_Y(L) = r_K(L) = \lim_{n\to\infty} \|L^n\|_K^{\frac{1}{n}}. \tag{2.26}$$

In (2.26), $r_Y(L)$ denotes the spectral radius of L as a map from Y to Y.

With these preliminaries we can prove our linear Krein-Rutman theorem.

(2.2) *THEOREM*

Let X be a Banach space, K a cone in X and $L:X \to X$ a bounded linear operator such that $L(K) \subset K$. If $\rho_K(L) = \nu$ and $r_K(L) = \mu$ are defined by equations (2.22) and (2.23), assume that $\nu < \mu$. Then there exists $x \in K-\{0\}$ such that $Lx = \mu x$.

Proof

Let s_n be a sequence of real numbers such that $\nu < s_n < \mu$ and $s_n \to \mu$ and define $g_n(x) = s_n^{-1}L(x)$. Our assumptions imply that

$$\rho_K(g_n) = s_n^{-1}\nu < 1 < s_n^{-1}\mu = r_K(g_n). \tag{2.27}$$

For notational convenience, fix n, write $g = g_n$, $s = s_n$ and select N so large that $\alpha_K(g^m) < 1$ for $m \geq N$. Just as in the proof of Proposition 7 in [18], define a generalized measure of noncompactness by

$$\beta(A) = \frac{1}{N}\sum_{j=0}^{N-1} \alpha(g^j(A)). \tag{2.28}$$

One can easily check that β is a generalized measure of noncompactness, that $\beta(\lambda A) = |\lambda|\beta(A)$ and that there exists a constant $c < 1$ such that

$$\beta(g(A)) \leq c\beta(A) \tag{2.29}$$

for every bounded set $A \subset K$.

If Y is the Banach space defined immediately before the statement of Theorem 2.2, we have seen that $r_K(g) = r_Y(g) =$ the spectral radius of g as a map of Y to Y. Since $r_K(g) > 1$, it follows that $|g^n|_Y$ is unbounded. The uniform boundedness principle on the Banach space Y implies that there exists $y \in Y$ such that $|g^n(y)|$ is unbounded. It follows that there exists $u \in K$ such that $|g^n u| = \|g^n u\|$ is unbounded. Theorem 2.1 now implies that there exists $t \geq 1$ and $x \in K$, $\|x\| = 1$, such that

$$g(x) = tx. \tag{2.30}$$

Of course $t = t_n$ and $x = x_n$ depend on n, and if we write $\mu_n = s_n t_n$ we have

$$L(x_n) = \mu_n x_n. \tag{2.31}$$

We must have $s_n \leq \mu_n \leq \mu$, so $\lim_{n\to\infty} \mu_n = \mu$. Exactly as in the last paragraph of the proof of Proposition 7 in [18] or as in the proof of Corollary 2.1, $\{x_n : n \geq 1\}$ has compact closure, so one can assume by taking a subsequence that $x_n \to x \in K$, $\|x\| = 1$, and taking the limit as $n \to \infty$ of (2.30) gives

$$Lx = \mu x \tag{2.32}$$

which is the desired result. ■

(2.3) *REMARK*

If $L|K$ is compact, so $\alpha_K(L) = 0$, Theorem 2.2 generalizes a result of Bonsall [1,2]. The original Krein-Rutman theorem deals with the case that L is compact as a map of X to X, K is total and $r(L) > 0$, and one might believe that, at least for total cones, Bonsall's result is equivalent to the Krein-Rutman theorem. However Bonsall gives a simple example which shows this hope is false.
Let $X = \{x \in C[0,1]: x(0) = 0\}$ and define $L:X \to X$ by $(Lx)(t) = x(\frac{1}{2}t)$. Bonsall constructs, for each $\gamma > 0$, a total cone K_γ such that $L(K_\gamma) \subset K_\gamma$, $L|K_\gamma$ is compact, and $r_{K_\gamma}(L) = (\frac{1}{2})^\gamma$. Perhaps most surprising is the fact that the cone spectral radius can vary for different total cones. Bonsall proves, however, that if L is compact as a map of X into itself, then $r_K(L) = r(L)$ for every total cone K such that $L(K) \subset K$. Our next proposition is a generalization of this fact, and the argument we give is a generalization of Bonsall's argument for the case L compact.

(2.2) *COROLLARY*

(Compare [1].) Let X be a Banach space and $L:X \to X$ a bounded linear map. Assume that $\rho(L) < r(L)$ ($\rho(L)$ and $r(L)$ defined by (2.18) and (2.20)). Then if K is any total cone in X such that $L(K) \subset K$ and $r_K(L)$ is defined by (2.23), one has

$$r_K(L) = r(L). \tag{2.33}$$

In particular, if $r = r(L)$, there exists $x \in K-\{0\}$ such that

$$Lx = rx. \tag{2.34}$$

If X^* is the dual space of X and $K^* = \{f \in X^* : f(x) \geq 0 \text{ for all } x \in K\}$, there exists $f \in K^*$ such that

$$L^*(f) = rf. \tag{2.35}$$

(2.4) *REMARK*

For reproducing cones, the latter half of Corollary 2.2 was proved in [7] by a linear argument like that used by Krein and Rutman.

Proof

Suppose we can prove that $r_K(L) = r(L)$. Since we clearly have $\rho_K(L) \leq \rho(L)$, it will then follow from Theorem 2.2 that there exists $x \in K$ satisfying (2.34). Since K is total, it is easy to see that K^* is a cone and $L^*:K^* \to K^*$. It is proved in [14] that $\rho(L^*) = \rho(L)$, so that $\rho_{K^*}(L^*) \leq \rho(L)$. If we can prove that $r_{K^*}(L^*) \geq r = r(L)$, the existence of f satisfying (2.35) is a consequence of Theorem 2.2 again. However, if x satisfies (2.34), a version of the Hahn-Banach theorem implies that there exists $g \in K^*$ with $\langle g,x\rangle > 0$ (where $\langle\ ,\ \rangle$ denotes the bilinear pairing between X^* and X) and one obtains

$$\langle (L^*)^n g,x\rangle = \langle g,L^n x\rangle = r^n\langle g,x\rangle. \tag{2.36}$$

Equation (2.36) easily implies that $r_{K^*}(L^*) \geq r$.

Thus to complete the proof it suffices to prove (2.33). We claim that there exists $x \in K$ such that

$$\limsup_{n\to\infty} \frac{\|L^n x\|}{\|L^n\|} > 0. \qquad (2.37)$$

If (2.37) holds for some x, it is easy to see (using the fact that K is total) that there exists $u \in K$ such that

$$\limsup_{n\to\infty} \frac{\|L^n u\|}{\|L^n\|} > 0. \qquad (2.38)$$

Inequality (2.38) implies that $r_K(L) \geq r(L)$, while the opposite inequality is immediate.

Thus it suffices to prove (2.37). We argue by contradiction and assume

$$\limsup_{n\to\infty} \frac{\|L^n x\|}{\|L^n\|} = 0 \quad \text{for all } x \in K.$$

Select numbers ρ_1 and ρ_2 such that

$$\rho = \rho(L) < \rho_1 < \rho_2 < r = r(L).$$

If $B = \{x: \|x\| \leq 1\}$ and α denotes the measure of noncompactness, select N_o such that

$$\alpha(L^n(B)) < \rho_1^n \alpha(B) \leq 2\rho_1^n, \quad n \geq N_o.$$

We can also assume that N_o is so large that

$$2^{(\frac{1}{N_o})} \rho_2 < r.$$

By definition of the measure of noncompactness, there exist sets S_i, $1 \leq i \leq m$, such that

$$L^{N_o}(B) = \bigcup_{i=1}^{m} S_i, \quad \text{diameter } (S_i) \leq 2\rho_1^{N_o}.$$

Select ε with $0 < \varepsilon < 2(\rho_2^{N_o} - \rho_1^{N_o})$ and for each i, $1 \leq i \leq m$, select $x_i \in S_i$ and N_i such that

$$\|T^n x_i\| < \varepsilon\|T^n\|, \; n \geq N_i.$$

Finally, select an integer K_o such that

$$N \stackrel{\text{def}}{=} k_o N_o \geq \max_{1 \leq i \leq m} N_i .$$

If $y \in B$, there exists an integer i, $1 \leq i \leq m$, such that

$$\|T^{N_o} y - x_i\| \leq 2\rho_1^{N_o} .$$

If $n \geq N$ we obtain that

$$\|T^{n+N_o} y\| \leq \|T^n x_i\| + \|T^n\| \|T^{N_o} y - x_i\|$$

$$\leq \varepsilon \|T^n\| + 2\rho_1^{N_o} \|T^n\| \tag{2.39}$$

$$\leq 2\rho_2^{N_o} \|T^n\| .$$

Inequality (2.39) implies that for $n \geq N$

$$\|T^{n+N_o}\| \leq 2\rho_2^{N_o} \|T^n\| . \tag{2.40}$$

Using (2.40) one sees that there exists a constant B independent of the integer $k \geq 1$ such that

$$\|T^{kN_o}\| \leq B(2\rho_2^{N_o})^k \tag{2.41}$$

and (2.41) implies that

$$\limsup_{k \to \infty} \|T^{kN_o}\|^{\left(\frac{1}{kN_o}\right)} \leq 2^{\frac{1}{N_o}} \rho_2 < r . \tag{2.42}$$

Inequality (2.42) contradicts the definition of r. ■

Many cones used in analysis are normal. We would like to close this paper by discussing a linear operator which is best studied in nonnormal cones. The operator L which we shall define below plays an important role in recent work by R. Bumby on the problem of finding the Hausdorff dimension of certain sets of real numbers defined by properties of their continued fraction expansions [4].

Let M denote a finite union of closed, bounded intervals of reals, let X_n, $n \geq 0$, denote the Banach space of real-valued maps $x: M \to \mathbb{R}$ which are n times continuously differentiable and let K_n denote the cone of nonnegative functions in X_n. If $x \in X_n$ define $\|x\|_n$ by

$$\|x\|_n = \sum_{j=0}^{n} \max_{t\in M} |x^{(j)}(t)|. \tag{2.43}$$

There is a natural map $J_n : X_n \to X_o$ by $J_n(x) = x^{(n)}$, the nth derivative of x. If S is a bounded set in X_n, we leave to the reader the straightforward proof that

$$\alpha(S) = \alpha(J_n(S)). \tag{2.44}$$

We use the same letter α in (2.44) to denote the measure of noncompactness in X_n on the left and in X_o on the right.

We now define a linear map L by

$$(Lx)(t) = \sum_{j=1}^{N} b_j(t)x(\tau_j(t)). \tag{2.45}$$

We shall always assume about the given functions b_j and τ_j in (2.45)

H1. The functions $b_j : M \to \mathbb{R}$ are C^∞ and nonnegative for $1 \le j \le N$ and $\sum_{j=1}^{N} b_j(t) > 0$ for all $t \in M$. The functions $\tau_j : M \to \mathbb{R}$ are C^∞, $\tau_j(M) \subset M$ and $\max_{1\le j\le N} (\max_{t\in M} |\tau_j'(t)|) = c < 1$.

Under assumption H1 L defines a bounded linear map of X_n to X_n for $n \ge 0$ and $L(K_n) \subset K_n$. In our next theorem we discuss the spectrum of $L : X_n \to X_n$; for reasons of length we shall not prove Theorem 2.3 here.

(2.3) *THEOREM*

(Compare [4]). Let L be defined by (2.45) and assume that H1 holds. Consider L as a map from $X_n = C^n(M)$ into itself and let ρ_n denote the essential spectral radius of L in X_n and $r_n = r_n(L)$ the spectral radius of L in X_n. Then if

$A = \max_{t\in M} \sum_{j=1}^{N} |b_j(t)|$, $B = \min_{t\in M} \sum_{j=1}^{N} b_j(t)$ and $c = \max_{j,t} |\tau_j'(t)| < 1$, one has

$$\rho_n < Ac^n \tag{2.46}$$

$$r_n \geq B. \tag{2.47}$$

Inequality (2.46) holds even if the functions b_j are not nonnegative. If $\rho_n < r_n$ (as will be true for large n), $r_n = r_o$; and if $\rho_n < r_n$ and D is a total cone in X_n such that $L(D) \subset D$, then $r_D(L) = r_n$. There exists a nonnegative, C^∞ function u, not identically zero, such that

$$Lu = r_o u. \tag{2.48}$$

(2.5) *REMARK*

If b_j is strictly positive for $1 \leq j \leq N$, Bumby [4] proves the existence of a continuous function which is strictly positive on M and satisfies (2.48). However, his proof does not apply under the weaker assumption H1, and indeed we do not prove here that u is positive on M under H1, although we conjecture that this is so. The fact that $r_n = r_o = r_D(L)$ when $\rho_n < r_n$ and the estimates on ρ_n appear to be new; if the b_j are strictly positive, we can prove that $r_n = r_o$ for all n.

(2.6) *REMARK*

If the hypotheses of Theorem 2.3 are weakened slightly, then r_o may fail to be a positive eigenvalue of L with corresponding eigenvector in K_o. For example, if $(Lx)(t) = tx(\frac{1}{2}t)$ for $x \in C[0,1]$, one can verify directly that the spectral radius r_o of L is 0; any C^∞ function with support in $[\frac{1}{2},1]$ is a C^∞ eigenfunction.

If $(Lx)(t) = b(t)x(t)$ for $x \in C[0,1]$, where $b(t)$ is not constant on any subinterval of $[0,1]$, one can prove directly that the spectral radius of L equals $\max_t |b(t)|$, but that L has no point spectrum.

(2.7) *REMARK*

The spectrum of L varies depending on what space L acts in. For example if $(Lx)(t) = x(ct)$, $0 < c < 1$, one can prove that the spectrum of L as a map of $C^n[0,1]$ into itself is $\{c^j: 0 \leq j \leq n-1\} \cup \{z: z \text{ complex}, |z| \leq c^n\}$; c^j $(0 \leq j \leq n-1)$ has algebraic multiplicity one and corresponds to the eigenvector $x_j(t) = t^j$. Furthermore if $|z| < c^n$, z is in the point spectrum of $L:C^n \to C^n$ and has infinite multiplicity; if $0 \leq z < c^n$, there are infinitely many nonnegative, linearly independent eigenvectors corresponding to z.

References

[1] BONSALL, F.F.: Linear operators in complete positive cones, Proc. London Math. Soc. 8 (1958), 53-75.

[2] BONSALL, F.F.: Positive operators compact in an auxiliary topology, Pacific J. Math. 10 (1960), 1131-1138.

[3] BONSALL, F.F.: Lectures on Some Fixed Point Theorems of Functional Analysis, Tata Institute of Fundamental Research, Bombay, 1962.

[4] BUMBY, R.T.: Hausdorff dimension of Cantor sets, to appear.

[5] DARBO, G.: Punti uniti in transformazioni a condiminio non compatto, Rend. Sem. Mat. Univ. Padova 24 (1955), 353-367.

[6] DUGUNDJI, J.: An extension of Tietze's theorem, Pacific J. Math. 1 (1951), 353-367.

[7] EDMUNDS, D.A. and POTTER, A.J.B.and STUART, C.A: Noncompact positive operators, Proc. Royal Soc. London A328 (1972), 67-81.

[8] KRASNOSEL'SKIǏ, M.A.: Positive Solutions of Operator Equations, P. Noordhoff Ltd., Groningen, The Netherlands, 1964.

[9] KREIN, M.G. and RUTMAN, M.A: Linear operators leaving invariant a cone in a Banach space (Russian), Uspehi Mat. Nauk 3, No. 1, 23 (1948), 3-95. English translation, A.M.S. translation 26.

[10] KURATOWSKI, C.: Sur les espaces complets, Fund. Math. 15 (1930), 301-309.

[11] LERAY, J.: Théorie des points fixes: indice total et nombre de Lefschetz, Bull. Soc. Math. France, 87 (1959), 221-233.

[12] LERAY, J. and SCHAUDER, J.: Topologic ct équations fonctionnelles, Ann. Sci. Ecole Norm. Sup. 51 (1934), 45-78.

[13] MASSABO, I. and STUART, C.A: Positive eigenvectors of k-set-contractions, Nonlinear Analysis, T.M.A. 3 (1979), 35-44.

[14] NUSSBAUM, R.D.: The radius of the essential spectrum, Duke Math. J. 37 (1970), 473-478.

[15] NUSSBAUM, R.D.: Some asymptotic fixed point theorems, Trans. Amer. Math. Soc. 171 (1972), 349-375.

[16] NUSSBAUM, R.D.: The fixed point index for local condensing maps, Ann. Mat. Pura. Appl. 89 (1971), 217-258.

[17] NUSSBAUM, R.D.: Generalizing the fixed point index, Math, Ann. 228 (1977), 259-278.

[18] NUSSBAUM, R.D.: Integral equations from the theory of epidemics, p. 235-255 in Nonlinear systems and applications, edited by V. Laksmikantham, Academic Press, New York, (1977).

[19] NUSSBAUM, R.D.: Periodic solutions of some nonlinear integral equations, p. 221-249, in Dynamical systems, Proceedings of a University of Florida International Symposium, edited by A.R. Bednarek and L. Cesari, Academic Press, New York, (1977).

[20] RABINOWITZ, P.: Théorie du degré topologique et applications à des problèmes aux limites non linéaires, notes by H. Berestycki from a course at Université Paris VI, Spring, (1973).

[21] SCHAEFER, H.H.: On nonlinear positive operators, Pacific J. Math. 9 (1959), 847-860.

[22] SCHAEFER, H.H.: Some spectral properties of positive linear operators, Pacific J. Math. 10 (1960), 1009-1019.

AN ε-PERTURBATION OF BROUWER'S DEFINITION OF DEGREE

BY

HEINZ-OTTO PEITGEN AND HANS-WILLI SIEGBERG

Fachbereich Mathematik
Forschungsschwerpunkt "Dynamische Systeme"
Universität Bremen
Bibliothekstrasse
Postfach 330 440
2800 Bremen 33
West Germany

0. INTRODUCTION.

Degree theories play a central role in many fields of modern mathematics. In each of these fields one knows a typical approach close to the "nerves" of that field. For example, one has a setting of degree theory in *differential topology*, for example [26], [28], [35], [41], *differential geometry*, for example [14], [25], [37], [48], *algebraic topology*, for example [2], [11], [13], [20], [47], *singularity theory*, for example [18], *analysis*, for example [12], [27], [36], [45], and in *linear* and *nonlinear functional analysis*, for example [5], [19], [21].

The aim of this note is to present an approach

- which one can consider to be the numerical analyst's approach,
- which is most elementary with respect to its definition and the proofs of its basic properties, and,
- which can be viewed to be only an ε-variation of Brouwer's original definition [7], [8].

This ε-variation, however, seems to be an essential observation, because it both provides a completely constructive approach to degree as well as a solid background for recent developments in numerical analysis: PL- and simplicial methods for solving nonlinear equations (for a survey we recommend [3] and [15]). In particular, it turns out that the generalized homotopy invariance of degree (see (4.3.2)) which is crucial in global bifurcation theory and which is the hardest to get in any other definition of degree, comes out surprisingly elementarily. Here, this property will be a consequence of a fundamental principle (the Door in/Door out Principle, see for example [15], [16], [23], [39], [50]) which is the heart of all recent PL- and simplicial algorithms.

Since the fundamental paper of Leray and Schauder [32] degree theories have become most important and useful in the context of existence proofs for linear and nonlinear operator equations. Solving these problems numerically it has become popular to make use of degree arguments as well, see for example [1], [4], [10], [31], [39], [40]. It therefore seems to be adequate to develop a purely numerical approach to degree theory, its basic properties and those important principles (Leray-Schauder Continuation, Global Bifurcation in the sense of Krasnosel'skii-Rabinowitz, Borsuk-Ulam Theorem etc.) which are essentially due to degree arguments.

Our approach will follow Brouwer's original definition [7], [8] (in a modern language of course; for a survey of the history of degree theory see [46]): Degree will be obtained by "PL-approximations" and reduction to "regular values". The difference to Brouwer's approach consists in the fact that we will

- provide "regular values" in a constructive and algorithmic way,
- have an algorithmic proof for the generalized homotopy invariance, and,
- deduce all other basic properties (Additivity, Solution Property, etc.) from this property.

We emphasize the affinity of our approach to Pontryagin's most elegant one in differential topology [41], see also [5], [19], [26], [28], [35]. In fact, the ingredients used in the present paper are "PL-substitutes" of differential topological frames as

- C^{∞}-Mollifiers
- Brown-Sard Theorem
- Implicit Function Theorem
- Classification of Smooth 1-Manifolds, and,
- Pontryagin Construction.

However, the PL-substitutes will turn out to be elementary facts which are easy to handle.

The first three chapters are purely preparatorial, and provide the framework which we need. The content of these chapters is essentially well-known from recent PL- and simplicial algorithms. In chapter 4 we give a definition of degree in terms of "$\overline{\varepsilon}$-regular" simplices, and deduce the properties characterizing the Brouwer degree. In the last chapter we prove a reduction property of the degree which is the crucial step in defining the Leray-Schauder degree. We conclude with some remarks concerning a degree for multivalued mappings.

1. PRELIMINARIES

We start with an axiomatic characterization of the Brouwer degree. According to results of Amann and Weiss [6], Führer [22], Nussbaum [38], and Zeidler [51] the Brouwer degree is uniquely determined by the following properties:

(1.1) *AXIOMS FOR A BROUWER DEGREE*

For every bounded open set $U \subset R^n$ let

$$M(U) := \{f : \overline{U} \to R^n : f \text{ continuous}, 0 \notin f(\partial U)\} .$$

The Brouwer degree in R^n is a collection of maps

$$d \equiv \{d(\cdot, U, 0) : M(U) \to Z : U \subset R^n \text{ bounded, open}\}$$

satisfying the following axioms:

NORMALIZATION: For every bounded open subset $U \subset R^n$ with $0 \in U$.

$$d(Id_{\overline{U}}, U, 0) = 1 .$$

ADDITIVITY: For every nonempty bounded open subset $U \subset R^n$, every pair of disjoint open subsets U_1, U_2 of U, and for every $f \in M(U)$ with $0 \notin f(\overline{U} \setminus (U_1 \cup U_2))$,

$$d(f,U,0) = d(f,U_1,0) + d(f,U_2,0).$$

HOMOTOPY INVARIANCE: For every nonempty bounded open subset $U \subset R^n$, and for every continuous map $h : \overline{U} \times [0,1] \to R^n$ with $0 \notin h(\partial U \times [0,1])$,

$$d(h(\cdot, t), U, 0) \equiv \text{const.}$$

as a function of t.

An easy caculation shows that *ADDITIVITY* implies the solution property see for example [6]:

SOLUTION PROPERTY: For every bounded open subset $U \subset R^n$, and for every $f \in M(U)$ with $0 \notin f(\bar{U})$,

$$d(f,U,o) = 0.$$

(1.2) *TRIANGULATION*

A k-*simplex* $\sigma = [a^o,\ldots,a^k]$ in R^n $(0 \le k \le n)$ is the convex hull of $k+1$ affinely independent points $a^o,\ldots,a^k$ in R^n, which are called the *vertices* of σ. A simplex $\tau \subset \sigma$ is called a *face* of σ if, and only if all its vertices are contained in $\{a^o,\ldots a^k\}$.

(1.2.1) *DEFINITION*

Let $T_n := \{\sigma_1,\sigma_2,\ldots\}$ be a (possibly infinite) set of n-simplices in R^n. The set

$$M := \bigcup_{\sigma \in T_n} \sigma$$

is called a *triangulable set of homogeneous dimension* n provided

1) for all $\sigma,\sigma' \in T_n$ the intersection $\sigma \cap \sigma'$ is empty or a common face of both σ and σ';

2) for every $x \in M$ there is a neighborhood U_x of x intersecting only a finite number of simplices $\sigma \in T_n$.

If 1) and 2) are satisfied the collection T of all faces of simplices in T_n is called a *triangulation* of M. The set of all k-simplices in T $(0 \le k \le n)$ is called the k-*skeleton* of T and denoted by T_k.

For any open subset $U \subset M$ define

$$M_T(U) := \bigcup_{\substack{\sigma \in T_n \\ \sigma \subset \overline{U}}} \sigma$$

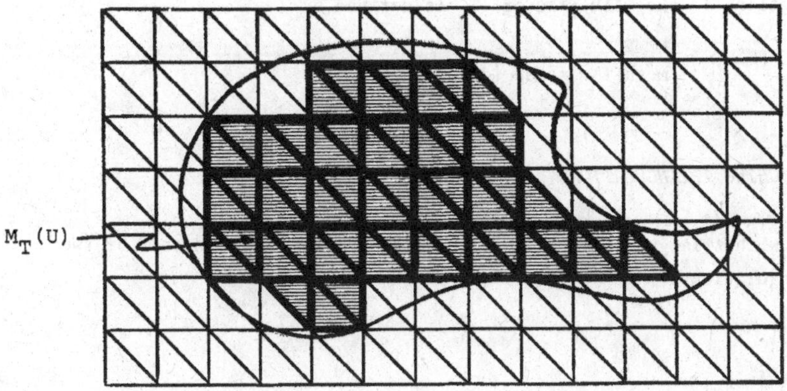

(1.2.2) Every (n - 1)-simplex in the triangulation T of M is either the face of exactly two n-simplices in T_n or the face of exactly one n-simplex in T_n. In the latter case the (n - 1)-simplex is called a *boundary simplex* of the triangulation.

(1.2.3) For numerical purposes it is important to remark that several triangulations of R^n - for example Kuhn's triangulation or Todd's "Union Jack" triangulation, see [49] - are easy to implement on a computer, see for example [49]:

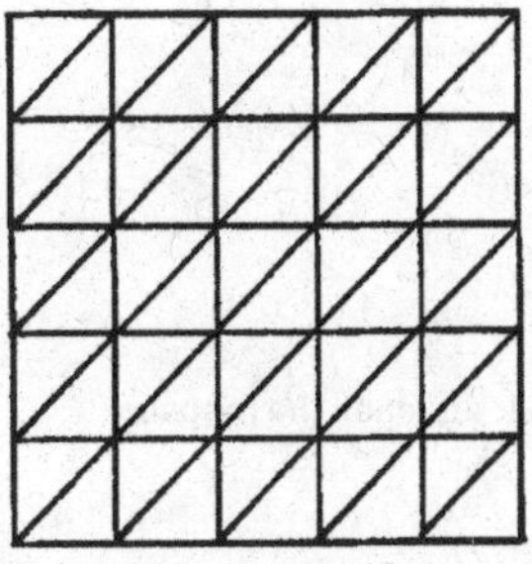

Kuhn's triangulation (R^2)

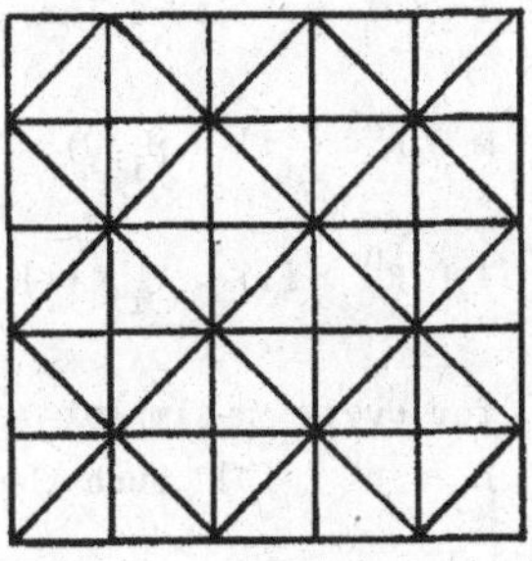

Todd's triangulation (R^2)

Throughout the paper the ℓ_∞-norm $|x| := \max\{|x_i| : i = 1,\dots,n\}$ is used.

For a bounded subset $B \subset R^n$ the *diameter* of B is defined by

$$\mathrm{diam}(B) := \sup \{|x-y| : x, y \in B\} .$$

For a triangulation T the *mesh-size* of T is defined by

$$\mathrm{mesh}(T) := \sup \{\mathrm{diam}(\sigma) : \sigma \in T\} .$$

(1.3) *MATCHING TRIANGULATIONS*

To work with triangulations one often needs a "common denominator" of two different triangulations:

(1.3.1) *LEMMA*

Let $U \subset R^n$ be a bounded open subset of R^n, and let T_o, T_1 be two triangulations of R^n. Let $f : \overline{U} \to R^n$ be continuous such that

$$0 \notin f(\overline{U \setminus M_i(U)}) \quad \text{where} \quad M_i(U) := M_{T_i}(U), \quad i = o,1.$$

Assume that for every boundary simplex τ of $M_i(u)$, $i = o,1$, there exists an open halfspace $H_\tau \subset R^n \setminus \{0\}$ such that $f(\tau) \subset H_\tau$.

Then there exists a triangulable set $M \subset \overline{U} \times [0,1]$ of homogeneous dimension $n+1$ and a triangulation T of M such that

1) $f^{-1}(o) \times [0,1] \subset \mathrm{int}\ M$

2) $M \cap R^n \times \{i\} = M_i(U) \times \{i\}, \quad i = o,1$

3) $T \cap R^n \times \{i\} = T_i \times \{i\}, \quad i = o,1$

4) for every n-simplex $\sigma \subset \partial M$ there exists an open halfspace $H_\sigma \subset R^n \setminus \{0\}$ such that

$$h(\sigma) \subset H_\sigma ,$$

where $h(x,t) := f(x)$ for $(x,t) \in M$.

Proof

Let $M := M_o(U) \times [0, 0.5] \cup M_1(U) \times [0.5, 1]$.

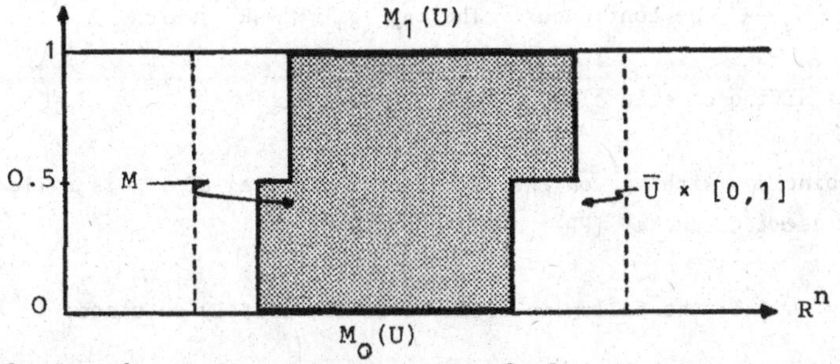

Let further $\lambda > 0$ be a Lebesque number [29] such that any subset $S \subset \overline{U \setminus M_i(U)}$, $i = o,1$, with $\mathrm{diam}(S) \le \lambda$ is mapped by f into some open half-space in $R^n \setminus \{0\}$.

Now choose a triangulation T of M which satisfies condition 3) such that for any n-simplex $\sigma \subset \partial M$ one of the following two alternatives holds:

i) σ is contained either in $\tau_o \times [0, 0.5]$ for some boundary simplex $\tau_o \subset M_o(U)$ or in $\tau_1 \times [0.5, 1]$ for some boundary simplex $\tau_1 \subset M_1(U)$;

or ii) $\mathrm{diam}(\sigma) \le \lambda$.

The construction of T requires standard techniques of PL-topology as described in for example [2], [30], [44]. By construction the triangulation T satisfies the desired properties. ■

2. PL-APPROXIMATION

In this chapter some basic PL-tools are developed which are in analogy with fundamental facts from differential topology.

In what follows M will always denote a bounded triangulable set of homogeneous dimension, and T will be a fixed triangulation of M. Moreover, mappings from n-space to n-space are always denoted by f, mappings from $(n+1)$-space to n-space by h, and mappings from m-space $(m \ge n)$ to n-space by g.

(2.1) *PL-MOLLIFIERS*

(2.1.1) *DEFINITION*

Let $g : M \to R^n$ be continuous. The map $g_T : M \to R^n$ which

1) is affine on each simplex $\sigma \in T$, and,

2) coincides with g on the set T_o of vertices of T is called the *piecewise linear* (PL-) *approximation* of g.

For any $\sigma \in T$ the following estimate is immediately obtained:

$$|g(x) - g_T(x)| \leq \text{diam}\, g(\sigma), \quad x \in \sigma. \tag{2.1.2}$$

Hence, the mapping g_T will converge uniformly to g provided mesh $(T) \to 0$.

(2.2) *THE $\bar{\varepsilon}$-PERTURBATION / PL-BROWN-SARD THEOREM*

The following ideas are crucial for our understanding of Brouwer degree. They may be interpreted as a regularization technique which is similar to the Brown-Sard theorem in differential topology.

Denote by $\gamma : [0,\infty) \to R^n$ the curve

$$\gamma(t) := (t, t^2, \ldots, t^n) \in R^n.$$

For small $\varepsilon > 0$ we will also use the abbreviation $\bar{\varepsilon} := \gamma(\varepsilon)$ in the following. The following definition is in analogy with differential topology.

(2.2.1) *DEFINITION*

Let $g : M \to R^n$ be continuous.

1) $0 \in R^n$ is called a *regular value* for g_T provided $g_T^{-1}(0) \cap T_{n-1} = \phi$;

2) $\sigma \in T_n$ is called a *regular* n-*simplex* for g_T provided $0 \in g_T(\sigma)$, and 0 is a regular value for $g_T|_\sigma$.

In differential topology it is a consequence of the Brown-Sard theorem that regularity is a generic quality: any smooth map $g : M \to R^n$ (M smooth manifold), no matter how bizarre its behavior, may be deformed by an arbitrary small amount into a map that has $0 \in R^n$ as a regular value [26, ch. 2]. This is achieved "with probability one", see for example [1], [10], simply by a translation $g + a$, $a \in R^n$ small. In view of this fact the following lemma is a PL-analogue of the celebrated Brown-Sard theorem.

(2.2.2) *LEMMA (PL-Brown-Sard Theorem)*

Let $g : M \to R^n$ be continuous. Then there exists $\varepsilon_o \equiv \varepsilon_o(g,T)$ such that for all ε, $0 < \varepsilon < \varepsilon_o$,

1) $0 \in R^n$ is a regular value for $g_T - \bar{\varepsilon}$, and,

2) $\sigma \in T_n$ is a regular n-simplex for $g_T - \bar{\varepsilon}$, once σ is a regular n-simplex for $g_T - \bar{t}$, for some fixed t, $0 < t < \varepsilon_o$.

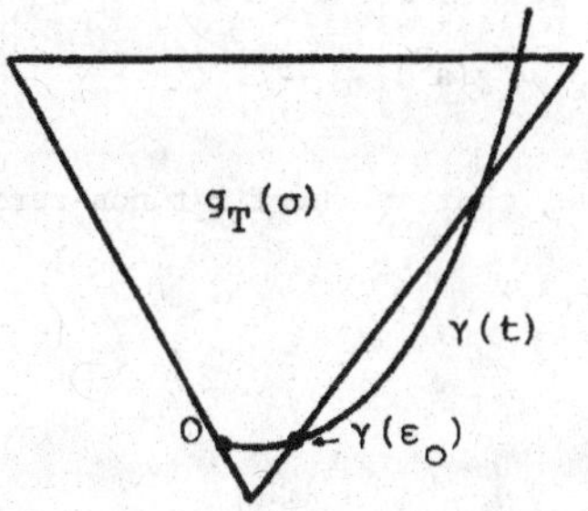

Proof

Observe that by the very definition of γ the points $\gamma(t_0), \gamma(t_1), \ldots, \gamma(t_n)$ are affinely independent provided $t_0, \ldots, t_n \in [0,\infty)$ are mutually different (van der Monde determinant). Hence, $\gamma([0,\infty))$ intersects any hyperplane of R^n in at most n points, and consequently, $g_T(T_{n-1})$ is intersected by γ in at most a finite number of points. Define

$$\varepsilon_o := \min \{\varepsilon > 0 : \bar{\varepsilon} \in g_T(T_{n-1})\} > 0.$$

Then 1) is satisfied by construction. To verify 2) observe that by definition of $\varepsilon_o\ \bar{\varepsilon}$ cannot escape from $g_T(\sigma \setminus \partial\sigma)$ provided $0 < \varepsilon < \varepsilon_o$. ■

In view of the previous lemma we weaken the notion of a regular simplex.

(2.2.3) *DEFINITION*

Let $g : M \to R^n$ be continuous. $\sigma \in T_n$ is called an $\bar{\varepsilon}$-*regular* n-simplex for g_T provided there exists $\varepsilon_o > 0$ such that σ is a regular n-simplex for $g_T - \bar{\varepsilon}$ for all ε with $0 < \varepsilon < \varepsilon_o$.

Obviously, any n-simplex $\sigma \in T_n$ which is regular for g_T is also $\bar{\varepsilon}$-regular for g_T.

Definition (2.2.3) is equivalent to the following numerical characterization which yields an immediate computational access:

(2.2.4) $\sigma = [a^o,\dots,a^n]$ is $\bar{\varepsilon}$-regular for g_T if and only if

$$\begin{bmatrix} 1 & \dots & 1 \\ g(a^o) & \dots & g(a^n) \end{bmatrix}^{-1}$$

exists, and is lexicographically positive, that is the first non-zero element in each row is positive.

<u>*Proof*</u>

Note that $\bar{\varepsilon} \in g_T(\sigma)$ if and only if there exist $\lambda_o,\dots,\lambda_n \geq 0$ such that

$$\begin{bmatrix} 1 & \dots & 1 \\ \\ g(a^o) & \dots & g(a^n) \end{bmatrix} \begin{bmatrix} \lambda_o \\ \vdots \\ \lambda_n \end{bmatrix} = \begin{bmatrix} 1 \\ \varepsilon \\ \vdots \\ \varepsilon^n \end{bmatrix}$$ ■

Using this characterization we can prove the following useful fact, see chapter 5:

(2.2.5) Let $\tau := [a^o,\dots,a^{n-1}] \subset R^{n-1}$ be any (n-1)-simplex in R^{n-1}. τ is $\bar{\varepsilon}$- regular for Id_τ if and only if $\sigma := [(a^o,o),\dots,(a^{n-1},o),(o,o,\dots,o,1)] \subset R^n$ is $\bar{\varepsilon}$-regular for Id_σ.

Proof

Observe that

$$\begin{bmatrix} 1 & \dots & 1 & 1 \\ a^o & \dots & a^{n-1} & 0 \\ 0 & 0 \,.\, 0 & 0 & 1 \end{bmatrix}^{-1} = \begin{bmatrix} A & v \\ 0 & 1 \end{bmatrix}, \quad v \in R^n ,$$

with

$$A = \begin{bmatrix} 1 & \dots & 1 \\ a^o & \dots & a^{n-1} \end{bmatrix}^{-1} \qquad \blacksquare$$

(2.2.2) together with the following lemma though being only a consequence of elementary linear algebra arguments can be considered to be the corner stone of our approach. They are well-known in PL-algorithms, and there they form the heart of such an algorithm. In view of the differential topology approach to degree the following lemma is essentially the implicit function theorem, see (2.3).

(2.2.6) *LEMMA (None or two)*

Let $h : M \to R^n$ be continuous, and let $\sigma \in T_{n+1}$ be any (n+1)-simplex in T. Then σ has

-either two $\bar{\varepsilon}$-regular n-faces for h_T
-or no $\bar{\varepsilon}$-regular n-face for h_T.

Proof

Assume there is an $\bar{\varepsilon}$-regular n-face $\tau \subset \sigma$ for h_T. Choose ε_o as in lemma (2.2.2), and fix ε, $0 < \varepsilon < \varepsilon_o$. Since $h_T|_\sigma$ is affine, $h_T^{-1}(\bar{\varepsilon}) \cap \sigma$ is a line intersecting $\partial\sigma$ in exactly two points: one point contained in τ, the other point contained in some other n-face $\tau' \subset \sigma$. τ' is by construction $\bar{\varepsilon}$-regular for h_T, and it is clear that τ and τ' are the only $\bar{\varepsilon}$-regular n-faces for h_T. $\blacksquare$

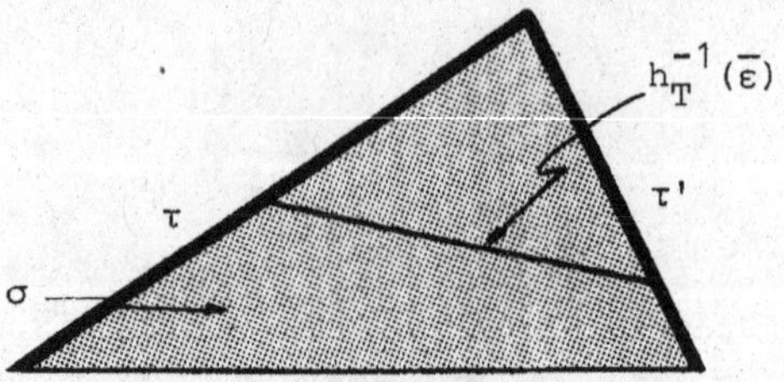

(2.2.7) Again it is important to emphasize the computational feature of this lemma:

Given one $\bar{\varepsilon}$-regular n-face of σ, then the second $\bar{\varepsilon}$-regular n-face of σ can actually be computed via a linear programming step, see for example [3, lemma (3.11)].

(2.3) *PL-IMPLICIT FUNCTION THEOREM*

As a consequence of the previous lemma we get a principle representing a PL-analogue to the implicit function theorem.

(2.3.1) *THEOREM (Door in / Door out principle)*

Let $h : M \to R^n$ be continuous, and let $\tau_o \in T_n$ be any n-simplex of T which is $\bar{\varepsilon}$-regular for h_T. Then τ_o determines a unique chain

$$ch_h(\tau_o) := \ \dots\ \tau_{-1}, \tau_o, \tau_1, \tau_2, \dots$$

of n-simplices which are $\bar{\varepsilon}$-regular for h_T.

Any such chain

- either starts and stops in a boundary simplex
- or is cyclic.

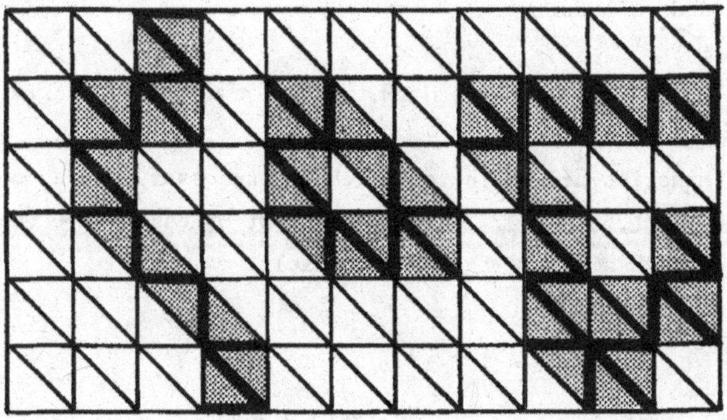

chains
of
$\bar{\varepsilon}$-regular
simplices

Proof

The proof follows immediately from (2.2.6). ■

(2.3.2) To each chain $ch_h(\tau_o)$ one can associate in a unique way the point set (piecewise linear manifold)

$$m_h(\tau_o) := h_T^{-1}(\bar{\varepsilon}) \cap \bigcup_{\tau \in ch_h(\tau_o)} \tau$$

(ε sufficiently small, that is $0 < \varepsilon < \varepsilon_o$, see (2.2.2)).

Because no "bifurcations" can occur the set $m_h(\tau_o)$ is isomorphic

- -either to [0,1]
- -or to S^1 (the cyclic case).

Observe further that

$$\partial m_h(\tau_o) = m_h(\tau_o) \cap \partial M.$$

In formulation (2.3.2) we have obtained a complete PL-substitute of

- -the implicit function theorem, and
- -the classification of compact smooth 1-manifolds

as they are used in differential topology, see [26], [28], [35]:

Let M be a compact smooth $(n+1)$-manifold with boundary, and let $h: M \to R^n$ be a smooth map such that $0 \in R^n$ is a regular value, both for h and $h|_{\partial M}$. Then $h^{-1}(o)$ is a compact submanifold of M of dimension 1 (hence, a collection of smooth copies of [0,1] and S^1) with

$$\partial h^{-1}(0) = h^{-1}(0) \cap \partial M.$$

Up to now all preliminaries are collected which are necessary for a "mod 2" degree theory, see for example [26], [35]. With the support of chapter 4 the reader will be able to develop such a degree theory by himself.

3. Orientation and Curve Index

To unfold the full power of the previous considerations one introduces orientation.

(3.1) *ORIENTATION*

Brouwer degree in this approach will be defined by counting $\bar{\varepsilon}$-regular simplices with a certain orientation.

(3.1.1) *DEFINITION*

Let $f : M \to R^n$ be continuous, and let $\sigma = [a^0,\dots,a^n] \in T_n$ be $\bar{\varepsilon}$-regular for f_T. The number

$$\mathrm{or}(\sigma) := \mathrm{sign}\det\begin{bmatrix} 1 & \dots & 1 \\ a^0 & \dots & a^n \end{bmatrix} \cdot \mathrm{sign}\det\begin{bmatrix} 1 & \dots & 1 \\ f(a^0) & \dots & f(a^n) \end{bmatrix}$$

is called the *orientation number* of σ.

It is easily seen that $\mathrm{or}(\sigma) \in \{-1,1\}$ is the sign of the determinant of the linear part of $f_T|_\sigma$.

(3.2) *CURVE INDEX*

In (2.3.1) it was described how $\bar{\varepsilon}$-regular boundary simplices are connected by chains of $\bar{\varepsilon}$-regular simplices. In order to relate the orientation numbers of the boundary simplices one defines the curve index, see [15], [16], and also [50].

(3.2.1) *DEFINITION*

Let $h : M \to R^n$ be continuous, let $\sigma = [a^o,\dots,a^{n+1}] \in T_{n+1}$, and let $\tau = [a^o,\dots,a^n] \subset \sigma$ be $\bar{\varepsilon}$-regular for h_T. The number

$$\operatorname{ind}_h(\tau \subset \sigma) := \operatorname{sign}\det\begin{bmatrix}1 & \dots & 1\\ a^o & \dots & a^{n+1}\end{bmatrix} \cdot \operatorname{sign}\det\begin{bmatrix}1 & \dots & 1\\ h(a^o) & \dots & h(a^n)\end{bmatrix}$$

is called the *curve index of* τ *in* σ (with recpect to h).

The following fundamental properties of the curve index are crucial:

(3.2.2) Let $\tau = [a^o,\dots,a^n]$ be contained in $R^n \times \{a\}$ or in $R^n \times \{b\}$, and let the vertex a^{n+1} of σ be contained in $R^n \times \{c\}$, where $a < c < b$. Then

$$\operatorname{ind}_h(\tau \subset \sigma) = \begin{cases} -\operatorname{or}(\bar{\tau}), & \tau \subset R^n \times \{b\}\\ \operatorname{or}(\bar{\tau}), & \tau \subset R^n \times \{a\}\end{cases}$$

($\bar{\tau} := \{x \in R^n : (x,a)$ respectively $(x,b) \in \tau\}$).

<u>*Proof*</u>

Let $d \in \{a,b\}$. Then

$$\operatorname{sign}\det\begin{bmatrix}1 & \dots & 1 & 1\\ a^o & \dots & a^n & a^{n+1}\end{bmatrix} = \operatorname{sign}\det\begin{bmatrix}1 & \dots & 1 & 1\\ \bar{a}^o & \dots & \bar{a}^n & \bar{a}^{n+1}\\ d & \dots & d & c\end{bmatrix}$$

$$= \operatorname{sign}(c-d)\cdot\operatorname{sign}\det\begin{bmatrix}1 & \dots & 1\\ \bar{a}^o & \dots & \bar{a}^n\end{bmatrix}. \blacksquare$$

The following fact is useful.

(3.2.3) *LEMMA*

Let $a^o,\dots,a^k$ be $k+1$ affinely independent points in R^{k+1}, and let H be the hyperplane spanned by $a^o,\dots,a^k$. The two halfspaces of R^{k+1} induced by H are determined by the sign of the functional $\ell : R^{k+1} \to R$,

$$\ell(z) := \det \begin{bmatrix} 1 & \dots & 1 & 1 \\ a^o & \dots & a^k & z \end{bmatrix}.$$

Proof

It is easy to find two points in $R^{k+1} \setminus H$ where ℓ has a different sign. Since ℓ vanishes if and only if $z \in H$, and since ℓ is continuous, ℓ has a constant sign on each of the two halfspaces. ■

(3.2.4) Let $h : M \to R^n$ be continuous, and let τ_1, τ_2 be the two $\bar{\varepsilon}$-regular n-faces of $\sigma \in T_{n+1}$. Then

$$\mathrm{ind}_h\,(\tau_1 \subset \sigma) + \mathrm{ind}_h(\tau_2 \subset \sigma) = 0.$$

Proof

Let $\sigma = [a^o,\dots,a^{n+1}]$ and $\tau_1 = [a^o,\dots,a^n]$, $\tau_2 = [a^1,\dots,a^{n+1}]$. Observe that $\bar{\varepsilon} \in h_T(\tau_1) \cap h_T(\tau_2)$ for $0 < \varepsilon < \varepsilon_o$ implies that $h(a^o)$ and $h(a^{n+1})$ are contained in the same halfspace induced by $h(a^1),\dots,h(a^n)$. Hence,

$$\mathrm{ind}_h(\tau_1 \subset \sigma) := \mathrm{sign}\det \begin{bmatrix} 1 & \dots & 1 \\ a^o & \dots & a^{n+1} \end{bmatrix} \cdot \mathrm{sign}\det \begin{bmatrix} 1 & \dots & 1 \\ h(a^o) & \dots & h(a^n) \end{bmatrix}$$

$$= -\,\mathrm{sign}\det \begin{bmatrix} 1 & \dots & 1 & 1 \\ a^1 & \dots & a^{n+1} & a^o \end{bmatrix} \cdot \mathrm{sign}\det \begin{bmatrix} 1 & \dots & 1 & 1 \\ h(a^1) & \dots & h(a^n) & h(a^o) \end{bmatrix}$$

$$= -\,\mathrm{sign}\det \begin{bmatrix} 1 & \dots & 1 & 1 \\ a^1 & \dots & a^{n+1} & a^o \end{bmatrix} \cdot \mathrm{sign}\det \begin{bmatrix} 1 & \dots & 1 & 1 \\ h(a^1) & \dots & h(a^n) & h(a^{n+1}) \end{bmatrix}$$

$$=: -\,\mathrm{ind}_h(\tau_2 \subset \sigma).\ ■$$

(3.2.5) Let $h : M \to R^n$ be continuous, and let $\tau \in T_n$ be a common $\bar{\varepsilon}$-regular face of σ_1 and $\sigma_2 \in T_{n+1}$. Then

$$\mathrm{ind}_h(\tau \subset \sigma_1) + \mathrm{ind}_h(\tau \subset \sigma_2) = 0.$$

Proof

Let $\tau = [a^o,\dots,a^n]$ and $\sigma_1 = [a^o,\dots,a^n,b]$, $\sigma_2 = [a^o,\dots,a^n,c]$. (3.2.3) implies that

$$\text{sign}\det\begin{bmatrix}1 & \dots & 1 & 1\\ a^o & \dots & a^n & b\end{bmatrix} + \text{sign}\det\begin{bmatrix}1 & \dots & 1 & 1\\ a^o & \dots & a^n & c\end{bmatrix} = 0,$$

which proves the assertion. ■

Combining (3.2.4) and (3.2.5) with the Door in/ Door out principle (2.3.1) we obtain the following PL-analogue of the Pontryagin construction.

(3.2.6) *THEOREM (Pl-Pontryagin construction)*

Let $h : M \to R^n$ be continuous, and let $\tau_o,\dots,\tau_k \in T_n$ be a finite chain of n-simplices which are $\bar{\varepsilon}$-regular for h_T. Assume τ_o and τ_k are boundary simplices. Then

$$\text{ind}_h(\tau_o \subset \sigma_o) + \text{ind}_h(\tau_k \subset \sigma_k) = 0,$$

where σ_o,σ_k are the corresponding adjoining (n+1)-simplices in T_{n+1}.

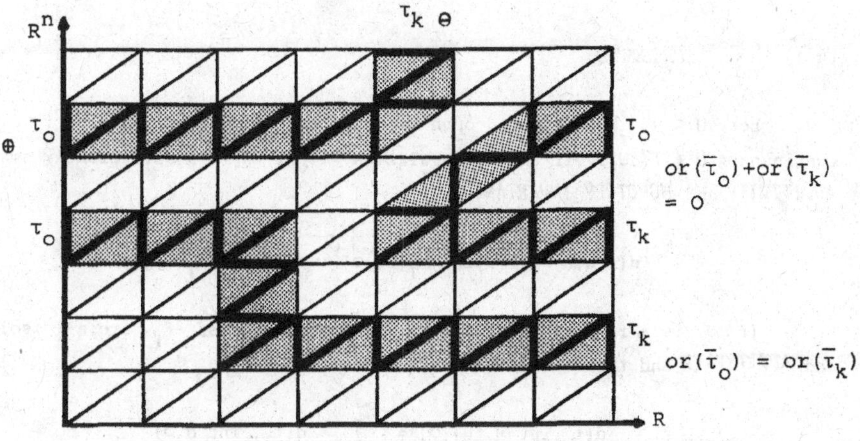

Proof

Let $\sigma_{ij} \in T_{n+1}$ be the $(n+1)$-simplex in T_{n+1} with the $\bar{\varepsilon}$-regular faces τ_i, τ_j. Then,

$$\begin{aligned} \mathrm{ind}_h(\tau_o \subset \sigma_{01}) &= -\,\mathrm{ind}_h(\tau_1 \subset \sigma_{01}) && \text{by (3.2.4)} \\ &= \mathrm{ind}_h(\tau_1 \subset \sigma_{12}) && \text{by (3.2.5)} \\ &= \ldots = \mathrm{ind}_h(\tau_{k-1} \subset \sigma_{k-1,k}) = -\,\mathrm{ind}_h(\tau_k \subset \sigma_{k-1,k}). \blacksquare \end{aligned}$$

If, in particular, $\tau_o, \tau_k \subset R^n \times\{a\} \cup R^n \times\{b\}$, $a < b$, then (3.2.2) implies

$$\mathrm{or}(\bar{\tau}_o) + \mathrm{or}(\bar{\tau}_k) = 0, \quad \text{in case } \tau_o, \tau_k \subset R^n \times\{a\} \text{ or } \tau_o, \tau_k \subset R^n \times\{b\}$$

and

$$\mathrm{or}(\bar{\tau}_o) = \mathrm{or}(\bar{\tau}_k) \quad , \text{ otherwise.}$$

4. Definition and Properties of Brouwer Degree.

First, to give a motivation, we will show how one is led to the definition of Brouwer degree given in this approach by employing the axioms from chapter 1 and the $\bar{\varepsilon}$-perturbation.

(4.1) *HEURISTICS*

Let $U \subset R^n$ be a bounded open set, and let $f \in M(U)$, that is $f : \bar{U} \to R^n$ continuous, $0 \notin f(\partial U)$. If the triangulation T of R^n is sufficiently small, *ADDITIVITY* and *HOMOTOPY INVARIANCE* imply

$$d(f,U,o) = d(f,\mathrm{int}\ M_T(U),o) = d(f_T,\mathrm{int}\ M_T(U),o).$$

If 0 is a regular value for f_T the solutions of f_T are all isolated, and *ADDITIVITY* and the *SOLUTION PROPERTY* imply:

$$d(f_T,\mathrm{int}\ M_T(U),o) = \sum_{\substack{\sigma \in T_n \\ \sigma \subset \bar{U}}} d(f_T, \mathrm{int}\ \sigma,o)$$

$$= \sum_{\substack{\sigma \in T_n \\ 0 \in f_T(\sigma)}} d(f_T, \text{int } \sigma, 0) .$$

Next one can compute - by the Leray-Schauder theorem, see for example [6] - for $\sigma \in T_n$ with $0 \in f_T(\sigma)$ $d(f_T, \text{int } \sigma, 0) = \text{sign det } A_\sigma = \text{or}(\sigma)$, where A_σ is the linear part of $f_T|_\sigma$.

This is the definition of degree as introduced by Brouwer in [7] (in the regular case), see also [46]. This is due to the observation that

$$\sum_{\substack{\sigma \in T_n \\ 0 \in f_T(\sigma)}} \text{sign det } A_\sigma = \#\{\sigma \in T_n \mid 0 \in f_T(\sigma),\ f_T|_\sigma \text{ is orientation preserving}\}$$
$$- \#\{\sigma \in T_n \mid 0 \in f_T(\sigma),\ f_T|_\sigma \text{ is orientation reversing}\}.$$

To extend this definition to the general case Brouwer used certain nonconstructive approximation steps. Here, we will leave Brouwer's route, and apply the $\bar{\varepsilon}$-perturbation technique as described in (2.2), to obtain a fully constructive (that is algorithmic) approach: for sufficiently small $\varepsilon > 0$ we have from the discussion in (2.2) that 0 is a regular value for $f_T - \bar{\varepsilon}$, therefore the same arguments as above imply

$$d(f_T, \text{int } M_T(U), 0) = d(f_T - \bar{\varepsilon}, \text{int } M_T(U), 0) = .. = \sum_{\substack{\sigma \in T_n \\ \bar{\varepsilon} \in f_T(\sigma)}} \text{or}(\sigma).$$

Observe that $\{\sigma \in T_n \mid \bar{\varepsilon} \in f_T(\sigma)\}$ is the set of $\bar{\varepsilon}$-regular simplices for f_T which are contained in $\bar{U}$ (provided ε is sufficiently small). In the following this set is abbreviated by $S_T^f(U)$ or briefly by S if the context is clear.

(4.2) *BROUWER DEGREE*

In the previous section it was motivated by relating $d(f,U,0)$ with $d(f_T, \text{int } M_T(U), 0)$ how to define the Brouwer degree via PL-approximations and the $\bar{\varepsilon}$-perturbation. To make arguments precise we need a crucial bound for the mesh-size of T:

(4.2.1) Let $\Omega \subset R^n \times [a,b]$ be a bounded open subset of $R^n \times [a,b]$ where $a \le b$, and let $g : \overline{\Omega} \to R^n$ be continuous with $0 \notin g(\partial\Omega)$. (closure and boundary relative to the topology of $R^n \times [a,b]$).

Let $\alpha := \mathrm{dist}(g^{-1}(0), \partial\Omega)$, that is $g(x) = 0$ then $|x - a| \ge \alpha > 0$ for all $a \in \partial\Omega$.

Let further $K := \{x \in \overline{\Omega} : \exists\, a \in \partial\Omega$ such that $|x - a| \le \alpha/2\}$ be a collar of $\partial\Omega$.

Since K is compact there is a Lebesgue number $\lambda > 0$ [29] such that for any subset $S \subset K$ with $\mathrm{diam}(S) \le \lambda$ the image $g(S)$ is contained in some open halfspace H_S of $R^n \setminus \{0\}$.

Let $r(g) := \min\{\alpha/2, \lambda\}$. Then for any triangulation T of $R^n \times [a,b]$ with $\mathrm{mesh}(T) \le r(g)$ one claims:

1) $0 \notin g(\overline{\Omega \setminus M_T(\Omega)})$

2) any simplex $\sigma \in T$ which is contained in $\overline{\Omega \setminus M_T(\Omega)}$ is mapped by g in some open halfspace $H_\sigma \subset R^n \setminus \{0\}$.

(4.2.2) *THEOREM AND DEFINITION*

Let $f \in M(U)$, and let T be a triangulation of R^n with $\mathrm{mesh}(T) \le r(f)$. Then

$$\deg(f,U,0) := \sum_{\sigma \in S^f_T(U)} \mathrm{or}(\sigma)$$

is independent of the triangulation T of R^n, and satisfies all the axioms of the Brouwer degree.

(4.2.3) Definition (4.2.2) is independent of the triangulation T of R^n.

Proof

Let T_0 and T_1 be two triangulations of R^n with $\mathrm{mesh}(T_0)$, $\mathrm{mesh}(T_1) \le r(f)$.

Then by the matching lemma (1.3.1) there exists a triangulable set $M \subset \overline{U} \times [0,1]$ and a triangulation T of M such that there is no $\overline{\varepsilon}$-regular n-simplex for h_T on ∂M, where $h(x,t) := f(x)$, and such that $T \cap R^n \times \{i\} = T_i \times \{i\}$, $i = 0,1$. Thus, applying (3.2.2) and (3.2.6) one obtains

$$\sum_{\sigma \in S^f_{T_0}(U)} or(\sigma) = \sum_{\sigma \in S^f_{T_1}(U)} or(\sigma) \blacksquare$$

(4.3) *PROPERTIES OF BROUWER DEGREE*

(4.3.1) *NORMALIZATION*

If 0 is contained in the bounded open set $U \subset R^n$, then $\deg(Id_{\overline{U}}, U, 0) = 1$.

Proof

$0 \in U$ is a regular value for $Id - \overline{\varepsilon}$ if and only if $\overline{\varepsilon} \notin T_{n-1}$. Hence, there is a unique $\sigma \in T_n$ with $\overline{\varepsilon} \in \sigma \setminus \partial\sigma$. $\blacksquare$

A surprising byproduct of our approach here is the fact that a direct and elementary proof of the so-called generalized homotopy invariance is obtained. The proof is entirely based on (3.2.6).

(4.3.2) *GENERALIZED HOMOTOPY INVARIANCE*

For every nonempty bounded open subset $\Omega \subset R^n \times [a,b]$, $a < b$, and for every continuous map $h : \overline{\Omega} \to R^n$ with $0 \notin h(\partial\Omega)$,

$$\deg(h(\cdot,t), \Omega_t, 0) \equiv \text{const.}$$

as a function of $t \in [a,b]$. ($\Omega_t := \{x \in R^n \mid (x,t) \in \Omega\}$.)

Proof

Let T be a triangulation of R^{n+1} with $\text{mesh}(T) \leq r(h)$ such that the restrictions

$$T^a := T|_{R^n \times \{a\}}, \quad T^b := T|_{R^n \times \{b\}}$$

are also triangulations. Due to the definition of $r(h)$ there are no $\bar{\varepsilon}$-regular n-simplices for h_T on $\partial M_T(\Omega)$, and, moreover, any $\bar{\varepsilon}$-regular n-simplex in $M_{T^a}(\Omega_a)$ respectively $M_{T^b}(\Omega_b)$ has an adjoining (n+1)-simplex in $M_T(\Omega)$. Hence, (3.2.2) and (3.2.6) yield the theorem. ■

We demonstrate the full power of this property by showing how the additivity and the solution property follow from the generalized homotopy invariance in a general setting.

(4.3.3) *SOLUTION PROPERTY*

Let $f \in M(U)$. If $0 \notin f(\bar{U})$ then $\deg(f,U,0) = 0$.

Proof

Set $\Omega := U \times [0,1) \subset R^n \times [0,2]$ and define $h : \bar{\Omega} \to R^n$ by $h(x,t) := f(x)$. Then $0 \notin h(\partial\Omega)$, and (4.3.2) implies

$$\deg(f,U,0) = \deg(h(\cdot,0),\Omega_0,0) = \deg(h(\cdot,2), \Omega_2,0) = 0,$$

because $\Omega_2 = \phi$. ■

(4.3.4) *ADDITIVITY*

Let $f \in M(U)$, $U \neq \phi$, and let U_1,U_2 be disjoint open subsets of U with $0 \notin f(\bar{U} \setminus (U_1 \cup U_2))$. Then

$$\deg(f,U,0) = \deg(f,U_1,0) + \deg(f,U_2,0).$$

Proof

Set $\Omega := U \times [0,1) \cup U_1 \times [0,2] \cup U_2 \times [0,2] \subset R^n \times [0,2]$

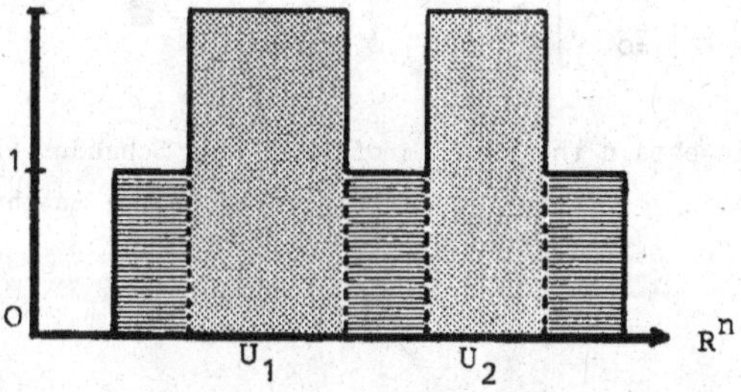

and define $h : \overline{\Omega} \to R^n$ by $h(x,t) := f(x)$. Then $0 \notin h(\partial\Omega)$, and (4.3.2) implies

$$\deg(f,U,o) = \deg(h(\cdot,0),\Omega_0,o) = \deg(h(\cdot,2),\Omega_2,o) = \deg(f,U_1 \cup U_2,o)$$

$$= \deg(f,U_1,o) + \deg(f,U_2,o). \blacksquare$$

(4.3.5) *REMARK*

1) Usually the generalized homotopy invariance is obtained as a consequence of the (usual) homotopy invariance and the additivity, see for example [5], [32].

2) In view of the constructive and numerical background of this approach we also present an alternative, constructive proof of the solution property:

CONSTRUCTIVE PROOF OF THE SOLUTION PROPERTY: Assume $\deg(f,U,o) \neq 0$, and let $\{T^k\}_{k\in N}$ be a sequence of triangulations of R^n such that $\mathrm{mesh}(T^k) \leq r(f)$ and $\mathrm{mesh}(T^k) \to 0$. For every $k \in N$ there is at least one $\overline{\varepsilon}$-regular simplex $\sigma_k = [a_k^o,\dots,a_k^n] \in T^k$ for f_{T^k}:

$$(*) \qquad 0 = \sum_{j=o}^{n} \lambda_j^k f(a_k^j), \quad (\sum \lambda_j^k = 1, \lambda_j^k \geq 0, j = 0,\dots,n)\ k \in N .$$

We may assume (take a subsequence, if necessary) that

$$(\lambda_o^k,\dots,\lambda_n^k) \to (\lambda_o,\dots,\lambda_n)$$

and

$$a_k^j \to a \in \overline{U}, \quad j = o,1,\dots,n$$

for $k \to \infty$. Since $f(a_k^j) \to f(a)$, we obtain from (*)

$$f(a) = \left(\sum_{j=0}^{n} \lambda_j\right) f(a) = \sum_{j=0}^{n} \lambda_j f(a) = 0. \blacksquare$$

Finally, we want to obtain the theorem of Leray and Schauder which provides the well-known representation of degree (= definition of degree in the context of differential topology):

(4.3.6) *THEOREM (Leray-Schauder)*

Let $f \in M(U)$ be of class C^1, and let $0 \in R^n$ be a regular value for f (that is $f(x) = 0 \Rightarrow \det f'(x) \neq 0$). Then

$$\deg(f,U,0) = \sum_{f(x)=0} \text{sign} \det f'(x) .$$

Proof

First observe that $f^{-1}(0)$ is a finite set, say $f^{-1}(0) = \{x_1,\dots,x_k\}$. Choose a triangulation T with $\text{mesh}(T) \leq r(f)$ such that $f^{-1}(0) \cap T_{n-1} = \phi$ and let $\sigma_i \in T_n$ be the n-simplex containing x_i.

If the triangulation is chosen sufficiently small, we have by definition (4.2.2)

$$\deg(f,\text{int }\sigma_i,0) = \text{or}(\sigma_i).$$

Since $f_T|_{\sigma_i}$ and $f'(x_i)(\cdot - x_i)|_{\sigma_i}$ are close for small mesh-size of T,

$$|f_T(x) - f'(x_i)(x - x_i)| \leq |f_T(x) - f(x)| + |f(x) - f'(x_i)(x - x_i)|, \quad x \in \sigma_i,$$

the linear parts of $f_T|_{\sigma_i}$ and $f'(x_i)(\cdot - x_i)$ are contained in the same component of $GL(n)$. Hence,

$$\text{or}(\sigma_i) = \text{sign} \det f'(x_i) ,$$

and, thus,

$$\deg(f,U,0) = \sum_{i=1}^{k} \deg(f,\text{int }\sigma_i,0) = \sum_{i=1}^{k} \text{sign} \det f'(x_i). \blacksquare$$

(4.3.7) *REMARK*

We emphasize again the fact that the proof of the generalized homotopy invariance (4.3.2) is constructive, and can be realized numerically by PL-algorithms. This is discussed in [31], [39], [42], [43]. There it is shown that the Leray-Schauder continuation method and global bifurcation in the sense of Krasnosel'skii-Rabinowitz can be obtained in the framework of PL-algorithms as a consequence of the background presented here.

5. EXTENSIONS

In this chapter two extensions of the Brouwer degree are sketched:

- Leray-Schauder degree [32]
- degree for multivalued mappings, for example [9], [24], [34].

(5.1) *REDUCTION PROPERTY*

In order to develop the full power of degree theory one has to extend Brouwer degree to mappings between infinite dimensional spaces. This was done in the classical paper of Leray and Schauder [32] for compact perturbations of the identity of normed vector spaces (of arbitrary dimensions). This extension is due to the following facts:

- any compact map admits arbitrary close finite dimensional approximations
- the degrees of these finite dimensional approximations stabilize.

To prove the latter fact one needs the so-called reduction property of Brouwer degree. The aim here is to give a proof of that property in the framework of our PL-approach.

(5.1.1) *THEOREM (Reduction property)*

Let $f \in M(U)$ and assume $(Id - f)(\overline{U}) \subset R^{n-1} \times \{0\}$. Then

$$\deg(f,U,o) = \deg(\overline{f},U_o,o),$$

where $U_o := \{x \in R^{n-1} \mid (x,0) \in U\}$ and $\overline{f} : \overline{U}_o \to R^{n-1}$, $\overline{f}(x) := f(x,0)$.

Proof

Throughout the proof $R^{n-1} \times \{0\}$ is identified with R^{n-1}. First, observe that f is "level preserving" on the last coordinate, that is $f(x,a) = (y,b) \in R^n$, for $a,b \in R$, then $a = b$. Hence, the map $\overline{f}$ is well defined. Choose a triangulation T of R^n such that $T^o := T|_{R^{n-1}}$ is also a triangulation, and such that $\mathrm{mesh}(T) \le r(f)$ and $\mathrm{mesh}(T^o) \le r(\overline{f})$.

The idea of the proof is to relate $\overline{\varepsilon}$-regular simplices for f_T with $\overline{\varepsilon}$-regular simplices for $\overline{f}_{T^o}$ using the machinery developed in chapters 2 and 3 . Since f is level preserving f_T is also level preserving. Thus, one obtains from the definition that

- any $\overline{\varepsilon}$-regular n-simplex $\sigma \in T_n$ for f_T is contained in the half-space $\{x \in R^n \mid x_n \ge 0\}$, and,
- any $\overline{\varepsilon}$-regular n-simplex $\sigma \in T_n$ for f_T has a nonempty intersection with R^{n-1}.

Define $M := M_T(U) \cap \{x \in R^n \mid x_n \ge 0\}$, and denote by $K \subset R^{n+1}$ the cone over M with vertex $p := (v,1,1)$, for some $v \in U_o$.

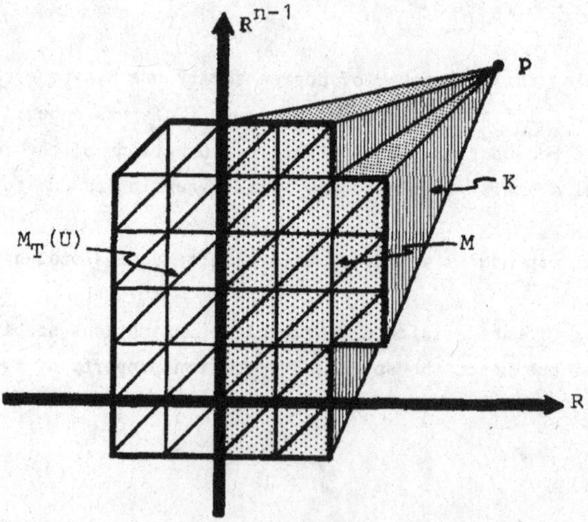

The triangulation T induces in a natural way a triangulation $\tilde{T}$ of K: if $\sigma \in T \cap M$, then the convex hull of σ and p is a simplex of $\tilde{T}$. Now, let $h : K \to R^n$ be an auxiliary map such that

1) $h(p) = (0,\dots,0,1)$;

2) $h(x,0) = f(x)$ for $x \in M$.

To outline the proof let $\sigma \in \tilde{T}_n$ be any boundary simplex of $\tilde{T}$ which is $\bar{\varepsilon}$-regular for $h_{\tilde{T}}$. Assume σ is not contained in M, that is $\sigma = [b^o,\ldots,b^{n-1},p]$.

If there is a vertex in $\{b^o,\ldots,b^{n-1}\}$ which is not contained in R^{n-1}, then $[b^o,\ldots,b^{n-1}]$ is a boundary simplex in $M_T(U)$, and the fact that σ is $\bar{\varepsilon}$-regular for $h_{\tilde{T}}$ would imply that there is a zero of f_T in $[b^o,\ldots,b^{n-1}]$ which was excluded.

Thus, the set of boundary simplices of $\tilde{T}$ which are $\bar{\varepsilon}$-regular for $h_{\tilde{T}}$, splits into two parts

$$S_1 := \left\{\sigma \in \tilde{T}_n \;\middle|\; \begin{array}{l}\sigma \text{ is } \bar{\varepsilon}\text{-regular for } h_{\tilde{T}}\,;\\ \sigma = [b^o,\ldots,b^{n-1},p], \quad b^o,\ldots,b^{n-1} \in R^{n-1}\end{array}\right\}$$

and

$$S_2 := \{\sigma \in \tilde{T}_n \mid \sigma \text{ is } \bar{\varepsilon}\text{-regular for } h_{\tilde{T}}\,;\ \sigma \subset M\} = S_T^f(U)\,.$$

We will prove

1) $\sigma \in S_1 : \mathrm{ind}_h(\sigma \subset \rho) = -\,\mathrm{or}(\tau)$,
where ρ is the adjoining $(n+1)$-simplex of σ in $\tilde{T}_{n+1}$, and
$\tau = [b^o,\ldots,b^{n-1}]$

2) $\sigma \in S_2 : \mathrm{ind}_h(\sigma \subset \rho) = \mathrm{or}(\sigma)$,
where ρ is the adjoining $(n+1)$-simplex of σ in $\tilde{T}_{n+1}$.

Using 1) and 2) and theorem (3.2.6) one obtains:

$$\begin{aligned}\deg(\bar{f},U_o,o) &:= \sum_{\sigma \in S} \mathrm{or}(\sigma) = -\sum_{\sigma \in S_1} \mathrm{ind}_h(\sigma \subset \rho)\\ &= \sum_{\sigma \in S_2} \mathrm{ind}_h(\sigma \subset \rho) \qquad (3.2.6)\\ &= \sum_{\sigma \in S} \mathrm{or}(\sigma) =: \deg(f,U,o).\end{aligned}$$

Proof of 1)

To be precise let $\tau := [b^o,\ldots,b^{n-1}] \subset R^{n-1}$, $\sigma = [(b^o,o,o),\ldots,(b^{n-1},o,o),p]$ and $\rho = [(b^o,o,o),\ldots,(b^{n-1},o,o),p,q]$.

(2.2.5) implies that τ is $\bar{\varepsilon}$-regular for $\bar{f}_{T^o}$ if and only if σ is $\bar{\varepsilon}$-regular for $h_{\tilde{T}}$. Because the n-th coordinate q_n of q is positive, say $q = (\bar{q}, q_n, o)$, we have

$$\mathrm{ind}_h(\sigma \subset \rho)$$

$$:= \text{sign det}\begin{bmatrix} 1 & \dots & 1 & 1 & 1 \\ b^o & & b^{n-1} & v & \bar{q} \\ o & & o & 1 & q_n \\ o & & o & 1 & o \end{bmatrix} \text{sign det}\begin{bmatrix} 1 & \dots & 1 & 1 \\ h(b^o,o,o), & \dots, & h(b^{n-1},o,o) & h(p) \end{bmatrix}$$

$$= - \text{sign det}\begin{bmatrix} 1 & \dots & 1 & 1 \\ b^o & \dots & b^{n-1} & \bar{q} \\ o & \dots & o & q_n \end{bmatrix} \text{sign det}\begin{bmatrix} 1 & \dots & 1 & 1 \\ f(b^o,o) & \dots & f(b^{n-1},o) & h(p) \end{bmatrix}$$

$$= - \text{sign det}\begin{bmatrix} 1 & \dots & 1 \\ b^o & \dots & b^{n-1} \end{bmatrix} \text{sign det}\begin{bmatrix} 1 & \dots & 1 \\ \bar{f}(b^o) & \dots & \bar{f}(b^{n-1}) \end{bmatrix}$$

$$=: - \mathrm{or}(\tau).$$

Proof of 2)

Let $\sigma := [(a^o,o),\dots,(a^n,o)] \in S_2$, $\rho = [(a^o,o),\dots,(a^n,o),p]$. Then

$$\mathrm{ind}_h(\sigma \subset \rho) := \text{sign det}\begin{bmatrix} 1 & \dots & 1 & 1 \\ a^o & \dots & a^n & \\ & & & p \\ o & \dots & o & \end{bmatrix} \text{sign det}\begin{bmatrix} 1 & \dots & 1 \\ h(a^o,o) & \dots & h(a^n,o) \end{bmatrix}$$

$$= \text{sign det}\begin{bmatrix} 1 & \dots & 1 \\ a^o & \dots & a^n \end{bmatrix} \text{sign det}\begin{bmatrix} 1 & \dots & 1 \\ f(a^o) & \dots & f(a^n) \end{bmatrix}$$

$$=: \mathrm{or}(\sigma). \blacksquare$$

(5.2) *MULTIVALUED MAPPINGS*

It is a surprising fact that the simplicial approach to Brouwer degree carries over verbatim ("mutatis mutandis") to multivalued mappings (for a degree theory for multivalued mappings see also [33]). We give some concluding remarks for this case.

Let $F : \overline{U} \to 2^{R^n}$, U bounded open subset of R^n, be an upper semicontinuous map such that

1) $F(x)$ is compact, convex and nonempty for $x \in \overline{U}$

2) $0 \notin F(x)$ for $x \in \partial U$.

Again we can associate to F a mesh-size $r(F)$ which guarantees that for any triangulation T of R^n with $\mathrm{mesh}(T) \le r(F)$

1) $0 \notin F(\overline{U \setminus M_T(U)})$

2) any simplex $\sigma \in T$ which is contained in $\overline{U \setminus M_T(U)}$ is mapped by F in some open halfspace $H_\sigma \subset R^n \setminus \{0\}$.

In order to define a degree for F (one can prove that there is a unique degree satisfying the axioms of chapter 1) choose a triangulation T of R^n with $\mathrm{mesh}(T) \le r(F)$ and any selection $f : T_o \to R^n$ (that is $f(x) \in F(x)$ for any vertex $x \in T_o$), and define

$$\deg(F,U,o) \quad := \sum_{\sigma \in S_T^{f_T}(U)} \mathrm{or}(\sigma) .$$

The proof that this definition is independent of the selection f and the triangulation T, and the proofs of the axioms of degree (Normalization, Additivity, Generalized Homotopy Invariance, Solution Property) are parallel to the proofs in the single valued case.

(5.3) In view of the important applications of Borsuk's theorem in nonlinear analysis (for example "Ljusternik-Schnirelman category" of the real projective spaces), see for example [5], [12], [19], we give for reasons of completeness a short proof of that theorem in the context of our approach.

In the following T will always denote a symmetric triangulation of R^n (that is $T = -T$), for example the Union Jack triangulation (1.2.3). For a PL-mapping $f_T : M \to R^n$ let

$$B(f_T) := \{\sigma \in T_n \mid \sigma \text{ is } \overline{\varepsilon}\text{-regular for } f_T;\ o \in f_T(\partial\sigma)\} .$$

Borsuk's theorem is an easy consequence of the following lemma.

(5.3.1) *LEMMA*

Let $M \subset R^n$ be a symmetric (that is $M = -M$) triangulable subset of homogeneous dimension n, and let $f_T : M \to R^n$ be an odd PL-mapping (that is $f_T(-x) = -f_T(x)$) such that $o \notin f_T(\partial M)$.

Let $\sigma = [a^o,\dots,a^n] \in \mathcal{B}(f_T)$ such that $[a^o,\dots,a^k] \subset \partial M$ for some k, $o \le k \le n-1$ $(a^o,\dots,a^n \ne o)$. Then there exists a PL-mapping $f_o : M \to R^n$ such that

- f_o is odd

- $f_o|_{\partial M} = f_T|_{\partial M}$

- $\#\mathcal{B}(f_o) < \#\mathcal{B}(f_T)$.

Proof

The proof is based essentially on the same argument as used in the proof of (2.2.2). Let $a^p \ne 0$, $k+1 \le p \le n$, be a vertex of σ such that

$$o \notin [f_T(a^o),\dots,f_T(\hat{a}^p),\dots,f_T(a^n)],$$

and let $\rho_1,\dots,\rho_\ell \in T_{n-1}$ be the set of all (n-1)-simplices having a^p as a vertex. Denote by H_j the affine subspace of R^n which is spanned by

$$f_T(\rho_j), \quad 1 \le j \le \ell.$$

There exists $\varepsilon_o > o$ such that for all ε, $o < \varepsilon < \varepsilon_o$, and for all j, $1 \le j \le \ell$, the following properties are satisfied:

i) $f_T(a^p) + \bar{\varepsilon} \notin H_j$

ii) if $o \notin f_T(\rho_j) = co\{f_T(b^1),\dots,f_T(a^p),\dots,f_T(b^{n-1})\}$,
then $o \notin co\{f_T(b^1),\dots,f_T(a^p)+\bar{\varepsilon},\dots,f_T(b^{n-1})\}$,
where $\rho_j = [b^1,\dots,a^p,\dots,b^{n-1}]$.

Define f_o to be the PL-mapping which maps a^p onto $f_T(a^p) + \bar{\varepsilon}$, $-a^p$ onto $-f_T(a^p) - \bar{\varepsilon}$, and which coincides with f_T on all other vertices of T. Then $\sigma \notin \mathcal{B}(f_o)$, and by definition of f_o it is clear that $\mathcal{B}(f_o) \setminus \mathcal{B}(f_T) = \phi$, which proves the lemma. ■

(5.3.2) *THEOREM (Borsuk)*

Let $f \in M(U)$, where U is a symmetric neighborhood of the origin. If f is odd, then $\deg(f,U,o)$ is odd.

Proof

Let T be a symmetric triangulation of R^n. The proof is easy if the situation is regular in the following sense: if $\sigma \in T_n$ is $\bar{\varepsilon}$-regular for f_T and if $o \notin \sigma$ then, since f_T is odd, σ is regular for f_T if and only if $-\sigma$ is regular for f_T.

Thus, if all $\bar{\varepsilon}$-regular simplices except one (which contains the origin) are regular for f_T the theorem follows immediately from the definition of degree. The idea of proof is to perturb f_T in such a way that this regular situation arises.

If the mesh-size of T is sufficiently small the set of simplices in T containing the origin and the set of simplices in T which have a nonempty intersection with $\partial M_T(U)$ are disjoint. Moreover, subdividing T symmetrically (if necessary) we can obtain that there is no n-simplex in T whose vertices are all contained in $\partial M_T(U)$.

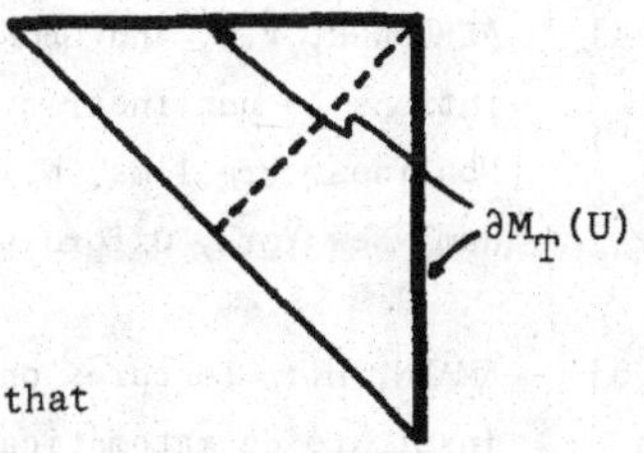

Now let $\bar{f}_T$ be the PL-extension of the following vertex map

$$v \in T_o \to \begin{cases} f_T(v), & v \in \partial M_T(U) \\ v, & \text{otherwise} \end{cases}$$

and apply lemma (5.3.1) to the map $\bar{f}_T$.
By induction we finally obtain a PL-mapping f_o such that

- f_o is odd
- $f_o|_{\partial M} = f_T|_{\partial M}$
- f_o is the identity in a neighborhood of the origin
- all $\bar{\varepsilon}$-regular simplices except the (unique) one which contains the origin are regular for f_o.

Since f_o and f_T are homotopic, the theorem is proved. ■

(5.3.2) *REMARK*

There are also easy and short proofs in the context of our approach for the multiplication property and the cartesian product property (see, for example [12], [19]) of Brouwer degree. (These proofs provide, for instance, elementary proofs for the n-dimensional Jordan-Brouwer separation theorem and for the "invariance of domain".) Because of reasons of length these proofs are omitted.

REFERENCES

[1] ALEXANDER, J.C. and YORKE, J.A.: The homotopy continuation method: Numerically implementable topological prodédures, Trans. AMS, 242 (1978), 271-284.

[2] ALEXANDROFF, P. and HOPF, H.: Topologie, berichtigter Reprint, Berlin, Heidelberg, New York, Springer-Verlag (1974), Die Grundlehren der mathematischen Wissenschaften, Bd. 45.

[3] ALLGOWER, E.L. and GEORG, K.: Simplicial and continuation methods for approximating fixed points and solutions to systems of equations, SIAM Review, 22 (1980), 28-85.

[4] ALLGOWER, E.L. and GEORG, K.: Homotopy methods for approximating several solutions to nonlinear systems of equations, in: Numerical Solution of Highly Nonlinear Problems, W. Forster (Ed), North-Holland Publishing Company-Amsterdam, New York, Oxford (1980), 253-270.

[5] AMANN, H.: Lectures on some fixed point theorems, Monografias de Matemática, Instituto de matemática pura e aplicada, Rio de Janeiro (1974).

[6] AMANN, H. and WEISS, S.A.: On the uniqueness of the topological degree, Math. Z. 130 (1973), 39-54.

[7] BROUWER, L.E.J.: Beweis der Invarianz der Dimensionszahl, Math. Ann. 70 (1911), 161-165.

[8] BROUWER, L.E.J.: Über Abbildung von Mannigfaltigkeiten, Math. Ann. 71 (1912), 97-115.

[9] CELLINA, A. and LASOTA, A.: A new approach to the definition of topological degree for multi-valued mappings, Atti Accad. Naz. Lincei Rend. Cl. Sci. Fis. Mat. Natur. 47 (1969), 434-440.

[10] CHOW, S.N., MALLET-PARET, J. and YORKE, J.A.: Finding zeros of maps: Homotopy methods that are constructive with probability one, Math. Comp. 32 (1978), 887-899.

[11] CRONIN, J.: Fixed points and topological degree in nonlinear analysis, Providence, R.I., Amer. Math. Soc. (1964).

[12] DEIMLING, K.: Nichtlineare Gleichungen und Abbildungsgrade, Berlin, Heidelberg, New York, Springer-Verlag (1974).

[13] DOLD, A.: Lectures on algebraic topology, Berlin, Heilderberg, New York, Springer-Verlag (1972), Die Grundlehren der mathematischen Wissenschaften, Bd. 200.

[14] DE RHAM, G.: Variétés différentiables, Paris, Hermann (1955).

[15] EAVES, B.C.: A short course in solving equations with PL-homotopies, SIAM-AMS Proceedings 9 (1976), 73-143.

[16] EAVES, B.C. and SCARF, H.: The solution of systems of piecewise linear equations, Math. of Op. Res. 1 (1976), 1-27.

[17] EILENBERG, S. and STEENROD, N.: Foundations of algebraic topology, Princeton, Univ. Press (1952).

[18] EISENBUD, D. and LEVINE, H.I.: The topological degree of a finite C^∞-map germ, in: Structural Stability, the Theory of Catastrophes, and Applications in the Sciences, Springer-Verlag (1976), Lectures Notes in Mathematics 525, 90-98.

[19] EISENACK, G. and FENSKE, C.C.: Fixpunkttheorie, Mannheim, Wien, Zürich, Bibliographisches Institut (1978).

[20] EPSTEIN, D.B.A.: The degree of a map, Proc. London Math. Soc. 16 (1966), 369-383.

[21] FENSKE, C.C.: Analytische Theorie des Abbildungsgrades in Banachräumen,

Math. Nachr. 48 (1971), 279-290.

[22] FÜHRER, L.: Ein elementarer analytischer Beweis zur Eindeutigkeit des Abbildungsgrades im R^n, Math. Nachr. 54 (1972), 259-267.

[23] GARCIA, C.B.: Computation of solutions to nonlinear equations under homotopy invariance, Math. of Op. Res. 2 (1977), 25-29.

[24] GRANAS, A.: Sur la notion du degré topologique pour une cartaine classe de transformations multivalentes dans les espaces de Banach, Bull. Acad. Polon. Sci. Sér. Sci. Mat. Astronom. Phys. 7 (1959), 191-194.

[25] GREUB, W., HALPERIN, S. and VANSTONE, R.: Connections, Curvature, and Cohomology, vol I, Academic Press, New York, London (1972).

[26] GUILLEMIN, V. and POLLACK, A.: Differential topology, Prentice Hall, Inc. New Jersey (1974).

[27] HEINZ, E.: An elementary analytic theory of degree of mappings in n-dimensional spaces, J. Math. Mech. 8 (1959), 231-247.

[28] HIRSCH, M.W.: Differential topology. Berlin, Heidelberg, New York, Springer-Verlag (1976), Graduate Texts in Mathematics 33.

[29] HOCKING, J.G. and YOUNG, G.S.: Topology, Addison-Wesley Publishing Company, Inc. (1961).

[30] HUDSON, J.F.P.: Piecewise linear topology, Benjamin, New York, (1969).

[31] JÜRGENS, H., PEITGEN, H.O. and SAUPE, D.: Topological perturbations in the numerical study of nonlinear eigenvalue and bifurcation problems, to appear in: Proc. Symposium on Analysis and Computation of Fixed Points, Madison, May 7-8, 1979, S.M. Robinson (Ed.), Academic Press, New York.

[32] LERAY, J. and SCHAUDER, J.P.: Topologie et équations fonctionnelles, Ann. Ecole Norm. Sup. (3) 51 (1934), 45-78.

[33] LLOYD, N.G.: Degree theory, Cambridge University Press (1978).

[34] MA, T.W.: Topological degree for set-valued compact vector fields in locally convex spaces, Dissertationes Math. 92 (1972).

[35] MILNOR, J.: Topology from the differentiable viewpoint, 2nd printing, Charlottesville, The University of Virginia Press (1969).

[36] NAGUMO, M.: A theory of degree of mapping based on infinitesimal analysis, Amer. J. Math. 73 (1951), 485-496.

[37] NIREMBERG, L.: Topics in nonlinear functional analysis, Courant Institute of Mathematical Sciences, New York University (1974).

[38] NUSSBAUM, R.D.: On the uniqueness of the topological degree for k-set contractions, Math. Z. 137 (1974), 1-6.

[39] PEITGEN, H.O. and PRÜFER, M.: The Leray-Schauder continuation method is a constructive element in the numerical study of nonlinear eigenvalue and bifurcation problems, in: Functional Differential Equations and Approximation of Fixed Points, H.O. Peitgen and H.O. Walther (Eds), Springer-Verlag (1979), Lectures Notes in Mathematics 730, 326-409.

[40] PEITGEN, H.O., SAUPE, D. and SCHMITT, K.: Nonlinear elliptic boundary value problems versus finite difference approximations: numerically irrelevant solutions, to appear in: J. Reine Angew. Math.

[41] PONTRYAGIN, L.S.: Smooth manifolds and their applications in homotopy theory, AMS Translations, Ser. 2, II (1959), 1-114.

[42] PRÜFER, M.: Calculating global bifurcation, in: Continuation Methods, H.J. Wacker (Ed), Academic Press, New York, San Francisco, London (1978), 187-213.

[43] PRÜFER, M.: Simpliziale Topologie und globale Verzweigung, Dissertation, University of Bonn (1978).

[44] ROURKE, C.P. and SANDERSON, B.J.: Introduction to piecewise-linear topology, Berlin, Heidelberg, New York, Springer-Verlag (1972).

[45] SCHWARTZ, J.: Nonlinear functional aanalysis, Gordon and Breach, New York, (1969).

[46] SIEGBERG, H.W.: Brouwer degree: history and numerical computation, in: Numerical Solution of Highly Nonlinear Problems, W. Forster (Ed), North-Holland Publishing Company Amsterdam, New York, Oxford (1980), 389-411.

[47] SPANIER, E.: Algebraic topology, New York, Mc Graw-Hill, (1966).

[48] SPIVAK, M.: A comprehensive introduction to differential geometry, vol I, Boston: Publish or Perish Inc. (1970).

[49] TODD, M.J.: The computation of fixed points, Berlin, Heidelberg, New York, Springer-Verlag (1976), Lectures Notes in Economics and Mathematical Systems 124.

[50] TODD, M.J.: Orientation in complementary pivoting, Math. of Op. Res. 1 (1976), 54-66.

[51] ZEIDLER, E.: Existenz, Eindeutigkeit, Eigenschaften und Anwendungen des Abbildungsgrades in R^n, in: Theory of Nonlinear Operators, Proceedings of a summer school held in Oct. 1972 at Neuendorf, Akademie-Verlag, Berlin (1974), 259-311.

Acknowledgement: this research was supported by the "Deutsche Forschungsgemeinschaft" (DFG-Projekt: 'Multiple Bifurcation').

FIXED POINTS AND SURJECTIVITY THEOREMS VIA THE A-PROPER MAPPING THEORY WITH APPLICATION TO DIFFERENTIAL EQUATIONS

By

W.V. PETRYSHYN*

Department of Mathematics
Rutgers University, New Brunswick, N.J.

0. Introduction

The purpose of these lectures is to show how the A-proper mapping theory can be used to obtain general approximation-solvability and/or existence theorems for nonlinear partial and ordinary differential equations involving operators which need not be coercive and which sometimes are of the form to which other abstract theories need not apply.

For the sake of clarity and completeness in Section 1 we introduce the notion of an A-proper and P_1-compact mappings, give some examples and state (with indication of the proofs) those facts from the A-proper mapping theory which are relevant to the subject matter of this conference and which will be needed in Section 2. Some of the results in Section 1 appear here for the first time.

In section 2 we first obtain variational approximation-solvability and/or existence theorems for elliptic BV Problems involving not necessarily coercive operators of order 2m on a domain R^n in generalized divergence form

$$A(u) = \sum_{|\alpha|\leq m} (-1)^{|\alpha|} D^{\alpha} A_{\alpha}(x,u,\ldots,D^{m}u), \tag{1}$$

where A_α's are of polynomial growth in $(u,\ldots,D^m u)$. Theorem 2.1 below extends to equations of the form (1) the constructive linear result obtained by the author in [32]. Our Theorem 2.2 unifies and extends for equations of the form (1) some of the earlier existence results of Višik [40], Browder [1, 2], Leray and Lions [15], Pohožayev [34] and others. (For details see [32].) In Theorem 2.4 we provide an al-

*Supported in part by the NSF Grant MCS-8003002.

ternative proof of a recent surjectivity result of Sanches [36] concerning a second order OD Equation appearing in the dynamics of wires. Theorem 2.5 establishes the existence of solutions to an OD Equation to which the abstract results of other authors cited in this article are not applicable.

1. SOME PRELIMINARY RESULTS ON A-PROPER MAPPINGS

For the sake of clarity and completeness, in this lecture we outline briefly the definition of the A-proper mapping, give some of its examples and state (with proofs in some cases) those facts from the theory of A-proper mappings which are relevant to the subject matter of this conference and which we shall use in our study of the solvability of nonlinear ordinary and partial differential equations. Some results in this section appear for the first time. For a complete survey of the theory of A-proper mappings developed by various authors up to 1975 see [26].

Let X and Y be real Banach spaces with an *admissible* scheme

$$\Gamma = \{X_n, V_n ; E_n, W_n\}, \text{ i.e., } \{X_n\} \text{ and } \{E_n\}$$

are sequences of oriented finite dimensional spaces with $\{X_n\} \subset X$ such that

$$\dim X_n = \dim E_n$$

for each n, $\text{dist}(x, X_n) \to 0$ for each $x \in X$, V_n is an injection of X_n into X and W_n is a linear map of Y into E_n with $\{W_n\}$ uniformly bounded. We use the same symbol $\|\cdot\|$ to denote the norm in X, Y and E_n, and "$\to$" and "$\rightharpoonup$" to denote the strong and the weak convergence respectively. The simplest special cases (to be used later) are the *injective* scheme $\Gamma_I = \{X_n; V_n; X_n^*, V_n^*\}$ for (X, X^*) and the *projective* scheme $\Gamma_\alpha = \{X_n, V_n; Y_n; Q_n\}$ for (X, Y) with $\alpha = \sup_n \|Q_n\| < \infty$.

Let D be a subset of X, $D_n = D \cap X_n$ for each n, $T:D \to Y$ a not necessarily linear map and $T_n = W_n T\big|_{D_n} : D_n \to E_n$ for each n. For a given f in Y, the following notion for

$$Tx = f \quad (x \in D,\ f \in Y) \tag{1.1}$$

will prove to be useful.

(1.1) *DEFINITION*

Equation (1.1) is said to be *strongly* (respectively *feebly*) *approximation-solvable* (*a-solvable*, for short) w.r.t. Γ if there exist $n_f \in Z_+$ such that

$$T_n(x) = W_n f \quad (x \in D_n,\ W_n f \in E_n) \tag{1.2}$$

has a solution $x_n \in D_n$ for each $n \geq n_f$ such that $x_n \to x_o \in D$ (respectively $x_{n_j} \to x_o \in D$) and x_o is a solution of (1.1).

The notion of an *Approximation-proper* (A-proper, for short) mapping arose as the answer to the following

Question: What conditions should be imposed on T so that (1.1) is strongly (or at least feebly) a-solvable w.r.t Γ ?

To answer this question the speaker was led in [23,24] to the following class of mappings.

(1.2) *DEFINITION*

Let $T:D \to Y$ and $g \in Y$. Then T is said to be *A-proper at* g w.r.t. Γ iff $T_n:D_n \to E_n$ is continuous for each $n \in Z_+$ and condition (H) holds, where:

(H) If $\{x_{n_j} | x_{n_j} \in D_{n_j}\}$ is any bounded sequence and that $T_{n_j}(x_{n_j}) - W_{n_j}(g) \to 0$, then there exist a subsequence $\{x_{n_{j(k)}}\}$ and an $x \in D$ such that $x_{n_{j(k)}} \to x$ as $k \to \infty$ and $Tx = g$.

If T is A-proper at g w.r.t. Γ for any $g \in Y$, then T is called *A-proper* w.r.t. Γ or simply A-proper.

Thus, for example, it was shown in [23,24] that if $D = X$ and $T:X \to Y$ is linear, then (1.1) is uniquely a-solvable w.r.t. Γ if T is A-proper w.r.t. Γ and one-to-one. The converse is also true if $\Gamma = \Gamma_\alpha$. Analogous characterization result was obtained in [23] when $T:X \to Y$ is nonlinear. This fact provided the initial motivation for the study of A-proper mappings.

It is easy to see show that if $T:D \to Y$ is A-proper and $C:D \to Y$ compact, then $T + C$ is also A-proper. This last fact, together with the Sobolev Imbedding Theorem, is particularly useful when the A-proper mapping theory is used to solve differential equations.

(1.1) *Examples and the fixed point for P_1-compact mappings and its consequences.*

It is easy to show that if $D \subset X$ is closed, $C:D \to X$ compact and $S:X \to X$ is ℓ-Lipschitz with $\ell < 1$, then $\lambda I-S-C$ is A-proper w.r.t. $\Gamma_1 = \{X_n, P_n\}$ for each $\lambda \geq 1$, i.e., $S + C$ is P_1-*compact* in the sense of [23]. If $F:D \to X$ is ball-condensing (i.e. $\beta(F(Q)) < \beta(Q)$ for $Q \subset D$ with $\beta(Q) \neq 0$ where $\beta(Q)$ denotes the ball-measure of noncompactness of Q), then $F + C$ is also P_1-compact. Another important example of P_1-compact maps is given by the following lemma whose proof can be found in [30,18, 38].

(1.1) *LEMMA*

Let $A:X \to X$ *be continuous and bounded accretive map,* F *ball-condensing and* C *compact. Then* $F + C - A$ *is* P_1*-compact w.r.t. to a nested scheme* Γ_1.

As was shown in [38], the above class of A-proper maps allows us to study semi-linear elliptic BV Problems when the underlying domain $Q \subset R^n$ is unbounded.

Note that a map G is P_1-compact iff $I-\mu G$ is A-proper for each $\mu \in (0,1]$. Hence Lemma 1.1 can also be deduced from Lemma 1.2 whose proof is based on the following known fact

$$\beta(\{Q_n y_n\}) \leq \beta(\{y_n\}) \sup_n \|Q_n\| \tag{1.3}$$

for any bounded sequence $\{y_n\} \subset Y$.

(1.2) *LEMMA*

Let $\Gamma = \Gamma_\alpha$, $D \subset X$ *closed,* $T:D \to Y$ *continuous and*

(a) $\begin{cases} \textit{There exist } \mu_o > 0 \textit{ and } n_o \in Z_+ \textit{ such that } \beta(\{Q_n Tx_n\}) \geq \mu_o \beta(\{x_n\}) \\ \text{for any bounded sequence } \{x_n | x_n \in D_n\} \text{ with } n \geq n_o. \end{cases}$

If $F:D \to X$ *is k-ball-contractive, then* $T_\mu \equiv T + \mu F$ *is A-proper w.r.t.* Γ_α

for each $\mu \in (-\mu_0 k^{-1}, \mu_0 k^{-1})$. *If* $\mu_0 = \alpha = 1$, *the same holds when* F *is ball-condensing and* $\mu \in [-1,1]$.

Proof

Note first that $T_{\mu n}: D_n \to Y_n$ is continuous for each $n \in Z_+$ and

$$\mu \in (-\mu_0 k^{-1}, \mu_0 k^{-1}).$$

Thus, it suffices to show that T_μ satisfies condition (H) for each fixed

$$\mu \in (-\mu_0 k^{-1}, \mu_0 k^{-1}).$$

So let $\{x_{n_j} | x_{n_j} \in D_{n_j}\}$ be any bounded sequence such that $Q_{n_j} T_\mu (x_{n_j}) - Q_{n_j} g \to 0$ for some g in Y, where without loss of generality we assume that $n_j \geq n_0$ for each $j \in Z_+$. Since $Q_{n_j} g \to g$ in Y, we see that

$$g_{n_j} \equiv Q_{n_j} T(x_{n_j}) + \mu Q_{n_j} F(x_{n_j}) \to g \text{ in } Y.$$

Since the assertion is trivial when $\mu = 0$, so we may assume that $\mu \neq 0$. Since $\{g_{n_j}\}$ is precompact and $Q_{n_j} Tx_{n_j} = g_{n_j} - \mu Q_{n_j} Fx_{n_j}$ for each $j \in Z_+$, it follows from (a) and (1.3) that

$$\mu_0 \beta(\{x_{n_j}\}) \leq \beta(\{Q_{n_j} Tx_{n_j}\}) \leq |\mu| \beta(\{Q_{n_j} Fx_{n_j}\}) \leq |\mu| k\alpha\beta(\{x_{n_j}\}).$$

Thus $\beta(\{x_{n_j}\}) = 0$ and so $\{x_{n_j}\}$ has a convergent subsequence $\{x_{n_{j(k)}}\}$ with $x_{n_{j(k)}} \to x$ for some $x \in D$. Hence, by the continuity of T and F and the completeness of Γ_α, we see that $Tx + \mu Fx = g$, i.e., T_μ is A-proper for each

$$\mu \in (-\mu_0 k^{-1}, \mu_0 k^{-1}).$$

The second part of Lemma 1.2 is proved similary. ■

PROBLEM

Show that (a) of Lemma 1.2 holds when $T: X \to Y$ is a continuous surjective map such that $\|T_n(x)-T_n(y)\| \geq \mu_o\|x-y\|$ for all $x,y \in X_n$ and $n \geq n_o$.

Since this workshop is devoted essentially to the fixed point theory, we shall now show how one can use extremely simple arguments, together with the finite dimensional Leray-Schauder or Brouwer fixed point theorems, to establish (in some cases constructively) a fixed point theorem for a map $F:D \to X$ which is P_1-compact at 0. A slightly weaker version of this theorem has been established earlier in [23] (see also [33]). But some of its consequences stated below appear to be new.

(1.1) *THEOREM*

Let $D \subset X$ *be a bounded and open subset with* $0 \in D$, $F:\bar{D} \to X$ *a bounded map such that* $\lambda I-F$ *is* A-*proper at* 0 *w.r.t.* $\Gamma_\alpha = \{X_n, P_n\}$ *for each* $\lambda \geq 1$ (i.e. F *is* P_1-*compact at* 0). *Suppose further that*

$$Fx \neq \lambda x \quad \text{for all} \quad x \in \partial D \quad \text{and} \quad \lambda > 1. \tag{1.4}$$

Then F *has a fixed point in* $\bar{D}$. *Moreover, the equation* $x-Fx = 0$ *is uniquely a-solvable w.r.t.* Γ_α *if* F *has a unique fixed point in* D. (i.e. *the Galerkin method converges*).

<u>Proof</u>

We may assume that F has no fixed points on ∂D. We claim that there exists $n_o \in Z_+$ such that

$$F_n(x) \neq \lambda x \quad \text{for all} \quad x \in \partial D_n,\ n \geq n_o \quad \text{and} \quad \lambda \geq 1. \tag{1.5}$$

If not, then there exist sequences $\{x_{n_j} | x_{n_j} \in \partial D_{n_j}\}$ and $\{\beta_j\}$ with $\beta_j \geq 1$ such that $F_{n_j}(x_{n_j}) = \beta_j x_{n_j}$ for each $j \in Z_+$. Now, since F is bounded and

$$\|x_{n_j}\| \geq \delta > 0 \quad \text{for some} \quad \delta > 0 \quad \text{and all} \quad j \in Z_+,$$

it follows that $\{\beta_j\}$ is bounded. Hence we may assume that $\beta_j \to \beta \geq 1$ and note that

$$F_{n_j}(x_{n_j})-\beta x_{n_j} = (\beta_j-\beta)(x_{n_j}) \to 0 \quad \text{as} \quad j \to \infty.$$

Hence, since $F-\beta I$ is A-proper at 0, there exist a subsequence $\{x_{n_{j(k)}}\}$ and an $x_o \in \bar{D}$ such that $x_{n_{j(k)}} \to x_o$ and $Fx_o-\beta x_o = 0$ with $x_o \in \partial D$, in contradiction to (1.4). Thus (1.5) holds for each $n \geq n_o$ and so, by the finite dimensional Leray-Schauder fixed point theorem, there exists an $x_n \in D_n$ such that $F_n(x_n)-x_n = 0$. Since $x_n-F_n(x_n) = 0 \to 0$ as $n \to \infty$ and I-F is A-proper at 0, there exist a subsequence $\{x_{n_j}\}$ and $x_o \in \overline{D}$ such that $x_{n_j} \to x_o$ and $x_o-F(x_o) = 0$ with $x_o \in D$. To prove the second assertion of Theorem 1.1, note that if x_o is the unique fixed point of F in D, then the entire sequence $\{x_n\}$ converges to x_o since otherwise there would exist a subsequence $\{x_{n_k}\}$ of $\{x_n\}$ such that $\|x_{n_k}-x_o\| \geq \varepsilon > 0$ for all $k \in Z_+$ and $F_{n_k}(x_{n_k})-x_{n_k} = 0$ for each $k \in Z_+$. Hence, again by the A-properness of I-F at 0, there exist a subsequence $\{x_{n_{k(i)}}\}$ and $\overline{x} \in D$ such that $x_{n_{k(i)}} \to \overline{x}$ and $F\overline{x} - \overline{x} = 0$ with $\overline{x} \neq x_o$, in contradiction to our assumption. The last fact also shows that the Galerkin method converges in this case. ■

(1.1) *REMARK*

Condition (1.4) is implied by any one of the following

$$\|Fx-x\|^2 \geq \|Fx\|^2 - \|x\|^2 \quad \text{for all} \quad x \in \partial B_r \equiv \partial B(0,r). \tag{1.6}$$

$$(Fx, w) \leq (x,w) \quad \text{for all} \quad x \in \partial B_r \quad \text{and some} \quad w \in Jx, \quad \text{where}$$
$$J \text{ is the normalized duality map of } X \text{ to } 2^{X^*}. \tag{1.7}$$

$$Fx \in \overline{B}_r \quad \text{for all} \quad x \in \partial B_r. \tag{1.8}$$

In view of Remark 1.1 and the discussion at the beginning of this section, we see that Theorem 1.1 includes the fixed point theorems of Schauder, Rothe, Leray-Schauder, Altman and Krasnoselskii when F is compact, Kaniel when F is quasicompact, Petryshyn where F is P_γ-compact, Sadovskii when F is ball-condensing and others (see [26] for details and references).

Let us add that if F is P_1-compact, then obviously F is P_1-compact at 0 but the converse need not be true. Thus, for example, when $F:\overline{B}_r \to \overline{B}_r$ is ℓ-Lipschitzian with $\ell \in (0,1)$, then F is P_1-compact at 0 but it is unknown whether F is P_1-compact. For other examples see [26].

Using Schauder fixed point theorem, it was shown by Schaeffer [37] that if $C:X \to X$ is compact, then either C has a fixed point in X or the set

$$\{x : x-\lambda Cx = 0, \ 0 < \lambda < 1\}$$

is unbounded. This theorem is known to be important in various applications. Martelli and Vignoli [17] extending this result to condensing mappings, while the author [25] extended it further to 1-set contractive maps satisfying condition (c). Using Theorem 1.1 we now show that the results in [37,17] admit an extension to maps which are P_1-compact at 0.

(1.2) *THEOREM*

Let $F:X \to X$ *be bounded, continuous and* P_1*-compact at* 0. *Then either* F *has fixed point in* X *or the set* $\{x: x-\lambda Fx = 0,\ 0 < \lambda < 1\}$ *is unbounded.*

Proof

The proof of Theorem 1.2 is based on Theorem 1.1 and the following lemma whose proof can be given in the same way as that of Lemma 4 in [22].

(1.3) *LEMMA*

Let R *be a radial retraction of* X *onto* $\bar{B}_r$. *If* $F:X \to X$ *is* P_1*-compact at* 0, *then* $R\cdot F$ *is also* P_1*-compact at* 0.

Proof of Theorem 1.2

Suppose $x-Fx = 0$ has no solution in X. For $k \in Z_k$, let R_k be the radial retraction of X onto $\bar{B}_k$. Then $F_k \equiv R_k\cdot F$ maps $\bar{B}_k$ into $\bar{B}_k$ and, by lemma 1.3, F_k is P_1-compact at 0. Hence, by Theorem 1.1, there exists $x_k \in \bar{B}_k$ such that $F_k(x_k) = x_k$ for each $k \in Z_k$. By our supposition, $Fx_k \in X \setminus \bar{B}_k$ for each k because if $Fx_k \in \bar{B}_k$ some $k = p$, then x_p would be such that $x_p = F(x_p)$, which is not possible. Hence $\|Fx_k\| > k$ for $k \geq 1$ and $x_k-\lambda_k Fx_k = 0$ where $\lambda_k = k/\|Fx_k\| < 1$. This proves Theorem 1.2. ■

In view of Lemma 1.1, Theorem 1.2 implies

(1.1) *COROLLARY*

Let $A:X \to X$ *be continuous, bounded and accretive,* $F:X \to X$ *ball-condensing and* $C:X \to X$ *compact. Then either the equation* $x-Fx-Cx+Ax = y$ *has a solution in* X *or the set* $\{x:x-\lambda\,[(F+C-A)\,(x)-y] = 0,\ 0 < \lambda < 1\}$ *is unbounded.*

Let us add that to the best of our knowledge Corollary 1.1, which is important when applied to elliptic BV Problems on unbounded domains in R^n (see [38]), cannot be obtained by any other existing abstract theory except by the A-proper mapping theory.

(1.2) *REMARK*

If $D = B_r$ and condition (1.8) holds on ∂B_r, then the following new result shows that one can weaken further the condition on F in Theorem 1.1.

(1.3) *THEOREM*

Let $F:\overline{B}_r \to X$ *be such that* $I-F$ *is A-proper at* 0 *w.r.t.* Γ_1 *and* $F(\partial B_r)\subset\overline{B}_r$. *Then the equation* $x-Fx = 0$ *is feebly a-solvable w.r.t.* Γ_1*; it is strongly a-solvable if the fixed point is unique.*

Proof

Since $\|P_n\| = 1$ and $\|Fx\| \le r$ if $\|x\| = r$, it follows that $F_n(x) \in \overline{B}_{rn}$ for $x \in \partial B_{rn}$ for all $n \in Z_+$. Consider the radial retraction $R_n:X_n \to \overline{B}_{rn}$. Then

$$A_n \equiv R_n \cdot F_n$$

is a continuous map of $\overline{B}_{rn}$ into $\overline{B}_{rn}$ and hence, by the Brouwer fixed point theorem, there exists $x_n \in \overline{B}_{rn}$ such that $A_n(x_n) = x_n$. Since $F_n(\partial B_{rn}) \subseteq \overline{B}_{rn}$, it follows easily that $\|x_n\| \le r$ and so $A_n(x_n) = F_n(x_n) = x_n$. Hence $x_n-F_n(x_n) = 0 \to 0$ as $n \to \infty$ and so, since $I-F$ is A-proper at 0, there exists a subsequence $\{x_{n_j}\}$ and $x \in \overline{B}_r$ such that $x_{n_j} \to x$ and $x-Fx = 0$. The proof of the second part follows as before. ■

It is known that $P_n \to I$ uniformly on the set $K \subset X$ iff K is precompact. In view of this, an immediate corollary of Theorem 1.3 is the following fixed point theorem of Frum-Ketkow [10] with a correct proof for X with a $\Gamma_1 = \{X_n, P_n\}$ scheme given by Nussbaum [20].

(1.2) *COROLLARY*

Let $F:\overline{B}_r \to X$ *be continuous and* $F(\partial B_r) \subset \overline{B}_r$. *Suppose there exist a compact set* $K \subset X$ *and* $k \in (0,1)$ *such that* $\mathrm{dist}(F(x),K) \leq k\ \mathrm{dist}(x,K)$ *for all* $x \in \overline{B}_r$. *Then the conclusions of Theorem* 1.3 *hold.*

Proof

It suffices to show that I-F is A-proper at 0. So let $\{x_{n_j} | x_{n_j} \in \overline{B}_{rn_j}\}$ be such that $x_{n_j} - P_{n_j} F(x_{n_j}) \to 0$. Since K is compact, the condition on F implies the existence of $y_{n_j} \in K$ for each j such that $\|Fx_{n_j} - y_{n_j}\| \leq k\|x_{n_j} - y_{n_j}\|$. It is now easy to show that $\{x_{n_j}\}$ has a convergent subsequence and so the continuity of F and the completeness of Γ_1 imply the A-properness of I-F at 0.■

We complete this section with the following.

(1.3) *REMARK*

The main result in [39] asserts that "if $F:X \to X$ is a continuous mapping, then the equation $F(x) = kx$ has a solution when $|k|$ is sufficiently large".

A counterexample has been given in [13] to show that the above assertion is false. Indeed, if $X = C[0,1]$ and $F:X \to X$ is a continuous mapping defined by

$$F(x)\ (t) = (\max\{t, |x(t)-x(0)|\})^{1/2},$$

then $F(x) = kx$ has no solution in X for any $k \in R$ as shown in [13].

However, it follows easily from Theorem 1.1 that

(1.3) *COROLLARY*

If $F:X \to X$ *is continuous, bounded,* P_1*-compact at* 0 *and* $F(0) \neq 0$, *then the equation* $F(x) = kx$ *has a nonzero solution for each sufficiently large* $k > 0$.

(1.2) *Other examples of A-proper mappings and surjectivity theorems.*

Let $L \in L(X\ Y)$ be Fredholm of index 0. Then there exists a compact map $C \in L(X, Y)$ such that $K \equiv L+C$ is bijective. Let $\{Y_n, Q_n\}$ be a complete projective scheme for Y and $X_n = K^{-1}(Y_n)$. Then $\Gamma_K = \{X_n, V_n; KX_n, Q_n\}$ is an admissible scheme for (X, Y) and, as was shown in [29], L is A-proper w.r.t. Γ_K.

An immediate consequence of Theorem 1.2 is the following surjectivity theorem. (cf. Proposition 3.1 in [8]).

(1.4) *THEOREM*

Suppose L, C *and* Γ_L *are as above. Suppose* $N:X \to Y$ *is a continuous and bounded nonlinear map such that* $L-\mu N$ *is A-proper w.r.t.* Γ_K *for each* $\mu \in (0,1]$ *and*

(i) *If* $\{x_j\} \subset X$ *and* $\{t_j\} \subset (0,1)$ *are any sequences such that*

$\{Lx_j + t_j Nx_j + (1-t_j)\,Cx_j\}$ *is bounded, then* $\{x_j\}$ *is bounded.*

Then $(L-N)(X) = Y$.

Proof

It is easy to see that $y \in (L-N)(X)$ iff the map $F_y \equiv [(N+C)K^{-1}+y]$ has a fixed point in Y. On the other hand, since CK^{-1} is compact, it is not hard to show that $L-\mu N$ is A-proper w.r.t. Γ_K for each $\mu \in (0,1]$ iff F_y is P_1-compact w.r.t. $\Gamma = \{Y_n, Q_n\}$. Now we claim that the set $\{w: w-\lambda F_y w = 0,\ 0 < \lambda < 1\}$ is bounded in Y. Indeed, if not, then there would exist sequences $\{w_j\} \subset Y$ and $\{\lambda_j\} \subset (0,1)$ such that $\{w_j\}$ is unbounded and $w_j - \lambda_j F_y(w_j) = 0$ for each j. But then $\{x_j\} \equiv \{K^{-1}w_j\}$ is unbounded in X and $\{Lx_j - \lambda_j Nx_j + (1-\lambda_j)\,Cx_j\} = \{\lambda_j y\}$ bounded, in contradiction to (i). Thus, by Theorem 1.2, F_y has a fixed point in Y for each y in Y, i.e., $(L-N)(X) = Y$. ■

In [31] the author studied the conditions on N which would ensure that $L-\mu N$ is A-proper w.r.t. Γ_K for each $\mu \in (0,1]$. Thus, for example, it was shown that this is the case if one of the following holds:

(i) Either N or the generalized inverse of L is compact.

(ii) N is k-ball-contractive with $k \in [0, \ell(L))$ and $\|Q_n\| = 1$, where $\ell(L) = \sup\{r > 0 : r\beta(Q) \le \beta(L(Q))$ for each bounded $Q \subset X\}$.

(iii) NK^{-1} is ball-condensing and $\|Q_n\| = 1$.

(iv) X and Y are Hilbert spaces and $N(x) = B(x, x)$, where $B: X \times X \to Y$ is continuous, bounded and such that

(1a) There is $\alpha \in (0,1)$ and function $\varphi : X \to R$ with $\varphi(0) = 0$, φ is weakly upper semicontinuous at 0 and

$$(B(x, x)-B(x, y),\ Lx-Ly) \le \alpha\|Lx-Ly\|^2 + \varphi(x-y) \quad \forall x, y \in X.$$

(2a) For each fixed $x \in X$ and any bounded sequence $\{x_n : x_n \in X_n\}$, $\{B(x_n, x)\}$ is precompact in Y.

Another class of A-proper mappings which is important in applications to PDE's are the maps of type (S) and type (S_+) introduced by Browder [4]. Let $K: X \to Y^*$ be a linear homeomorphism such that $Q_n^* Kx = Kx$ for all $x \in X_n$. A map $T: X \to Y$ is said to be of *type* (KS) (respectively (KS_+)) if $x_j \rightharpoonup x_o$ in X and

$$\lim_j (Tx_j - Tx_o,\ Kx_j - Kx_o) = 0$$

(respectively $\limsup (Tx_j - Tx_o, Kx_j - Kx_o) \le 0$), then $x_j \to x_o$ in X. It was shown in [29] (see also [6] when $Y = X^*$ and $K = I$) that if X is reflexive, and $T: X \to Y$ is demicontinuous, semibounded and of type (KS), then T is A-proper w.r.t. $\Gamma_\alpha = \{X_n, V_n; Y_n, Q_n\}$. In particular, every semibounded map $T: X \to X^*$ of type (S) is A-proper w.r.t. Γ_I. Since every monotone map is semibounded, firmly monotone (and, in particular, strongly monotone) maps are A-proper. Using this, one shows that if X is reflexive with X^* strictly convex and $T: X \to X^*$ is demicontinuous, semibounded and *quasimonotone* (i.e. $\lim_j \sup(Tx_j, x_j - x_o) \ge 0$ whenever $x_j \rightharpoonup x_o$ in X), then $T + \mu J$ is A-proper w.r.t. Γ_I for each $\mu > 0$. It is this fact which allows us to apply the A-proper mapping theory to equations involving mappings of monotone type. (see [30, 28]).

We complete this section with the following theorem which we shall need to obtain existence results for quasilinear PDE's and whose proof is given in the same way as that of Theorem 1.1 in [30].

First, we recall that $T:X \to Y$ is said to satisfy *condition* (+) if $\{x_j\} \subset X$ is bounded whenever $Tx_j \to g$ in Y for some g in Y. The above condition is known to play an important role in establishing the surjectivity of T (see [30, 9, 35, 14]).

(1.5) *THEOREM*

Let $\Gamma = \{X_n, V_n; E_n, W_n\}$ *be admissible for* (X, Y), K *some map of* X *into* Y^* *and suppose there exists a bounded, odd map* $G:X \to Y$ *such that*

(H1) G *is A-proper w.r.t.* Γ *and* $(Gx, Kx) = \|Gx\| \, \|Kx\| > 0$ *for* $x \neq 0$.

(H2) $T_\mu = T+\mu G$ *is A-proper w.r.t.* Γ *for each* $\mu > 0$.

(H3) $Tx \neq \gamma Gx$ *for all* $x \in X\text{-}B(0, r)$, *all* $\gamma < 0$ *and some* $r \geq 0$.

(H4) T *satisfies condition* (+) *or, more generally, to each* $f \in Y$ *there are* $r_f \geq r$ *and* $\alpha_f > 0$ *such that* $\|Tx-tf\| \geq \alpha_f$ *for* $x \in \partial B_{r_f}$, $t \in [0,1]$.

(A1) *If* T *is A-proper w.r.t.* Γ *and bounded, then the equation* $Tx = f$ *is feebly a-solvable to each* $f \in Y$, *and strongly a-solvable if it is uniquely solvable.*

(A2) *If* T *is odd on* $X\text{-}B(0, r)$ *then the conclusions hold without the conditions* (H1) - (H3).

2. Solvability of quasilinear elliptic BV Problems and OD Equations.

The prupose of these lectures is to show how the theory of A-proper mappings developed in Section 1 can be used to obtain general variational a-solvability and/or existence theorems for not necessarily coercive elliptic BV Problems involving operators of order $2m$ on a domain in R^n in generalized divergence form

$$A(u) = \sum_{|\alpha| \leq m} (-1)^{|\alpha|} D^\alpha A_\alpha(x, u, \ldots, D^m u), \tag{2.1}$$

where A_α's are of polynomial growth in $(u, \ldots, D^m u)$. The existence theorems for equations of the form (2.1) were first obtained by Višik [40] using compactness argument and a priori estimates on the (m+1)-derivatives. The theory of coercive monotone operators was first applied to equations of the form (2.1) by Browder [1] and by Leray-Lions [15] to the case when $A(u)$ gives ríse to a special case of pseudomonotone operator. Odd operators $A(u)$ satisfying strong monotonicity conditions were first studied by Pohožayev [34] and later by Browder [2, 3] in case of monotone and semi-monotone maps. The generalized degree for pseudomonotone maps, which is based on the degree theory for A-proper maps of Browder-Petryshyn [7], were first applied in [5]. Subsequently, the solvability of equations of the form (2.1) with $A(u)$ giving rise the operators $T:X \to X^*$ of monotone type have been studied by many authors (see [21,16] for references).

The direct application of the A-proper mapping theory to the solvability of linear and quasilinear differential equations was initiated by the author in [26,27].

In this section we use the theory of A-proper maps and their uniform limits to study the solvability of equations of the form (2.1). Theorem 2.1 below is a constructive result which is an extension of Theorem 1.1(A) in [32], while Theorem 2.2 is a unification and generalization of the corresponding results of [2, 5, 1, 15, 34] and others for operators which need not be coercive or monotone (see [32]).

(2.1) *Formulation and solution of the variational BV Problem for* (2.1).

Let $Q \subset R^n$ be a bounded domain with boundary ∂Q so smooth that the Sobolev Imbedding Theorem holds on Q. Let $C_o^\infty(Q)$ be a family of infinitely differentiable functions with compact support in Q. For a multiindex $\alpha = (\alpha_1,\dots,\alpha_n)$ we denote by D^α the generalized derivative $D^\alpha = D_1^{\alpha_1}\dots D_n^{\alpha_n}$ of order $|\alpha| = \alpha_1+\dots+\alpha_n$. For $m \in Z_+$ and $p \in (1,\infty)$, the Sobolev space

$$W_p^m(Q) = \{u \mid u \in L_p(Q),\ D^\alpha u \in L_p(Q) \quad \text{for } |\alpha| \le m\}$$

is a uniformly convex and separable Banach space with respect to the norm

$$\|u\|_{m,p} = \{\sum_{|\alpha|\le m} \|D^\alpha u\|_p^p\}^{1/p},$$

where $\|\cdot\|_p$ denotes the norm in L_p. We let $\mathring{W}_p^m$ denote the closure of $C_o^\infty(Q)$ in W_p^m. Let $\langle u,v\rangle = \int_Q uv\,dx$ denote the pairing between $u \in L_p$ and $v \in L_q$ with $q = p(p-1)^{-1}$. When $p = 2$, $\langle .\rangle$ denotes the inner product in $L_2(Q)$ and W_2^m is the Hilbert space with the inner product $(u, v)_m = \sum_{|\alpha|\le m} \langle D^\alpha u, D^\alpha v\rangle$.

For a given $m \in Z_+$, we let $\xi = \{\xi_\alpha : |\alpha| \le m\}$ and set

$$\zeta = \{\zeta_\alpha : |\alpha| = m\},\ \eta = \{\eta_\beta : |\beta| \le m-1\},$$

where each ξ_α, ζ_α, η_β is an element of R. The set of all ξ of the above form is an Euclidean space R^{s_m}, and correspondingly, $\zeta \in R^{s'_m}$, $\eta \in R^{s_{m-1}}$. We also set

$$\begin{cases} D^m u = \{D^\alpha u : |\alpha| = m\},\ \delta u = \{D^\alpha u : |\alpha| \le m-1\},\ \xi = (\eta,\zeta) \text{ and} \\ A_\alpha(x,\xi) = A\,(x,\eta,\zeta) \quad \text{for} \quad |\alpha| \le m. \end{cases}$$

We shall assume the following concerning $A(u)$ in (2.1):

(a1) For each α, $A_\alpha : Q \times R^{s_m} \to R$ is such that $A_\alpha(x,\xi)$ is measurable in x for fixed ξ and continuous in ξ for fixed x. For a given $p \in (1,\infty)$, there exist $c > 0$ and $k(x) \in L_q$ such that

$$|A_\alpha(x,\eta,\zeta)| \le c[|\eta|^{p-1} + |\zeta|^{p-1} + k(x)], \quad p^{-1} + q^{-1} = 1. \tag{2.2}$$

To define the variational BV Problem for (2.1), we assume that we are given a closed subspace V of W^m_p with $\overset{\circ}{W}{}^m_p \subseteq V$ and with A represented by (2.1) we associate the generalized form defined on W^m_p by

$$a(u,v) = \sum_{|\alpha| \le m} \langle A_\alpha(x,u,\dots,D^m u),\ D^\alpha v\rangle \qquad (u,v \in W^m_p). \tag{2.3}$$

It is known (see [21]) that, in view of (a1), the form $a(u,v)$ is well defined on W^m_p and by Hölder's inequality

$$|a\,(u,v)| \le c_o(\|u\|^{p-1}_{m,p} + \|k\|_q)\ \|v\|_{m,p} \qquad \forall u,v \in W^m_p. \tag{2.4}$$

For w in the dual space V^* and $v \in V$, the value of w at v is denoted by (w,v). We now define the *variational* BV Problem corresponding to (A,V):

(2.1) *DEFINITION*

Let $F \in V^*$. Then u is said to be a variational solution of the BV Problem $Au = F$ corresponding to V if $u \in V$ and

$$a(u,v) = (F,v) \qquad \forall\, v \in V. \tag{2.5}$$

It follows from (2.4) that for each fixed $u \in V$, $a(u,v)$ is a bounded linear functional of v in V. Hence there exists a nonlinear bounded and continuous mapping

$T: V \to V^*$ such that

$$a(u,v) = (Tu,v) \quad \forall v,u \in V. \tag{2.6}$$

Thus the solvability of (2.5) is equivalent to that of

$$Tu = F. \tag{2.7}$$

To apply the A-proper mapping theory to the solvability (2.5) or (2.7) we choose a sequence $\{X_n\}$ of finite dimensional subspaces of V such that

$$\operatorname{dist}(u,X_n) \to 0 \quad \text{for each } u \in V$$

and let V_n be a linear injection of X_n into V. Then $\Gamma_I = \{X_n, V_n; X_n^*, V_n^*\}$ is an admissible scheme for (V, V^*).

(2.2) *DEFINITION*

For a given $F \in V^*$, (2.5) is said to be *strongly* (respectively *feebly*) *a-solvable* iff there exists $n_F \in Z_+$ such that the finite dimensional problem

$$a(u_n,v) = (F,v) \quad \forall v \in X_n, \quad n \geq n_o, \tag{2.8}$$

has a solution $u_n \in X_n$ for each $n \geq n_o$ such that $u_n \to u_o$ (respectively $u_{n_j} \to u_o$) in V and u_o satisfies (2.5).

Since $u_n \in X_n$ is a solution of (2.5) iff u_n is a solution of

$$V_n^*T(u_n) = V_n^*F \qquad (u_n \in X_n, \; V_n^*F \in X_n^*), \tag{2.9}$$

Definition (2.2) is equivalent to Definition (1.1) when $\Gamma = \Gamma_I$.

(2.3) *DEFINITION*

The form $a(u,v)$ is said to satisfy *condition* (+) iff $\{u_j\} \subset V$ is such that $a(u_j,v) \to (g,v)$ for some g in V^*, uniformly w.r.t. $v \in S_1 \equiv \partial B(0,1) \subset V$, then $\{u_j\}$ is bounded.

It is easy to see that $a(u,v)$ satisfies condition (+) iff T does. Condition (+) is known (see [9, 35, 28 , 14, 32]) to be important in proving the surjectivity theorems for various classes of nonlinear maps. Thus, for example, in improving some surjectivity results in Browder [1] and Minty [19], it was shown by Rockafeller [35] (see also [9]) that for a demicontinuous monotone map $T:V \to V^*$ the condition (+) is not only sufficient for $T(V) = V^*$ but also necessary. However, to prove the surjectivity results for other classes of maps some further condition has to be imposed. It seems that our condition (a3) below is the weakest one among such additional conditions.

Our first theorem in this section extends the constructive linear result in [32] to nonlinear variational problem (A, V).

(2.1) *THEOREM*

Let V *be a closed subspace of* W_p^m *with* $V \supseteq \mathring{W}_p^m$ *and suppose that the functions* A_α *satisfy* (a1) *and*

(a2) *There exists a constant* $\mu_o > 0$ *such that*

$$\sum_{|\alpha|=m} [A_\alpha(x,\eta,\zeta)-A_\alpha(x,\eta,\zeta')] \ (\zeta_\alpha-\zeta'_\alpha) \geq \mu_o \sum_{|\alpha|=m} |\zeta_\alpha-\zeta'_\alpha|^p$$

for $x \in Q$ (a.e.), $\eta \in R^{s_{m-1}}$ *and* $\zeta,\zeta' \in R^{s'_m}$.

(a3) $Tu \neq \gamma\, Ju$ *for all* $u \in V\text{-}B(0,r)$, *all* $\gamma < 0$ *and some* $r > 0$, *with* $J:V \to V^*$ *a duality map given by* $(Ju,u) = \|u\|^2$ *and* $\|Ju\| = \|u\|$.

Then, if a(u,v) *satisfies condition* (+), (2.5) *is feebly a-solvable for each* $F \in V^*$ *(and, in particular,* $T(V) = V^*$*), and* (2.5) *is strongly a-solvable if it is uniquely solvable for given* F.

If T *is odd on* V-B(0,r) (i.e. $T(-u) = -Tu$ for $u \in V\text{-}B(0,r)$), *then the conclusions hold without condition* (a3).

(2.1) *REMARK*

Condition (+) holds if one of the following holds:

(1+) $a(u,u)/\|u\|_V \to \infty$ as $\|u\|_V \to \infty$ (i.e. a(u,u) is coercive).

(2+) $\|Tu\|+a(u,u)/\|u\|_V \to \infty$ as $\|u\|_V \to \infty$.

(3+) There exists a continuous function $\psi: R^+ \to R^+$ such that $\|u\|_V \le \psi(\|F\|)$ for each solution u of (2.5).

(2.2) *REMARK*

(a3) holds if, for example, one of the following holds:

($a3_1$) $a(u,u) \ge 0$ for all $u \in V-B(0,r)$ and some $r \ge 0$.

($a3_2$) $\|Tu\|+a(u,u)/\|u\|_V \ge 0$ for all $u \in V-B(0,r)$ and some $r \ge 0$.

Proof of Theorem 2.1

To deduce Theorem 2.1 from Theorem 1.5 we set

$$X = V,\ Y = X^*,\ K = I,\ G = J,\ E_n = X_n^*,\ W_n = V_n^*$$

and note that $\Gamma_I = \{X_n, V_n; X_n^*, V_n^*\}$ is an admissible scheme for (V,V^*) and the hypotheses (H1)-(H4) are verified. Indeed, by the definition of J,

$$(Ju,u) = \|Ju\|\ \|u\| > 0 \quad \text{for} \quad u \ne 0,$$

J is continuous, bounded, odd and A-proper w.r.t. Γ_I; hence (H1) holds. Now, (H3) holds because of condition (a3), while T satisfies condition (+) because $a(u,v)$ does. Thus, to complete the proof, we must verify (H2) and prove that T is A-proper w.r.t. Γ_I. This follows from

(2.1) *LEMMA*

If A_α*'s satisfy* (a1) *and* (a2), *then* $T: V \to V^*$ *given by* (2.6) *is A-proper w.r.t.* Γ_I. *The map* $T + \mu J: V \to V^*$ *is also A-proper w.r.t.* Γ_I *for each* $\mu > 0$.

For the proof of Lemma 2.1 see [32].

(2.3) *REMARK*

Note that when $p = 2$ and when we let

$$A_\alpha(x,u,\dots,D^m u) = \sum_{|\beta|\le m} A_{\alpha\beta}(x)D^\beta u$$

with $A_{\alpha\beta}(x) \in L_\infty(Q)$ for $|\alpha| \le m$ and $|\beta| \le m$, then the form $a(u,v)$ in (2.3) coincides with bilinear form $B[u,v]$ in [32] and the condition (a2) reduces to the assumption

$$(a2_L) \qquad \sum_{|\alpha|=|\beta|=m} A_{\alpha\beta}(x)\zeta_\alpha\zeta_\beta \ge \mu_0 \sum_{|\alpha|=m} |\zeta_\alpha|^2 \quad \text{for } x \in Q(\text{a.e.})$$

and all $\zeta = \{\zeta_\alpha | (\alpha) = m\} \in R^{s'_m}$ which is the same as the assumption (c2) in [32] for the linear case when all functions are real. Since the linear operator $T(=L)$ is odd and T satisfies (+) iff T has a bounded inverse which, because T is A-proper, is the case iff T is one-to-one, we see that (A1) of Theorem 1.1 in [32] is a special case of Theorem 2.1 when T is odd. Observe that the linear result is deduced without the use of Garding inequality even when $V = \mathring{W}_2^m$.

It should be pointed out, however, that the strong ellipticity assumption $(a2_L)$ is different from the classical condition which requires that

$$(*) \qquad \sum_{|\alpha|=|\beta|=m} A_{\alpha\beta}(x)\xi^\alpha\xi^\beta \ge \mu_0 \sum_{|\alpha|=m} |\xi^\alpha|^2 \quad \text{for } x \in Q(\text{a.e.})$$

and for all $\xi = (\xi_1,\dots,\xi_n) \in R^n$.

We now show how the A-proper mapping approach can still be used to establish the solvability of (2.5) for each $F \in V^*$ when instead of condition (a2) we assume a much weaker condition

$$\sum_{|\alpha|=m} [A_\alpha(x,\eta,\zeta)-A_\alpha(x,\eta,\zeta')] \ (\zeta_\alpha-\zeta'_\alpha) \ge 0 \tag{2.10}$$

for $x \in Q(\text{a.e.})$, $\eta \in R^{s_{m-1}}$ and $\zeta,\zeta' \in R^{s'_m}$, provided we also assume that $a(u,v)$ satisfies condition (++) given by

(2.4) *DEFINITION*

The form $a(u,v)$ is sais to satisfy *condition* (++) iff $\{u_j\} \subset V$ is bounded and such that $a(u_j,v) \to (g,v)$ for some $g \in V$, uniformly w.r.t. $v \in S_1$, then there is $u_0 \in V$ such that $a(u_0,v) = (g,v)$ for all $v \in S_1$.

(2.2) *THEOREM*

Let A_α's *satisfy condition* (a1) *and* (2.10) *and suppose that* T *is either odd on* V-B(0,r) *or condition* (a3) *of Theorem* 2.1 *holds. Suppose further that* a(u,v) *satisfies condition* (++). *Then, if* a(u,v) *satisfies* (+), (2.5) *is solvable for each* $F \in V^*$ (i.e., $T(V) = V^*$).

Theorem 2.2 will be deduced from the following surjectivity result for $T:X \to Y$ which is *not* A-proper but a uniform limit of A-proper maps and which satisfies *condition* (++) (i.e. If $\{x_j\} \subset X$ is bounded and $Tx_j \to g$ for some g in Y, then there exists $x_o \in X$ such that $Tx_o = g$).

(2.3) *THEOREM*

Let $\Gamma, T, G:X \to Y$ **and** $K:X \to Y^*$ *be such that* (H1)-(H4) *of Theorem* 1.5 *hold. Then:*

(A3) *If* T *is bounded and satisfies condition* (++), then $T(X) = Y$.

(A4) *If* T *is odd on* X-B(0,r), *then* $T(X) = Y$ *without the hypotheses* (H1) *and* (H3) *and the boundedness condition on* T.

The validity of Theorem 2.3 follows from Theorem 1.5 since it is easy to show that for each fixed $\lambda > 0$ the map $T_\lambda = T+\lambda G$ is A-proper w.r.t. Γ and satisfies conditions (H1)-(H4). For details and various special cases of Theorem 2.3 see [32].

Proof of Theorem 2.2

It suffices to show that the hypotheses of Theorem 2.2 imply those of Theorem 2.3 when $X = V$, $Y = V^*$, $K = I$, $G = J$, $\Gamma = \Gamma_I = \{X_n, V_n; X_n^*, V_n^*\}$ and $T:V \to V^*$ is given by $(Tu,v) = a(u,v)$. Now, as in the proof of Theorem 2.1, one checks that (H1), (H3) and (H4) hold. The condition (H2), i.e., the requirement that $T+\mu J$ be A-proper w.r.t. Γ_I for each $\mu > 0$, follows from the same argument as that used to prove Lemma 2.1 since (2.10) and the definition of J imply for $T_\mu = T+\mu J$ the inequality

$$(T_\mu(u)-T_\mu(v),u-v) \geq \mu(\|u\|-\|v\|)^2 + \sum_{|\alpha|=m} \langle A_\alpha(x,\delta u,D^m v)-A_\alpha(x,\delta v,D^m v),D^\alpha u-D^\alpha v\rangle$$

$$+ \sum_{|\alpha|<m} \langle A_\alpha(x,\delta u,D^m u)-A_\alpha(x,\delta v,D^m v),D^\alpha u-D^\alpha v\rangle \tag{2.11}$$

for each $\mu > 0$ and all $u,v \in V$. Finally T satisfies condition (++) since $a(u,v)$ does. Thus all conditions of Theorem 2.3 are verified and so (2.5) is solvable for each $F \in V^*$, i.e., $T(V) = V^*$. ■

In applying Theorem 2.2 to the solvability of (2.5), it is important to find some analytic conditions on $A(x,\eta,\zeta)$ which would ensure that $a(u,v)$ satisfies conditions (++).

As our first corollary to Theorem 2.2 we show that if in (2.10) we require inequality to be strictly greater than 0, then Theorem 2.2 remains valid without condition (++).

(2.1) *COROLLARY*

Suppose that in addition to (a1) *the functions* $A_\alpha(x,\eta,\zeta)$ *satisfy the condition:*

$$\sum_{|\alpha|=m} [A_\alpha(x,\eta,\zeta)-A_\alpha(x,\eta,\zeta')]\,(\zeta_\alpha-\zeta'_\alpha) > 0 \tag{2.12}$$

for $\zeta,\zeta' \in R^{s'_m}$ *with* $\zeta \neq \zeta'$, *all* $\eta \in R^{s_{m-1}}$ *and* $x \in Q$(a.e.). *Suppose further*

(H3.1) *Either* T *is odd on* $V-B(0,r)$ *or* $Tu \neq \gamma Ju$ *for* $u \in V-B(0,r)$ *and* $\gamma < 0$.

Then, if $a(u,v)$ *satisfies condition* (+), $T(V) = V^*$.

Proof

Since (2.12) implies (2.10), it follows that (2.11) holds for each $\mu > 0$. This implies that $T+\mu J$ is A-proper w.r.t. Γ_I for each $\mu > 0$. Thus, all we need is to show that $a(u,v)$ satisfies condition (++) if (2.12) holds.

To indicate the proof of it, as in [15], define $G:V\times V \to V^*$ by

$$(G(u,w),v) = \sum_{|\alpha|=m} <A_\alpha(x,\delta u,D^m w),D^\alpha v> + \sum_{|\alpha|<m} <A_\alpha(x,\delta u,D^m u),D^\alpha v>. \quad (2.13)$$

It follows from (a1) tant G is bounded, continuous and $G(u,u) = T(u)$ for $u \in V$. By (2.12)

$$(G(u,u)-G(u,w),u-w) = \sum_{|\alpha|=m} <A_\alpha(x,\delta u,D^m u)-A_\alpha(x,\delta u,D^m w),D^\alpha u-D^\alpha w> \geq 0. \quad (2.14)$$

Using (2.12) one then proves the following claim:

If $u_j \rightharpoonup u_o$ *in* V *and* $(G(u_j,u_j)-G(u_j,u_o),u_j-u_o) \to 0$, *then*

$$(G(u_j,w),u_j-w) \to (G(u_o,w),u_o-w)$$

for each $w \in V$. For the details of the proof of this claim see [33].

We now show that $a(u,v)$ satisfies condition (++). So, let $\{u_j\} \subset V$ be bounded and such that $a(u_j,v) \to (g,v)$ for some $g \in V^*$, uniformly w.r.t. $v \in S_1$. Hence $Tu_j \to g$ in V^*. This and the nonrestrictive assumption that $u_j \rightharpoonup u_o$ for some $u_o \in V$ imply that $(G(u_j,u_j)-G(u_j,u_o), u_j - u_o) \to 0$ and hence, by the Claim, $(G(u_j,w),u_j-w) \to (G(u_o,w),u_o-w)$ for each $w \in V$. Since, by (2.14),

$$(G(u_j,u_j)-G(u_j,w),u_j-w) \geq 0$$

for each $w \in V$, the passage to the limit in the last inequality yields

$$(g-G(u_o,w),u_o-w) \geq 0$$

for each $w \in V$. It follows from this and Minty's lemma that $g = G(u_o,u_o) = Tu_o$, i.e., condition (++) holds. ■

(2.4) *REMARK*

Corollary 2.1 extends a surjectivity theorem of Leray and Lions [15] proved by them for for the case when $a(u,v)$ is coercive and the following condition (2_2) also holds:

(2_2) $\sum_{|\alpha|=m} A_\alpha(x,\eta,\zeta)\zeta_\alpha/|\zeta|+|\zeta|^{p-1} \to \infty$ as $|\zeta| \to \infty$ uniformly for bounded η and $x \in Q$(a.e.).

In the original formulation the writer also imposed condition (2_2) to prove the Claim. I am grateful to Prof. J.P. Gossez who suggested a variant of an argument due to R. Landes which was used here to establish the Claim using only (2.12) and not (2_2). I am also thankful to J.P. Gossez for sending me Landes' preprint "On the Galerkin method in the existence theory of quasilinear elliptic equations" which also contains Corollary 2.1 when $a(u,u)$ is coercive and $V = \overset{\circ}{W}{}^m_p$ but whose proof is totally different from that of ours.

As a second corollary to Theorem 2.2 we show that if (2.10) is required to hold for all $|\alpha| \leq m$, then Theorem 2.2 remains valid without condition (++).

(2.2) *COROLLARY*

Suppose A_α's *satisfy* (a1) *and*

$$\sum_{|\alpha|\leq m} [A_\alpha(x,\eta,\zeta)-A_\alpha(x,\eta,\zeta')]\,(\zeta_\alpha-\zeta'_\alpha) \geq 0 \quad \textit{for} \quad x \in Q(\text{a.e.}), \tag{2.15}$$

$\eta \in R^{s_{m-1}}$ *and* $\zeta,\zeta' \in R^{s'_m}$. *Suppose also that* (H3.1) *of Corollary* 2.1 *holds. Then, if* $a(u,v)$ *satisfies condition* (+), $T(V) = V^*$.

<u>*Proof*</u>

By theorem 2.2, it suffices to show that $a(u,v)$ satisfies condition (++) if (2.15) holds.

As in [2], write $a(u,v)$ in another notation, i.e., let us define $a(u,w;v)$ on V by

$$a(u,w;v) = \sum_{|\alpha|\leq m} \langle A_\alpha(x,\delta u,D^m w)D^\alpha v\rangle \quad (u,v,w \in V). \tag{2.16}$$

It follows from (a1) that for fixed $u,w \in V$, $a(u,w;w)$ is a bounded linear functional in $v \in V$. Hence there exists a unique $S(u,w) \in V^*$ such that

$$a(u,w;w) = (S(u,w),v) \quad \forall v \in V \tag{2.17}$$

and the previous operator T is the given by $Tu = S(u,u)$. It follows from (2.15) and (2.16)-(2.17) that

$$(S(u,u)-S(u,v),u-v) \geq 0 \quad \forall u,v \in V. \tag{2.18}$$

Furthermore, as in [2], one proves that in view of (2.16)-(2.18) and the Sobolev imbedding theorem one has:

(a) For fixed $u \in V$, $S(u_j,u) \to S(u_o,u)$ if $u_j \rightharpoonup u_o$ in V.

(b) For fixed $u \in V$, $S(u,u_j) \to S(u,u_o)$ if $u_j \to u_o$ in V.

This shows that $T(u) = S(u,u)$ is semimonotone in the sense of Browder [2, 3]. Now let $\{u_j\} \subset V$ be bounded and such that $a(u_j,v) \to (g,v)$ for some $g \in V^*$, uniformly for $v \in S_1$. Then $tu_j \to g$ in V^* and we may assume that $u_j \rightharpoonup u_o$ in V. Since, by (2.18),

$$(S(u_j,u_j)-S(u_j,u),u_j-u) \geq 0 \quad \forall u \in V, \tag{2.19}$$

the passage to the limit gives $(g-S(u_o,u),u_o-u) \geq 0$ for all $u \in V$ which, on account of Minty's lemma, implies that $g = S(u_o,u_o) = Tu_o$, i.e., condition (++) holds. ∎

(2.5) *REMARK*

When T is coercive Corollary 2.2 was proved by Browder [5] by applying the abstract theorem for semimonotone operators. Equations of type (2.1) involving pseudo-monotone odd operators satisfying condition (3+) have been studied in [5].

(2.2) *A note on an OD Equation of the dynamics of wires.*

Using the continuation principle for stably-solvable maps of Furi-Martelli-Vignoli [11, 12], it was shown by Sanches [36] that a differential equation which appears in the study of the dynamics of wires has a solution for each $f(t) \in L^2(0,b)$. Here we indicate how theorem 1.5, when T is odd, can be used to deduce the result of [36] in a simpler and sometimes constructive way. A perturbation problem to which the abstract results in [11, 12] cannot be applied will also be treated.

Let $b > 0$ and consider the nonlinear OD Equation

$$\left.\begin{aligned} &u''+r(t)u-p(t)u^3+q(t)u'+|u'|u' = f(t) \\ &u(0) = u(b), \quad u'(0) = u'(b), \end{aligned}\right\} \tag{2.20}$$

where $f \in L_2(0,b)$ and the functions r, q and p satisfy the following conditions:

(b1) $r,q \in L_\infty(0,b)$

(b2) $p \in W^1_\infty(0,b)$, $p(0) = p(b)$ and there exists $\alpha > 0$ such that $p(t) \geq \alpha$ for $t \in [0,b]$.

To apply Theorem 1.5 to the solvability of (2.20) we let $Y = L_2(0,b)$, $X = \{u \in W^2_2(0,b) \,|\, u(0) = u(b),\ u'(0) = u'(b)\}$ and define the map $T: X \to Y$ by

$$Tu = u''+r(t)u-p(t)u^3+q(t)u'+|u'|u'. \tag{2.21}$$

Let $\{Y_n\} \subset L_2(0,b)$ be finite dimensional subspaces such that $\operatorname{dist}(g,Y_n) \to 0$ for each $g \in L_2(0,b)$. It is easy to see that $L \in L(X,Y)$ defined by $Lu = u''$ is Fredholm of index 0 and that $K \in L(X,Y)$ defined by $Ku = u''-u$ is a linear homeomorphism of X onto Y. Hence, if for each n, we choose X_n in X to be such that $Y_n = K(X_n)$ and let Q_n be the orthogonal projection of Y onto Y_n, then $\Gamma_K = \{X_n, V_n; Y_n, Q_n\}$ is an admissible scheme for (X,Y), the operator L is A-proper w.r.t. Γ_K and the operator $T: X \to Y$ defined by (2.21) is also A-proper w.r.t. Γ_K since the map $N: X \to Y$ given by $Nu = r(t)u-p(t)u^3+q(t)u'+|u'|u'$ is compact, by the Sobolev imbedding theorem. Since T is also odd, to apply Theorem 1.5 to (2.20), all we need is to show that T satisfies condition (+), i.e., if $\{u_j\} \subset X$ is any sequence such that $Tu_j \to g$ for some g in Y, then $\|u_j\|_{2,2} \leq M$ for all j and some constant $M > 0$. Thus, as in [36], we are led to derive certain a-priori estimates, i.e., it suffices to show that there exist a continuous function $\psi: R^+ \to R^+$ such that if $u \in X$ is a solution of

$$u''+r(t)u-p(t)u^3+g(t)u'+|u'|u' = f \tag{2.22}$$

for any given $f \in L_2(0,b)$, then $\|u\|_{2,2} \leq \psi(\|f\|)$. To prove this we will need the following inequalities for any positive a and b:

$$ab \leq \frac{\varepsilon}{2}\, a^2 + \frac{1}{2\varepsilon}\, b^2, \quad \varepsilon > 0 \tag{2.23}$$

$$ab \leq \varepsilon a^{1/\alpha} + \left(\frac{\alpha}{\varepsilon}\right)^{\alpha/(1-\alpha)} (1-\alpha) b^{1/(1-\alpha)}, \quad \varepsilon > 0, \quad \alpha \in (0,1). \tag{2.24}$$

To get an estimate for $\|u''\|_2$ we first note that

$$\int_0^b |u'|u'u''dt = 0 \quad \text{since} \quad |u'|u'' = \left(\tfrac{1}{2}|u'|u'\right)'$$

and

$$\int_0^b |u'|u'u''dt = \frac{1}{2}|u'|^3\Big|_0^b - \frac{1}{2}\int_0^b |u'|u'u''dt.$$

Thus multiplying (2.22) by u'', integrating over $[0,b]$ and using Cauchy-Schwarz inequality one gets

$$\|u''\|_2^2 \le R\|u\|_2\|u''\|_2 + P\|u\|_6^3\|u''\|_2 + Q\|u'\|_2\|u''\|_2 + \|f\|_2\|u''\|_2,$$

where R, P, Q denote the $L_\infty(0,b)$-norms of r, p, q and $\|\cdot\|_p$ is the $L_p(0,b)$-norm. Hence, in view of (2.23),

$$\|u''\|_2^2 \le c\,(\|u\|_6^6 + \|u'\|_2^2 + \|f\|_2 + 1), \tag{2.25}$$

where here and afterward c denotes various constants which are independent of u. Next multiply (2.22) by u' and integrate to get

$$\int_0^b ruu'dt + \frac{1}{4}\int_0^b p'u^4dt + \int_0^b q(u')^2dt + \int_0^1 |u'|^3 = \int_0^b fu'dt,$$

where we have used the facts that

$$\int_0^b u''u'dt = 0 \quad \text{and} \quad \int_0^b pu^3u'dt = -\frac{1}{4}\int_0^b p'u^4dt.$$

Again, as above, we get from the preceeding equality that

$$\|u'\|_3^3 \le c\{\|u'\|_2^2 + \|u\|_4^4 + \|f\|_2^2 + 1\}. \tag{2.26}$$

Finally multiply (2.22) by u and integrate to obtain

$$\|u'\|_2^2 + \alpha\|u\|_4^4 \le R\|u\|_2^2 + Q\|u\|_2\|u'\|_2 + \|f\|_2\|u\|_2 + \int_0^b \|u'\|^2|u|dt, \tag{2.27}$$

where we used the inequality $p(t) \ge \alpha$ and the fact that $\int_0^b u''udt = -\|u'\|_2^2$. Now, using (2.24) with $\alpha = \frac{3}{4}$ we get

$$|u'|^2|u| \le \varepsilon|u'|^{8/3} + \left(\frac{1}{\varepsilon}\,\frac{3}{4}\right)^3 \left(\frac{1}{4}\right)|u|^4 \quad \text{for any } \varepsilon > 0$$

and therefore

$$\int_0^b |u'|^2|u|dt \le \varepsilon\|u'\|_{8/3}^{8/3} + \frac{1}{4}\left(\frac{3}{4\varepsilon}\right)^3 \|u\|_4^4.$$

In view of this, it follows from (2.27) that

$$\|u'\|_2^2+\|u\|_4^4 \le c\{\|u'\|_{8/3}^{8/3}+\|f\|_2^2+1\}. \tag{2.28}$$

Since

$$\|u'\|_{8/3}^{8/3} = \int_0^b |u'|^{8/3}dt \le c(\int_0^b |u'|^3)^{8/9} = c(\|u\|_3^3)^{8/9},$$

it follows from (2.28) and (2.26) that

$$\|u'\|_2^2+\|u\|_4^4 \le c\{[\|u'\|_2^2+\|u\|_4^4+\|f\|_2^2+1]^{8/9}+\|f\|_2^2+1\}. \tag{2.29}$$

This obviously implies that $\|u'\|_2^2+\|u\|_4^4 \le M_o$, M_o being a constant independent of u, and by virtue of Sobolev theorem, we also have $\|u\|_6$ bounded. This and (2.25) imply that $\|u''\|_2$ is also bounded. Hence $\|u\|_{2,2} \le \psi(\|f\|)$, where $\psi(t) = c\{t_2^2+1\}$.

The above discussion implies the validity of

(2.4) *THEOREM*

For any functions p, q *and* r *satisfying* (b1) *and* (b2), *the BV Problem* (2.20) *is feebly a-solvable w.r.t.* Γ_K *for each* f *in* $L_2(0,b)$ *and, in particular,* $T(X) = L_2$. *If, for some* f *in* $L_2(0,b)$, *the problem* (2.20) *has a unique solution, then* (2.20) *is strongly a-solvable.*

Using the above result we now use again Theorem 1.5 to solve a perturbation problem to which the abstract results in [11, 12] are not applicable.

(2.5) *THEOREM*

Suppose the functions p, q *and* r *satisfy conditions* (b1) *and* (b2) *and suppose that*

(b3) $g:[0,b] \times R^3 \to R$ *is continuous and bounded and such that there exists* $d \in [0,1)$ *such that*

$$[g(t,s,r,z_1)-g(t,s,r,z_2)](z_1-z_2) \ge -d|z_1-z_2|^2 \quad \text{if } t \in [0,b],\ s,r,z_1,z_2 \in R$$

and also suppose that

$$g(t,s,r,z) = -g(t,-s,-r,-z) \quad \textit{if} \quad t \in [0,1],\ s,r,z \in R.$$

Then the conclusions of Theorem 2.4 *hold for*

$$\left.\begin{aligned} -u'' + g(t,u,u',u'') + N(u) &= f(t) \\ u(0) = u(b), \quad u'(0) &= u'(b) \end{aligned}\right\} \tag{2.30}$$

for each $f \in L_2(0,b)$ *with* $N(u)$ *the same as in Theorem* 2.4.

Proof

Let $G:X \to Y$ be the mapping defined by $G(u) = g(t,u,u',u'')$. Condition (b3) implies that G is continuous, has a bounded range, odd and, in view of Lemma 3.1 in [29], $L+G:X \to Y$ is A-proper w.r.t. Γ_K. Furthermore, since $N:X \to Y$ is compact, it follows that $L+G+N$ is A-proper. Since $G(X)$ is bounded it follows $L+G+N$ satisfies condition (+). This and the oddness of $L+G+N$ allows us to invoke the assertion (A2) of Theorem. 1.5. ■

References

[1] BROWDER, F.E.: Nonlinear elliptic BV Problems, Bull. AMS, 69 (1963), 862-874.

[2] BROWDER, F.E.: Existence and uniqueness theorems for solutions of nonlinear BV Problems, Proc. Symp. in Appl. Math., vol XVII, AMS, (1965), 24-49.

[3] BROWDER, F.E.: Mapping theorems for noncompact nonlinear operators in Banach spaces, Proc. Nat. Acad. Sci., USA, 54 (1965), 337-342.

[4] BROWDER, F.E.: Nonlinear eigenvalue problems and Galerkin approximations, Bull., AMS, 74 (1968), 651-656.

[5] BROWDER, F.E.: Nonlinear elliptic BV Problems and the generalized topological degree, Bull., AMS, 78 (1970), 999-1005.

[6] BROWDER, F.E.: Nonlinear operators and nonlinear equations of evolution in Banach spaces, in Proc. Symp. in Pure Math,. Vol. 18, AMS, Providence, R.I., (1976).

[7] BROWDER, F.E. and PETRYSHYN, W.V.: Approximation methods and the generalized topological degree for nonlinear mappings in Banach spaces, J. Funct. Anal. 3 (1969), 217-245.

[8] FITZPATRICK, P.F.: On nonlinear perturbation of linear second order elliptic BV Problems, Math. Proc. Camb. Phil. Soc. 84 (1978), 143-157.

[9] FITZPATRICK, P.M.: Surjectivity results for nonlinear mappings from a Banach space to its dual, Math. Ann. 204 (1973), 177-188.

[10] FRUM-KETKOV, R.L.: On mappings of the sphere of a Banach space, Soviet Math. Dokladi, 8 (1967), 1004-1006.

[11] FURI, M, MARTELLI, M. and VIGNOLI, A.: Stably-solvable operators in Banach spaces, Atti Accad. Naz. Lincei Rend. Cl. Sci. Fiz. Mat. Nat., 60 (1976), 21-26.

[12] FURI, M., MARTELLI, M. and VIGNOLI, A.: On the solvability of nonlinear operator equations in Banach spaces, Ann. di Mat. Pura. Appl. (to appear).

[13] GOEBEL, K. and RZYMOWSKI, W.: An equation $f(x) = kx$ possessing no solution, Proc., AMS, 51 (1975).

[14] GOSSEZ, J.P.: Surjectivity results for pseudomonotone mappings in complementary systems, J. Math. Anal. Appl. 53 (1976), 484-494.

[15] LERAY, J. and LIONS, J.L.: Quelques résultats de Višik sur les problèmes elliptiques nonlinéaires par les méthodes de Minty-Browder, Bull. Soc. Math. France 93 (1965), 97-107.

[16] LIONS, J.L.: Quelques méthodes de résolution des problèmes aux limites non linéaires, Dunod Gauthier-Villons, Paris (1970).

[17] MARTELLI, M. and VIGNOLI, A.: Eigenvectors and surjectivity for α-Lipschitz mappings in Banach spaces, Amer. Mat. Pura Appl. (4) 94 (1972), 1-9.

[18] MILOJEVIČ, P.S.: A generalization of the Leray-Schauder Theorem and surjec-

tivity results for multivalued A-proper and pseudo-A-proper mappings, Nonlin. Anal., Theory, Methods and Appl. 1 (1977), 263-276.

[19] MINTY, G.J.: On a "monotonicity" method for the solution of nonlinear equations in Banach spaces, Proc. Nat. Acad. Sci., USA, 50 (1963), 1038-1041.

[20] NUSSBAUM, R.D.: The fixed point index and fixed point theorems for k-set-contractions, PH. D. Dissertation, Univ. of Chicago, Chicago, (1969).

[21] PASCALI, D. and SBURLAN, S.: Nonlinear Mappings of Monotone Type, Sijthoff and Noordhoff Intern. Publ., Alphen ann den Riju, The Netherlands (1978).

[22] PETRYSHYN, W.V.: Construction of fixed points of demicompact mappings in Hilbert space, J. Math. Anal. Appl. 14 (1966), 276-284.

[23] PETRYSHYN, W.V.: On the approximation-solvability of nonlinear functional equations in normed linear spaces, Num. Anal. of PDE's (C.I.M.E. 2°, Ciclo, Ispra, (1967)), Edizioni Cremonese, Roma (1968), 343-355. See also Math. Ann. 177 (1968), 156-164.

[24] PETRYSHYN, W.V.: On projectional-solvability and the Fredholm alternative for equations involving linear A-proper mappings, Arch. Rat. Mech. Anal. 30 (1968), 270-284.

[25] PETRYSHYN, W.V.: Generalization of Schaeffer's Theorem to 1-set-contractive operators, Dopovidi, Ukr. Acad. Sci. Ser. A. No. 10 (1973), 889-891.

[26] PETRYSHYN, W.V.: The approximation-solvability of equations involving A-proper and pseudo-A-proper mappings, Bull. AMS, 81 (1975), 223-312.

[27] PETRYSHYN, W.V.: Fredholm alternative for nonlinear A-proper mappings with applications to nonlinear elliptic BV Problems, J. Funct. Anal., 18 (1975), 288-317.

[28] PETRYSHYN, W.V.: On the relationship of A-properness to maps of monotone type with application to elliptic equations, in "Fixed point theory and its appl." (ed. S. Swaminatham), Acad. Press, N.Y., (1976), 149-174.

[29] PETRYSHYN, W.V.: Existence theorems for semilinear abstract and differential equations with noninvertible linear part and noncompact perturbations, in

"Nonlinear Equations in Abstract Spaces", ed. L. Lakshmikantham, Acad. Press, N.Y. (1978), 275-316.

[30] PETRYSHYN, W.V.: On the solvability of nonlinear equations involving abstract and differential equations, in "Funct. Anal.Methods in Num. Anal.", (M.Z. Nashed, ed.), Lecture Notes in Math., No. 701, Springer-Verlag, Berlin (1979), 209-247.

[31] PETRYSHYN, W.V.: Using degree theory for densely defined A-proper maps in the solvability of semilinear equations with unbounded and noninvertible linear part, Nonlin. Anal., Theory, Methods and Appl. 4 (1980), 259-281.

[32] PETRYSHYN, W.V.: Solvability of linear and quasilinear elliptic BV Problems via the A-proper mapping theory (submitted).

[33] PETRYSHYN, W.V. and TUCKER, T.S.: On the functional equations involving non linear generalized P-compact operators, Trans., AMS, 135 (1969), 343-373.

[34] POHOŽAYEV, S.I.: The solvability of nonlinear equations with odd operators, Funktion. Analis i Priloz. 1 (1967), 66-73.

[35] ROCKAFELLAR, R.T.: Local boundedness of nonlinear maximal monotone operators, Mich. Math. J. (1969), 397-407.

[36] SANCHES, L.: A note on a differential equation of the dynamics of Wires, Bollettino, UMI, (5) 16-A (1979), 391-397.

[37] SCHAEFFER, H.: Über die Methode der a priori Schranken, Math. Ann. 129 (1955), 415-416.

[38] TOLAND, J.F.: Global bifurcation theory via Galerkin method, Nonlin. Anal., Theory, Methods and Appl. 7 (1977), 305-317.

[39] VENKATESWARAN, S.: The existence of a solution of $f(x) = kx$ for a continous not necessarily linear operator, Proc., AMS, 36 (1972), 313-314.

[40] VIŠIK, M.I.: Quasilinear strongly elliptic systems of differential equations in divergence form, Trudy Moskov. Mat. Obšč. 12 (1963), 125-184.

AN EXISTENCE THEOREM AND APPLICATION TO A NON-LINEAR ELLIPTIC BOUNDARY VALUE PROBLEM

By

A. J. B. POTTER*

Department of Mathematics
University of Aberdeen
Aberdeen, Scotland

0. INTRODUCTION

Let U and V be real Banach spaces. In this paper we are concerned with the solvability of the equation

$$Lu + Nu = f \tag{0.1}$$

where f is a given element of V, $L : D \subset U \to V$ and $N : U \to V$ are certain non-linear mappings.

Our method of solution is to first consider the approximate equations

$$Lu + \lambda Cu + Nu = f \tag{0.2}$$

where $C : U \to V$ is a mapping such that L + C has nice enough properties to allow the solvability of (0.2). The solvability of (0.1) is then established by considering the behavior of solutions of (0.2) as $\lambda \to 0$.

Such a method was used by Petryshyn in [5]. He assumed that L was a bounded linear Fredholm operator of index zero. He was able to apply his results to the linear boundary value problem

$$\begin{aligned} -\Delta u + g(x,\nabla u,\Delta u) + h(x,u) &= f \quad \text{on } \Omega \\ \partial u/\partial n &= 0 \quad \text{on } \Gamma \end{aligned} \tag{0.3}$$

---

*This work was completed while the author was visiting Rutgers University. He is grateful for the generous hospitality given him by the Mathematics Department at Rutgers.

(here Ω is a bounded open subset of R^m with smooth boundary Γ and g and h particular non-linear mappings).

In this paper we show similar techniques can be used in the case L is non-linear. We apply our results to equation (0.3) but with the linear boundary condition replaced by

$$-\partial u/\partial u \in \beta(u) \tag{0.4}$$

where β is a maximal monotone graph in R^2.

1. EXISTENCE RESULT

Throughout this section U and V denote real Banach spaces and $L : U \to V$ and $N : U \to V$ denote bounded non-linear mappings. Let D be a subset of U and put $A = L|_D$ (that is A is the restriction of L to D). The norms in U and V are denoted by $||\cdot||$ and $|\cdot|$ respectively. We make the following hypotheses concerning A and N.

(H1) There exists a bounded mapping $C : U \to V$ such that $A + C$ is one-one and onto Y ($A + C$ is considered as a mapping from D into Y) and $C \circ (A+C)^{-1} : Y \to Y$ is compact and non-expansive.

(H2) There exists a constant $K > 0$ such that

$$||u|| \leq K(|Au + Cu| + 1)$$

for all $u \in D$.

(H3) $0 \in D$ and $A\,0 = 0$.

(H4) $N \circ (A+C)^{-1} : V \to V$ is a k-set contraction ($k < 1$).

(1.1) *REMARKS*

(i) Although hypotheses (H1), (H2), (H3) and (H4) are made with an eye on our application, it is interesting to note that they are satisfied by many operators both linear and non-linear. If A is a single-valued m-accretive operator on a

Banach space X with compact resolvent then A satisfies (H1) (put $U = V = X$, $D = D(A)$ and C = identity on X). If L is a bounded linear Fredholm operator of index zero from U to V then L satisfies (H1), (H2) and (H3) provided V is renormed suitably (put $C = M \circ P$ where P is a projection of U onto $N(L)$, the null-space of L, and M is a linear isomorphism of $N(L)$ onto a complementary subspace of $R(L)$, the range of L; the new norm on V is defined by

$$|||v||| = |v_1| + |v_2|$$

where $v = v_1 + v_2$, $v_1 \in V_1$ and $v_2 \in R(L)$, V_1 being the chosen complementary subspace of $R(L)$).

(ii) (H2) is an estimate satisfied by many operators arising in the study of elliptic boundary value problem. (H3) is a technical condition.

(iii) For a bounded subset B of a metric space X, the set-measure of non-compactness is defined by

$$\gamma(B) = \inf \left\{ d > 0 \;\middle|\; \begin{array}{l} B \text{ is contained in the union of a finite} \\ \text{number of sets of diameter } \leq d \end{array} \right\}.$$

We say $T : X \to X$ is a k-set contraction if T is continuous and $\gamma(T(B)) \leq k\gamma(B)$ for all bounded subsets B of X. Properties of k-set contractions $(k < 1)$ and in particular a fixed point index theory for such mappings can be found in the paper of Nussbaum [4]. It will be clear from our argument that the compactness assumption in (H1) can be dropped if we assume $N \circ (A + C)^{-1}$ is compact.

Let us now consider the equation

$$Au + Nu = f. \tag{1.1}$$

To do this we first consider the equations

$$Au + \lambda Cu + Nu = f \tag{1.2}$$

for $\lambda > 0$. For simplicity we put $B = (A + C)^{-1}$. Clearly to establish the solvability of (1.2) it is sufficient to prove the existence of a solution $v_\lambda \in V$ of the equation

$$v - (1 - \lambda)\, CBv + NBv = f\,. \tag{1.3}$$

For each $t \in [0,1]$ define $T_t : V \to V$ by

$$T_t v = (1-\lambda)\, CBv - tNBv + tf. \tag{1.4}$$

We must show T_1 has a fixed point. In order to do this it is sufficient to show that the set

$$S = \{v \in V : T_t v = v \text{ for some } t \in [0,1]\}$$

is bounded in V. For then if R is sufficiently large

$$\mathrm{ind}_V(T_1, B_R) = \mathrm{ind}_V(T_0, B_R)$$

($B_R = \{v \in V : |v| < R\}$). This follows from the homotopy invariance of the fixed point index of k-set contractions ($k < 1$) (it should be noted that the map $(t,v) \to T_t v$ of $[0,1] \times V \to V$ is a 'permissible' homotopy of k-set contractions). But $\mathrm{ind}_V(T_0, B_R) \neq 0$ since $J - T_0 : V \to V$ is a homeomorphism (T_0 being a strict-contraction by (H1)).

Thus we conclude that if S is bounded then T_1 has a fixed point.

(1.2) *REMARK*

If we replaced (H4) by the hypothesis that $I - T_t$ is an A-proper homotopy with respect to an approximation scheme for V (see [6] for definitions) then the same conclusion would result (this would require using A-proper degree theory). This in turn would allow slightly more general assumptions in our application but we choose to use k-set contraction theory for simplicity.

In order to establish the boundedness of S we need further hypotheses. For instance

(H5) $N(u) = o(||u||)$ as $||u|| \to \infty$

or

(H5') V is a Hilbert space and A and N satisfy the following

(i) $(Au, Cu) \geq 0$ for all $u \in D$

(ii) $(Nu, Cu) \geq K_1|Cu| + K_2$ for all $u \in D$

(iii) $|Nu| \le K_3|Cu| + K_4$ for all $u \in D$

where K_1, K_2, K_3, K_4 are constants and $(\cdot,\cdot)$ is the inner product on V.

(1.3) *THEOREM*

Let $L : U \to V$ and $N : U \to V$ be bounded mappings satisfying (H1), (H2), (H3) and (H4). Moreover, suppose L and N satisfy either (H5) or (H5'). Then for each $\lambda > 0$ there exists $u_\lambda \in D$ such that

$$Au_\lambda + \lambda Cu_\lambda + Nu_\lambda = f.$$

Proof

It is sufficient to prove S is bounded. The proof in the case (H5) is satisfied is trivial so we consider the case (H5') is satisfied. Assume (H5') and suppose $v \in S$ so $T_t v = v$ for some $t \in [0,1]$. Let $u = Bv$. Then

$$Au + \lambda Cu + tNu = tf.$$

Taking the inner product with Cu gives

$$(Au,Cu) + \lambda|Cu|^2 + t(Nu,Cu) = (tf,Cu).$$

Hence

$$\lambda|Cu|^2 \le (t||f|| - tK_1)|Cu| - tK_2.$$

So $|Cu|$ is bounded for $v \in S$. It follows from (H2) and (H5'(iii)) that $|v|$ is bounded for $v \in S$. ■

(1.4) *COROLLARY*

Suppose (λ_k) is a sequence of positive real numbers such that $\lambda_k \to 0$ as $k \to \infty$. If a corresponding sequence (u_k) of solutions of (1.2) is bounded in U, then there exists $u \in D$ such that

$$Au + Nu = f.$$

Proof

Put $v_k = (A+C)u_k$. So (v_k) is a bounded sequence in V and

$$v_k - (1-\lambda_k)C(A+C)^{-1}v_k + N(A+C)^{-1}v_k = f.$$

Since $C(A+C)^{-1}$ is compact it follows there is a subsequence (v_{k_j}) such that

$$\gamma((v_{k_j})) = \gamma((N(A+C)^{-1}v_{k_j})) \leq k\gamma((v_{k_j})).$$

So we may assume, passing to a subsequence if necessary, that $v_{k_j} \to v$ as $j \to \infty$. Thus by continuity,

$$v - C(A+C)^{-1}v + N(A+C)^{-1}v = f.$$

Putting $u = (A+C)^{-1}v$ we get $Au + Nu = f$. ■

2. APPLICATION

Let Ω be a bounded open subset of R^n with smooth boundary Γ. We consider the non-linear boundary value problem

$$\begin{aligned} -\Delta u + g(x,\nabla u,\Delta u) + h(x,u) &= f \quad \text{on } \Omega \\ -\partial u/\partial n &\in \beta(u) \quad \text{on } \Gamma. \end{aligned} \tag{2.1}$$

We make the following assumptions on g, h and β

(A1) β is a maximal monotone graph in R^2 with $0 \in \beta(0)$.

(A2) $g : \Omega \times R^n \times R \to R$ is continuous and satisfies

(i) if $(x,u,s) \in \Omega \times R^n \times R$ then g is continuous in u uniformly with respect to $(x,s) \in \Omega \times R$;

(ii) there is a constant $k \in (0,1)$ such that

$$|g(x,u,s_1) - g(x,u,s_2)| \leq k|s_1 - s_2| \quad \text{for } (x,u) \in R \times R^m,\ s_1,s_2 \in R$$

(iii) there exists $\varphi \in L^2(\Omega)$ such that

$$|g(x,u,s)| \le \varphi(x) \quad \text{for} \quad (x,u,s) \in \Omega \times R^m \times R.$$

(A3) $h : \Omega \times R \to R$ is continuous and satisfies

(i) there is $a \in L^2(\Omega)$ and a constant b such that

$$|h(x,u)| \le a(x) + b|u| \quad \text{for} \quad (x,u) \in \Omega \times R$$

(ii) there is a $T \in L^2(\Omega)$ such that

$$h(x, u)u \ge 0 \quad \text{for} \quad |u| \ge T(x)$$

for almost all $x \in \Omega$.

Let $U = W^{2,2}(\Omega)$, $W = W^{1,2}(\Omega)$ and $V = L^2(\Omega)$ (all the spaces being equipped with usual norms). Let E_1, E_2 denote the embeddings of U into W and W into V. Put $C = E_2 \circ E_1$. Since Γ is assumed smooth the embedding theorems give that E_1 and E_2 are compact. We consider Δ and ∇ as bounded linear mappings from U into V and W into V respectively. Let $L = -\Delta$ and

$$D = \{u \in W^{2,2}(\Omega) : -\partial u/\partial n \in \beta(u) \quad \text{a.e. on } \Gamma\}$$

(the trace theorems show D is well-defined). It is well known (see p. 63 [1]) that $C \circ (A+C)^{-1}$ is a non-expansion and compact (if one considers $A = L|_D$ as an operator in V then it is maximal monotone) and also that there is a constant K such that

$$||u|| < K(|Au + Cu| + 1) \quad \text{for} \quad u \in D. \tag{2.2}$$

Thus A satisfies (H1), (H2) and (H3).

Define $N : U \to V$ by

$$Nu(x) = g(x,\nabla u(x),\Delta u(x)) + h(x,u(x)) \quad \text{for a.a.} \quad x \in \Omega.$$

It is well known that N is well defined bounded and continuous $N \circ (A+C)^{-1}$ is also continuous. [It should be remarked at this stage that a simple argument shows that $E_1 \circ (A+C)^{-1} : V \to W$ is continuous (in fact it is a compact mapping), however, it is not clear that $(A+C)^{-1} : V \to U$ is continuous. This difficulty is overcome by noticing that $\Delta(A+C)^{-1}$ is continuous because $\Delta(A+C)^{-1}v = -v + C(A+C)^{-1}v$]. It is now routine to check that $N \circ (A+C)^{-1}$ is a k-set contraction (same k as in ((A2) (ii)). For further details we refer the reader to [3] where similar assertions are proved. Thus A and N satisfy (H4).

Again using the assumptions on g and h and the fact that A is monotone (when considered as an operator in V) it is almost trivial to check that hypothesis (H5') is satisfied.

We are now in a position to apply Theorem (1.3). We deduce that for each $\lambda > 0$ there exists $u_\lambda \in W^{2,2}(\Omega)$ such that

$$\begin{aligned} -\Delta u_\lambda + \lambda u_\lambda + g(x,\nabla u_\lambda,\Delta u_\lambda) + h(x,u_\lambda) &= f \quad \text{a.e. on } \Omega \\ -\partial u_\lambda/\partial n &\in \beta(u_\lambda) \quad \text{a.e. on } \Gamma. \end{aligned} \tag{2.3}$$

In order to establish the existence of a solution of (2.1) we must show that if $\lambda_k \to 0$ with $\lambda_k > 0$ and (u_k) a corresponding sequence of solutions of (2.3), then (u_k) is bounded. This will follow from the next theorem. But first we need some more notation.

For $t \in R$ let $\beta^0(t)$ be the element in $\beta(t)$ of least absolute value if $\beta(t) \neq \phi$ and put $\beta^0(t) = \pm\infty$ if $\beta(t) = \phi$ (and $t > 0$, $t < 0$ respectively). Put $\beta_\pm = \lim_{t\to\pm\infty} \beta^0(t)$ (in the extended sense). Further let

$$h_+(x) = \liminf_{t\to\infty} h(x,t) \quad\text{and}\quad h_-(x) = \limsup_{t\to-\infty} h(x,t).$$

(2.1) *THEOREM*

Let (λ_k) be a sequence of positive real numbers such that $\lambda_k \to 0$ and let (u_k) be a corresponding sequence of solutions of (2.3). Moreover, suppose that

$$\int_\Gamma \beta_- d\gamma + \int_\Omega (h_-(x) + \varphi(x))dx < \int_\Omega f(x)dx < \int_\Gamma \beta_+ d\gamma + \int_\Omega (h_+(x) - \varphi(x))dx. \tag{2.4}$$

Then the sequence (u_k) is bounded. Consequently, there is a $u \in W^{2,2}(\Omega)$ satisfying (2.1). ('$d\sigma$' denotes Lebesgue measure on Γ.)

Proof

Suppose not, then going to a subsequence if necessary, we may assume $||u_k|| \to \infty$. Using the estimate (2.2) we see

$$||u_k|| \le K(|(1-\lambda_k)u_k - g(x,\nabla u_k,\Delta u_k) - h(x,u_k) + f| + 1).$$

By our assumptions on g and h there are constants α_1 and α_2 such that

$$||u_k|| < \alpha_1(|u_k| + \alpha_2)$$

(note at this stage we are not distinguishing between u_k, $E_1 u_k$, Cu_k, etc.). Thus $(u_k/|u_k|)$ is a bounded sequence in U. Put $v_k = u_k/|u_k|$. By the compactness of the embeddings, we may assume, going to a subsequence if necessary, that

$v_k \rightharpoonup v$ in U ('$\rightharpoonup$' means weak convergence)

$v_k \to v$ in W

and $v_k \to v$ in V.

Also by the trace theorems $v_k|_\Gamma \to v|_\Gamma$ in $L^2(\Gamma)$ and $\partial v_k/\partial n \rightharpoonup \partial v/\partial n$ in $L^2(\Gamma)$. Multiply (2.3) by u_k and integrate. Then applying Green's theorem we get

$$\int_\Omega \nabla u_k^2 dx \le -\int_\Gamma \beta(u_k)u_k d\sigma - \int_\Omega h(u_k)u_k dx - \int_\Omega g(x,\nabla u_k,\Delta u_k)u_k dx + \int_\Omega f u_k dx \,. \tag{2.5}$$

Dividing by $|u_k|^2$ and letting $k \to \infty$ we see that

$$\int_\Omega \nabla v_k^2 dx \to 0$$

and so $\int_\Omega \nabla v^2 dx = 0$. Thus v = constant. This constant is non-zero since $|v| = 1$. Returning to (2.5) we have

$$\int_\Gamma \beta(u_k)u_k d\sigma + \int_\Omega g(x,\nabla u_k,\Delta u_k)u_k dx + \int_\Omega h(x,u_k)u_k dx \le \int_\Omega f u_k dx.$$

Divide by $|u_k|$ to get

$$\int_\Omega f v_k dx \ge \int_\Gamma \beta(u_k)v_k d\sigma + \int_\Omega g(x,\nabla u_k,\Delta u_k)v_k dx + \int_\Omega h(x,u_k)v_k dx \,.$$

Thus going to the limit as $k \to \infty$

$$\int_\Omega f v dx \geq \underline{\lim}\{\int_\Gamma \beta(u_k)v_k d\sigma + \int_\Omega g(x,\nabla u_k,\Delta u_k)v_k dx + \int_\Omega h(x,u_k)v_k dx\}$$

$$\geq \begin{cases} \int_\Gamma \beta_+ v d\sigma - \int_\Omega \varphi|v|dx + \int_\Omega h_+ v dx, & \text{if } v > 0 \\ \int_\Omega \beta_- v d\sigma - \int_\Omega \varphi|v|dx + \int_\Omega h_- v dx, & \text{if } v < 0 \end{cases}$$

(by Fatou's lemma, note $u_k \to \pm\infty$ pointwise a.e. on Ω (depending on whether $v>0$ or $v<0$); or at least some subsequence does).

Cancelling the constant v gives the required contradiction and the theorem is proved. ■

In the case $g \equiv 0$ the condition (2.4) reduces to the condition of Theorem 1 p. 24 [2]. That theorem applies to problems with more general boundary conditions than those in this paper also to solutions in L^p-spaces. However, at least in the case $p = 2$ and with our restricted boundary condition we have shown that we can allow non-linearities with some dependency on Δu and ∇u.

In the case $\beta \equiv 0$ our theorem generalizes theorem (3.7) p. 176 [3]. In [5] Petryshyn considers (2.1) with $\beta \equiv 0$ but with a more general non-linearity g. Using the theory of A-proper mapping we could treat his type of non-linearity using our method.

REFERENCES

[1] BARBU, V.: Non-linear semigroups and differential equations in Banach spaces, Noordhoff (Leyden) (1976).

[2] CALVERT, B.D. and GUPTA, C.P.: Non-linear elliptic boundary value problems in L^p-spaces and sums of ranges of accretive operators, Non-lin Anal. 2 (1978), 1-26.

[3] FITZPATRICK, P.M.: Existence results for equations involving non-compact perturbations of Fredholm mappings, J. Math. Anal. and Appl. 66 (1978), 151-177.

[4] NUSSBAUM, R.D.: The fixed point index for local condensing mappings, Ann. Mat. Pura Appl. 89 (1971), 217-258.

[5] PETRYSHYN, W.V.: Existence theorems for semilinear abstract and differential equations with non-invertible linear parts and non-compact perturbations, Proceedings of the Symposia on Non-linear equations in Abstract spaces, Academic Press (1978).

[6] PETRYSHYN, W.V.: The approximation solvability of equations involving A-proper and pseudo A-proper mappings, Bull. Amer. Math. Soc. 81 (1975), 223-312.

NONEXPANSIVE MAPPINGS WITH PRECOMPACT ORBITS

By

WILLIAM O. RAY
Department of Mathematics
Iowa State University
Ames, Iowa 50010

AND

ROBERT C. SINE
Department of Mathematics
University of Rhode Island
Kingston, Rhode Island 02881

§0.

Since the results of Browder [3], Göhde [10], and Kirk [13] of 1965 there has been great interest in determining when a nonexpansive map T of a closed bounded nonempty convex set K into itself has a fixed point. Counterexamples have been known for some time. Lim [17] recently produced a weak* compact counterexample in a renormed ℓ_1 and even more recently Alspach [1] has given a weak compact counterexample in L_1. Most positive results involve assumption of nice geometric properties of the ambient Banach space X such as uniform convexity, Opial's condition, normal structure, or asymptotic normal structure [19]. Here we consider spaces C(E) where E is compact Hausdorff and Stonian (= extremally disconnected). While such spaces are far from the previously mentioned special classes of Banach space they do have other geometric properties (existence of Chebyshev centers and extension properties) which we exploit here to obtain certain ideal fixed points (which have the virtue of always existing). This existence will be established in Section 1. Then in Section 2 we apply the results and methods to show the fixed point set is a nonexpansive retract and to prove that Krasnosel'skiǐ averaging converges in an arbitrary space. We then extablish a common fixed point result for Lipschitzian semigroups with precompact orbits in certain settings.

§1.

Let K be a nonempty convex closed bounded set in X and T a nonexpansive map of K into itself. In this section we show the existence of ideal fixed points which are fixed under certain extensions of T.

PROPOSITION A

Let X be an arbitrary Banach space. Then there is a compact Hausdorff Stonian space E so that X is linearly isometric to a subspace of $C(E)$.

REMARKS

This extension of Banach's classical universality theorem can be established in essentially the same way. The fact that any compact Hausdorff space is the continuous image of a Stonian space is needed and can be put together from results in [16, p. 41]. We note that if X is a dual space the embedding can be arranged so that $C(E)$ is also a dual space and the embedding map is weak* continuous. The easy technique for this last refinement was suggested by the second part of the example in [11, p. 225]. Thus $C(E)$ contains both Lim's and Alspach's counterexamples.

PROPOSITION B

Let E be Stonian and A a subset of $C(E)$. Suppose $T : A \to A$ is nonexpansive. Then for any order interval J which contains A there is a nonexpansive extension $\hat{T}$ of T which maps J into J.

REMARK

This proposition is an immediate corollary of a reworking by Wells and Williams [22] of a theorem of Aronszajn and Panitchpakdi [2].

PROPOSITION C

Let J be a nonempty closed bounded order interval in $C(E)$. If $T : J \to J$ is nonexpansive then T has a fixed point in J.

REMARK

This result was obtained independently by P. Soardi [21] and by the second author [20].

PROPOSITION D

Let A be a bounded set in $C(E)$ with E Stonian. Then there is a central point z in $C(E)$ so that $d(f,z) = (1/2)\mathrm{dia}(A)$ for all f in A. Moreover z can be taken to be $(1/2)(\overline{f} + \underline{f})$ where $\overline{f} = \vee(A)$ and $\underline{f} = \wedge(A)$.

REMARK

This result is a trivial consequence of the Binary Ball Intersection property in $C(E)$. If E is only compact Hausdorff but A is compact the result still holds. We will need this last case as well which apparently was first observed by Lorentz [18] in the case E is compact metric (an easy topological quotient argument reduces the compact Hausdorff case to the compact metric case).

If we now combine all of these results we see for any convex closed bounded nonempty set K in X and any nonexpansive map T of K that there is a fixed point p for an extension of the linear isometric copy of (X, K, T) in some $C(E)$.

§2.

Krasnosel'skiĭ [15] established his result for uniformly convex spaces in 1955. It was extended to strictly convex spaces by Edelstein [8] in 1966. The result as stated below for arbitrary spaces came as a corollary of work on asymptotic regularity both by Ishikawa [12] and by Edelstein and O'Brien [9]. The arguments of both of these papers are considerably less geometric than that presented here.

THEOREM 1

Let K be a nonempty closed convex set and C a compact subset of K. Suppose T is a nonexpansive map of K into C. If $S \equiv (1/2)(I + T)$ then for each x in K the iterates $\{S^n x\}$ converge to a fixed point of T.

Proof

For a fixed x in K let $K_0 = \overline{\mathrm{co}}\{x, C\}$ to get a compact invariant set. It is clear that $\mathrm{Fix}(S) = \mathrm{Fix}(T)$ and $\{S^n x\}$ has a nonempty compact ω-limit set G. We need only show G is a singleton. So pick v in G and let A be the norm closure of $\{S^n v : n \geq 0\} \cup \{TS^n v : n \geq 0\}$. With A regarded as a subset of $C(E)$ we let J be the minimal order interval over A and $\hat{T}$ be the nonexpansive extension of T to J. Let p be a fixed point of $\hat{T}$ in J. If v itself is fixed we are done. If not we can assume without loss of generality the p is the zero function and $\mathrm{dist}(p,v) = 1$. Let $M(0) = \{t \text{ in } E : |v(t)| = 1\}$ and in general $M(n+1) = \{t \text{ in } E : |S^{n+1}v(t)| = 1\}$. It is easy to show that $M(n+1)$ is a nonempty compact subset of $M(n)$. For a point t_0 in $\cap M(n)$ we have

$$TS^n v(t_0) = -S^n v(t_0) = v(t_0).$$

Now $\Lambda(A)$ and $V(A)$ satisfy the same condition at t_0 (since A is compact these lattice extrema are pointwise limits). But p is itself a point in the lattice hull so we would have $p(t_0) = v(t_0)$ thus $0 = |p(t_0)| = 1$ giving a contradiction which finishes the proof. ■

THEOREM 2

Let T be a nonexpansive map of a nonempty closed bounded nonempty order interval J in $C(E)$ where E is Stonian. Then $\mathrm{Fix}(T)$ is a (nonempty) nonexpansive retract of J.

Proof

One shows that $\mathrm{Fix}(T)$ is metrically convex and has the Binary Ball Intersection Property quite easily from Proposition C. These facts together with the machinery that gave us Proposition B imply that the identity map of $\mathrm{Fix}(T)$ has a nonexpansive extension to a map π from J into $\mathrm{Fix}(T)$. This map is the required retraction map. ■

REMARK

The fixed point set of a nonexpansive map need not be convex [7] nor a nonexpansive retract [5, Example 1] in general. Bruck [4] has shown that the fixed point set is a nonexpansive retract whenever a conditional fixed point property holds. But the conditional fixed point property does not hold in $C(E)$ for we need only embed Alspach's example in $C(E)$ and extend the map to the order hull of the embedding.

COROLLARY

Let J be a nonempty closed bounded order interval in $C(E)$ where E is Stonian. Then any countable abelian family of nonexpansive maps have a nonempty common fixed point set (which is a nonexpansive retract of J).

Proof

If T_1 and T_2 commute then T_2 maps $Fix(T_1)$ into itself. But $Fix(T_1)$ is a nonexpansive retract of J, a set with the fixed point property for nonexpansive maps. Thus T_2 has a fixed point in $Fix(T_1)$. Continuing in this fashion we obtain a descending sequence $H_1 \supset H_2 \supset \dots$ where the points of H_n are the common fixed points of $\{T_1, T_2, \dots, T_n\}$. Next we use a key lemma of Bruck [5, p. 61] to claim that $\cap H_n$ is $Fix(S)$ for some nonexpansive map S of J into itself. Since $Fix(S)$ is nonempty by Proposition C we are done. ■

Let Φ be a one-parameter semigroup (defined either on Z_+ or R_+) mapping K into K. We call Φ (uniformly) γ-Lipschitzian if for all x and y in K

$$||\varphi_t(x) - \varphi_t(y)|| \le \gamma ||x - y||.$$

THEOREM 3

Let Y be a compact Hausdorff space and J a nonempty closed bounded order interval in $C(Y)$. Suppose Φ is γ-Lipschitzian with precompact orbits defined on J. If $\gamma < \sqrt{2}$ then Φ has a nonempty common fixed point set.

Proof

For each x and y in J we define

$$\rho(x,y) = \lim_{t\to\infty} \sup_{s\ge t} ||\phi_s(x) - y||$$

and $d(x) = \rho(x,x)$.

Now for a fixed x in J and u in Y set

$$A(u) = \lim_{t\to\infty} \sup \{\varphi_s(x)(u) : s \geq t\}.$$

The Ascoli-Arzela theorem together with the fact that x has a nonempty compact ω-limit set can be used to show $A(u)$ is continuous. We do the same for the function B defined with infima in place of suprema. Set $2r = ||A - B||$ and $z = (1/2)(A + B)$. Then it can be shown that

$$\rho(x, z(x)) \leq r \tag{1}$$

$$\gamma d(x) \geq 2r \tag{2}$$

and

$$d(z(x)) \leq \gamma\rho(x, z(x)). \tag{3}$$

Combination of estimates (1), (2), and (3) yields

$$d(z(x)) \leq (1/2)\gamma^2 \, d(x).$$

Now we define the function ψ by

$$\psi(x) = (2+\gamma)(2-\gamma^2)^{-1} \, d(x).$$

This function is clearly continuous and satisfies

$$\psi(x) - \psi(z(x)) \geq ||x - z(x)||.$$

Thus Caristi's fixed point theorem [6] can be applied to the mapping z to conclude that z has a fixed point w in J. Since $\gamma < \sqrt{2}$ we see $d(w) = 0$ so $w = \lim \varphi_t w$. But $\varphi_s(w) = \lim_{t\to\infty} \varphi_s\varphi_t(w) = w$ so w is fixed under each φ_s in Φ. ■

REMARK

For Φ a semigroup generated by a single map and $\gamma = 1$ this was obtained by the second author [20]. The result also generalizes a theorem of Kirk and Torrejon [14] who assumed an symptotic nonexpansivity condition.

REFERENCES

[1] ALSPACH, D.E.: A fixed point free nonexpansive map, preprint.

[2] ARONSZAJN, N. and PANITCHPAKDI, P.: Extensions of uniformly continuous transformations and hyperconvex metric spaces, Pac. J. Math. 6 (1956), 405-439.

[3] BROWDER, F.: Nonexpansive nonlinear operators in a Banach space, Proc. Nat. Acad. Sci. 54 (1965), 1041-1044.

[4] BRUCK, R.E., Jr.: Properties of fixed point sets of nonexpansive mappings in Banach spaces. Trans. Amer. Math. Soc., 179 (1973), 251-262.

[5] BRUCK, R.E., Jr.: A common fixed theorem for a commuting family of nonexpansive mappings. Pac. J. Math., 53 (1974), 59-71.

[6] CARISTI, J.V.: Fixed point theorems for mappings satisfying inwardness conditions, Trans. Amer. Math. Soc. 215 (1976), 241-251.

[7] DeMARR, R.: Common fixed points for commuting mappings, Pac. J. Math. 13 (1963), 1139-1141.

[8] EDELSTEIN, M.: A remark on a theorem of Krasnosel'skiǐ, Amer. Math. Monthly, 73 (1966), 509-510.

[9] EDELSTEIN, M. and O'BRIEN, R.C.: Nonexpansive mappings, asymptotic regularity and successive approximations, to appear in J. London Math. Soc.

[10] GÖHDE, D.: Zum Prinzip der kontractiven Abbildung, Math. Nachr., 30 (1965), 251-258.

[11] HOLMES, R.B.: Geometric Functional Analysis and its Applications, Springer Verlag, (1975).

[12] ISHIKAWA, S.: Fixed points and iteration of a nonexpansive mapping in a Banach space, Proc. Amer. Math. Soc., 59 (1976), 65-71.

[13] KIRK, W.A.: A fixed point theorem for mappings which do not increase distance, Amer. Math. Monthly 72 (1965), 1004-1006.

[14] KIRK, W.A. and TORREJON, R.: Asymptotically nonexpansive semigroups in Banach spaces. J. Nonlinear Anal., Theory Meth. and Appl. 3 (1979), 111-121.

[15] KRASNOSEL'SKIǏ, M.A.: Two remarks on the method of successive approximation, Uspehi Mat. Nauk 10 (1955) No. 1 (63) 123-127.

[16] LACEY, H.E.: The Isometric Theory of Classical Banach Spaces, Springer-Verlag, (1974).

[17] LIM, T.-C.: Asymptotic centers and nonexpansive mappings in some conjugate Banach spaces, preprint.

[18] LORENTZ, G.G.: Approximation of Functions, Holt, Rinehart and Winston, (1966).

[19] SCHÖNEBERG, R.: Asymptotic normal structure and fixed points of nonexpansive mappings, preprint.

[20] SINE, R.: On nonlinear contraction semigroups in sup norm spaces, J. Nonlinear Anal., Theory, Meth. and Appl., 3 (1979), 885-890.

[21] SOARDI, P.: Existence of fixed points of nonexpansive mappings in certain Banach lattices. Proc. Amer. Math. Soc., 73 (1979), 25-29.

[22] WELLS, J.H. and WILLIAMS, L.R.: Embeddings and Extensions in Analysis, Springer-Verlag, (1975).

FIXED POINT SETS OF CONTINUOUS SELFMAPS

BY

HELGA SCHIRMER*

Department of Mathematics
Carleton University
Ottawa, Ontario, Canada

1. THE COMPLETE INVARIANCE PROPERTY

In 1967 H. Robbins [16] investigated the set of fixed points of a continuous function or a homeomorphism of a closed n-ball B^n. It turned out that it is not easy to find necessary and sufficient conditions for a subset A of B^n so that A can be the fixed point set of a homeomorphism and the problem is still not completely solved (see these proceedings, Problem # 11). However, for B^n, and indeed for a rather general class of spaces X which we will describe, the problem for a continuous function has a simple solution. Namely, any non-empty closed subset A of X is realizable as the fixed point set of a self map $f: X \to X$. (The case of A being empty is excluded because X may have the fixed point property.) In other words, using the following definition such spaces X have the "complete invariance property".

DEFINITION 1 (Ward [23])

A topological space X has the complete invariance property (CIP) if every closed and nonempty subset A of X is the fixed point set of a continuous selfmap of X.

During the last decade several papers have appeared which deal with the CIP, giving it the features of a less important but healthy younger brother of the fixed point property. As in the case of the fixed point property, research has been mainly concerned with two topics:

I Which spaces have the CIP?

---

* **The research for this article was supported in part by NSERC Grant A 7579.**

II How does the CIP behave with regard to geometric constructions?

While the problems are similar, the answers are not. The fixed point property is comparatively rare, but spaces in several large and important classes (including manifolds and polyhedra) have the CIP. Hence it may be surprising that its behavior under geometric constructions is even more pathological than that of the fixed point property. None of the operations investigated so far preserve the CIP, not even those of taking wedges and of retraction which preserve the fixed point property. This phenomenon is related to the anomalies which occur in continua, especially to the failure of higher dimensional Peano continua to have the CIP.

This paper is mainly expository. Its only new feature is the emphasis on path fields as the most useful tool known so far in establishing the CIP. As in the case of the fixed point property no method of proof has been found which covers all existing results.

I would like to thank John Martin for some helpful discussions.

2. PATH FIELDS

Robbin's proof that a ball has the CIP is in essence based on the fact that a ball admits a vector field with one singularity. The concept of a field of vectors generalizes to that of a field of paths which never return to their initial point.

DEFINITION 2

Let X be a topological space, I the unit interval, and X^I the path space of X with the compact-open topology. A *path field* on X is a map $\beta : X \to X^I$ so that $\beta(x)$ is either a path $p : I \to X$ with $p(0) = x$ and $p(t) \neq x$ for $0 < t \leq 1$, or the constant path at x. If $\beta(x)$ is the constant path at x, then x is called a *singularity* of β.

The existence of path fields on X can imply the CIP of X.

THEOREM 1

Let (X,d) be a metric space. If for every $a \in X$ there exists a path field on X which has no singularities on $X - \{a\}$, then X has the CIP.

Proof

We can assume that d is a bounded metric with $d < 1$. If A is a given closed and nonempty subset of X, choose $a \in A$ and a path field $\beta : X \to X^I$ which has no singularities on $X - \{a\}$. Then define a map $f : X \to X$ with fixed point set A by

$$f(x) = \beta(x)(t), \quad \text{where} \quad t = d(x,A). \blacksquare$$

Note that it is immaterial whether the path field β actually has a singularity at the point a or not. Hence all metric spaces which admit a path field without singularities have the CIP. Theorem 1 is related to Theorem 1 in [23], which has been called "Ward's Lemma" and has sometimes been applied to solve problems concerning the CIP. Ward's Lemma uses a homotopy between a selfmap and the identity, and clearly a path field β establishes a homotopy between the map which assigns to every $x \in X$ the endpoint of $\beta(x)$ and the identity map. Path fields have an advantage over Ward's Lemma as they are geometrically concrete, and as their existence has been studied by R.F. Brown and E. Fadell on topological manifolds [4], [6] and on certain polyhedra [7].

A path field on X is independent of a metric, but Theorem 1 requires that X is endowed with one. This requirement is typical, as all existing proofs of results which establish the CIP on a space X assume that X is metrizable, even if path fields are not used in the proof. An example by L.E. Ward, Jr. [23] p. 555 shows that the CIP does not readily extend to non-metric settings, as $I \times I$ in the dictionary order topology is a "long" but non-metrizable interval which does not have the CIP. No example of a non-metrizable Hausdorff space with the CIP has yet been described.

3. Spaces with the Complete Invariance Property

(i) Convex sets

Theorem 1 implies at once that a ball has the CIP. Ward observed a generalization to linear spaces.

THEOREM 2 (*L.E. Ward, Jr.* [23] *Corollary* (1.1)).

Let X be a convex subset of a normed linear space. Then X has the CIP.

(ii) TOPOLOGICAL MANIFOLDS

The following simple theorem is sometimes useful.

THEOREM 3

If every component of a locally connected space X has the CIP, then X has the CIP.

Proof

A map f with the given closed and nonempty fixed point set A can be obtained by constructing f on each component of X which intersects A, and mapping the other components to a point in A. ■

Hence it is only necessary to consider connected manifolds. The existence of path fields on connected topological manifolds without boundary was used by H. Schirmer [18], Theorem 4 to establish the CIP for such spaces. The proof of Theorem (4.2) in [6] shows that even a connected and compact topological n-manifolds with boundary admits a path field with at most one singularity which lies in its interior, and the homogeneity of manifolds ensures that this singularity can be moved to any given interior point. It is easy to see that the singularity can also be moved to any boundary point of the manifold.

THEOREM 4

Let X be a compact topological n-manifold with or without boundary. Then X has the CIP.

(iii) POLYHEDRA

Path fields on polyhedra are generated by maps which are small deformations of the identity. We denote by $|\sigma|$ an open simplex of the polyhedron $|K|$, by $|\bar{\sigma}|$ its closure, and by $|\varkappa(x)|$ the carrier of the point $x \in |K|$. A map $f : |K| \to |K|$ is called a *proximity map* [5] p. 124, if $|\bar{\varkappa}(x)| \cap |\bar{\varkappa}(f(x))| \neq \phi$ for all $x \in |K|$. The track of a point $x \in |K|$ under a homotopy between the identity and a proximity map f can be chosen as a broken line segment, and these segments will determine a path field on $|K|$ whose singularities are precisely the fixed points of f. (See [20], Lemma (1.1), [5] p. 124, or [7] Lemma (2.1)). It was first shown by F. Wecken [24], Satz 1, that proximity maps with at most one fixed point exist on all finite polyhedra which satisfy a connectedness condition modelled on manifolds of dimension ≥ 2. Such polyhedra are called *2-dimensionally connected*, and are defined by the property that every maximal simplex has dimension ≥ 2, and that for every two maximal simplexes σ and σ' of K there exists a sequence $\sigma_0 = \sigma, \sigma_1, \ldots, \sigma_r = \sigma'$ so that $|\sigma_i| \cap |\sigma_{i+1}|$ has dimension ≥ 1 for $i = 0, 1, \ldots, r-1$. (They have also been called *polyhedra of type* W, or said to satisfy the *Wecken condition*.) Hence 2-dimensionally connected finite polyhedra admit a path field with at most one singularity (see [7], Observation (3.2)). In order to apply Theorem 1 it is still necessary to move this singularity to an arbitrary point $a \in |K|$; this is a technical detail which can be found in [19] p. 223.

Lately Shi [22] showed that proximity maps without fixed points exist on all polyhedra which are infinite but locally finite, 2-dimensionally connected and have the weak topology, hence such polyhedra have the CIP also.

Finally Boju Jiang and H. Schirmer [9] used Shi's work [21] on selfmaps of polyhedra which are homotopic to the identity to obtain maps with a given fixed point set on polyhedra which are not necessarily 2-dimensionally connected. These maps are no longer deformations, but their construction makes use of the existence of path fields on 2-dimensionally connected subpolyhedra. Here is their result.

THEOREM 5 *(Boju Jiang and H. Schirmer* [9]*).*

Let $|K|$ be a locally finite simplicial complex. Then $|K|$ has the CIP.

A simplicial complex with the weak topology is locally finite if and only if it is metrizable, therefore a positive answer can be expected to

QUESTION 1

Does there exist an infinite simplicial complex with the weak topology which does not have the CIP?

The next question is of greater interest, but is difficult to attack at present, even for the class of ANR (metric).

QUESTION 2

Do all ANR's have the CIP?

(iv) TOPOLOGICAL GROUPS

A. Gleason [8], Theorem 1, proved that every locally compact and not totally disconnected topological group $(G, \cdot)$ contains an arc $q : I \to G$. After translating $q(0)$ to the identity element of G one can define a path field without singularities $\beta : G \to G^I$ by $\beta(g)(t) = g \cdot q(t)$. Hence such a group has the CIP if it is metrizable. This result extends to a totally disconnected group, as a purely topological argument [11], p. 1028 shows that each coset gH which meets a closed and nonempty subset A of the group admits a retraction onto $gH \cap A$. The construction of a selfmap of the group with fixed point set A is then analogous to the one given in the proof of Theorem 3. Note that the map is not a group morphism.

THEOREM 6 (*J.R. Martin and S.B. Nadler, Jr.* [11], *Theorem* (5.4)).

Let X be a locally compact metrizable topological group. Then X has the CIP.

Similar to Question 1 is

QUESTION 3

Does there exist a topological group which is either not metrizable or not locally compact, and which does not have the CIP?

(v) CONTINUA

Several papers have appeared which investigate the CIP for *Peano continua*, that is for compact, metric, connected and locally connected spaces. One special class of Peano continua admits path fields with one singularity. These are the *dendrites* (or acyclic curves), which are Peano continua which contain no simple closed curve. Every two points x,y of a dendrite D are connected by a unique arc $[x,y]$ with x and y as its endpoints [25], p. 89. As Bing [1], Theorem 6 has shown that every Peano continuum has a convex metric [15], p. 38, we can endow D with a metric d which is convex and bounded. Then there exists for every $x,y \in D$ and $t \in I$ a unique point $z \in [x,y]$ with $d(x,z) = t\,d(x,y)$, and we can define, for any $a \in D$, a path field $\beta : D \to D^I$ with a as its only singularity by $\beta(x)(t) = z$, where $z \in [x,a]$ and $d(x,z) = t\,d(x,a)$. Hence Theorem 1 shows that a dendrite has the CIP [17], Theorem (3.1).

J.R. Martin and E.D. Tymchatyn [14] used a decreasing sequence of partitionings [1], p. 545, of a 1-dimensional Peano continuum X to construct a sequence $B_1 \subset B_2 \subset B_3 \subset \ldots$, where each B_k is a finite acyclic graph, so that B_k reaches with increasing k towards all points of a given closed and nonempty subset A of X. There exists a retraction $r : X \to B$, where $B = A \cup \bigcup_{k=1}^{\infty} B_k$. As $B - A$ is essentially acyclic, B admits a path field with A as its set of singularities. This path field determines a selfmap g of B with fixed point set A, and so a selfmap f of X with fixed point set A is obtained as $f = g \circ r$. This proves the following

THEOREM 7 (*J.R. Martin and E.D. Tymchatyn* [14]).

Let X be a 1-dimensional Peano continuum. Then X has the CIP.

That this result does not extend to higher-dimensional Peano continua was shown by J.R. Martin [10]. He described, for each $n = 1, 2, \ldots,$ an $(n+1)$-dimensional LC^{n-1} [3], p. 30 continuum which has the Čech homology of a point, and which contains an n-sphere which cannot be the fixed point set of a selfmap.

But it is not known whether compactness can be replaced by local compactness. R.L. Wilder [26] p. 76, defines a Peano space as a locally compact, metric, connected and locally connected space. As the CIP for polyhedra has been extended from compact to locally compact ones, we ask

QUESTION 4

Does every 1-dimensional Peano space have the CIP?

More intriguing might be

QUESTION 5

Does every chainable continuum have the CIP? In particular, does the pseudo-arc have the CIP?

It is of interest to note that Peano continua behave quite differently under multivalued maps. Work in progress by J.T. Goodykoontz and S.B. Nadler, Jr. shows for example that every closed and nonempty subset of a Peano continuum of arbitrary dimension can be the fixed point set of a continuum-valued continuous multifunction.

4. BEHAVIOUR OF THE COMPLETE INVARIANCE PROPERTY WITH REGARD TO GEOMETRIC CONSTRUCTIONS.

Finally we deal with topic II. Several papers by J.R. Martin, S.B. Nadler, Jr., L. Oversteegen, E.D. Tymchatyn and L.E. Ward, Jr. [11], [12], [13], [23] contain examples of pathologies. As all polyhedra have the CIP, these examples are not taken from this class of spaces, but from that of continua. The search for nice spaces (for example locally contractible continua) with bad behavior still continues.

(i) PRODUCTS

THEOREM 8

There exists a space X with the CIP such that the product $X \times I$ does not have the CIP.

X can be chosen as a 1-dimensional planar Peano continuum which is constructed from a nullsequence of Hawaian earrings located inside a circle. This construction can be modified to obtain X as an n-dimensional LC^{n-1} $(n > 1)$ continuum, or as an LC^{∞} [3], p. 30 continuum. The proof uses the fact that X contains a sequence of points which are *homotopically stable*, that is which remain fixed under every deformation of X [13], examples (3.5) and (3.6).

(ii) CONES

THEOREM 9

There exists a space Y with the CIP such that the cone C(Y) does not have the CIP.

The simplest candidate for Y is the Cantor set [23], p. 556, [11] p. 1029. The Cantor set is not a continuum, but there also exists an example for Y which is a 1-dimensional planar Peano continuum, an n-dimensional LC^{n-1} continuum $(n > 1)$ or an LC^{∞} continuum. The construction of Y is then similar to that of X, but somewhat trickier [13], examples (4.1) and (4.2).

(iii) WEDGES

THEOREM 10

There exists a space Z with the CIP such that the wedge $Z \vee Z$ does not have the CIP.

As the fixed point property is preserved by wedging [2], Theorem 6, [5], p. 147, this anomaly is surprising. The wedge of two 1-dimensional Peano continua is again a 1-dimensional Peano continuum, therefore Theorem 7 shows that Z cannot be found in this class of spaces. But Z can be chosen as a 2-dimensional Peano continuum, namely as the product of the Hawaiian earring with I [11], example (3.1), or as a 1-dimensional contractible (but not locally connected) planar continuum [12].

(iv) DEFORMATION RETRACTS

The fixed point property is invariant under a retraction [2], Theorem 2, but the CIP is not even invariant under a strong deformation retraction.

THEOREM 11

There exists a strong deformation retract W_o of a space W such that W has the CIP, but W_o has not.

In [11], example (4.3), W is obtained from the disjoint union of a cone over the Cantor set and a cone over a circle with a spiral approaching it from inside by identifying the two vertices, and W_o is the cone over the Cantor set.

Further geometric constructions have not yet been considered, but a negative answer to the final question can be expected.

QUESTION 6

Is the CIP invariant under other geometric constructions, such as the suspension, the product $X \times X$ and the join?

References

[1] BING, R.H.: Partitioning continuous curves, Bull. Amer. Math. Soc. 58 (1952), 536-556.

[2] BING, R.H.: The elusive fixed point property, Amer. Math. Monthly 76 (1969), 119-132.

[3] BORSUK, K.: Theory of Retracts, Polish Scientific Publishers, Warsaw, (1967).

[4] BROWN, R.F.: Path fields on manifolds, Trans. Amer. Math. Soc. 118 (1965), 180-191.

[5] BROWN, R.F.: The Lefschetz Fixed Point Theorem, Scott, Foresman and Co., Glenview, Ill., (1971).

[6] BROWN, R.F. and FADELL, E.: Nonsingular path fields on compact topological manifolds, Proc. Amer. Math. Soc. 16 (6) (1965), 1342-1349.

[7] FADELL, E.: A remark on simple path fields in polyhedra of characteristic zero, Rocky Mount. J. Math. 4 (1974), 65-68.

[8] GLEASON, A.M.: Arcs in locally compact groups, Proc. Nat. Acad. Sci. 36 (1950), 663-667.

[9] JIANG, Boju (Po-chu Chiang) and SCHIRMER, H.: Fixed point sets of continuous

selfmaps of polyhedra, these proceedings.

[10] MARTIN, J.R.: Fixed point sets of Peano continua, Pac. J. Math. 74 (1978), 163-166.

[11] MARTIN, J.R. and NADLER, S.B., Jr.: Examples and questions in the theory of fixed point sets, Can. J. Math. 31 (1979), 1017-1032.

[12] MARTIN, J.R. and NADLER, S.B. Jr.: A note on fixed points sets and wedges, Can. Math. Bull. (to appear).

[13] MARTIN, J.R., OVERSTEEGEN, L.G. and TYMCHATYN, E.D.: Fixed point sets of products and cones, preprint.

[14] MARTIN, J.R. and TYMCHATYN, E.D.: Fixed point sets of 1-dimensional Peano continua, Pac. J. Math. (to appear).

[15] NADLER, S.B., Jr.: Hyperspaces of Sets, Marcel Dekker Inc., New York, (1978).

[16] ROBBINS, H.: Some complements to Brouwer's fixed point theorem, Israel J. Math. 5 (1967), 225-226.

[17] SCHIRMER, H.: Properties of fixed point sets of dendrites, Pac. J. Math. 36 (1971), 795-810.

[18] SCHIRMER, H.: Fixed point sets of homeomorphisms of compact surfaces, Israel J. Math. 10 (1971), 373-378.

[19] SCHIRMER, H.: Fixed point sets of polyhedra, Pac. J. Math. 52 (1974), 221-226.

[20] SHI GEN HUA : On the least number of fixed points and Nielsen numbers, Chinese Math. 8 (1966), 234-243.

[21] SHI GEN HUA : The least number of fixed points of the identity mapping class, Acta Math. Sinica 18 (1975), 192-202.

[22] SHI GEN HUA : On the least number of fixed points for infinite complexes, preprint.

[23] WARD, L.E., Jr.: Fixed point sets, Pac. J. Math. 47 (1973), 553-565.

[24] WECKEN, F.: Fixpunktklassen III, Math. Ann. 118 (1942), 544-577.

[25] WHYBURN, G.T.: Analytic Topology, Amer. Math. Soc., Providence, R.I., (1942).

[26] WILDER, R.: Topology of Manifolds, Amer. Math. Soc., Providence, R.I., (1949).

WHAT IS THE RIGHT ESTIMATE FOR THE LJUSTERNIK-SCHNIRELMANN COVERING PROPERTY?

By

H. STEINLEIN
Mathematisches Institut
der Ludwig-Maximilians-Universität
D 8 München 2,
Theresienstrasse 39
West Germany

In [2,3], we described how a special variant (if it is true) of the Ljusternik-Schnirelmann covering theorem would yield a positive answer to the longstanding question in asymptotic fixed point theory, whether each continuous map on a nonempty closed convex subset K of a normed space with some compact iterate (that is $f^m(K)$ is relatively compact for some $m \in \mathbb{N}$) has a fixed point. To be able to state this variant, we need the notion of the genus (or sectional category, see [1]) in the sense of A. S. Švarc [4,5]:

DEFINITION

Let M be a normal space, p a prime number and $f : M \to M$ a free $\mathbb{Z}_p$-action (that is f is continuous, $f^p = id$ and $f(x) \neq x$ for all $x \in M$). Then the genus $g(M,f)$ is defined by

$$g(M,f) := \min \left\{ \operatorname{card} \mathcal{U} \;\middle|\; \begin{array}{l} \mathcal{U} = \{H_i \mid i \in I\}, \text{ where all } H_i \subset M \text{ are closed,} \\ \bigcup\limits_{i \in I} \bigcup\limits_{j=0}^{p-1} f^j(H_i) = M \text{ and } H_i \cap f^j(H_i) = \phi \\ \text{for } i \in I \text{ and } j = 1,\ldots,p-1 \end{array} \right\}.$$

For example it is well known that for any free $\mathbb{Z}_p$-action f on a k-dimensional sphere S^k, $g(S^k,f) = k + 1$.

The above mentioned variant of the Ljusternik-Schnirelmann covering property can be formulated as the following problem:

PROBLEM

Let $k \in \mathbb{N}$ and p be a prime number. What is the minimal number $r_{k,p}$ such that for any normal space M and any free $\mathbb{Z}_p$-action $f : M \to M$ we have $g(M,f) \leq r_{k,p}$ whenever there exist closed sets $M_1,\dots,M_k$ with $\bigcup_{i=1}^{k} M_i = M$ and

$$M_i \cap f(M_i) = \phi \quad \text{for} \quad i = 1,\dots,k.$$

$r_{k,p}$ is explicitly known only in some special cases. We have

$$r_{k,2} = k - 1 \quad \text{for all} \quad k \in \mathbb{N} \qquad \text{(see [6,3])}$$

and

$$r_{3,p} = \begin{cases} 1 \text{ if } & p = 3 \\ 2 \text{ if } & p > 3 \end{cases} \qquad \text{(see [3])}.$$

Furthermore, trivially $r_{k,p} = 0$ for $k = 1$ or $k = 2$ and $p \geq 3$. In [3], the following estimate was proven:

$$r_{k,p} \leq \frac{p-1}{2}(k-3) + \begin{cases} 1 & \text{if } p = 3 \\ 2 & \text{if } p > 3. \end{cases}$$

In order to prove the above mentioned conjecture in asymptotic fixed point theory, one would need $r_{k,p} = o(p)$ for all $k \in \mathbb{N}$ instead of the above $O(p)$-estimate (see [2]).

In [3], we formulated the conjecture that $r_{k,p} = k - s_{k,p}$ for every $k \in \mathbb{N}$, p prime, where $s_{k,p} \in \{1,2,3\}$. It is the purpose of this short note to give a simple example, which shows that this conjecture is wrong.

THEOREM

For any prime number $p \geq 5$, $r_{4,p} \geq 4$. More explicitly, let $a \in \mathbb{N}$ with $\frac{1}{6} < \frac{a}{p} < \frac{1}{3}$ and $f_p : S^3 \subset \mathbb{C}^2 \to S^3$, $f_p(z_1,z_2) := (e^{\frac{a}{p}2\pi i} z_1, e^{\frac{-a}{p}2\pi i} z_2)$.

Then S^3 can be covered by 4 closed sets $M_1,\dots,M_4$ with $M_i \cap f_p(M_i) = \phi$ for $i = 1,\dots,4$, and therefore $r_{4,p} \geq g(S^3,f_p) = 4$ (see the remark after the definition). In particular, $r_{4,5} \geq 4$, and on the other hand, by the estimate

$$r_{k,p} \leq \frac{p-1}{2}(k-3) + 2,$$

we have $r_{4,5} \leq 4$, so $r_{4,5} = 4$, but we even do not know whether $r_{4,7} = 4$ or $r_{4,7} = 5$ (which are, by the same argument, the only possibilities). Observe that for $p = 2$ and $p = 3$ the theorem is not valid: $r_{4,2} = 3$ and $r_{4,3} = 2$.

Proof of the theorem

Let

$$D_{i,1} := \{(z_1,z_2) \in S^3 : |z_1| \geq |z_2|, \ \frac{i-1}{3}\pi \leq \arg z_1 \leq \frac{i}{3}\pi\}$$

and

$$D_{i,2} := \{(z_1,z_2) \in S^3 : |z_2| \geq |z_1|, \ \frac{i-1}{3}\pi \leq \arg z_2 \leq \frac{i}{3}\pi\}$$

for $i = 1,\dots,6$, and let

$$N_i := D_{i,1} \cup D_{i+3,1} \cup D_{i,2} \cup D_{i+3,2}$$

for $i = 1,2,3$. With f_p as in the statement of the theorem, we have for $i = 1,2,3$ and $j = 1,2$

$$(D_{i,j} \cup D_{i+3,j}) \cap f_p(D_{i,j} \cup D_{i+3,j}) = \phi$$

and hence

$$\bigcup_{i=1}^{3} N_i \cap f_p(N_i) = \bigcup_{i=1}^{3} (D_{i,1} \cup D_{i+3,1}) \cap f_p(D_{i,2} \cup D_{i+3,2})$$

$$\cup \bigcup_{i=1}^{3} (D_{i,2} \cup D_{i+3,2}) \cap f_p(D_{i,1} \cup D_{i+3,1})$$

$$\subset \{(z_1,z_2) \in S^3 : |z_1| = |z_2|\} =: T,$$

where T is a torus. Figure 1 shows the case of $p = 5$.

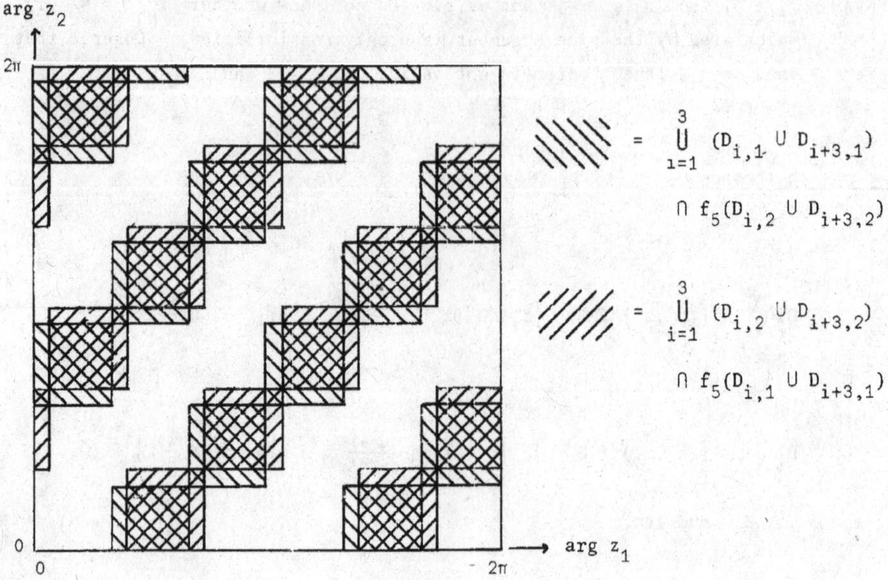

Figure 1

Now, let

$$N_4 := \bigcup_{i=1}^{6} D_{i,1} \cap \bigcup_{\substack{j \in \{1,\dots,6\} \\ j \equiv i-1 \pmod 6 \\ \text{or } j \equiv i-2 \pmod 6}} D_{j,2}$$

(see figure 2).

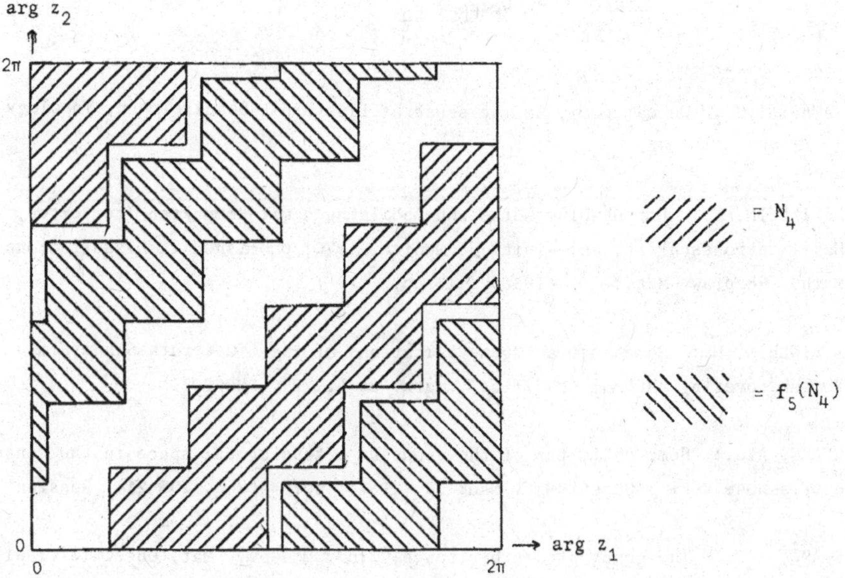

Figure 2

It is easy to see that

a) $N_4 \cap f_p(N_4) = \phi$,

b)
$$\bigcup_{i=1}^{3} N_i \cap f_p(N_i) = \bigcup_{i=1}^{6} (D_{i,1} \cap f_p(D_{i,2})) \cup (D_{i,2} \cap f_p(D_{i,1}))$$
$$\cup \bigcup_{\substack{i=1 \\ j\equiv i+3 (\mathrm{mod}\ 6)}}^{6} (D_{i,1} \cap f_p(D_{j,2})) \cup (D_{i,2} \cap f_p(D_{j,1}))$$
$$\subset N_4 \cup f_p(N_4).$$

Let M_4 be a closed neighborhood of N_4 in S^3 with $M_4 \cap f_p(M_4) = \phi$, and define $M_i := \overline{N_i \setminus M_4}$ for $i = 1,2,3$. Then it is obvious that the sets $M_1,\ldots,M_4$ have the desired properties. ■

It would be important to compute some more numbers $r_{k,p}$, for example, $r_{4,7}$ and $r_{5,5}$, to get hints for the general result, in particular, to see whether one can expect $r_{k,p} = o(p)$ or not.

References

[1] JAMES, I.M.: On category, in the sense of Ljusternik-Schnirelmann, Topology 17 (1978), 331-348.

[2] STEINLEIN, H.: Borsuk-Ulam Sätze und Abbildungen mit kompakten Iterierten, Habilitationsschrift, University of Munich, 1976, published in Dissertationes Math. (Rozprawy Mat.) 177 (1980), 116 pp.

[3] STEINLEIN, H.: Some abstract generalizations of the Ljusternik-Schnirelmann Borsuk covering theorem, Pacific J. Math. 83 (1979), 285-296.

[4] ŠVARC, A.S.: Some estimates of the genus of a topological space in the sense of Krasnosel'skiĭ, Uspehi Mat. Nauk 12 (1957), no. 4 (76), 209-214 (Russian)

[5] ŠVARC, A.S.: The genus of a fiber space, Trudy Moskov. Mat. Obšč. 10 (1961), 217-272 and 11 (1962), 99-126; (Russian), English translation in Amer. Math. Soc., Translat., II. Ser. 55 (1966), 49-140.

[6] YANG, Chung-Tao.: On theorems of Borsuk-Ulam, Kakutani-Yamabe-Yujobô and Dyson, I, Ann. Math. 60 (1954), 262-282.

Added in proof.

It turned out that

$r_{k,7} \leq 2(k - 2)$ for $k \geq 2$,

in particular $r_{4,7} \leq 4$, and hence, by the above theorem, $r_{4,7} = 4$. Of course, this increases the chances for the desired o(p)-estimate for $r_{k,p}$.

ON A CONJECTURE OF HOPF FOR α-SEPARATING MAPS FROM MANIFOLDS INTO SPHERES

By

FRIEDRICH WILLE
Fachbereich Mathematatik
Universität Kassel
Heinrich -Plett Strasse 41
3500 Kassel, West Germany.

1. INTRODUCTION

Let $f: M \to X$ be a continuous map from a metric space M into a topological space X. Assume that there exists a real number $\alpha > 0$ satisfying

$$(x_1, x_2 \in M \text{ and } d(x_1,x_2) = \alpha) \Rightarrow f(x_1) \neq f(x_2). \tag{1}$$

(d denotes the metric of M). A map with this property will be called a *α-separating map.*

In this paper we study the following case: assume $X = S^n = \{x \in \mathbb{R}^{n+1} : |x| = 1\}$ being the n-dimensional sphere and $M = M^n$ a smooth compact connected oriented n-dimensional manifold with a Riemannian metric d. Furthermore let α be a positive real number such that for any two points $x_1, x_2 \in M^n$ with $d(x_1,x_2) = \alpha$ there is a unique minimal geodesic from x_1 to x_2. Considering this α let

$$f : M^n \to S^n$$

be a α-separating map. We will prove the following.

THEOREM

Assuming f as above, the topological degree of f does not vanish: $\deg f \neq 0$.

This theorem was conjectured by H. Hopf [4, p. 136-137] in 1945. Especially he noted that even in the case $M^n = S^n$ $(n \geq 2)$ the result is still unknown. He remarked that the theorem is obviously true for $n = 1$, $M^1 = S^1$. In [2, 3] G. Hirsch proved $\deg f \neq 0$ under the strong additional assumption that

$$f(x_1) \neq -f(x_2)$$

if $d(x_1,x_2) = \alpha$. In [6, 7, 8] the writer proved the theorem for $M^n = S^n$, n even, and in the case $\alpha = \pi/2$ for all n. In a recent paper [1] T. tom Dieck and L. Smith gave the following result: under the assumptions above the Euler characteristic of M^n is even and the following congruences are true: $\deg f \equiv \chi(M^n)/2$, mod 2, if n is even, and $\deg f \equiv \chi_{1/2}(M^n)$, mod 2, if n is odd and $n \neq 1, 3, 7$. (χ denotes the Euler characteristic and $\chi_{1/2}$ the Kervaire semi-characteristic). This gives an affirmative answer to the conjecture in the case $M^n = S^n$ for all n except $n = 3$ and $n = 7$. In the following sections the theorem will be proved solving the problem of Hopf.

2. BASIC CONDITIONS

Let us fix α and f as above and assume $n \geq 2$. Defining the ball

$$K_{p,\alpha} := \{x \in M^n : d(x,p) \leq \alpha\}, \qquad p \in M^n,$$

one obtains

LEMMA 1

The topological degree

$$\delta := \deg(f, K_{p,\alpha}, f(p))$$

is odd and independent from $p \in M^n$.

Proof

The oddness of δ follows from [8] (proof of Lemma 3). The independence of p, we get from the homotopy invariance of the topological degree in the following way: we move p on M^n continuously and reformulate this moving as a homotopy by use of the exponential mapping. (The execution of this concept needs some technical arguments, but it is not difficult in principle, [5].) ■

We restrict our investigations to *spherical piecewise linear mappings* $f:M^n \to S^n$, which are defined in the following way: let $T = (K,\tau)$ be a triangulation of M^n, where K is a simplicial complex, $|K|$ its geometric realization and $\tau:|K| \to M^n$ a homeomorphism. Let $F:|K| \to \mathbb{R}^{n+1}$ be a map which agrees with $f \circ \tau$ at any vertex v of K: $F(v) = f(\tau(v))$. Moreover let F be linear on every geometric simplex of $|K|$. Then f has the form $f(x) = F(\tau^{-1}(x))/|F(\tau^{-1}(x))|$, assuming $F(z) \neq 0$ for all $z \in |K|$. Additionally we assume that $f^{-1}(y)$ is a finite set for every $y \in S^n$. (If this condition does not hold we get this by arbitrary small changes of $f(v')$, $v' = \tau(v)$, v vertex).

Hence without loss of generality we assume that

(a) $f: M^n \to S^n$ is spherical piecewise linear, based on a triangulation $T = (K,\tau)$ of M^n, and

(b) $f^{-1}(y)$ is a finite set for each $y \in S^n$.

If the theorem is proved for these mappings the whole statement follows by approximation.

Let Δ^k be a k-dimensional geometric simplex of $|K|$ and $\sigma^k = \tau(\Delta^k)$ the assigned topological simplex on M^n $(0 \leq k \leq n)$. $x \in M^n$ shall be called a *regular point of* f if x belongs to the interior of some simplex σ^n. $y \in S^n$ is called a *regular value of* f if all points of $f^{-1}(y)$ are regular points of f and if $f^{-1}(y)$ is not empty.

By (b) the image $f(\sigma^n)$ of any n-dimensional simplex σ^n is a nondegenerate spherical simplex. Since M^n is triangulated by a finite number of simplices the image $f(M^n)$ is covered by finitely many nondegenerate simplices $f(\sigma^n)$. Hence there exist some elements $y \in f(M^n)$ in the interior of some simplices $f(\sigma^n)$, which do not belong to any $f(\sigma^{n-1})$. These elements y are regular values. Hence regular values of f exist.

Let y be a regular value of f and $\{x_1,x_2,\ldots,x_m\} = f^{-1}(y)$ its preimage. Hence

$$\delta = \deg(f,K_{x_i,\alpha},y) = \sum_{d(x_k,x_i)<\alpha} j(x_k)\,, \tag{3}$$

where $j(x)$ denotes the index of x corresponding to f (that is $j(x) = \deg(f,\overline{V},f(x))$ where V is an open neighborhood of x such that $f(z) \neq f(x)$ for all $z \in \overline{V}$, $z \neq x$. x regular $\Rightarrow |j(x)| = 1$). Defining

$$a_{ik} := \begin{cases} j(x_i)j(x_k) & \text{if } d(x_i,x_k) < \alpha, \\ 0 & \text{if } d(x_i,x_k) > \alpha, \end{cases} \tag{4}$$

it follows from (3)

$$\sum_{k=1}^{m} a_{ik} = j(x_i)\delta \neq 0 \tag{5}$$

and

$$\sum_{i,k=1}^{m} a_{ik} = \delta \deg f , \tag{6}$$

using $\deg f = \sum_{i=1}^{m} j(x_i)$. The symmetrical matrix $A(y) = (a_{ik})_{m,m}$ will be called a *distance matrix* of y (with respect to f). The distance matrix $A(y)$ is uniquely determined by y, disregarding permutations of rows and columns respectively

3. CHARACTERISTIC MATRIX

First we define the *reduction* of a symmetrical square matrix A: assume $A_i = -A_\nu$ where A_i, A_ν denote the i-th and the ν-th row vector of A. Hence $A^i = -A^\nu$ where A^i and A^ν are the i-th and the ν-th column vector of A. The deleting of the rows A_i, A_ν and the columns A^i, A^ν will be called a *reduction* of A. A symmetrical square matrix which cannot be reduced in this sense is said to be *irreducible*.

Now we reduce the distance matrix $A(y)$ if possible. The remaining matrix will be reduced too if possible, etc. From (5) it follows that $A(y)$ does not completely vanish by successive reductions (because the last two vanishing rows A_i, A_ν would satisfy $\sum_{k=1}^{m} a_{ik} = 0$, $\sum_{k=1}^{m} a_{\nu k} = 0$ contradicting (5)). The repeated reducings stop when there remains an irreducible matrix $\overline{A}(y)$. We call it a *characteristic matrix* of y (with respect to f).

Showing the uniqueness of the characteristic matrix (refrained from permutations of rows and columns) we select a sequence of reductions of $A(y)$ so that a characteristic matrix $\bar{A}(y)$ remains. Without loss of generality we assume that by the first reduction the last two rows A_m, A_{m-1} will be deleted just as the last two columns A^m, A^{m-1}. By further reductions the last two rows and columns of each of the remaining matrices will by likewise deleted. Hence the characteristic matrix $\bar{A}(y)$ is placed in the "upper left corner" of $A(y) = (a_{ik})_{m,m}$, that is $\bar{A}(y) = (a_{ik})_{q,q}$ with $q \le m$, see figure 1.

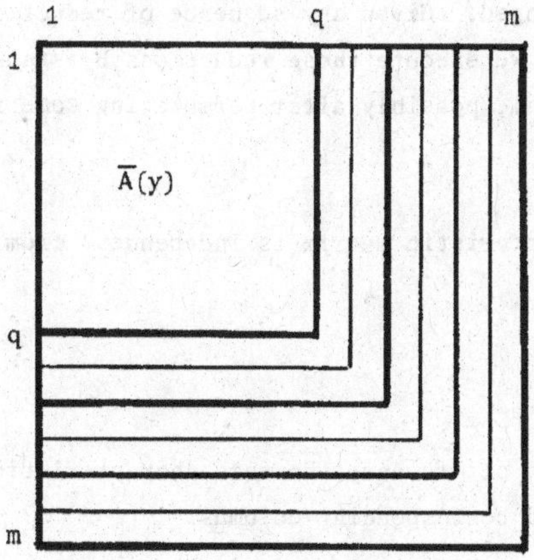

Figure 1. Distance matrix $A(y)$.

Therefore "outside of $\bar{A}(y)$" one has

$$a_{i,k-1} = -a_{ik} \qquad \text{if } i \le k = q + 2v \tag{7}$$

and by symmetry

$$a_{k-1,i} = -a_{ki} \qquad \text{if } i \le k = q + 2v \tag{8}$$

for all $v \in \{1,2,\ldots,(m-q)/2\}$.

Consider any reduction of $A(y)$ (which has the form of figure 1), that is deletion of two rows A_ν, A_μ ($A_\nu = -A_\mu$) and the corresponding columns A^ν, A^μ. Assume $\nu < \mu$. We define the *standard reduction step* of this reduction by the following procedure:

(a) If $\mu = q + 2v$ (v positive integer) replace A_ν by $A_{\mu-1}$. Then replace $A_{\mu-1}$ by $A_{\mu+1}$, A_μ by $A_{\mu+2}$, $A_{\mu+1}$ by $A_{\mu+3}$ etc. . Finally replace A_{m-2} by A_m. The columns will be handled analogously.

(b) If $\mu = q + 2v - 1$ replace A_ν by $A_{\mu+1}$. Then replace A_μ by $A_{\mu+2}$, $A_{\mu+1}$ by $A_{\mu+3}$ etc. until replacing A_{m-2} by A_m. Analogous replacings of columns will be added.

The square matrix of m-2 rows and columns in the left upper corner is the result of this procedure. It has a form like figure 1 also and the characteristic matrix $\overline{A}(y)$ has been unchanged. Given any sequence of reductions so that an irreducible matrix B remains we execute these reductions by standard reduction steps. Hence B and $\overline{A}(y)$ are equal, possibly after permutating some rows and columns respectively.

Furthermore the characteristic matrix is independent from y.

LEMMA 2

All regular values y of f have the same characteristic matrix, disregarding permutations of rows and corresponding columns.

Proof

Let y and y' be any two regular values of f and let $c:[0,1] \to S^n$ be a path connecting y and y'. Assume that $c([0,1])$ does not intersect any simplex $f(\sigma^{n-2})$ where σ^{n-2} denotes an (n-2)-dimensional simplex $\tau(\Delta^{n-2})$. Furthermore assume that $c(t)$ meets (n-1)-dimensional simplices $f(\sigma^{n-1})$ at most in a finite number of times, that is there exist $t_1,\ldots,t_s \in [0,1]$ such that $c(t)$ is regular for all $t \neq t_i$, $i=1,\ldots,s$. Let t increase from 0 to 1. The distance matrix $A(c(t))(t \neq t_i)$ only changes when t crosses any t_i (without regarding row and column permutations). These changes consist of reductions or "reverse reductions", that is extendings of $A(c(t))$ by two row vectors (whose sum is zero) and by two corresponding column vectors. Using standard reduction steps and the reverse of them the characteristic matrix does not change, which proves the lemma.■

4. THE RANK OF A BASE POINT

Let y be a regular value of f and let $A(y) = (a_{ik})_{m,m}$ be a distance matrix which has the form of figure 1. By the definition (4) of the distance matrix, every point $x_i \in f^{-1}(y)$ corresponds to a row A_i of $A(y)$ and to a column A^i respectively. The first q points $x_1,\dots,x_q$ provide the characteristic matrix $\overline{A}(y)$, by (4). There may be other subsets $\{x_1',\dots,x_q'\}$ of $f^{-1}(y)$ generating $\overline{A}(y)$ by (4) (disregarding permutations of rows and columns). Any set

$$\{x_1',\dots,x_q'\} \subset f^{-1}(y)$$

of this type we call a *characteristic set* of y (with respect to f), and any point x_i' of this set is called a *base point* of $f^{-1}(y)$.

Let $\{x_1,\dots,x_q\}$ be any characteristic set of y. Without loss of generality it corresponds to the first q rows $A_1,\dots,A_q$ of $A(y)$ and to the columns $A^1,\dots,A^q$ respectively. Let x_i be any base point of $f^{-1}(y)$. Then the *rank* $r(x_i)$ of x_i is defined by the number of points $x_1,\dots,x_q$ which belong to $K_{x_i,\alpha}$, that is

$$r(x_i) := \sum_{k=1}^{q} |a_{ik}|. \tag{9}$$

We remark that in this definition x_i may be one of the points $x_1,\dots,x_q$ or not.

$r(x_i)$ is independent of the selected characteristic set $\{x_1,\dots,x_q\}$. To prove this, consider a characteristic set of y containing x_i. Without loss of generality this set may be the set $\{x_1,\dots,x_q\}$ and $x_i = x_1$. Let $\{x_1',\dots,x_q'\}$ be another characteristic set of y which determines the submatrix $\overline{A}'(y)$ of $A(y)$. We get $\overline{A}'(y)$ from $A(y)$ by a certain sequence of reductions. Executing these reductions by standard reduction steps the points $x_1,\dots,x_q$ which differ from the points x_k' will be replaced by these points successively. The replacing of one of these points by use of a standard reduction step does not change (9). The proof of this fact by use of (7) and (8) is elementary. Hence after complete execution of all standard reduction steps (9) remains unchanged proving the required independence.

LEMMA 3

Let $\{x_1,\dots,x_q\}$ be a characteristic set of a regular point y. Then all points x_j of this set have the same rank.

Proof

The idea is to shift x_1 continuously to any other point x_j $(1 < j \le q)$ and to prove that the corresponding sum on the right hand side of (9) remains unchanged.

Assume $q \ge 2$ (the case $q = 1$ is trivial). Let $c : [0,1] \to M^n$ be a path connecting x_1 and x_j $(1 < j \le q) : x_1 = c(0)$, $x_j = c(1)$. Let $g := f \circ c$ be the corresponding path on S^n and assume that $g([0,1])$ does not intersect any simplex $f(\sigma^{n-2})$. Furthermore assume that there exists a finite number of values

$$t_1, \ldots, t_s \in [0,1]$$

such that $g(t_i) \in f(\sigma^{n-1})$ and $g(t)$ is a regular value for each

$$t \ne t_i, \; i \in \{1, \ldots, s\} .$$

Without regarding permutations of rows and columns the distance matrix $A(f(c(t)))$ only changes when t crosses any t_i. We assume that $A(f(c(t)))$ has the form of figure 1 for $t \ne t_i$.

Let t increase from 0 to 1. If t crosses any t_i only the following cases may occur (since f satisfies (a) and (b)).

Case 1. The correspondence of $c(t)$ changed from a row A_β to A_γ where $A_\beta = -A_\gamma$. Since these changings begin with $\beta = 1$ we obtain from (7), (8)

$$a_{\gamma,k-1} = -a_{\gamma k} \; , \; a_{k-1,\gamma} = -a_{k\gamma} \tag{10}$$

for all $k = q + 2v$, $v = 1,2 \ldots$, ($k \le$ number of rows).

Case 2. Reduction happens. We execute this reduction by a standard reduction step. After execution equations (10) are still true and $\sum_{k=1}^{q} |a_{\gamma k}|$ remains unchanged (the proof is elementary using (10)).

Case 3. Reverse reduction happens, that is two rows will be added below and two corresponding columns on the right. The sum of the rows and the columns respectively are zero. Hence (10) remains true. The considered sum remains untouched.

Case 4. $A(f(c(t)))$ and the correspondence of $c(t)$ to a row remain unchanged.

In all cases the sum $\sum_{k=1}^{q} |a_{\gamma k}|$ corresponding to $c(t)$, $t \neq t_i$, has the same value. Hence $r(x_1) = r(x_j)$, proving the lemma. ■

5. PROOF OF THE THEOREM

Assume $\deg f = 0$. Let y be a regular value of f and let $A(y)$ be a distance matrix which has the form of figure 1. Let $\{x_1,\dots,x_m\} = f^{-1}(y)$ where x_i corresponds to the i-th row A_i of $A(y)$ for each $i = 1,\dots,m$. Let the points $x_1,\dots,x_q$ $(q \le m)$ determine the characteristic matrix by (4). From figure 1 it follows $j(x_q+2v) = -j(x_q+2v-1)$ for all $v = 1,\dots,(m-q)/2$. Hence

$$\sum_{i=1}^{q} j(x_i) = \sum_{i=1}^{m} j(x_i) = \deg f = 0,$$

Since $j(x_i)$ is 1 or -1 it follows that q is even. We arrange the points $x_1,\dots,x_q$ so that $j(x_i) = 1$ for $i = 1,\dots,\frac{q}{2}$ and $j(x_i) = -1$ for $i = \frac{q}{2}+1,\dots,q$. Therefore the characteristic matrix $\overline{A}(y) = (a_{ik})_{q,q}$ may be written in the form

$$\overline{A}(y) = \begin{pmatrix} B & C \\ C' & D \end{pmatrix}$$

where B, D, C, C' are square matrices with $q/2$ rows. The elements of B and D are equal to 1 or 0 and the elements of C and C' equal -1 or 0. From (5) we derive

$$\sum_{k=1}^{q} a_{ik} = j(x_i)\delta \neq 0$$

using (7). Denoting $h := q/2$ it follows

$$\sum_{i=1}^{h} \sum_{k=1}^{q} a_{ik} = \sum_{i=1}^{h} \sum_{k=1}^{h} |a_{ik}| - \sum_{i=1}^{h} \sum_{k=h+1}^{q} |a_{ik}| = h\delta$$

and

$$\sum_{i=h+1}^{q} \sum_{k=1}^{q} a_{ik} = - \sum_{i=h+1}^{q} \sum_{k=1}^{h} |a_{ik}| + \sum_{i=h+1}^{q} \sum_{k=h+1}^{q} |a_{ik}| = -h\delta .$$

Subtraction of these equations provides

$$\sum_{i=1}^{h} \sum_{k=1}^{h} |a_{ik}| - \sum_{i=h+1}^{q} \sum_{k=h+1}^{q} |a_{ik}| = 2h\delta \neq 0 \tag{11}$$

using the symmetry of $\bar{A}(y)$. Furthermore let $\rho = r(x_i) = \sum_{k=1}^{q} |a_{ik}|$ be the common rank of all x_i, $i = 1,\ldots,q$ (see Lemma 3). Hence

$$\sum_{i=1}^{h} \sum_{k=1}^{h} |a_{ik}| + \sum_{i=1}^{h} \sum_{k=h+1}^{q} |a_{ik}| = h\rho ,$$

$$\sum_{i=h+1}^{q} \sum_{k=1}^{h} |a_{ik}| + \sum_{i=h+1}^{q} \sum_{k=h+1}^{q} |a_{ik}| = h\rho .$$

By subtraction we obtain

$$\sum_{i=1}^{h} \sum_{k=1}^{h} |a_{ik}| - \sum_{i=h+1}^{q} \sum_{k=h+1}^{q} |a_{ik}| = 0$$

contradicting (11). This completes the proof of the theorem. ■

6. Remarks

I. First examples for α-separating mappings $f : S^n \to S^n$ are given by maps satisfying $f(x) \neq f(-x)$ for all $x \in S^n$. By continuity there is some α_o such that f is a α-separating map for all $\alpha \in (\alpha_o,\pi]$ (using the angle metric). Another example of a α-separating map $f : S^1 \to S^1$ is given by

$$f(z) := \frac{\hat{f}(z)}{|\hat{f}(z)|} \quad \text{with} \quad \hat{f}(z) := \begin{cases} z^2 & \text{if } \operatorname{Re} z \geq 0 \\ \frac{1}{3}(z^{-2} - 2) & \text{if } \operatorname{Re} z < 0 \end{cases} , \text{ for } |z| = 1,$$

(z complex number), see figure 2.

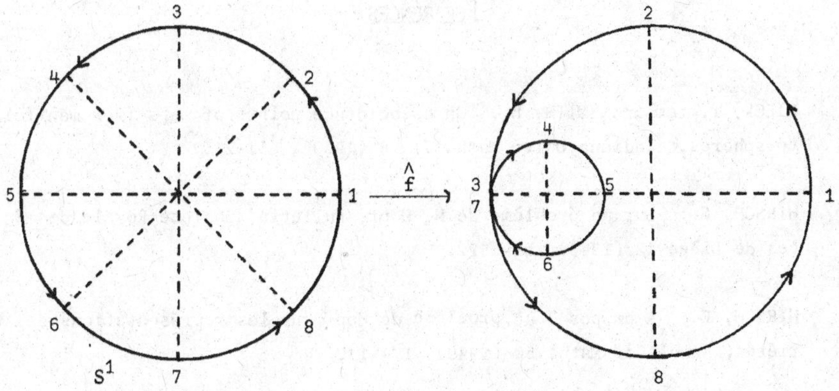

Figure 2. Example of a α-seperating map : $f = \hat{f}/|\hat{f}|$, $\alpha = \frac{3}{4}\pi$.

This map is α-seperating for $\alpha = 3\pi/4$, but not for $\alpha = \pi$! Similar examples will be found for $n \geq 2$.

II. The proved theorem leads to statements about the solvability of nonlinear operator equations.

Let $\|\cdot\|$ denote a strict convex norm on $\mathbb{R}^{n+1}$ and let $S^n = \{x \in \mathbb{R}^{n+1} : \|x\| = 1\}$ be the sphere based on this norm. Using the metric $d(x_1,x_2) = \|x_1 - x_2\|$ on S^n the proved theorem can be extended to α-separating maps $f : S^n \to S^n$, [8]. (Since S^n is homeomorphic to an euclidean sphere the degree of f can be defined in a natural way.) Furthermore by approximation our theorem can be extended to spheres $S^n \subset \mathbb{R}^{n+1}$ based on any norm. Hence by well known arguments of the Leray-Schauder theory we get the following fixed point theorem:

COROLLARY

Let X be a real Banach space, $B = \{x \in X : \|x\| \leq 1\}$ the unit ball and $G : B \to X$ a compact continuous map satisfying

$$\inf_{\substack{\|x-y\| = \alpha \\ \|x\|=\|y\| = 1}} \left\| \frac{x - G(x)}{\|x - G(x)\|} - \frac{y - G(y)}{\|y - G(y)\|} \right\| > 0 \qquad \text{for some } \alpha \in (0,2],$$

and $G(x) \neq x$ if $\|x\| = 1$. Then G has a fixed point.

REFERENCES

[1] DIECK, T. tom and SMITH, L.: On coincidence points of maps from manifolds to spheres. Indiana Univ. Math. J. 28 (1979), 251-255.

[2] HIRSCH, G.: Sur un problème de H. Hopf, Bulletin, Société Royale des Sciences de Liège 12 (1943), 514-522.

[3] HIRSCH, G.: A propos d'un problème de Hopf sur les représentations des variétés, Annals of Math. 50 (1949), 174-179.

[4] HOPF, H.: Eine Verallgemeinerung bekannter Abbildungsund Überdeckungssätze, Portug. Math. 4 (1945), 129-139.

[5] SEEMANN, R.: Verallgemeinerung von Sätzen der Abbildungsgradtheorie auf Sphären mit strikt konvexer Norm und Riemannsche Mannigfaltigkeiten, thesis, Univ. of Kassel, (1980).

[6] WILLE, F.: Ein neuer Fixpunktsatz, Zeitschr. Angew. Math. Mech. 56 (1976) T 269.

[7] WILLE, F.: Über eine Vermutung von H. Hopf zur Abbildungsgradtheorie, preprint, FB Math. Univ. of Kassel, (1976).

[8] WILLE, F.: Ein Analogon zum Borsukschen Antipodensatz für rechtwinklige Punktepaare, preprint, FB Math. Univ. of Kassel, (1978).

THE LERAY-SCHAUDER CONDITION IS NECESSARY FOR THE EXISTENCE OF SOLUTIONS

By

T.E. WILLIAMSON, JR.
Montclair State College
Upper Montclair N.J.
U.S.A. 07043

0. Introduction

The Leray-Schauder boundary condition, when reformulated, is a necessary condition for the existence of solutions for a variety of Hilbert space problems. These include existence of fixed points for nonexpansive mappings, and existence of zeros for strongly monotone and strictly monotone mappings. One consequence is that, when combined with a theorem of F.E. Browder and W.V. Petryshyn, the Leray-Schauder condition is a necessary and sufficient condition for the existence of fixed points of nonexpansive mappings whose domain is a nonempty closed bounded convex subset of a Hilbert space.

§1.

We define a Leray-Schauder condition, and show its equivalence to those of Browder-Petryshyn and Cramer-Ray for convex domains, and to the traditional Leray-Schauder boundary condition when the domain is a ball. For all of these definitions D is a subset of a Hilbert space H, and $T : D \to H$ is a mapping. Actually these definitions are relationships between pairs of distinct points in the space, but for clarity we view them through the mapping T.

DEFINITION 1

Let $x \in D$ with $x \neq Tx$, define $LS(x,Tx) = \{y \in H : \mathrm{Re}(Tx - x, y - x) > 0\}$. T satisfies the Leray-Schauder condition (LS) at x relative to D if and only if $LS(x,Tx) \cap D \neq \phi$.

REMARK 1

If $Tx \in D$ or $x \in \text{interior}(D)$ then T satisfies LS at x relative to D.

DEFINITION 2 *(Leray-Schauder* (1934) [2]*).*

Let $D = B(0,r)$ and $x \in \text{boundary}(D)$ with $x \neq Tx$. T satisfies the Leray-Schauder boundary condition (LSB) at $x \Leftrightarrow Tx \neq \lambda x$, for every $\lambda > 1$.

DEFINITION 3 *(F.E. Browder-W.V. Petryshyn* (1967) [1]*).*

(We restate their definition in an equivalent form.) Let $x \in D$ with $x \neq Tx$, define $BP(x,Tx) = \text{open ball } B(Tx, \|x - Tx\|)$. T satisfies BP at x relative to $D \Leftrightarrow BP(x,Tx) \cap D \neq \phi$.

DEFINITION 4 *(W.J. Cramer-W.O. Ray* (1980) [4]*).*

Let $x \in D$ with $x \neq Tx$. T satisfies CR at x relative to $D \Leftrightarrow$

$$\liminf_{h \to 0^+} h^{-1} d((1-h)x + hTx,\ D) < \|x - Tx\| .$$

PROPOSITION 1

Let D be convex and $x \in D$ with $x \neq Tx$. T satisfies LS at $x \Leftrightarrow$ T satisfies BP at x.

Proof

$\Leftarrow$) It suffices to prove $BP(x,Tx) \subseteq LS(x,Tx)$. Let $y \in BP(x,Tx)$. Since $\|y - Tx\| < \|x - Tx\|$, the equality

$$\|Tx - x\|^2 + \|y - x\|^2 - 2\text{Re}(Tx - x, y - x) = \|Tx - y\|^2 \qquad (\dagger)$$

implies $\|y - x\|^2 < 2\mathrm{Re}(Tx - x, y - x)$. Hence $y \in LS(x,Tx)$.

$\Rightarrow$) Let $y \in LS(x,Tx) \cap D$ and define $q = x + a(y - x)$ where

$$a = 2\|y - x\|^{-2}\,\mathrm{Re}(Tx - x, y - x).$$

Note that $a > 0$ and $q \in LS(x,Tx)$. We claim that $\{y,q\} \cap BP(x,Tx) \cap D \neq \phi$.

Case 1. If $a > 1$, $2\,\mathrm{Re}(Tx - x, y - x) > \|y - x\|^2$ which implies $y \in BP(x,Tx)$, by (†). Since $y \in D$ we have $y \in D \cap BP(x,Tx)$.

Case 2. If $a \leq 1$, the convexity of D implies that $q \in D$. We note that $\mathrm{Re}(Tx - q, q - x) = 0$. Using this, the fact that $x \neq q$ and the equality $\|x - Tx\|^2 = \|x - q\|^2 + \|q - Tx\|^2 + 2\mathrm{Re}(x - q, q - Tx)$ we see that $q \in BP(x,Tx)$, which completes the proof. ■

PROPOSITION 2

Let D be convex and $x \in D$ with $x \neq Tx$. T satisfies LS at $x \Leftrightarrow T$ satisfies CR at x.

LEMMA 1

Let D be convex and $x \in D$ with $x \neq Tx$. T satisfies CR at $x \Leftrightarrow$ there exist $y \in D$ and h, $0 < h \leq 1$, such that

$$h^{-1}\|(1 - h)x + hTx - y\| < \|x - Tx\| .$$

Proof of the lemma

$\Rightarrow$) Obvious. $\Leftarrow$) Without loss of generality, for such y and h choose $0 < k < 1$ such that

$$h^{-1}\|(1 - h)x + hTx - y\| \leq k\|x - Tx\| .$$

It suffices to show that for every a, $0 < a < 1$, $z_a = x + a(y - x)$, which lies in D, satisfies

$$(ah)^{-1}\|(1-ah)x+ahTx-z_a\| \le k\|x-Tx\|.$$

Since $(ah)^{-1}\|(1-ah)x+ahTx-z_a\| = h^{-1}\|(1-h)x+hTx-y\| \le k\|x-Tx\|$, the lemma is proved. ■

Proof of Proposition 2

⇐) Let $y \in D$ and h as in Lemma 1. It suffices to show $y \in D \cap LS(x,Tx)$. Note first that

$$h^{-1}\|(1-h)x+hTx-y\| = h^{-1}\|h(Tx-x)+(x-y)\|$$

and

$$h^{-2}\|h(Tx-x)+(x-y)\|^2 = \|Tx-x\|^2 + h^{-2}\|x-y\|^2 - 2h^{-1}\mathrm{Re}(Tx-x,y-x).$$

Hence $2\mathrm{Re}(Tx-x,y-x) > h^{-1}\|x-y\|^2$ which implies $y \in LS(x,Tx)$.

⇒) Just reverse the above steps after choosing $h > 0$ small enough. ■

PROPOSITION 3

Let $D = B(0,r)$ and $x \in \mathrm{boundary}(D)$ with $x \neq Tx$. T satisfies LSB at $x \Leftrightarrow T$ satisfies BP at x.

Proof

⇐) Let $y \in D \cap BP(x,Tx)$. Then

$$\|Tx\| \le \|y\| + \|y-Tx\| < \|x\| + \|x-Tx\|.$$

This implies that x does not lie on the line segment between 0 and Tx, that is, that $Tx \neq \lambda x$, for any $\lambda > 1$.

⇒) Let $Tx \neq \lambda x$ for any $\lambda > 1$. By Remark 1 we may assume $\|Tx\| > r$. Claim $y = r\|Tx\|^{-1}Tx \in D \cap BP(x,Tx)$. Clearly $y \in D$. Now $\|y-Tx\| = \|Tx\|-\|y\|$. Since x does not lie on the line segment between 0 and Tx, $\|Tx\| < \|x\|+\|x-Tx\|$, by the strict convexity of H. Combining these last two gives

$$\|y - Tx\| = \|Tx\| - \|y\| < (\|x\| - \|y\|) + \|x - Tx\|.$$

Since $\|x\| = \|y\| = r$, we get that $y \in BP(x,Tx)$. ■

§2.

We define Leray-Schauder geometric mappings for arbitrary domains $D \subseteq H$ and show that this class of mappings is large. In addition we show that any such mapping must necessarily satisfy LS at each x in the domain or else have empty fixed point set on that domain.

DEFINITION 5

T is a Leray-Schauder geometric mapping $\Leftrightarrow$ for every $x \in D$ with $x \neq Tx$, the fixed point set of T on D, $F_D(T)$ is contained in $LS(x,Tx)$.

THEOREM 1

Let T be a LS geometric mapping. If for some $x \in D$, with $x \neq Tx$, T fails to satisfy LS at x relative to D, then $F_D(T) = \phi$.

Proof

$\phi = D \cap LS(x,Tx) \supseteq D \cap F(T) = F_D(T)$, completing the proof. ■

Some members of this class of LS geometric mappings can be identified through their geometric estimators [5, 6].

THEOREM 2

Let $T : D \to H$, $D \subseteq H$ be nonexpansive (for every $x,y \in D$ $\|Tx - Ty\| \leq \|x - y\|$). Then T is a LS geometric mapping.

Proof

If $x \in D$ with $x \neq Tx$, define the geometric estimator

$$\Gamma(x,Tx) = \{y \in H : \|Tx - y\| \leq \|x - y\|\}$$

and note that $F_D(T) \subseteq \Gamma(x,Tx)$. Hence it suffices to show $\Gamma(x,Tx) \subseteq LS(x,Tx)$. From (†), if $y \in \Gamma(x,Tx)$, $\|Tx-x\|^2 \leq 2\mathrm{Re}(Tx - x, y - x)$. Hence $y \in LS(x,Tx)$, completing the proof. ■

THEOREM 3

Let $T : D \to H$, $D \subseteq H$ with $(-T)$ monotone (for every $x,y \in D$ $\mathrm{Re}(Tx - Ty, x - y) \leq 0$). Then T is a LS geometric mapping.

Proof

If $x \in D$ with $x \neq Tx$ define

$$\Gamma(x,Tx) = \{y \in H : \mathrm{Re}(Tx - y, x - y) \leq 0, \quad y \neq x\}$$

and note that $F_D(T) \subseteq \Gamma(x,Tx)$. Choose $y \in \Gamma(x,Tx)$ and note that $\mathrm{Re}(Tx - x, y - x) \geq \|x - y\|^2$. Since $y \neq x$, $y \in LS(x,Tx)$. ■

THEOREM 4

Let $T : D \to H$, $D \subseteq H$ with $(I - T)$ strictly monotone (for every $x,y \in D$, $\mathrm{Re}((x - Tx) - (y - Ty), x - y) > 0$). Then T is a LS geometric mapping.

Proof

If $x \in D$ with $x \neq Tx$ define

$$\Gamma(x,Tx) = \{y \in H : \mathrm{Re}(x - Tx, x - y) > 0\}$$

and note that $F_D(T) \subseteq \Gamma(x,Tx)$. Since $\Gamma(x,Tx) = LS(x,Tx)$, the proof is complete. ■

REMARK 2

The usual (equivalent) formulations of the problems addressed in the last two theorems are: finding zeros of a strongly monotone operator M (Theorem 3), and finding zeros of a strictly monotone operator M (Theorem 4). For these formulations the usual statement of the LS boundary condition (when $D = B(0,r)$) is: for $x \in \text{boundary}(D)$ $Mx \neq \gamma x$ for $\gamma < 0$. By Theorem 1 we see that, in fact, this condition is necessary for the existence of zeros in these problems. We also note that the domains are arbitrary in all of the above theorems, and that this analysis applies equally as well to multivalued mappings.

THEOREM 5

Let D be a closed bounded convex nonempty subset of a Hilbert space H and $T : D \to H$ be nonexpansive on D. Then $F_D(T) \doteq \phi \Leftrightarrow$ there exists $x \in D$ with $x \neq Tx$ such that T fails to satisfy LS at x relative to D.

Proof

$\Rightarrow$) Browder-Petryshyn [1, Theorem 9], and Proposition 1.

$\Leftarrow$) Theorems 1 and 2. ■

REMARK 3

The above theorem characterizes existence of fixed points in a manner different from the recent characterization of nonexpansive self-mappings by W.O. Ray [3]. The two, when viewed together, give a rather firm picture of the solution to the existence question for nonexpansive mappings in Hilbert space.

REMARK 4

We could say T satisfies graph-LS at x if and only if T satisfies LS at x relative to $D \cap T(D)$, then note that this is a necessary condition for the existence of fixed points of LS geometric mappings. This approach can be modified to handle other mappings by changing the LS condition appropriately. For example, if T is expanding (for $x,y \in D$ $\|Tx - Ty\| \geq \|x - y\|$) then for $x \in D$ with $x \neq Tx$

define $LS(Tx,x) = \{y \in H : Re(x - Tx, y - Tx) > 0\}$, $\Gamma(x,Tx) = \{y \in H : \|Tx - y\| \geq \|x - y\|\}$ and note that if $T(D) \cap LS(Tx,x) = \phi$ then $F_D(T) = \phi$.

REFERENCES

[1] BROWDER, F.E. and PETRYSHYN, W.V.: Construction of fixed points of nonlinear mappings in Hilbert space, J. Math. Anal. Appl. 20 (1967), 197-228.

[2] LERAY, J. and SCHAUDER, J.: Topologie et équations fonctionnelles, Ann. Sci. Ecole Norm. Sup. (3) 51 (1934), 45-78.

[3] RAY, W.O.: The fixed point property and unbounded sets in Hilbert space, Trans. Amer. Math. Soc. 258 (1980), 531-538.

[4] RAY, W.O. and CRAMER, W.J., Jr: Some remarks on the Leray-Schauder boundary condition, Talk delivered by Cramer at Fixed Point Workshop, Univ. de Sherbrooke, Canada, June 2-20, (1980).

[5] WILLIAMSON, T.E., Jr.: Geometric estimation of fixed points of Lipschitzian mappings II, J. Math. Anal. Appl. 62 (1978), 600-609.

[6] WILLIAMSON, T.E., Jr.: Geometric estimation of the solution to $x + Tx = 0$ for unbounded densely defined monotone operator T in Hilbert space, Proc. Amer. Math. Soc. 74 (1979), 278-284.

[7] WILLIAMSON, T.E., Jr: One-step estimates of the solution to $x + Tx = 0$ for accretive operator T in Banach space, submitted.

A PRIMER ON CONNECTIVITY

by

J. C. ALEXANDER*

Department of Mathematics
University of Maryland
College Park, Maryland

0. INTRODUCTION

Fixed point theory, in functional analysis, is a method of proving the existence of solutions of operators on (usually Banach) spaces. The standard method is to develop a degree (or fixed point index) theory for a class of operators, and show that a non-zero degree implies a solution. This is usually done by approximating the operators in the class by "simpler" operators which are known to have a workable degree theory.

Suppose it is already known the operator has some solutions of two types, so that the known solutions fall into two pieces A,B. For example A could be the zero solution and B could be a solution at infinity. Or the operator could be parametrized by an interval [a,b] and A (resp. B) could be the solutions for the parameter a (resp. b). If it can be shown that A and B are connected to each other in the set of all solutions, the existence of solutions other than A and B will be established.

In both philosophy and practice, using connectivity in this way has a lot in common with fixed point theory. And like fixed point

*Partially supported by N.S.F.

theory, it has been used for some time. However, neither its philosophy nor its practice are as standardized as in fixed point theory. Some easy, useful results seem to be unknown. The purpose of this paper is to survey and organize some results on connectivity useful in analysis, and to illustrate with some examples using connectivity.

The abstract (soft) form of fixed point theory is essentially a context in which to work with explicit problems. To make it work for a particular problem, a priori estimates and controls must be established (which of course is the essence of the problem). The abstract theory is a conceptual framework, showing what kind of estimates are required. The same is true of using connectivity. The discussion here is abstract and topological. The object is to codify somewhat the framework and show some "tricks of the trade."

1. POINT SET TOPOLOGY

First let us develop the concepts. All topological spaces are Hausdorff. (This is a real restriction, because one would like to be able to prove results using weak topologies, but it is not clear how this should be done.)

A *separation* of a space X is a pair of non-empty open (hence also closed) subsets with

$$U \cap V = \emptyset, \quad U \cup V = X .$$

A space is *connected* if it does not admit a separation. We need to consider how subsets lie with respect to a separation. Two sets $A,B \subset X$ are *connected (to each other) in* X if there exists a connected set Y with $A \cap Y \neq \emptyset$, $B \cap Y \neq \emptyset$. (Note that A,B need not themselves be connected.) Two non-empty sets A,B are *separated (from each other) in* X if there is a separation U,V of X with $A \subset U$, $B \subset V$. As a matter of convention, to say A,B are separated or not separated in X requires them to be non-empty.

These last two definitions are not opposites. If A,B are separated in X, they certainly are not connected in X, but the converse assertion is false. The standard example is the following. In the plane, let C_r be the circle around the origin of radius $1-r^{-1}$. Let $a = (1,0)$ and $b = (-1,0)$. Let

$$X = \bigcup_{r=2,3,\ldots} C_r \cup \{a,b\} .$$

Then a,b are not connected in X, but neither are they separated.

The set of points connected to x is called the <u>component</u> of x; the set of points that cannot be separated from x is called the <u>quasi-component</u> of x. Two sets $A,B \subset X$ are connected in X if and only if they both intersect some one component, i.e. if some point in A is connected to some point in B. The analogous statement for separation is false. Let X be the subset of the plane:

$$X = \{(x,y) : x = 0 \ \text{ or } \ x = \frac{1}{n}, n = 1,2,\ldots, \ 0 < y < 1\} .$$

Let

$$A = \{(\frac{1}{n},\frac{1}{n}) : n \ \text{ odd}\} ,$$

$$B = \{(\frac{1}{n},\frac{1}{n}) : n \ \text{ even}\} .$$

Then A and B are closed in X; they intersect no common quasi-component, but they cannot be separated from each other.

Connectivity is a "better" concept, in that it is stronger and closer to our intuition (as the above examples show). However (and this is one of the main philosophical points of this paper), the standard approximation techniques of functional analysis yield non-separation results. To get connectivity, one must fall back on some point set topology; the relevant topology is now discussed.

The following result is the main topological tool to handle

approximations.

PROPOSITION 1.

Suppose X is normal and A,B are closed in X. Let C be a closed subset of X, and C_α a family of closed subsets of X with the properties:

i) $A \cap C$, $B \cap C$ are not separated in any C_α,

ii) for each neighborhood N of C, there is a $C_\alpha \subset N$ (the C_α approximate C). Then $A \cap C$, $B \cap C$ are not separated in C.

Condition ii is equivalent to

ii') if $x_\alpha \in C_\alpha$ and a net in the x_α converges, then $\lim x_\alpha \in C$, and

ii") if $x_\alpha \in C_\alpha$, there exists a convergent net in the x_α.

Proof.

First we show $A \cap C \neq \emptyset$. $B \cap C \neq \emptyset$. If say, $A \cap C = \emptyset$, then $X - A$ is a neighborhood of C_n. Then there exists $C_\alpha \subset X - A$ so $A \cap C_\alpha = \emptyset$ which contradicts i). Suppose U,V separate $A \cap C$, $B \cap C$ in C; say $A \cap C \subset U$, $B \cap C \subset V$. Since U,V are closed in C, hence in X, there exist disjoint open U_1, V_1 on X with $U \subset U_1$, $V \subset V_1$. Let

$$U_2 = U_1 \cap (X \setminus B),\ V_2 = V_1 \cap (X \setminus A) .$$

Then U_2, V_2 are disjoint, open in X, $U = U_2 \cap C$, $V = V_2 \cap C$, and $U_2 \cap B = \emptyset$, $V_2 \cap A = \emptyset$. Since $U_2 \cap V_2$ is a neighborhood of C, there is some $C_\alpha \subset U_2 \cup V_2$. But then $U_2 \cap C_\alpha$, $V_2 \cap C_\alpha$ separate $A \cap C_\alpha$, $B \cap C_\alpha$ in C_α. This contradicts i). Thus the proposition is proved.

Sometimes the difference between connectivity and separation is immaterial. For example, a typical use of connectivity is the following. Let $[a,b]$ be a closed interval, B a Banach space. Let $F : [a,b] \times B \to B$ be an operator. Suppose it is known that the zeros

Z of F (or fixed points) connect the "ends" $\{a\} \times B$, $\{b\} \times B$. Then there is a zero of F for each y , $a \leq y \leq b$. This simple idea uses the two facts:

i) the continuous image of a connected set is connected,

ii) the only connected sets in the real R are points and intervals.

In this particular case, non-separability works as well as connectivity. That is

i') if A,B are not separated in X and $f : X \to Y$ is continuous, then $f(A)$, $f(B)$ are not separated in Y ,

ii') if a,b are not separated in $Y \subset R$, then $Y \supset [a,b]$.

Thus for subsets of R , components and quasi-components are the same. The most useful condition on X that allows equating components and quasi-components is compactness. In fact considerably more is true.

Given A,B closed in X , consider all closed sets C such that $A \cap B$, $B \cap C$ are not separated in C . The collection C of all such C is partially ordered by inclusion. A minimal such set (if one exists) is called irreducible (between A and B).

PROPOSITION 2.

Any irreducible C is connected. Thus if an irreducible C between A and B exists, A and B are connected in X .

Proof.

Suppose U,V separate C . Note that U,V are closed in C , hence in X . If $A \cap U \neq \emptyset$, $B \cap U \neq \emptyset$, then C is not minimal. If $A \cap V = \emptyset$, $B \cap U \neq \emptyset$, then A,B are separated in C . So C cannot admit a separation.

PROPOSITION 3.

If X is compact, A and B closed and not separated in X , then an irreducible C exists.

Proof.

Use Zorn's lemma. To make it work, we need to show that if $C_0 \subset C$ is totally ordered, then $\bigcap_{C \in C_0} C = C_0$ does not separate $A \cap C_0$, $B \cap B_0$. But this follows from Proposition 1, since condition ii there is automatic for compact X (and any compact space is normal).

COROLLARY 4.

If X is compact, A and B closed and not separated in X, then A and B are connected in X.

Hence in particular, the quasi-components of a compact X are components. This last statement has come to be known to analysts as "Whyburn's Lemma", but it goes back to the early days of topology. A complete compendium with attribution on the subject of connectedness in Chapter V of Kuratowski's book [K].

The above proof of Corollary 4 is different from ones usually given. This proof emphasizes irreducible sets, and it may well be possible to produce an irreducible set (which probably amounts to handling condition ii") of Proposition 1) without compactness.

There is further refinement which is quite useful in applications. Often, A and B are "trivial" solutions of a problem and they are shown to be not separated in the set X of all solutions. It is sometimes desirable to find a connected set of non-trivial solutions between A and B. That is, one would like to say something about the components of $X - (A \cup B)$. In general, nothing can be said. There exist connected spaces such that the removal of one point (called a dispersion point) leaves a totally disconnected space. However in the compact case, the best possible result is available.

PROPOSITION 5.

Suppose A and B are closed and not separated in a compact X. Then there exists a connected D in $X - (A \cup B)$ such that $\overline{D} \cap A \neq \emptyset$, $\overline{D} \cap B \neq \emptyset$.

Proof.

Let C be irreducible between A and B. We may assume $A \subset C$, $B \subset C$. The claim is that $D = C \setminus (A \cup B)$ is the desired set. First we show that we can assume that A and B are each a single point. For let C' be the quotient of C where A and B are identified to separate single points a, b resp. Because A and B are closed, the set D is homeomorphic to $D' = C' \setminus \{a,b\}$. Because C is compact the following is true

*) if a is in the closure of D', then $\overline{D} \cap A \neq \emptyset$ (and similarly for B).

Thus we may assume $A = \{a\}$, $B = \{b\}$. Suppose D admits a separation U, V. If a, b are in the closure of U in C, then C is not minimal. (It is easily checked that V is not contained in the closure of $U \cup \{a,b\}$ in C.) If $a \in \overline{U}$, $b \in \overline{U}$, $a \in \overline{V}$, $b \in \overline{V}$, then $U \cup \{a\}$, $V \cup \{b\}$ separate a and b in C. All other possibilities are symmetries in notation. The result is proved.

Remark.

Kuratowski proves this result under different hypotheses. If X is not compact, fact (*) of the proof can fail. In this case, it likely is better to weaken the conclusion accordingly; the weakened result is probably good enough for the application.

2. FIXED POINTS

To emphasize the similarities between fixed points and connectivity, I would like to quickly run through the rudiments of fixed point theory. Suppose $F : D \to B$ is defined on some subset D of a Banach space to another Banach space B. We desire to show that F has a solution (zero of fixed point). The standard method is to approximate F in some sense or other by $F_n: D \to B$. The F_n belong to a class of operators which are known to have a good fixed point theory. Three standard examples (for fixed points) are the following:

Class of F	Class of F_n
compact	finite dimensional
condensing (contracting)	(Leray-Schauder) compact
multi-valued (compact, upper-semi-continuous, convex point images)	single-valued.

Suppose that D is open. There are two types of controls needed on the approximations. One is near the boundary of D. These controls are needed essentially to show the degree theory is well-defined -- in particular that solutions do not "escape" across the boundary. The other type of control is at the solutions -- in particular to make sure the solutions of the approximations approximate those of F. Controls of the first type can be rather loose; essentially they determine what kind of homotopies can be used. Controls of the second type are tighter and more geometric. In the first example above -- from finite-dimensional to compact -- the approximations are uniform on bounded subsets, and nothing moves very far. But in the second example -- from compact to condensing -- the approximating F_n are considerably different from F near the boundary. The boundary is allowed to move a considerable distance. But at the fixed points, the F_n are quite close to F. In fact there is a set containing the fixed points on which F and the F_n are equal.

Here we are interested in the second type of control. Let $S(S_n)$ denote the fixed points of F (resp. F_n). All controls boil down to the following requirements on the S_n :

1. S and the S_n are closed in D (which is usually automatic),
2. if $x_n \in S_n$ and $\{x_n\}$ converges, then $\lim x_n \in S$ (this is usually straightforward),
3. If $x_n \in S_n$, there exists a convergent subsequence (this requires care in choosing the F_n).

It is clear that if each F_n has a fixed point, then so does F. Also, if each S_n is compact, so is $S_\infty = \lim S_n \subset S$. In the next section we discuss analogues of these conditions for connectivity.

<u>Remarks</u>. 1. If D is not open, but is for example, a cone, there may be somewhat different controls on the F_n to ensure the index or degree is well-defined, but the controls at the solutions are the same. This will also be true for connectivity.

2. the above conditions refer to the strong topology. They also work in weak topologies, and effective fixed point theories relying on weak topologies have been developed for, for example, pseudo-monotone operators. It would be extremely useful to carry over to connectivity results arguments based on weak topologies, but it is not clear it can be meaningfully done.

3. APPROXIMATIONS AND CONNECTIVITY

Suppose, as before, that $F : D \to B$ is defined on some subset D of a Banach space to another Banach space B. Let $F_n: D \to B$ be a sequence of operators. Let $S(S_n)$ be the zeros (or fixed points) of F (resp. F_n). Suppose A and B are two closed subsets and that the following conditions are satisfied:

0. $A \cap S_n$, $B \cap S_n$ are not separated in each S_n,
1. S and each S_n are closed in D,
2. if $x_n \in S_n$ and $\{x_n\}$ converges, then $\lim x_n \in S$,
3. if $x_n \in S_n$, then there exists a convergent subsequence.

PROPOSITION 6.

Under these conditions, $A \cap S$ and $B \cap S$ are not separated in S. Moreover, if each S_n is compact, so is $S_\infty = \lim S_n \subset S$.

Proof.

Let $C_n = S \cup \bigcup_{m \geq n} S_m$. It is routine to verify that $S = \cap C_n$ and the conditions of Proposition 1 are satisfied. This proves the result.

4. POINTS AT INFINITY

The one-point compactification is a useful technical tool in topology. In its simplest form, for Euclidean space, the construction consists of adjoining a point ∞ to Euclidean space. A neighborhood basis of ∞ consists of complements of bounded sets. The resulting space is sphere. A similar construction is useful in analysis. It can make statements of results and proofs simpler and more uniform. The simplest form of the construction is to adjoin a point ∞ to a Banach space B to get a space B^+. A neighborhood basis of ∞ is complements of bounded sets. This is not the one-point compactification because Banach space is not locally compact, but it serves a similar purpose.

Suppose that S is a closed subset of B such that the intersection of S with every bounded subset of B has compact closure. Let $S^+ = S \cup (\infty)$ be considered as a subset of B^+. Then S^+ is the true one-point compactification of S. Thus we can use all the machinery of Section 1 on S^+.

Here is a typical example. Suppose $F : B \to B$ is a compact operator with $F(0) = 0$, and we want to show there is an unbounded component of fixed points which contains 0. Suppose we can approximate F by compact (or finite-dimensional) operators F_n defined for x with $x \leq n$. These are to have the properties that $F_n(0) = 0$, that $F_n(x) - F(x) < \frac{1}{n}$ for $x < n$ and that the fixed-point set of F_n connects 0 and the sphere of radius n. From this we can conclude that the fixed point set of F has an unbounded component containing 0. We let

$$S_n^+ = (\text{fixed-point set of } F_n) \cup \{x: x \geq n\}$$
$$S^+ = (\text{fixed-point set of } F) \cup \{\infty\} .$$

We are assuming that 0 and ∞ are connected in each S_n^+. Condition 4 of Proposition 6 is true if $\{x_n\}$ is bounded, and if $\{x_n\}$ is unbounded, it converges to $m \in S^+$. Thus by Proposition 1, the points 0 and ∞ are connected in S^+. By Proposition 5, there is actually an unbounded component of non-zero solutions with 0 in its closure.

Roughly speaking, adjoining a point at ∞ makes some unbounded sets compact and approximations need be uniformly close only on bounded sets.

If F is defined only on $\{x: x < N\}$, there is a variant of the construction. Let B^+ be B with the set $\{x: x \geq N\}$ identified to a point called ∞. Similar arguments work. There is one minor point where one has to be careful. Let S be the fixed-point set of F and $S^+ = S \cup \{\infty\} \subset B^+$. Suppose S^+ connects 0 and ∞. This does not necessarily imply that $\overline{S}$ contains a point x of norm N.

There are two versions of compactness for such an operator F. The stronger definition is that F takes all of its domain to a set with compact closure. Then $\overline{S}$ does contain a point x of norm N. The weaker definition is that F takes sets $\{x\, x \leq M\}$ for $M < N$ to sets with compact closure. In this case we can only assert that S contains points of norm arbitrarily close to N.

There is another version of adjoining points at ∞ that can be useful. Suppose F is parametrized by the reals, so that $F : R \times D \to B$. It might be convenient to adjoin two points $\pm\infty$. A neighborhood basis of $-\infty$, say, is all points (r,x) with $r < -R$, $R = 1,2,\ldots$.

Obviously there are other variations on this theme. The use of points at ∞ does not make the estimates any easier, but the topological arguments are usually cleaner.

5. AN ALGEBRAIC TOPOLOGICAL CRITERION

Since non-separation is a topological concept which is preserved in the limit of approximations, it is not surprising there is an algebraic topological criterion. Occasionally, it is useful to have.

To reemphasize the analogue between fixed-point theory and connectivity methods, consider first the corresponding criterion for fixed points. Here the question is one of existence, so the analogue is a criterion for a space to be non-empty. For any space X (empty or not), let $\varphi : X \to p$ be the unique map to a point p. The criterion is that X is not empty if and only if the induced map on zero-dimensional cohomology $\varphi^* : H^0(p) \to H^0(X)$ is non-trivial.

For connectivity, the criterion is less trivial, and is a one-dimensional rather than zero-dimensional result. For Q a closed subspace of P, let $H^1(P,Q)$ denote the one-dimensional <u>Cech</u> cohomology group of the pair. Let X be a paracompact space, A and B two disjoint closed subsets of X. By Tietze's extension theorem, there is a continuous map $\varphi : X \to [0,1]$ with $\varphi(A) = \{0\}, \varphi(B) = \{1\}$. Since $[0,1]$ is convex, any two such φ are homotopic rel $A \cup B$ (i.e. with A and B not moving throughout the homotopy). Recall that $H^1([0,1], \{0,1\})$ is a copy of the integers Z.

PROPOSITION 7.

The following are equivalent.

1. A and B cannot be separated in X,

2. any such φ is surjective,

3. $\varphi^*: H^1([0,1],\{0,1\}) \to H^1(X, A \cup B)$ is non-trivial.

<u>Proof</u>.

$2 \Rightarrow 1$. If U, V separate A and B, define $\varphi(U) = \{0\}$, $\varphi(V) = \{1\}$. This is a non-surjective φ.

$3 \Rightarrow 2$. If some φ is not surjective, it can be deformed (rel. $A \cup B$) to a map that takes all of X to $\{0\} \cup \{1\}$. Then φ^* is the trivial homomorphism.

$1 \Rightarrow 3$. Note that $[0,1]$ with $\{0,1\}$ identified to a point s_0 is a circle S^1. Let $\hat{X} = X/A \cup B$ be X with A and B identified to a single point x_0. The map φ induces a map $\hat{\varphi}: \hat{X} \to S^1$ with $\hat{\varphi}(x_0) = s_0$. The circle is an Eilenberg-MacLane space $K(Z,1)$. This means in particular that if φ^* is trivial, φ is homotopic rel x_0 to the map that takes all of X to s_0.

Let $\hat{\Phi}_t: \hat{X} \times [0,1] \to S^1$ be the homotopy, thus $\hat{\Phi}_0 = \hat{\varphi}$, $\hat{\Phi}_1(X) = \{s_0\}$. Let $\Phi_t: X \times [0,1] \to S^1$ be the composition of $\hat{\Phi}_t$ with the projection $X \to \hat{X}$. Note that $\hat{\Phi}_t$ induces the trivial homomorphism on the fundamental group of each component; hence so does Φ_t. By the homotopy lifting property, Φ_t lifts to a map $\tilde{\Phi}_t: X \times [0,1] \to R$ (R = reals = universal cover of S^1) with $\tilde{\Phi}_0 = \Phi: X \to [0,1] \subset R$ and $\tilde{\Phi}_1(X) \subset Z \subset R$. Moreover $\tilde{\Phi}_t(A) = \{0\}$, $\tilde{\Phi}_t(B) = \{1\}$ for all t. Let $U = \tilde{\Phi}_1^{\,-1}(0)$, $V = \tilde{\Phi}_1^{\,-1}(Z-\{0\})$; then U, V is a separation of A and B in X.

6. CONTINUATION

Let I be a compact interval $[a,b]$ in the reals. Let D be a convex set in a Banach space and consider continuous $F: I \times D \to B$.

Let Σ denote the solution set of F . Let ∂D be the boundary of D; suppose there are bounds that guarantee $(I \times \partial D) \cap \Sigma = \emptyset$. Suppose F is of a class such that the approximation methods of § 2,3 work (e.g. F compact). Then for any $t \in [a,b]$ the degree of F_t is defined. This degree is independent of t and if it is non-zero the t-slice Σ_t of Σ is non-empty. Continuation asserts more; it asserts Σ_a and Σ_b cannot be separated in Σ .

The term "continuation" comes from the idea that the solutions can be continued across the interval. It is also true if I is not compact; this can be proved using Proposition 1 and the methods of § 4. It is also allowed that the size of D varies continuously in t.

I do not know who first observed the continuation property. It is explicitly discussed by Browder (Br 1960] for D a ball and by Dancer [Da 1973] for $I \times D$ a cone, and is implicit in work of Nussbaum and Rabinowitz. In the finite-dimensional case it has formed the basis of a method in numerical analysis. One numerically "follows" the solution across the interval. See, e.g. [Al-Ge 1980], [Wa 1978], [Pe-Wa 1979] and the references therein for more on this subject. It turns out a large number of topological invariants (but not variational invariants such as Morse index or Lusternik-Schnirelman degree) imply continuation. As a purely topological subject, this has been investigated by K. Alligood [A] 1980].

7. BIFURCATION

Perhaps the major use of connectivity in functional analysis has been to get global bifurcation branches. The seminal paper is [Ra 1971]. It has engendered a large number of generalizations and applications, and we can only touch upon them here.

The latest general version of global bifurcation runs something as follows. Let $F : R^n \times B \to B_1$ be a continuous operator where

$\lambda \in R^n$ is considered a parameter and B and B_1 are Banach spaces. Let S^{n-1} be a (small) sphere around the origin in R^n and let D be a (small) ball around the origin in B. Let $S = \partial D$. Let S_+^{n-1} be S^{n-1} thickened up slightly to an open annulus in R^n. Let S_0^{n-1} be the inner boundary of S_+^{n-1} and S_∞^{n-1} the outer boundary. Suppose as a hypothesis that F has no solutions on $S^{n-1} \times S$. Adjoin a point ∞ to $R^n \times B_1$ as in § 4. Let $\Sigma = F^{-1}(0) \cup \{\infty\}$ and $\hat{\Sigma} = \Sigma - (\Sigma \cap (S_+^{n-1} \times D))$. Global bifurcation is said to occur if $\hat{\Sigma} \cap (S_0^{n-1} \times D)$ and $\hat{\Sigma} \cap (S_\infty^{n-1} \times D)$ are non-empty and cannot be separated in $\hat{\Sigma}$.

In most applications, there are known "trivial" solutions with $x \in B$ equal to zero, and it is known no non-trivial solutions exist in $S_+^{n-1} \times D$. Usually Σ is compact (e.g. F = identity - compact). Then the standard form of the result obtains: there exists a connected subset Σ_0 of non-trivial solutions with the origin in the closure of Σ_0 and either i) Σ_0 is unbounded or ii) there is another trivial solution in the closure of Σ_0 (because of Proposition 5, this is slightly stronger than the usually stated result). If F is not defined on all $R^n \times B$, a third possibility must be added: the closure of Σ_0 contains a point in the complement of the domain of definition. The more general version stated first is convenient for extending the results to new classes of operators, because it works well with approximation methods.

Krasnoselski [Kr 1964) proved a local bifurcation result and showed that degree could be used to guarantee bifurcation. Rabinowitz [Ra 1971] made the conclusion global. For $n = 1$, the sphere S^{n-1} is two points $\lambda_\pm$, one on each side of the bifurcation point $\lambda = 0$. At these points, the local degree of $F_{\lambda_\pm}$, is defined, and if the degrees are different. global bifurcation is guaranteed. This is the most commonly used version of global bifurcation.

Such results have been extended to more general F and in another direction to $n > 1$. In the latter case, degree no longer suffices and more general topological invariants must be developed. In turn, the results for have been extended to more general F. And of course, there have been a large number of applications, some with their own ad hoc versions of global bifurcation. There have also been global bifurcation results using invariants other than topological degree and its generalizations. As two examples, Turner has used general position arguments and Dancer has used arguments from complex analysis. The reference list contains a (surely partial) list of all of these kinds of results.

8. THREE EXAMPLES

Here we illustrate some connectivity methods by three examples. Bifurcation results (non-linear eigenvalue problems) are fairly standard by now, so we illustrate other techniques. For reasons of space, the examples are ones that can be described in a page or two, and the topology is emphasized. For more details, in particular for the a priori estimates, the reader is referred to the original articles.

The first example emphasizes the point that variational arguments (the other main topological method in analysis) do not imply non-separation or connectivity. A free boundary problem is considered that naturally fits into a variational framework. In order to get families of solutions, however, it has to be recast as a parameterized fixed point problem. The second example is one where the approximation methods of § 1 explicitly come into play. The third is one where the existence of a problem is sought, not a connected family. Connectivity is used as a tool in the proof.

<u>Example 1</u>.

In [Au 1977], Auchmuty considers the shape of a self-gravitating rotating incompressible fluid, e.g. a planet. He uses variational techniques to get his result. The reader is referred to that paper for some details. In brief, he considers a (not necessarily connected) body G, an open subset of R^3 rotating around the z-axis. He considers bodies that have an S^1 symmetry around the z-axis (axisymmetric, so the body rotates as a system of infinitesimally thin shells), and a reflection symmetry through the plane $z = 0$. Using cylindrical coordinates (r,θ,z), he further assumes the intersection of G with a vertical line through $(r,\theta,0)$ is a segment $(-u(r), u(r))$ (independent of θ); this defines a height function $u(r)$ that completely describes the shape of G.

The tolal mass of G is

$$M = 4\pi \int_0^\infty ru(r)dr .$$

Let $m(r)$ be the proportion of mass within distance r of the z-axis:

$$m(r) = 4\pi M^{-1}\int_0^r su(s)ds .$$

The gravitational potential at a point $x = (r,0,z) \in R^3$ is

$$V(x) = V_u(s) = \int_G \frac{dy}{x-y} .$$

The physics of the body enters via the rotation law which is prescribed by a C^1 non-decreasing function $j:[0,1] \to [0,\infty]$ with $j(0) = 0$ and $\int_0^1 j(m)dm = 1$ which is the distribution of angular momentum per unit mass. If J is the total angular momentum let $j_J(m) = Jj(m)$. Thus $J = 0$ corresponds to no rotation.

For $x = (r,\theta,z)$, let

$$\varphi_G(x) = \varphi(r,z) = V(x) - \int_r^\infty s^{-3}j_J^2(m(s))ds .$$

This function is, up to a constant, the pressure at the point x. Thus it is not hard to believe (and it is true) that G is a possible shape if and only if there exists $\lambda > 0$ so that

$$\varphi_G(r,z) \quad \begin{cases} > \lambda & \text{if } 0 \leq z < u(r) \\ = \lambda & \text{if } z = u(r) \\ < \lambda & \text{if } z > u(r) \end{cases}$$

for $u(r) > 0$, and

$$\varphi_G(r,z) \leq 0 \quad \text{if} \quad u(r) = 0 \ .$$

In his article, Auchmuty shows that for any $J \geq 0$, there is such a u (and hence G) by a variational argument. In the classical case with

$$j_J(m) = \frac{8J}{5}[2-5(1-m) + 3(1-m)^{5/3}]$$

there is a connected family of solutions, the MacLaurin spheroids. It seems reasonable that for other j , there should be connected families of solutions. However, variational techniques cannot guarantee non-separation. Thus Auchmuty has recast the problem as a fixed-point problem (this is work in progress, and I am indebted to Auchmuty for sharing his notes).

Let X be the space of continuous bounded functions on $[0,\infty]$ with

$$\|u\|_1 = \int_0^\infty r\, u(r)\, dr < \infty \ .$$

Let

$$\|u\| = \max(\sup_r u(r) \ , \ \|u\|_1)$$

and C be the intersection of the positive cone of X with the hyperplane of u which satisfy

$$\int_0^\infty ru(r)dr = \frac{M}{4\pi} \quad M \ \text{fixed.}$$

Auchmuty defines a compact transformation

$$T(u,J) = T_J(u) : C \times [0,\infty) \to C \ .$$

Intuitively, T is the following transformation: Suppose G is a "rotating body" described by the height function u. It sets up a "pressure field" φ_G in R^3. Suppose G' is another body that "feels" this pressure field and thus takes on the shape so that its surface is an isobar of φ_G. If v is the height function of G', then $T_J(u) = v$. Clearly u is a solution to the problem if and only if it is a fixed point of T_J. Auchmuty shows $Tu = v$ is well-defined mathematically.

There are a priori bounds due to Friedman and Turkington (to appear in the Indiana Journal) which show the solutions are bounded in C. And presumably, the sphere is the only solution for $J = 0$ and the degree of T_0 is one. Then continuation applies and there is a global branch of solutions unbounded in J, and the connected family is established. (However, Auchmuty reports that in fact the classical work does not establish that the sphere is the only solution for $J = 0$; the classical arguments are not global. As of this writing, this point remains to be done.)

<u>Example 2</u>.

The following application of Nussbaum and Stuart [Nu-St 1976] considers a singular problem where continuation cannot be directly applied.They approximate the problem by non-singular problems and then apply approximation results as in Proposition 1.

The problem is to study the solutions (u,λ) for $\lambda \geq 0$ of the equation

$$(*) \qquad u'' + \lambda f(x,u(x),u'(x)) = 0 \quad 0 < x < 1$$

$$(*_0) \qquad u(0) = u(1) = 0 \quad .$$

Where f is continuous and has bounds:

there exist continuous $f_1 : (0,\infty) \to [0,\infty)$ with $\lim_{p\to 0} p^{-1} f_1(p) = 0$

and $f_2:(0,\infty) \to (0,\infty)$ such that $f_1(p) \leq f(x,p,q) \leq f_2(p)$ for all $(x,p,q) \in [0,1] \times (0,\infty) \times R$. Thus the problem can be singular whenever $u(x) = 0$.

The authors work in the Banach space $C[0,1]$ of functions continuous on $[0,1]$ with the sup norm. They effectively show there is an unbounded locally compact component S of solutions (u,λ) with $u(x) > 0$ and $u''(x)$ existing continuously for $0 < x < 1$ such that $(0,0) \in \overline{S}$. (Because of our Propositon 5, this result is a little stronger than the authors state. Also we arrange our continuation argument somewhat differently than the author's so as to fit in our context; they quote an equivalent result from [Ra 1973].)

The idea is to modify the boundary condition

$$(*_\varepsilon) \qquad u(0) = u(1) = \varepsilon > 0$$

and look for solutions in $C^2[0,1] \times [0,\infty)$. Nussbaum and Stuart transform the problem into an integral equation via a Green's function. Define

$$f_\varepsilon(x,p,\varepsilon) = \begin{cases} f(x,p,\varepsilon) & p \geq \varepsilon \\ f(x,\varepsilon,q)(>0) & p \leq \varepsilon \end{cases}$$

let

$$g(x,y) = \begin{cases} x(1-y) & 0 \leq x \leq y \leq 1 \\ x(1-x) & 0 \leq y \leq x \leq 1 \end{cases}$$

and consider the compact operator

$$F = F_\varepsilon : C^1[0,1] \to C^2[0,1] \subset C^1[0,1] ,$$

$$(F_\varepsilon u)(x) = \varepsilon + \int_0^1 g(x,y) f_\varepsilon(u,u(y),u'(y))dy .$$

Then it is straightforward to check (u,λ) satisfies $(*),(*_\varepsilon)$ if and only if $u = \lambda F_\varepsilon u$. Let $S_\varepsilon \subset C^1[0,1] \times [0,\infty)$ be the solutions of $u = \lambda F_\epsilon u$.

We work in $C^1[0,1] \times [0,\infty)$ with ∞ appended. Let $\Sigma_\epsilon = S_\epsilon \cup \{\infty\}$. Note Σ_ϵ is compact. The first claim is that $(u,\lambda) = (\epsilon,0)$ is not separated from ∞ in Σ_ϵ. Define

$$\alpha = \alpha_\epsilon : C^1[0,1] \times [0,\infty) \to [0,\infty)$$

by

$$\alpha(u,\lambda) = (\|u-\epsilon\|^2 + \lambda^2)^{1/2} .$$

For each $t > 0$, $\alpha^{-1}(t)$ is a closed "hemisphere" H_t and α gives $(C^1[0,1] \times [0,\infty)) - \{(\epsilon,0)\}$ a product structure $H \times (0,\infty)$. Note that $\{\alpha^{-1}(t,\infty)\} \cup \{\infty\}$ forms a neighborhood basis for ∞ . For $\lambda = 0$, the operator λF_ϵ is zero, so the degree of $I - \lambda F_\epsilon$ around the constant function ϵ is one, hence it is one on H_t for small t . Moreover, there are no solutions on the boundary of any H_t (the boundary of H_t consists of points with $\lambda = 0$) , so continuation can be invoked. For any t_1, t_2, $S_\epsilon \cap H_{t_1}$ and $S_\epsilon \cap H_{t_2}$ cannot be separated in S_ϵ . As $t_1 \to 0$, H_{t_1} gets inside any neighborhood of $(\epsilon,0)$ and as $t_2 \to \infty$, H_{t_2} gets inside any neighborhood of ∞ . Thus invoking Proposition 1, the points $(\epsilon,0)$ and ∞ are not separated in Σ_ϵ .

The authors then note (Cor. 1) that because there is a C^0 bound on u' for $(u,\lambda) \in S_\epsilon$ (via f_2) , the inclusion

$$C^1[0,1] \times [0,\infty) \to C[0,1] \times [0,\infty)$$

induces a proper map on S_ϵ . Thus the map extends to a homeomorphism

$$\Sigma_\epsilon \subset (C^1[0,1] \times [0,\infty)) \cup \{\infty\} \quad \text{to} \quad \Sigma_\epsilon \subset (C[0,1] \times [0,\infty)) \cup \{\infty\} .$$

Henceforth Σ_ϵ is considered as a subspace of $(C[0,1] \times [0,\infty)) \cup \{\infty\}$.

Next one pauses to observe that any solution of $(*)$, $(*_\epsilon)$ is concave down and non-negative on $(0,1)$. Thus S_ϵ lies in the cone

C of such functions.

Now the object is to let $\epsilon \to 0$ and see what happens to Σ_ϵ. The topological idea is to use something like our Proposition 1. This involves obtaining a priori estimates on elements of S_ϵ that are independent of ϵ. The reader is referred to the paper for the estimates; we indicate how the topology works out.

Let $A = \bigcup_{0 \leq \epsilon \leq \epsilon_0} \{(u,0)\ u = \epsilon\}$ and $B = \{\infty\}$. It has been shown that $A \cap \Sigma_\epsilon$ and B cannot be separated in each $\Sigma_\epsilon, \epsilon > 0$. The authors' Lemma 2 precisely proves condition (ii") of our Proposition 1. Their Lemma 3 precisely proves condtion (ii'). Therefore $\Sigma = X \cup \{\infty\}$ is compact and $a \cap \Sigma$ and B cannot be separated in Σ. Their Lemma 4 and Theorem 2 are a proof of our Proposition 1 for this particular case, and thus the result is established.

It is worth noting the significance ot the authors' Lemma 1. Without Lemma 1, it is established that the Σ_ϵ converge to some closed subspace $\hat{\Sigma}$ of $C \cup \{\infty\}$ that does not separate $\{(0,0)\}$ and $\{\infty\}$. However, there is the unpleasant possibility that $\hat{\Sigma}$ contains the identically zero function for some λ. Lemma 1 precludes this possibility by imposing a non-zero lower bound on the u for $(u,\lambda) \in S_\epsilon$ that is independent of ϵ.

<u>Example 3.</u>

Amann, Ambrosetti, and Mancini [Am-Am-Ma 1978] use continuation to study elliptic differential equations defined on a bounded domain $\overline{\Omega} \subset R^n$ (a similar use has been made by Shaw [Sh 1977]). Let L be a second order uniformly elliptic differential operator on $\overline{\Omega}$ and let B a combination of Dirichlet and Neumann boundary conditions. Suppose 0 is the principal eigenvalue of $(L,B=0)$. Let f be a bounded C^1 function defined on $\overline{\Omega} \times R \times R^n$. They study the resonant semi-linear system

$$(*) \qquad \begin{aligned} Lu &= f(x,u,\text{grad } u) \\ Bu &= 0 \, . \end{aligned}$$

A C^2 function v von $\overline{\Omega}$ is called a subsolution if

$$\begin{aligned} Lv &\leq f(x,u,\text{grad } v) \\ Bu &\leq 0 \end{aligned}$$

and a supersolution if the inequalities are reversed. In earlier work Amann has shown that under weaker hypotheses, (*) has a solution if there exist a subsolution $\overline{v}$ and a supersolution $\hat{v}$ with

$$(**) \qquad \overline{v} \leq \hat{v} \text{ on } \overline{\Omega} \, .$$

In this paper it is shown the boundedness of f allows the dropping of (**).

They work in the Sobolev space $W_p^2(\overline{\Omega})$ and let

$$X = \{u \in W_p^2(\overline{\Omega}) \; Bu = 0\} \, .$$

Then (*) defines a Fredholm operator of index 0:

$$L:X \to Y = L_p(\overline{\Omega}) \, .$$

(They have a regularity result which says any solution in X is a classical pointwise solution.) In fact the null space of L and its transpose L' are one-dimensional and are spanned by strictly positive mooth functions φ, ψ respectively. Also the Nemytskii operator $F:X \to Y$ induced by $-f$ is compact (by the Sobolev embedding theorem) and $F(X)$ is bounded (by the boundedness of f).

Write $X = R\varphi \oplus W$ and $Y = R\psi \oplus R_L$ (R_L = range L). Then $L\, W:W \to R_L$ has a bounded inverse $H_0:R_L \to W$. Let $Q_0:Y \to R_L$ be the projection parallel to $R\psi$. A Lyapunov-Schmidt reduction implies that for $u = t\varphi + \omega \in N \oplus W$, the equation $Lu + Fu = 0$ is equivalent to

$$\omega + H_0Q_0F(t\varphi+\omega) = 0$$

$$< \psi , F(t\varphi+\omega) > = 0 .$$

Let $\Sigma \subset R \times W$ be the set of (t,ω) such that $(t\varphi,\omega)$ satisfy the first of these Lyapunov-Schmidt equations. Since $F(X)$ is bounded, the set of ω occuring in Σ is bounded in X. A maximum principle implies there is a positive constant τ such that $-\tau\varphi \leq \omega \leq \tau\varphi$ for all ω occuring in Σ. Let $\bar{v}$ and $\hat{v}$ be sub- and supersolutions of (*). There is a constant $\alpha > \tau$ such that $(\alpha-\tau)\varphi \geq \bar{v}$, $-(\alpha-\tau)\varphi \leq \hat{v}$.

At this point the authors bring in continuation. Let

$$G(t,\omega) = \omega + H_0Q_0F(t\varphi + \omega) : R \times W \to W .$$

Since $F(X)$ is bounded, the zeros of G lie in some $R \times D$ where D is a large enough ball in W. Also G is a compact operator with Leray-Schauder degree 1. Thus by continuation, the slices $\Sigma_{-\alpha}$ and Σ_α of Σ cannot be separated in Σ (and by the compactness of Σ, they are connected in Σ).

Define $\gamma(t,\omega) = < \psi , F(t\varphi + \omega) > \in R$. Then $\gamma(\Sigma_{-\alpha})$ and $\gamma(\Sigma_\alpha)$ are connected by an interval I. If $0 \in I$, the second of the Lyapunov-Schmidt equations is satisfied and (*) has a solution. If $I \subset (0,\infty)$, then $L(t\varphi+\omega) + ((t\varphi+\omega) \geq 0$, so $\alpha t + \varphi \in X_\alpha$ is a supersolution v_1. Also $\hat{v}_1 \geq \bar{v}$. Therefore (**) is satisfied and (*) has a solution. If $I \subset (-\infty,0)$, a new subsolution $\bar{v}_1$ is obtained with $\hat{v} > \bar{v}_1$. In all cases (*) has a solution.

REFERENCES

General topological reference for connectivity:

[K] K. Kuratowski, Topology, vol. 2; Academic Press, N.Y. (1968).

Articles referred in the text:

Al 1980, K. Alligood, Topological conditions for the continuation of fixed points, to be published.

Al-Ge 1980, E. Allgower and K. Georg, Simplicial and continuation methods for approximating fixed points and solutions to systems of equations. SIAM Review 22 (1980), 28-85.

Am-Am-Ma 1978, H. Amann, A. Ambrosetti, and G. Mancini, Elliptic equations with non-invertible Fredholm linear part and bounded nonlinearities, Math.Z. 158 (1978), 179-194.

Au 1977, J. F. G. Auchmuty, Models of rotating self-gravitating liquids, to appear in Trends in Applications of Pure Mathematics to Mechanics, Pitman.

Br 1960, F. L. Browder, On continuity of fixed points under deformation of continuous mappings, Summa Br. Math. 4 (1960), 183-191.

Da 1973, E.N. Dancer. Global solution branches for positive mappings, Arch. Rat. Mech. and Anal. 52 (1973), 181-192.

Kr 1964, M.A. Krasnoselski, Topological Methods in the Theory of Nonlinear Integral Equations, MacMillan, N.Y., 1964.

Nu-St 1976, R. D. Nussbaum and C. A. Stuart, A non-singular bifurcation problem, J. London Math. Soc. 14 (1976), 31-38.

Ra 1971, P.H. Rabinowitz, Some global results for non-linear eigenvalue problems, J. Func. Anal. 7 (1971), 487-513.

Ra 1973, R. H. Rabinowitz, Some aspects of non-linear eigenvalue problems, Rocky Mountain Math. J. 3 (1973), 161-202.

Pe-Wa 1979, H-O. Peitgen and H-O. Walther, ed., Functional Differential Equations and Approximations of Fixed Points, Springer Lecture Notes # 730, 1979.

Sh 1977, H. Shaw, A non-linear elliptic boundary value problem at resonance, J. Diff. Eq. 26 (1977), 335-346.

Wa 1978, H-J. Wacker, Continuation Methods, Academic Press, 1978.

General reference list. (This list is surely incomplete. Apologies are offered to any omitted author.)

J. C. Alexander, Bifurcation of zeros of parametrized function, J. Func. Anal. 29 (1978), 37-53.

J. C. Alexander and J. F. G. Auchmuty, Global bifurcation of waves, Manus. Math. 27 (1979), 159-166.

J. C. Alexander and P. M. Fitzpatrick, The homotopy of certain spaces of nonlinear operators, and its relation to global bifurcation of the fixed points of parametrized condensing operators, J. Func. Anal. 34 (1979), 87-106.

J. C. Alexander and P. M. Fitzpatrick, Galerkin approximations in several parameter bifurcation problems, Math. Proc. Camb. Phil. Soc. 87 (1980), 489-500.

J. C. Alexander and P. M. Fitzpatrick, Global bifurcation for solutions of equations involving several parameter multivalued condensing mappings, to be published.

J. C. Alexander and J. A. Yorke, Global bifurcation of periodic orbits. Am. J. Math. 100 (1978), 263-292.

C. J. Amick and J. F. Toland, Finite amplitude solitary waves, to be published.

H. Amann, A. Ambrosetti, and G. Macini, Elliptic equations with noninvertible Fredholm linear part and bounded nonlinearities, Math. Z. 158 (1978), 179-194.

A. Ambrosetti and P. Hess, Positive solutions af asymptotically linear eigenvalue problems, to be published.

S. S. Antman, Bifurcation problems for nonlinearly elastic structures, in Symp. on Applications of Birfurcation Theory, P. M. Rabinowitz, ed., Academic Press, New York (1977), 73-124.

S. S. Antman, Buckled states of nonlinearly elastic plates, Arch. Rat. Mech. Anal. 67 (1978), 111-149.

S. S. Antman, Nonlinear eigenvalue problems for whirling elastic strings, Proc. Royal Soc. Edin. 85A (1980), 59-85.

S. S. Antman and Nachman, Large buckled states of rotating rods, Nonlinear Anal. 4 (1980), 303-327.

S. S. Antman and J. E. Dunn, Qualitative behavior of buckled nonlinearly elastic arches, J. of Elas. (1980), in press.

S. S. Antman, Geometric aspects of global bifurcation in nonlinear elasticity, in Geometric Methods in Physics, G. Kaiser and J. E. Marsden, eds., Springer-Lecture Notes in Math. #775 (1980), 1-29.

S. S. Antman, Global analysis of problems from nonlinear elastostatics, in Applications of Nonlinear Analysis, M. Amann, N. Bazley, K. Kirchgassner, eds., Pitman, to appear.

S. S. Antman and G. Rosenfeld, Global behavior of buckled states of nonlinearly elastic rods, SIAM Review 20 (1978), 513-566. Corrections and additions, ibid. 22 (1980), 186-187.

S. N. Chow and J. Mallet-Paret, The Fuller index and global Hopf bifurcation, J. Diff. Eq. 29 (1978), 66-85.

S. N. Chow, J. Mallet-Paret and J.A. Yorke, Global Hopf bifurcation from a multiple eigenvalue, Nonlinear Analysis 2 (1978), 753-763.

F. E. Browder, On continuity of fixed points under deformation of continuous mapping, Summa. Bras. Math. 4 (1960), 183-191.

M. G. Crandall and P. H. Rabinowitz, Nonlinear Sturm-Liouville eigenvalue problems and topological degree, J. Math. Mech. 19 (1970), 1083-1102.

E. N. Dancer, Global solution branches for positive mappings. Arch. Rat. Mech. Anal. 52 (1973), 181-192.

E. N. Dancer, Global structure of the solutions of nonlinear real analytic eigenvalue problems, Proc. London Math. Soc. 27 (1973), 747-765.

E. N. Dancer, On the structure of solutions on nonlinear eigenvalue problems, Ind. Univ. Math. J. 23 (1974), 1069-1076.

E. N. Dancer, Solution branches for mappings in cones, and applications, Bull. Aust. Math. Soc. 11 (1974), 131-143.

E. N. Dancer, A note on bifurcation from infinity, Quart. J. Math. Oxford, 25 (1974), 81-84.

G. Hetzer, Bifurcation theorems of Rabinowitz type, Nonlinear Anal. 1 (1977), 471-479.

J. Ize, Bifurcation theory for Fredholm operators, Mem. Am.Math. Soc. 174 (1976).

J. Ize. Periodic solutions of nonlinear parabolic equations, to be published.

G. Keady and J. Norbury, On the existence theory for irrotational water waves, Math. Proc. Camb. Phil. Soc. 83 (1978), 137-157.

A. Lev. Branching of solutions of equations in Banach spaces without multiplicity assumptions, Proc. London Math. Soc., 37 (1978), 306-341.

J. A. MacBain, Local and global bifurcation from normal eigenvalues I and II, Pac. J. Math. 63 (1976), 445-466, 74 (1978), 143-152.

R. J. Magnus, A generalization of multiplicity and the problem of bifurcation, Proc. London Math.Soc. 32 (1976), 251-278.

R. D. Nussbaum, A global bifurcation theorem with applications to functional differential equations, J. Func. Anal.19 (1975),319-339.

R. D. Nussbaum, Periodic solutions of some nonlinear integral equations, in Dynamical System, Academic Press, New York (1977), 221-249.

R. D. Nussbaum, A periodicity threshold theorem for some nonlinear integral equations. SIAM J. Math. Anal. 9 (1978) 356-376.

R. D. Nussbaum, A Hopf global bifurcation theorem for retarded functional differential equations, Trans. Am. Math. Soc. 238 (1978), 139-164.

R. D. Nussbaum and C. A. Stuart, A singular bifurcation problem, J. London, Math. Soc. 14 (1976), 31-38.

P. M. Rabinowitz, Some global results for non-linear eigenvalue problems. J. Func. Anal. 7 (1971), 487-513.

P.M. Rabinowitz, Some aspects of nonlinear eigenvalue problems, Rocky Mountain Math. J. 3 (1973) ,161-202.

P.M. Rabinowitz, On bifurcation from infinity, J. Diff. Eq. 14 (1973), 462-475.

P. M. Rabinowitz, A note on pairs of solutions of a nonlinear Sturm-Liouville problem, Manus. Math. 11 (1974), 273-282.

H. Shaw, A nonlinear elliptic boundary value problem at resonance, J. Diff. Eq. 26 (1977), 335-346.

C. A. Stuart, Some bifurcation theory for k-set contractions, Proc. London Math. Soc. 27 (1973), 531-550.

C. A. Stuart, Solutions of large norm for nonlinear Sturm-Liouville problems, Quart, J. Math. Oxford, 24 (1973), 129-139.

C. A. Stuart, Global properties of components of solutions of non-linear second order ordinary differential equations on the half line, Ann. Sc. Norm. Sup. Pisa.

C. A. Stuart, Existence theory for the Hartree equation, Arch. Rat. Mech. and Anal. 51 (1973), 60-69.

C. A. Stuart, Spectral theory of rotating chains, Proc. Royal Soc. Edin. 12 (1974/75), 199-214.

C. A. Stuart, Steadily rotating chains, in Math. Lecture Notes # 503 (1976), 490-499.

J. F. Toland, Asymptotic linearity in nonlinear eigenvalue problems, Quart. J. Math. Oxford, 24 (1973), 241-250.

J. F. Toland, Global bifurcation for k-set contractions without multiplicity assumptions, Quart. J. Math. Oxford 27 (1976), 199-216.

J. F. Toland, Global bifurcation theory via Galerkin's method, Nonlinear Anal. 1 (1977), 305-317.

J. F. Toland, On the existence of a wave of greatest height and Stoke's conjecture, Proc. Royal Soc. London 363 (1978), 469-485.

R. E. L. Turner, Nonlinear Sturm-Liouville problems, J. Diff. Eq. 10 (1971); 141-146.

R. E. L. Turner, Nonlinear eigenvalue problems with application to elliptic equations, Arch. Rat. Mech. Anal. 42 (1971), 184-193.

R. E. L. Turner, Transversality in nonlinear eigenvalue problems, in Contributions to Nonlinear Functional Analysis, Academic Press, New York (1971), 37-67.

R. E. L. Turner, Transversality and cone maps, Arch. Rat. Mech. Anal. 58 (1975), 151-179.

R. E. L. Turner, Superlinear Sturm-Liouville problems, J. Diff. Eq. 13 (1971), 151-171,

J. H.Wolkowisky, Existence of buckled states of circular plates, Comm. Pure Appl. Math. 20 (1967), 549-560.

J. H. Wolkowisky, Nonlinear Sturm-Liouville problems, Arch. Rat. al. 35 (1969), 299-320.

J. H. Wolkowisky,Branches of periodic solutions of the nonlinear Mill's equation, J. Diff. Eq. 11 (1972), 385-400.

FIXED POINT THEORY FOR NONEXPANSIVE MAPPINGS

By

W.A. KIRK(*)

Department of Mathematics
The University of Iowa
Iowa City, Iowa 52242

0. Introduction

Probably the fixed point theorem most frequently cited in analysis is the 'Banach contraction mapping principle', which asserts that if (M,ρ) is a complete metric space and if T is a self-mapping of M which satisfies for fixed $k < 1$ and all $x,y \in M$, $\rho(T(x),T(y)) \leq k\rho(x,y)$, then T has a unique fixed point in M, and moreover for each $x \in M$ the Picard iterates $\{T^n(x)\}$ converge to this fixed point. Within the context of complete metric spaces the asumption $k < 1$ is crucial even to the existence part of this result, but within more restrictive yet quite natural settings an elaborate fixed point theory exists for the case $k = 1$. Mappings in this wider class are called *nonexpansive*.

Our purpose here is to review the more fundamental aspects of the development of the nonexpansive theory and, in particular, to describe in precise terms what is currently known about the following central question. Given a Banach space X and a nonempty (and, generally, bounded closed convex) subset K of X, what further assumptions on K (or X) guarantee the existence of fixed points for every nonexpansive self-mapping of K? We also discuss a number of additional topics; principally ones which have evolved directly from the study of the above question. These include the existence of common fixed points for commuting families of nonexpansive mappings, the existence of fixed points for uniformly k-lipschitzian mappings for $k > 1$, and certain extensions of the theory to wider classes of spaces. However in this secondary respect our scope is limited. We do not treat the broad auxiliary theory in which the self-mapping assumption is replaced with various boundary or inwardness assumptions, nor do we discuss the relationship between the nonexpansive theory and the theory of accretive operators. And while we touch upon iterative techniques for appro-

*Research supported in part by National Science Foundation grant MCS-8001604.

ximating fixed points of non-expansive mappings, no attempt is made to document the vast literature on this subject.

For the most part we use conventional notation: $\overline{\text{conv}}\, S$ denotes the closed convex hull of a subset S of a Banach space X, $\text{diam}(S) = \sup\{\|x-y\| : x,y \in S\}$, and for $u,v \in X$ and $r > 0$, $B(x;r) = \{x \in X : \|u-x\| \le r\}$;

$$\text{seg}[u,v] = \{tu + (1-t)v : t \in [0,1]\}.$$

1. The Central Question

The study of the problem of determining those subsets of Banach spaces which have the fixed point property for nonexpansive self-mappings has its origins in four papers which appeared in 1965. In the first of these ([8]), F. Browder, drawing on concepts from the theory of monotone operators, proved that bounded closed convex subsets of Hilbert spaces have this property, and subsequently in [9], using a more direct argument much like the one given below, he extended this result to the much wider class of all uniformly convex spaces. At the same time this latter result was obtained independently by D. Göhde ([27]), while in [36] the present writer, exploiting a property shared by all uniformly convex spaces, obtained the same result for an even wider class of spaces. In order to describe these results in precise terms we need two definitions, the first of which describes a property shown by Clarkson to hold in all the standard ℓ^p and L^p spaces for $1 < p < \infty$, while the second is a concept introduced by Brodskii and Milman in connection with their study of fixed points of isometries.

DEFINITION 1

A Banach space X is said to be *uniformly convex* ([17]) if for each $\varepsilon > 0$ there exists $\delta(\varepsilon) > 0$ such that for $x,y \in X$, the conditions $\|x\| \le 1$, $\|y\| \le 1$, and $\|x-y\| \ge \varepsilon$ imply $(1/2)\|x+y\| \le 1 - \delta(\varepsilon)$.

DEFINITION 2

A convex subset K of X is said to have *normal structure* ([7]) if each bounded convex subset H of K for which $\text{diam}(H) > 0$ contains a point z such that

$\sup\{\|x-y\| : x \in H\} < \mathrm{diam}(H)$.

If H is a convex subset of a uniformly convex space with $\mathrm{diam}(H) = d \in (0,\infty)$, and if $z = (1/2)(u + v)$ where $u,v \in H$ are chosen so that $\|u-v\| \geq d/2$, then it follows routinely that for all $x \in H$, $\|z-x\| \leq d(1-\delta(d/2))$. Thus all convex subsets of uniformly convex spaces have normal structure (as do compact convex sets in arbitrary spaces ([7])). Certain generalizations of uniform convexity are also known to imply normal structure, for example 'uniform convexity in each direction' ([18]) and 'K-uniform rotundity' ([55]). Other standard assumptions are also known to imply normal structure, for example, Opial's condition; see [28].

We now suppose K is a nonempty weakly compact convex subset of a Banach space X and that K has normal structure. If $T : K \to K$ is a given nonexpansive mapping, then a routine application of Zorn's lemma implies that K contains a minimal T-invariant nonempty closed convex subset H. Suppose $d = \mathrm{diam}(H) > 0$. By normal structure there exists $r \in (0,d)$ for which

$$C = \{z \in H : H \subset B(z;r)\} \neq \phi.$$

Let $z \in C$. By nonexpansiveness of T, $T(H) \subset B(T(z);r)$; hence $\overline{\mathrm{conv}}\, T(H) \subset B(T(z);r)$. But $\overline{\mathrm{conv}}\, T(H)$ is also weakly compact and T-invariant, so by minimality of H, $\overline{\mathrm{conv}}\, T(H) = H$. Hence $H \subset B(T(z);r)$, proving $T(z) \in C$, that is, C is T-invariant. Also C is closed and convex (thus weakly closed), so again by minimality, $C = H$. But if $u,v \in C$, then $\|u-v\| \leq r$ from which $\mathrm{diam}(C) \leq r < d = \mathrm{diam}(H)$. This contradicts the assumption $d > 0$ and proves the following.

(1) ([36]; cf. [9],[27]). *Let* K *be a nonempty weakly compact convex subset of a Banach space* X, *and suppose also that* K *has normal structure. Then every nonexpansive mapping* $T : K \to K$ *has a fixed point.*

It is quite easy to find examples (see, for example, [9],[36]) of bounded closed convex subsets of Banach spaces for which fixed point free nonexpansive self-mappings exist. Also (see [8],[36]), if B is the closed unit ball in ℓ^2, $\varepsilon > 0$, $e_1 = (1,0,0,\ldots)$ and S the right shift operator, then the mapping $T : B \to B$ defined by

$$T(x) = \varepsilon(1-\|x\|)e_1 + S(x)$$

is a fixed point free mapping having Lipschitz constant $1+\varepsilon$. Such examples define in general terms the limits of the theory: nonexpansiveness of T is essential, and additional restrictions must be placed on K.

Motivated by Göhde's proof of [27], Browder in 1968 formalized a useful 'demi-closedness' principle for mappings U for which $T = I-U$ is nonexpansive. Its fixed-point-theoretic formulation is the following. (Notice that here T is not a self-mapping.)

(2) ([11]). *Let* K *be a bounded closed convex subset of a uniformly convex Banach space* X *and let* T *be a nonexpansive mapping of* K *into* X. *Suppose for* $\{x_n\}$ *in* K, $x_n - T(x_n) \to 0$ *(strongly) while* $x_n \rightharpoonup x_0$ *(weakly). Then* $T(x_0) = x_0$.

While (2), in its present setting, provides no new information concerning the *existence* of fixed points for self-mappings of K, it is the prototype of a result which could be used to establish existence for such mappings in more general spaces. This is because if K is a bounded closed convex subset of an arbitrary space and if $T: K \to K$ is nonexpansive, then it is always possible to uniformly approximate T on K with contraction mappings and to conclude that $\inf\{\|x-T(x)\| : x \in K\} = 0$. Thus if in addition X is known to be reflexive (or K weakly compact), it is possible to obtain a sequence $\{x_n\}$ in K which satisfies the hypothesis of (2). To date, however, (2) has not been extended beyond the framework of uniformly convex spaces, and indeed this may well be its natural setting. Browder's proof (see [11]) makes very strong use of the uniform convexity structure of the space in conjunction with a clever thinning of the sequence $\{x_n\}$.

Although a number of important peripheral results were obtained shortly after 1965, no essential progress was made on the central question until the period 1974-76 when Karlovitz obtained explicit positive results ([34],[35]), results which in fact triggered a renewed vigorous interest in the problem. Karlovitz's approach involved the introduction of a generalized notion of orthogonality.

If w and v are elements of a Banach space X, then w is said to be *orthogonal* to v if for all scalars λ, $\|w\| \le \|w+\lambda v\|$. Symmetry of orthogonality is known to characterize Hilbert spaces among Banach spaces of dimension greater than 2. But weaker forms of symmetry hold in certain spaces. Orthogonality in X is said to be *uniformly approximately symmetric* (for example ℓ^p, $1 < p < \infty$) if for each $x \in X$ and $\varepsilon > 0$ there exists a closed linear subspace $U = U(x,\varepsilon)$ of finite codimension and a number $\delta = \delta(x,\varepsilon) > 0$ such that for each $u \in U$ with $\|u\| = 1$,

$$\|u\| \le \|u+\lambda x\| - \delta \quad \text{when} \quad |\lambda| \ge \varepsilon.$$

If in addition X is a conjugate space and U can be chosen to be weak* closed, then orthogonality is said to be *weak* uniformly approximately symmetric* (for example, ℓ^1;

James's space J_0 [32]). In [34] Karlovitz establishes the fixed point property for nonexpansive self-mappings of nonempty bounded convex closed (respectively weak* closed) subsets of reflexive (respectively, duals of separable) spaces which possess these respective orthogonality properties. While this conclusion was already known in the special ℓ^p case for $1 < p < \infty$, the extension of the theory to nonreflexive spaces represented a signifiant development. The following is a special case of Karlovitz's general result.

(3) ([34]). *Nonempty bounded weak* closed convex subsets of* ℓ^1 *and of* J_0 *have the fixed point property for nonexpansive self-mappings.*

We now define a class of spaces which has recently been the object of very intensive study ([4]).

DEFINITION 3

For $\beta \geq 1$, let X_β be the Hilbert space ℓ^2 renormed by taking

$$\|x\| = \max\{\|x\|_2, \beta\|x\|_\infty\}, \quad x \in \ell^2.$$

Since $\|x\|_2 \leq \|x\| \leq (\beta+1)\|x\|_2$, the space X_β are not only reflexive but also superreflexive, and moreover it is known that for $\beta \geq \sqrt{2}$, X_β fails to have normal structure. (This observation is due to R.C. James; see [6].) The signifiance of the following, which is also due to Karlovitz, lies in the fact that at least in certain reflexive spaces, the normal structure assumption of Theorem 1 is not essential.

(4) ([35]). *Nonempty bounded closed convex subsets of* X_β *for* $\beta = \sqrt{2}$ *have the fixed point property for nonexpansive self-mappings.*

In 1977, K. Goebel and T. Kuczumow proved (see [25]) that certain closed convex, but non-weak* compact, subsets of ℓ^1 also have the fixed point property for nonexpansive self-mappings, and in addition they discovered some surprising pathology: it is possible in ℓ^1 to have a descending sequence $\{K_n\}$ of bounded closed convex sets with the property that for n odd, K_n has the fixed point property for nonexpansive self-mappings, while for n even, K_n fails to have this property, and moreover the sequence $\{K_n\}$ may be defined in such a manner that $\bigcap_{n=1}^{\infty} K_n$ is nonempty, and by pre-choice either does or does not have the fixed point property.

Research announced in 1978 included a noteworthy development within the context of Banach lattices. R. Sine and P. Soardi independently obtained results which imply (implicity) the following. (See also, Ray and Sine [47].)

(5) (cf. [53],[54]). *Let* S *be a compact stonian space (completely regular and extremally disconnected) and let* $C(S)$ *denote the space of all continuous real valued functions defined on* S. *Suppose* I *is a closed order interval in* $C(S)$. *Then every nonexpansive mapping of* I *into itself has a fixed point.*

The key to the proof of (5) amounts to showing that a bounded closed set M in $C(S)$ always has an optimal Chebyshev center, that is, there exists a point $x_0 \in C(S)$ such that M is contained in a closed ball centered at x_0 with radius $(1/2)\mathrm{diam}(M)$. (5) has the following corollary for classical (nonreflexive) spaces.

(6) ([53],[54]). *Closed balls in* $L^\infty(\Omega,\mu)$ *over a finite (or* σ*-finite) measure space have the fixed point property for nonexpansive self-mappings.*

Also in 1978, E. Odell and Y. Sternfeld obtained a positive result in c_0, a space whose norm seemingly fails to have any nice geometric properties. (Recall that c_0 is the space of all sequences $\{x_n\} = x$ of real numbers which converge to 0, with $\|x\| = \sup\{|x_n| : n \in \mathbb{N}\}$.)

(7) ([45]). *The closed convex hull of a weakly convergent sequence in* c_0 *has the fixed point property for nonexpansive self-mappings.*

The proof of the above is technically quite complex. It may be assumed (without loss) that $K = \overline{\mathrm{conv}}\{x_n : n \in \mathbb{N}\}$ where the sequence $\{x_n\}$ is weakly convergent to 0 and $\|x_n\| \le 1$ for each n. By approximating T with contraction mappings it is possible to obtain a sequence $\{y_n\}$ in K such that $\|y_n - T(y_n)\| \to 0$ as $n \to \infty$, and by passing to a subsequence it may be supposed that $y_n \rightharpoonup y_0$ (weakly) while $\|y_n - y_0\| \to r \ge 0$. The complexity arises in showing that by assuming $r > 0$ it is possible to construct a new set $\{w^\varepsilon\}_{\varepsilon>0}$ such that $\|w^\varepsilon - T(w^\varepsilon)\| \to 0$ as $\varepsilon \to 0$ and at the same time $w^\varepsilon \to z \in K$ (strongly). (Thus $T(z) = z$.)

In light of (2) there is another observation in [45] that is of interest. Namioka has shown ([43]) that in any weakly compact convex set K in a Banach space, the set $D = \{x \in K : \{x_n\}$ in K and $x_n \rightharpoonup x$ weakly $\Rightarrow \|x_n - x\| \to 0\}$ is a weakly dense G_δ subset of K. It is noted in [44] that if in addition K is a subset of

c_0, then the set D is in fact norm dense in K. Since it is always possible to obtain $\{y_n\}$ in such K for which $\|y_n - T(y_n)\| \to 0$ and $y_n \rightharpoonup y_0 \in K$ (weakly), y_0 would be a fixed point of T if it could be shown that $y_0 \in D$.

We now discuss several results which have been obtained within the past two years. The precise chronology is difficult to determine, so matters are taken up more or less as they came to the writer's attention.

In [4], Baillon and Schöneberg introduce the following.

DEFINITION 4

A subset K of a Banach space has *asymptotic normal structure* if for each bounded convex subset H of K with $\mathrm{diam}(H) > 0$ and each sequence $\{x_n\}$ in H for which $\|x_n - x_{n+1}\| \to 0$, there exists $x_0 \in H$ such that

$$\liminf_n \|x_n - x_0\| < \mathrm{diam}(H).$$

The class of spaces which normal structure is properly contained in the class just defined, as evidenced by the following facts ([4]).

(i) The space X_β has normal structure if and only if $\beta < \sqrt{2}$.

(ii) The space X_β has asymptotic normal structure if and only if $\beta < 2$.

(8) ([4]). *Nonempty weakly compact convex subsets of Banach spaces which have asymptotic normal structure have the fixed point property for nonexpansive self-mappings.*

The proof of (8) is a reasonably straightforward application of ideas which by now have become standard in the theory, while the proof of (ii), which decisively shows that (8) is a generalization of (1), is more complex. In view of (ii), (8) also properly includes (4). But in this direction slightly more is known, although the proof seems to require a major escalation in complexity.

(9) ([4]). *Nonempty bounded closed convex subsets of* X_β *for* $1 \le \beta \le 2$ *have the fixed point property for nonexpansive self-mappings.*

W.L. Bynum also has obtained results which include (4), but by a different approach. He proves in [16] that associated with any reflexive space X there is a number $WCS(X) \geq 1$ (the weakly convergent sequence coefficient of X) for which the following is true: nonempty bounded closed convex subsets of a Banach space X have the fixed point property for nonexpansive self-mappings if there exists a uniformly convex Banach space Y for which the Banach-Mazur distante $d(X,Y)$ satisfies $d(X,Y) \leq WCS(Y)$. Bynum also shows that for the ℓ^p spaces, $1 < p < \infty$, $WCS(\ell^p) = 2^{1/p}$. For $p = 2$ this yields the following fact (cf. (4)), which according to Bynum was already known to Baillon. *Let* $(X, \|\cdot\|_2)$ *be a Hilbert space and let* $\|\cdot\|_\infty$ *be any norm on* X *for which* $\|\cdot\|_\infty \leq \|\cdot\|_2$. *Renorm* X *according to:* $\|x\| = \max\{\|x\|_2, \sqrt{2}\|x\|_\infty\}$. *Then nonempty bounded closed convex subsets of* $(X, \|\cdot\|)$ *have the fixed point property for nonexpansive self-mappings.*

The proof given at the outset for Theorem (1) carries over under the assumption that K is a nonempty weak* compact convex subset of a conjugate space, with K having *weak* normal structure* in the sense that every weak* compact convex set $H \subset K$ with $\mathrm{diam}(H) > 0$ contains a point z such that $\sup\{\|z-x\| : x \in H\} < \mathrm{diam}(H)$. T.C. Lim proves in [41] that ℓ^1 has weak* normal structure, thus providing an alternate proof of the result (3). Lim also utilizes a renorming idea due to Bynum to devise the following counterexample.

EXAMPLE ([41]).

Let c_0 be the sequences of real numbers which converge to 0 with $\|x\|_\infty = \sup_{i\geq 1}|x_i|$ for $x = \{x_i\} \in c_0$. For each such x, let x^+ and x^- be the respective positive and negative parts of x, and renorm c_0 by taking

$$|x| = \|x^+\|_\infty + \|x^-\|_\infty.$$

The dual of $(c_0, |\cdot|)$ is isometrically isomorphic to $(\ell^1, \|\cdot\|)$ with the norm $\|\cdot\|$ defined by

$$\|x\| = \max\{\|x^+\|_1, \|x^-\|_1\}, \quad x \in \ell^1.$$

Define $K \subset (\ell^1, \|\cdot\|)$ by:

$$K = \{x = \{x_i\} \in \ell^1 : x_i \geq 0, \textstyle\sum x_i \leq 1\}.$$

Then K is weak* compact and convex, while the mapping $T : K \to K$ defined by

$$T(x) = (1-\sum x_i, x_1, x_2, \ldots)$$

is affine and isometric, yet fixed point free.

The question of whether all weakly compact convex subsets of c_0 have the fixed point property remains open, but there has been further progress. An additional class (cf. (7)) of weakly compact sets has been discovered for which the conclusion holds. These are the weakly compact coordinatewise star-shaped sets. A subset K of c_0 is said to be *coordinatewise star-shaped* ([30]) if there exists $z \in K$ such that for each $x \in K$ and $y \in c_0$, if $y_i \in [z_i, x_i]$, then $y \in K$. Such sets may fail to be convex (and conversely), but they are always star-shaped in the usual sense. Haydon and the others ([30]) prove that *weakly compact coordinatewise star-shaped subsets of* c_0 *have the fixed point property for nonexpansive self-mappings.* In this instance it is proved additionally that such a fixed point can be obtained in a constructive manner.

It might be noted at this point that in each of the results stated thus far, the domain of the mapping is assumed bounded. This assumption appears to be essential. Indeed, W.O. Ray has recently shown in [47] that a closed convex subset of ℓ^2 has the fixed point property for nonexpansive self-mappings *only if* it is bounded. This is somewhat remarkable in view of the fact that if K is a nonempty closed convex linearly bounded subset of ℓ^2, or more generally of ℓ^p, $1 < p < \infty$ ([46]), then for any nonexpansive $T : K \to K$, $\inf\{\|x-T(x)\| : x \in K\} = 0$. Thus unbounded sets may have the 'almost fixed point property', a fact first noticed by Goebel and Kuczumov in [24].

We remark also that a standard embedding procedure always leads to existence of fixed points for extensions of nonexpansive mappings to larger domains. In [48] Ray and Sine observe that if X is an arbitrary Banach space, then it follows routinely from classical theory that there exists an extremally disconnected compact Hausdorff space E such that X is isometrically isomorphic to a subspace of $C(E)$, the continuous real valued functions defined on E. Moreover if K is an arbitrary set in $C(E)$, if $T : K \to K$ is nonexpansive, and if J is any closed order interval of $C(E)$ which contains K, then there exists a nonnexpansive extension $\hat{T}$ of T which maps J into J. By (5), $\hat{T}$ has a fixed point in J.

The fundamental open question in the theory from the outset has been whether or not an arbitrary weakly compact convex set in a Banach space must have the fixed point property for nonexpansive self-mappings. In a recent dramatic development, D. Alspach has settled this question in the negative. Subsequently, additional examples have been found (Schechtman [52]).

EXAMPLE (Alspach [1])

Let X be the function space $L^1[0,1]$ and let

$$K = \{f \in X : \int_0^1 f = 1,\ 0 \le f \le 2 \text{ almost everywhere}\}.$$

Then K is a closed and convex subset of the order interval $[0,2]$, hence K is weakly compact. Define $T : K \to K$ by

$$Tf(t) = \begin{cases} \min\{2f(2t),2\} & , \quad 0 \le t \le 1/2 \\ \max\{2f(2t-1)-2,0\}, & 1/2 < t < 1. \end{cases}$$

Then the mapping T is in fact isometric on K, but T has no fixed points.

Finally, one additional positive result has recently been announced. Recall that a space X is *uniformly smooth* if for each $\eta > 0$ there exists $\xi(\eta) > 0$ such that if $x,y \in X$ and $\|x-y\| \le \xi(\eta)$, then $\|x+y\| \ge \|x\| + \|y\| - \eta\|x-y\|$. (Equivalent conditions: the norm of X is uniformly Fréchet differentiable (Smulyan); the dual X^* of X is uniformly convex.) S. Reich reports in [50] (we have no direct citation) that Baillon has proved the following: *every weakly compact convex subset of a uniformly smooth Banach space has the fixed point property for nonexpansive self-mappings.*

2. Related Results

1) Families of nonexpansive mappings.

In classical theorems concerning the existence of common fixed points for families of mappings, such as the Markov-Kakutani theorem ([42],[33]) and its well-known generalization due to Ryll-Nardzewski ([51]), the mappings of the family are usually assumed to be linear, or at least to be weakly continuous and affine ([44]). In the nonlinear theory weak continuity is not assumed, but stronger geometric structure is utilized. In particular, if K is a convex subset of a Banach space X whose norm is strictly convex, then the fixed point set $F(T)$ of any nonexpansive $T : K \to K$ must also be convex. For if $u,v \in F(T)$ and $m \in \operatorname{seg}[u,v]$, nonexpansiveness implies

$$\|u-T(m)\| + \|T(m)-v\| \le \|u-m\| + \|m-v\| = \|u-v\|,$$

whence by strict convexity, $T(m) \in \mathrm{seg}[u,v]$; thus $T(m) = m$. It follows (cf. Browder [9]) that if in addition to the assumptions on K in (1) the space X has strictly convex norm, and if $F = \{T_\alpha : \alpha \in A\}$ is any commutative family of nonexpansive self-mappings of K, then $T_\alpha(F(T_\beta)) \subset F(T_\beta)$ for any $\alpha, \beta \in A$, from which (since $F(T_\beta)$ is closed and convex) $F(T_\alpha) \cap F(T_\beta) \neq \phi$. An induction argument shows that the family $\{F(T_\alpha) : \alpha \in A\}$ has the finite intersection property. Since each of these sets is weakly compact, $\bigcap_{\alpha \in A} F(T_\alpha) \neq \phi$.

The question as to whether the strict convexity assumption is essential for the above conclusion proved difficult to resolve. Without this assumption weak compactness of the $F(T_\alpha)$ is not assured. It was already known in 1963 ([19]) that such an assumption is not necessary if K is assumed compact in the norm topology, and in 1966 Belluce and Kirk observed ([5]) that the same is true if F is finite (or finitely generated). But it was not until 1974 that the problem was completely settled. Using fundamentally different approaches, R.E. Bruck and T.C. Lim not only eliminated the strict convexity assumption, but weakened other assumptions as well. By a careful analysis of the deeper implications of normal structure, Lim obtained his result for left reversible topological semigroups F. (Thus each two closed right ideals in F have nonempty intersection, an assumption weaker than commutativity.) The *discrete* version of Lim's result is the following.

(10) ([39]); also see [40]). *Let* K *be a nonempty weakly compact convex subset of a Banach space, and suppose* K *has normal structure. Suppose* F *is a left reversible semigroup of nonexpansive self-mappings of* K. *Then* $\bigcap_{T \in F} F(T) \neq \phi$.

Bruck's more abstract approach shows that the common fixed point property holds for commuting families of nonexpansive mappings not only in the setting of (1), but also in the settings of (3), (4), (8), (9).

(11) ([13]). *Let* K *be a nonempty closed convex subset of a Banach space and suppose* K *is either weakly compact, or bounded and separable. Suppose also that* K *has the following property. If* $T : K \to K$ *is nonexpansive, then* T *has a fixed point in every nonempty bounded closed convex* T*-invariant subset of* K. *Suppose* F *is a commutative family of nonexpansive self-mappings of* K. *Then* $\bigcap_{T \in F} F(T)$ *is a nonempty nonexpansive retract of* K.

Lim has recently obtained ([41]) a common fixed point theorem for left reversible topological semigroups of nonexpansive self-mappings of weak* closed convex subsets of ℓ^1. This result is not included in (11), even for commutative families.

In [52] it is shown that there exists a weakly compact convex subset K of $L^1(0,1)$ and a sequence $\{T_n\}$ of commuting nonexpansive self-mappings of K such that any finite subcollection of the $\{T_n\}$ have a common fixed point, while $\bigcap_{n=1}^{\infty} F(T_n) = \phi$. (Thus in general the sets $F(T_n)$ are *not* weakly compact, even if the domain is.)

Finally, results are known which do not require even a reversibility assumption on the semigroup. Bruck has shown ([14]) that if K is a bounded closed convex subset of a strictly convex space and if S is a convex semigroup of nonexpansive self-mappings of K, and if either K is strongly compact or S is compact in the topology of weak pointwise convergence, then the assumption $\overline{\text{conv}}\ s_1(K) \cap \overline{\text{conv}}\ s_2(K) \neq \phi$ for all $s_1, s_2 \in S$ implies S has a common fixed point.

2) Uniformly lipschitzian families.

An example given in part (I) shows that (1) may fail to hold for the class of mappings T having Lipschitz constant $k > 1$, no matter how near to 1 we choose k. A class intermediate between these and the nonexpansive mappings is provided by the following. A mapping $T : K \to K$, $K \subset X$, is said to be *uniformly k-lipschitzian* $(k \geq 1)$ if for each $x, y \in K$,

$$\|T^n(x) - T^n(y)\| \leq k\|x-y\|, \quad n = 1,2,\cdots.$$

(12) ([22]). *Let X be a uniformly convex Banach space. Then there exists a constant $\gamma > 1$ such that if K is a nonempty bounded closed convex subset of X, and if $T : K \to K$ is uniformly k-lipschitzian for $k < \gamma$, then T has a fixed point in K.*

The constant γ of (12) is derived from the modulus of convexity of X; that is, the function $\delta : [0,2] \to [0,1]$ defined as follows:

$$\delta(\varepsilon) = \inf\{1-(1/2)\|x+y\| : x,y \in X, \|x\| \leq 1, \|y\| \leq 1, \|x-y\| \geq \varepsilon\}.$$

It was shown in [22] that the conclusion of (12) holds if γ is taken to be the solution of the equation $\gamma(1-\delta(\gamma^{-1})) = 1$. In Hilbert space this yields $\gamma = \sqrt{5}/2$. On the other hand, an example in [23] shows that there exist uniformly 2-lipschitzian mappings of closed convex subsets of the unit ball of ℓ^2 which are fixed point free.

The question as to the validity of (12) for X a Hilbert space, $\gamma \in (\sqrt{5}/2, 2)$, was subsequently taken up by others. Lifschitz shows in [38] that if (M,ρ) is any metric space, it is possible to associate whith M a constant $\kappa(M)$ as follows:

$$\kappa(M) = \sup\left\{\beta > 0 \,\middle|\, \begin{array}{l} \exists\, \alpha > 1 \text{ such that } \forall\, x,y \in M \text{ and } r > 0,\ \rho(x,y) > r \Rightarrow \\ \exists\, z \in M \text{ such that } B(x;\beta r) \cap B(y;\alpha r) \subset B(z;r) \end{array}\right\}.$$

In general $\kappa(M) \geq 1$. If X is a uniformly convex space, then

$$\inf\{\kappa(K) : K \subset X \text{ is nonempty and convex}\} > 1,$$

and moreover if X is a Hilbert space this infimum is $\geq \sqrt{2}$. It is further shown [38] that if (M,ρ) is a complete and bounded metric space and if $T : M \to M$ is uniformly k-lipschitzian for $k < \kappa(M)$, then T has a fixed point in M. Combined, these results improve the constant γ of (12), and in particular they establish the following.

(13) ([38]). *Let* K *be a nonempty bounded closed convex subset of a Hilbert space and suppose* $T : K \to K$ *is uniformly* k*-lipschitzian for* $k < \sqrt{2}$. *Then* T *has a fixed point in* K.

Recently Baillon ([3]) has found an example of a fixed point free uniformly $\pi/2$-lipschitzian mapping which leaves invariant a bounded closed convex subset of ℓ^2. The validity of (13) for $\sqrt{2} \leq k < \pi/2$ remains open.

(13) has been extended to a common fixed point theorem for left reversible uniformly k-lipschitzian semigroups, $k < \sqrt{2}$, in [20].

3) Extensions of the theory to non-normed spaces.

We begin with a description of the results of Goebel and the others [26]. Let B be the open unit ball in complex Hilbert space and let

$$F = \{f : B \to B,\ f \text{ holomorphic}\}.$$

Thus for each $f \in F$, the Fréchet derivative of f at $x \in B$, denoted $Df(x)$, exists as a complex bounded linear map from H to H. Define for $x,y \in B$,

$$\alpha(x,y) = \sup\{\|Df(x,y)\| : f \in F\};$$

$$\rho(x,y) = \inf_{\gamma} \int_0^1 \alpha(\gamma(t),\gamma'(t))dt;$$

where γ ranges over all piecewise differentiable curves joining x and y. Then ρ is a metric on B, known as the hyperbolic metric, and the following facts are generally known: (B,ρ) is unbounded and complete; the ρ-topology is equivalent to the norm topology on any ball $B_r = \{x \in B : \|x\| \le r\}$, $r < 1$; $\rho(0,x) = \tanh^{-1}\|x\|$; and $\rho(f(x),f(y)) \le \rho(x,y)$ for all $f \in F$. Also if for $f \in F$, $\sup\{\|f(x)\| : x \in B\} < 1$, then it is known ([21]) that there exists $k \in (0,1)$ such that $\rho(f(x),f(y)) \le k\rho(x,y)$, $x,y \in B$.

It is shown [26] that there exists a continuous function $\delta : (0,\infty) \times [0,2] \to [0,1]$ satisfying:

(i) $\delta(r,0) = 0$;

(ii) $\delta(r,\varepsilon)$ is increasing in ε;

(iii) For each $a,x,y \in B$, $r \in (0,\infty)$, and $\varepsilon \in [0,2]$, the conditions $\rho(a,x) \le r$, $\rho(a,y) \le r$, $\rho(x,y) > \varepsilon r$ imply $\rho(a, \frac{1}{2}[x,y]) \le (1-\delta(r,\varepsilon))r$, where $\frac{1}{2}[x,y]$ denotes the midpoint of the ρ-geodesic joining x and y.

Thus the space (B,ρ) has the essential geometric features of uniform convexity, a fact which is used to prove the following.

(14) ([26]). *A holomorphic mapping* $T : B \to B$ *has a fixed point if and only if there exists* $r < 1$ *such that* $T(x) \ne \lambda x$ *for all* $x \in B$ *with* $\|x\| = r$ *and* $\lambda > 1$.

COROLLARY ([26])

Suppose $T : B \to B$ *is holomorphic and has a continuous extension to the closed unit ball* $\overline{B}$. *Then* T *has a fixed point in* $\overline{B}$.

The proof of the corollary is accomplished as follows. Let $z(t)$ denote the unique fixed points of the mappings tT, $t \in (0,1)$ (which are contractions on (B,ρ)). Suppose the assumptions of the theorem fail. Then there exists a weakly convergent sequence $\{z(t_n)\}$ with $z(t_n) \rightharpoonup u \in \overline{B}$ as $t \to 1$. It is then shown that $\|u\|$ must equal 1 from which it follows from a property of the Hilbert space norm that $z(t_n) \to u$ strongly, yielding $T(u) = u$.

We mention one final extended result. In [37], Lami Dozo takes up the study of nonexpansive mappings in F-spaces. These are complete metric linear spaces with translation invariant metric. Using the fact that the asymptotic center technique and Opial's condition carry over in a natural way to separable F-spaces, Lami Dozo obtains the following as a special case of his results: *Closed balls in the spaces* $(\ell^p, \|\cdot\|_p)$, $0 < p \le 1$, $(\|x\|_p = \sum |x_i|^p, x = \{x_i\} \in \ell^p)$ *have the fixed point property for nonexpansive self-mappings.* The proof (which carries over to the case $p > 1$ with the usual ℓ^p norm) makes strong use of the fact that balls in such spaces are compact in the topology of coordinatewise convergence.

In view of the above and (1), (3), and (5), closed balls in all the ℓ^p spaces, $0 < p \le \infty$, have the fixed point property for nonexpansive self-mappings. The same is true of the L^p-spaces, $1 < p \le \infty$, but in view of Alspach's example, this is almost surely false for L^1.

(4) Iteration and approximation.

As indicated in the introduction, we make no attempt at a complete survey of this area. We mention only three basic results.

(a) (Browder [10]). Let H be a Hilbert space, K a nonempty bounded closed convex subset of H, and $T: K \to K$ nonexpansive. Suppose $0 \in K$, and for $t \in (0,1)$, let x_t be the fixed point of the contraction mapping tT in K. Then $\lim_{t \to 1^-} x_t$ exists and converges to a fixed point of T. (For another proof, see Halpern [29]. Reich subsequently extended this result in [49] to spaces X having Gâteaux differentiable norm and possessing a weakly sequentially continuous duality map.)

(b) (Ishikawa [31]). Let X be an arbitrary Banach space, K a closed convex subset of X, and $T: K \to K$ nonexpansive. Fix $x = x_1 \in K$, let $\{t_n\}$ in $\mathbb{R}$ satisfy $0 \le t_n \le b < 1$, $n = 1,2,\cdots$, suppose $\sum_{n=1}^{\infty} t_n = \infty$, and define the sequence $\{x_n\}$ by

$$x_{n+1} = (1-t_n)x_n + t_n T(x_n).$$

If $\{x_n\}$ is bounded, then $\|x_n - T(x_n)\| \to 0$ as $n \to \infty$. If in addition the range of T is precompact, then T has a fixed point in K and $\{x_n\}$ converges to this fixed point.

(c) (Baillon [2], Bruck [15]). Suppose X is uniformly convex with Fréchet differentiable norm, K a bounded closed convex subset of X, $T: K \to K$ nonexpansive, and $x \in K$. Let $\{S_n(x)\}$ denote the Césaro means of $\{T^n(x)\}$, that is,

$$S_n(x) = (1/n) \sum_{i=0}^{n-1} T^i(x).$$

Then $\{S_n(x)\}$ converges weakly to a fixed point of T.

Appendix

Summary of Part (I). Let X be a Banach space, K a *bounded closed convex* subset of X, $K \neq \phi$. Nonexpansive self-mappings of K always have fixed points under the following *additional* assumptions on X and K.

X	K	Reference
Hilbert space	—	[8]
uniformly convex	—	[9],[27]
—	weakly compact with normal structure	[36]
ℓ^1, J_0	weak* compact	[34]
$X_{\sqrt{2}}$	—	[35]
ℓ^1	certain non-weak* compact sets	[25]
$L^\infty(\Omega,\mu)$	closed balls	[53],[54]
c_0	closed convex hull of a weakly convergent sequence	[45]
—	weakly compact with asymptotic normal structure	[4]
X_β, $\beta \in [1,2]$	—	[4]
$d(X,Y) \leq WCS(Y)$ for Y uniformly convex	—	[16]
C(E) with E stonian	closed order interval	[48],[53],[54]
c_0	weakly compact coordinate-wise star-shaped (delete convex)	[30]
uniformly smooth	—	(Baillon) [50]

OPEN QUESTIONS

In each of the following, X is assumed to be a Banach space and K a nonempty bounded closed and convex subset of X.

1) Suppose in addition that K is weakly compact. For spaces X does K always have the fixed point property for nonexpansive self-mappings? That some restriction on X is necessary is now known ([1]). It is also known that normal structure, or more generally asymptotic normal structure, suffices ([36],[4]). What about the following cases?

(i) X is superreflexive; specifically, $X = X_\beta$, $\beta > 2$,

or

(ii) X is strictly convex and reflexive.

2) If X is uniformly convex and $T : K \to X$ nonexpansive, then the assumption $\inf\{\|x-T(x)\| : x \in K\} = 0$ implies that T has a fixed point in K ([11]). Does this result hold for a wider class of spaces; specifically, what if X is reflexive and has normal structure?

3) Suppose $X = \ell^2$ and suppose $T : K \to K$ satisfies

$$\|T^i(x)-T^i(y)\| \leq k\|x-y\|, \quad x,y \in K, \quad i = 1,2\cdots.$$

Then T has a fixed point in K if $k < \sqrt{2}$ ([38]) and need not have a fixed point in K if $k \geq \pi/2$ ([3]). What if $k \in [\sqrt{2},\pi/2)$?

References

[1] ALSPACH, D.: A fixed point free nonexpansive map, preprint.

[2] BAILLON, J.-B.: Comportement asymptotique des contractions et semi-groupes de contractions - Equations de Schrödinger nonlinéaires et divers, Thèse, Université Paris VI (1978).

[3] BAILLON, J.-B.: Personal communication to R. Schöneberg, (1979).

[4] BAILLON, J.-B. and SCHÖNEBERG, R.: Asymptotic normal structure and fixed points of nonexpansive mappings, Proc. Amer. Math. Soc. 81 (1981), 257-264.

[5] BELLUCE, L.P. and KIRK, W.A.: Fixed point theorems for families of contraction mappings, Pacific J. Math. 18 (1966), 213-217.

[6] BELLUCE, L.P., KIRK, W.A. and STEINER, E.F.: Normal structure in Banach spaces, Pacific J. Math. 26 (1968), 433-440.

[7] BRODSKII, M.S. and MILMAN, D.P.: On the center of a convex set, Dokl. Akad. Nauk SSSR 59 (1948), 837-840 (Russian).

[8] BROWDER, F.E.: Fixed point theorems for noncompact mappings in Hilbert space, Proc. Nat. Acad. Sci. U.S.A. 53 (1965), 1272-1276.

[9] BROWDER, F.E.: Nonexpansive nonlinear operators in a Banach space, Proc. Nat. Acad. Sci. U.S.A. 54 (1965), 1041-1044.

[10] BROWDER, F.E.: Convergence of approximants to fixed points of non-expansive nonlinear mappings in Banach spaces, Arch. Rational Mech. Anal. 24 (1967), 82-90.

[11] BROWDER, F.E.: Semicontractive and semiaccretive nonlinear mappings in Banach spaces, Bull. Amer. Math. Soc. 74 (1968), 660-665.

[12] BROWDER, F.E.: Nonlinear Operators and Nonlinear Equations of Evolution in Banach Spaces, Proc. Symp. Pure Math. 18, pt. 2, Amer. Math. Soc., Providence, R.I., (1976).

[13] BRUCK, R.E.: A common fixed point theorem for a commuting family of nonexpansive mappings, Pacific J. Math. 53 (1974), 59-71.

[14] BRUCK, R.E.: A common fixed point theorem for compact convex semigroups of nonexpansive mappings, Proc. Amer. Math. Soc. 53 (1975), 113-116.

[15] BRUCK, R.E.: A simple proof of the mean ergodic theorem for nonlinear contractions in Banach spaces, Israel J. Math. 32 (1979), 107-116.

[16] BYNUM, W.L.: Normal structure coefficients for Banach spaces, Pacific J. Math. 86 (1980), 427-436.

[17] CLARKSON, J.A.: Uniformly convex spaces, Trans. Amer. Math. Soc. 40 (1936), 396-414.

[18] DAY, M.M., JAMES, R.C. and SWAMINATHAN, S.: Normed linear spaces that are uniformly convex in every direction, Canad. J. Math. 23 (1971), 1051-1059.

[19] DeMARR, R.: Common fixed points for commuting contraction mappings, Pacific J. Math. 13 (1963), 1139-1141.

[20] DOWNING, D. and RAY, W.O.: Uniformly lipschitzian semigroups in Hibert space, preprint.

[21] EARLE, C.J. and HAMILTON, R.S.:: A fixed point theorem for holomorphic mappings, Proc. Symp. Pure Math. vol. 16, Amer. Math. Soc., Providence, R.I. (1970), 61-65.

[22] GOEBEL, K and KIRK, W.A.: A fixed point theorem for transformations whose iterates have uniform Lipschitz constant, Studia Math. 47 (1973), 135-140.

[23] GOEBEL, K, KIRK, W.A. and THELE, R.L.: Uniformly lipschitzian families of transformations in Banach spaces, Canad. J. Math. 26 (1974), 1245-1256.

[24] GOEBEL, K. and KUCZUMOW, T.: A contribution to the theory of nonexpansive mappings,

[25] GOEBEL. K and KUCZUMOW, T.: Irregular convex sets with the fixed point property for nonexpansive mappings, Colloquim Math. 40 (1979), 259-264.

[26] GOEBEL, K., SEKOWSKI, T. and STACHURA, A.: Uniform convexity of the hyperbolic metric and fixed points of holomorphic mappings in the Hilbert ball, Nonlinear Analysis. 4 (1980), 1011-1021.

[27] GÖHDE, D.: Zum Prinzip der kontraktiven Abbildung, Math. Nachr. 30 (1965), 251-258.

[28] GOSSEZ, J.P. and LAMI DOZO, E.: Some geometric properties related to the fixed point theory for nonexpansive mappings, Pacific J. Math. 40 (1972), 565-573.

[29] HALPERN, B.: Fixed points of nonexpanding maps, Bull. Amer. Math. Soc. 73 (1967), 957-961.

[30] HAYDON, R., ODELL, E. and STERNFELD, Y.: A fixed point theorem for a class of star-shaped sets in c_0, preprint.

[31] ISHIKAWA, S.: Fixed points and iteration of a nonexpansive mapping in a Banach space, Proc. Amer. Math. Soc. 59 (1976), 65-71.

[32] JAMES, R.C.: A separable somewhat reflexive Banach space with nonseparable dual, Bull. Amer. Math. Soc. 80 (1974), 738-743.

[33] KAKUTANI, S.: Two fixed-point theorems concerning bicompact convex sets, Proc. Imp. Acad. Tokyo 14 (1938), 242-245.

[34] KARLOVITZ, L.A.: On nonexpansive mappings, Proc. Amer. Math. Soc. 55 (1976), 321-325.

[35] KARLOVITZ, L.A.: Existence of fixed points for nonexpansive mappings in spaces without normal structure, Pacific J. Math. 66 (1976), 153-156.

[36] KIRK, W.A.: A fixed point theorem for mappings which do not increase distances, Amer. Math. Monthly 72 (1965), 1004-1006.

[37] LAMI DOZO, E.: Centres asymptotiques dans certains F-espaces, Boll. Un. Mat. Ital. (to appear).

[38] LIFSCHITZ, E.A.: Fixed point theorems for operators in strongly convex spaces, Voronez Gos. Univ. Trudy Mat. Fak. 16 (1975), 23-28 (Russian).

[39] LIM, T.C.: A fixed point theorem for families of nonexpansive mappings, Pacific J. Math. 53 (1974), 487-493.

[40] LIM, T.C.: Characterizations of normal structure, Proc. Amer. Math. Soc. 43 (1974), 313-319.

[41] LIM, T.C.: Asymptotic centers and nonexpansive mappings in some conjugate spaces, Pacific J. Math. 90 (1980), 135-143.

[42] MARKOV, A.: Quelques théorèmes sur les ensembles abéliens, Dokl. Acad. Nauk SSSR 10 (1936), 311-314.

[43] NAMIOKA, I.: Neighborhoods of extreme points, Israel J. Math. 5 (1967), 145-152.

[44] NAMIOKA, I. and ASPLUND, E: A geometric proof of Ryll-Nardzewski's fixed point theorem, Bull. Amer. Math. Soc. 73 (1967), 443-445.

[45] ODELL, E. and STERNFELD, Y.: A fixed point theorem in c_0, preprint.

[46] RAY, W.O.: Nonexpansive mappings on unbounded convex domains, Bull. Acad. Polon. Sci. 26 (1978), 241-245.

[47] RAY, W.O.: The fixed point property and unbounded sets in Hilbert space, Trans. Amer. Math. Soc. 258 (1980), 531-537.

[48] RAY, W.O. and SINE, R.: Nonexpansive mappings with precompact orbit, preprint.

[49] REICH, S.: Asymptotic behavior of contractions in Banach spaces, J. Math. Anal. Appl. 44 (1973), 57-70.

[50] REICH, S.: The fixed point property for nonexpansive mappings II, Amer. Math. Monthly 87 (1980), 292-294.

[51] RYLL-NARDZEWSKI, C.: On fixed points of semi-groups of endomorphisms of linear spaces, Proc. Fifth Berkeley Symp. on Statistics and Probability II, pt. 1,(1966).

[52] SCHECHTMAN, G.: Some remarks on commuting families of nonexpansive operators, preprint.

[53] SINE, R.: On nonlinear contractions in sup norm spaces, Nonlinear Analysis 3 (1979), 885-890.

[54] SOARDI, P.: Existence of fixed points of nonexpansive mappings in certain Banach lattices, Proc. Amer. Math. Soc. 73 (1979), 25-29.

[55] SULLIVAN, F.: A generalization of uniformly rotund Banach spaces, Canad. J. Math. 31 (1979), 628-636.

OPEN PROBLEMS

1. (R.F. Brown)

(A very old one). Hopf proved that if X is a finite polyhedron and $f : X \to X$ any map, then there exists a map g homotopic to f such that g has a finite number of fixed points. Is the same true if X is a compact ANR?

2. (R.F. Brown)

Let X be a space, X^X the space of maps of X to itself (compact-open topology), $x_0 \in X$ some point, and define $e : X^X \to X$ by $e(f) = f(x_0)$. Then e induces $e_* : \pi_1(X^X, 1) \to \pi_1(X, x_0)$, where 1 denotes the identity map. The space X is Jiang if e_* is surjective. Jiang spaces are important on fixed point theory because if X is Jiang and $f \in X^X$, then the Nielsen number $N(f)$ depends only on the homomorphism of the fundamental group of X induced by f. Aside from the trivial example of simply connected spaces, all known Jiang spaces: H-spaces, generalized lens spaces, and homogeneous spaces G/H where G is a compact topological group and H is a closed connected subgroup, were discovered by Jiang in 1964. The problem is to find a way of constructing more Jiang spaces. Jingyal Pak has suggested the following question: if $p : E \to B$ is a fibre space with fibre F that is orientable in a strong sense and B and F are Jiang, is E Jiang? It is known even whether E must have an abelian fundamental group in this case (as every Jiang spaces does). A good special case to consider is bundles $p : E \to S^n$, $n \geq 2$, with fibre F a Jiang space: must E then be Jiang?

3. (R.F. Brown)

(Essentially due to Ben Halpern). A theorem of Shub and Sullivan states that if M is a smooth compact manifold and $f : M \to M$ is a C^1 map such that the set $\{L(f^n)\}$ of Lefschetz numbers of iterates of f is unbounded, then the set of periodic points of f, that is $\{x \in M : f^n(x) = x$ for some $n\}$, is infinite. They also exhibit a map $f : S^2 \to S^2$ such that $L(f^n) = 2^n$ but the set of periodic points consists of just two points. The example, which depends crucially on suspension structure, can be extended to all n-spheres, $n \geq 2$. For the n-torus T^n, however, every continuous map $f : T^n \to T^n$ has at least $|L(f)|$ fixed points so the Shub-Sullivan result holds without requiring f to be C^1. (For $n = 2$ this follows from an old result of

Brouwer; for general n it comes from a paper of Brooks and others). A recent result of Halpern implies that the "non-C^1 version" of Shub-Sullivan is true on the Klein bottle. The problem is: can the Shub-Sullivan theorem be proved without assuming f is C^1, provided M is not an n-sphere, $n \geq 2$? A special case that might well lead to a general solution is when M is the closed orientable surface of genus 2 (the "two-holed torus"). Even if the answer turns out to be negative in general, a solution to the problem could throw more light on the role of differentiability hypotheses in fixed point theory.

4. (A. Dold)

The fixed point index $I(f)$ of a mapping $f : V \to E$ over B is an element of the 0-th stable cohomotopy group of B plus a point; $I(f) \in \pi_s^0(B^+)$. This assumes $p : E \to B$ to be an ENR_B, $V \subset E$ an open subset, and $Fix(f) \to B$ to be proper (see Dold, Inventiones math. 25 (1974), 281-297). Every element $\beta \in \pi_s^0(B^+)$ occurs as the index of some such f, and one can even arrange it so that $V = E = R^n \times B$ (see (3.5) and (3.6) l.c.). The question is whether one can arrange it so that p is a proper bundle, or f to be locally trivial? The mapping f is said to be locally trivial, if every $b \in B$ has a neighborhood $U \subset B$ such that $p^{-1}(U) \approx U \times Y$, $p^{-1}(U) \cap V \approx U \times Z$ over B, for some ENR Y and open subset $Z \subset Y$; and such that f under these homeomorphisms over U takes the form $f(u,z) = (u,\varphi(z))$ for some $\varphi : Z \to Y$, and all $u \in U$, $z \in Z$.

What we are asking is which elements $\beta \in \pi_s^0(B^+)$ occur as the indices $(\beta = I(f))$ of maps f in the following classes.

(LT) Locally trivial f, as above.

(PB) Maps $f : E \to E$ over B, where $p : E \to B$ is a bundle with fibre a compact ENR.

(Eul) Identity maps $f = id : E \to E$ of bundles as in (PB).

The last case asks, so to speak, which elements $\beta \in \pi_s^0(B^+)$ occur as "Euler-characteristics": it's a question which I was asked by T. tom Dieck.

5. (A. Dold)

If Y is a compact metric space we denote by $\chi(Y) = \sum_{j=0}^{\infty} (-1)^j \dim \check{H}^j(Y)$ its Euler characteristic, calculated in Čech cohomology $\check{H}(Y) = \check{H}(Y,Q)$ with rational coefficients. It assumes $\left(\sum_{j=0}^{\infty} \dim H^j(Y)\right) < \infty$; otherwise $\chi(Y)$ is not defined. Motivated by the Vietoris mapping theorem and analogous results, we ask whether the following is true: "if $f : Y \to Z$ is a continuous map between compact metric spaces such that

$\chi(f^{-1}(z)) = 1$ for all $z \in Z$ then $\chi(Y) = \chi(Z)$".

The question is of interest also for simpler spaces Y, Z, say compact CW-spaces or manifolds. On the other hand, the question loses much of its interest if the continuous map f is further restricted (for example simplicial or fibration). Answers for restricted classes of maps will therefore not be counted towards a solution!

A little more general, one can ask whether $\chi(Y) = k\chi(Z)$ if $\chi(f^{-1}(z))$ has the same value $k \in Z$ for all $z \in Z$. (These questions originate from a course in algebraic topology and were brought up by students).

6. (G. Fournier)

If K is a compact subspace of V an open subspace of a linear normed space E, does there exists a basis $\{W_m\}$ of neighbourhoods of K in V such that (i) $W_n \subset W_m$ if $n \geq m$ and (ii) $i_{n,m*} : H(W_n) \to H(W_m)$ is an epimorphism for any $n \geq m$ where $i_{n,m} : W_n \to W_m$ is the inclusion?

7. (R.D. Nussbaum)

Recall that a cone K in a Banach space X is a closed subset of X such that (a) if x and y are any elements of K and r and s are any nonnegative real numbers, then $ax + sy \in K$ and (b) if $x \in K \setminus \{0\}$ then $-x \notin K$. Walter Petryshyn has defined a cone to be "quasinormal" if there exists $u \in K \setminus \{0\}$ and $\gamma > 0$ such that

$$\|x + u\| \geq \gamma \|x\| \tag{1}$$

for all $x \in K$; and E. Lami-Dozo has pointed out to me that a simple argument shows that for any $u \in K \setminus \{0\}$, K an arbitrary cone, there exists a $\gamma > 0$ such that (1) is satisfied. If $\gamma(u)$ denotes the supremum of $\gamma > 0$ for which (1) is satisfied for all $x \in K$, define $\gamma(K)$ by

$$\gamma(K) = \sup \{\gamma(u) : u \in K \setminus \{0\}\}. \tag{2}$$

It is clear that $\gamma(K) \leq 1$.

<u>QUESTION.</u>

For what common cones in analysis does one have $\gamma(K) = 1$? When does there exists

$u \in K \setminus \{0\}$ such that $\gamma(u) = 1$?

If Ω is a bounded region in $\mathbb{R}^n$ and K is the cone of nonnegative functions in $W^{m,p}(\Omega)$, it is clear that $\gamma(K) = 1$ (just take u to be the function identically 1). However, it is not even clear what the value of $\gamma(K)$ is when K is the set of non-negative functions in $W_0^{m,p}(\Omega) = X$.

Knowing that $\gamma(K) = 1$ is sometimes convenient in the theory of fixed points of non-linear operators in cones.

8. (R.D. Nussbaum)

Fix a number $\alpha > 1$ and consider the integral equation

$$U(x) = 1 + \lambda \int_x^1 u(y)^\alpha\, u(y-x)^\alpha\, dy, \qquad 0 \le x \le 1, \quad \lambda \ge 0. \tag{1}$$

Let $S = \{(u,\lambda) : \lambda \ge 0,\ u \in C[0,1],\ u$ and λ solve (1)$\}$. It is not hard to prove that there exists a positive number $\lambda(\alpha) \le \frac{1}{2}$ such that equation (1) has no positive continuous solutions for $\lambda > \lambda(\alpha)$ and at least one positive solution for each λ with $0 \le \lambda < \lambda(\alpha)$.

QUESTION.

Does there exists $\varepsilon > 0$ (ε dependent on α) such that $\{\|u\| : (u,\lambda) \in S$ and $\lambda \ge \varepsilon\}$ is unbounded? Really, we conjecture that a graph of the points $(\lambda, \|u\|)$ for $(u,\lambda) \in S$ has the following appearance

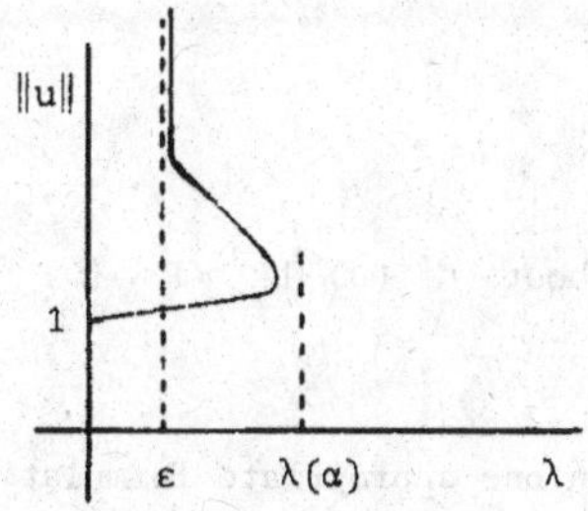

with a vertical asymptote at some $\varepsilon > 0$.

9,10. (H.-O. Peitgen)

Let $F : R^n \times R \to R^n$. We are interested in a *global numerical study* of the general non-linear eigenvalue problem

$$F(x,\lambda) = 0$$

One possible way of studying the solution set $F^{-1}(0)$ numerically is to pass to a suitable global PL-unfolding of $F^{-1}(0)$: let T be a triangulation of R^{n+1} and let F_T denote the associated PL-approximation to F. Furthermore let

$$\bar{\varepsilon} := (\varepsilon, \varepsilon^2, \ldots, \varepsilon^n), \quad 0 < \varepsilon \le \varepsilon_0.$$

One can show that for ε sufficiently small $\bar{\varepsilon} \in R^n$ is a regular value for $F_T : R^n \times R \to R^n$ in the PL sense. Define

$$G(x,\lambda,\varepsilon) := F_T(x,\lambda) - \bar{\varepsilon} .$$

Then one may consider $G^{-1}(0)$ to be a "global numerical unfolding" of $F^{-1}(0)$. Recent studies (see [1], [2], [3]) have shown the numerical relevance of this approach. However the following problems arise:

PROBLEM 9:

How is $F^{-1}(0)$ related with $G^{-1}(0) \cap R^n \cap \{0\}$? (structurally)

Concrete computer computations in this approach are done for $G^{-1}(0) \cap R^n \times R \times \{\varepsilon\}$. Therefore,

PROBLEM 10:

What structural information about $G^{-1}(0) \cap R^n \times R \times \{0\}$ can be obtained from $G^{-1}(0) \cap R^n \times R \times \{\varepsilon\}$?

There is some ividence that in one appropriate formulation one may obtain some partial answers by exploiting continuity properties of Čech-cohomology.

[1] JÜRGENS, H., PEITGEN, H.-O. and SAUPE, S.: Topological perturbations in the numerical study of nonlinear eigenvalue- and bifurcation problems, Proc. Conf. on Analysis and Computation of Fixed Points, Madison (1979), S.M. Robinson ed., Academic Press, N.Y., (1980).

[2] PEITGEN, H.-O, and PRÜFER, M.: The Leray-Schauder continuation method is a constructive element in the numerical study of nonlinear eigenvalue and bifurcation problems, Proc. Conf. Functional Differential Equations and Approximation of Fixed Points, H.-O. Peitgen and H.O. Walther, eds., Springer Lecture Notes in Mathematics, 730, (1979), 326-409.

[3] PEITGEN, H.-O., SAUPE, D. and SCHMITT, K.: Nonlinear elliptic boundary value problems versus their finite difference approximations: numerically irrelevant solutions, submitted.

11. (H. SCHIRMER)

Let A be contained in the interior of a closed $(2n+1)$-dimensional ball B^{2n+1}. Find necessary and sufficient conditions so that A can be the fixed point set of a homeomorphism.

Vol. 728: Non-Commutative Harmonic Analysis. Proceedings, 1978. Edited by J. Carmona and M. Vergne. V, 244 pages. 1979.

Vol. 729: Ergodic Theory. Proceedings, 1978. Edited by M. Denker and K. Jacobs. XII, 209 pages. 1979.

Vol. 730: Functional Differential Equations and Approximation of Fixed Points. Proceedings, 1978. Edited by H.-O. Peitgen and H.-O. Walther. XV, 503 pages. 1979.

Vol. 731: Y. Nakagami and M. Takesaki, Duality for Crossed Products of von Neumann Algebras. IX, 139 pages. 1979.

Vol. 732: Algebraic Geometry. Proceedings, 1978. Edited by K. Lønsted. IV, 658 pages. 1979.

Vol. 733: F. Bloom, Modern Differential Geometric Techniques in the Theory of Continuous Distributions of Dislocations. XII, 206 pages. 1979.

Vol. 734: Ring Theory, Waterloo, 1978. Proceedings, 1978. Edited by D. Handelman and J. Lawrence. XI, 352 pages. 1979.

Vol. 735: B. Aupetit, Propriétés Spectrales des Algèbres de Banach. XII, 192 pages. 1979.

Vol. 736: E. Behrends, M-Structure and the Banach-Stone Theorem. X, 217 pages. 1979.

Vol. 737: Volterra Equations. Proceedings 1978. Edited by S.-O. Londen and O. J. Staffans. VIII, 314 pages. 1979.

Vol. 738: P. E. Conner, Differentiable Periodic Maps. 2nd edition, IV, 181 pages. 1979.

Vol. 739: Analyse Harmonique sur les Groupes de Lie II. Proceedings, 1976–78. Edited by P. Eymard et al. VI, 646 pages. 1979.

Vol. 740: Séminaire d'Algèbre Paul Dubreil. Proceedings, 1977–78. Edited by M.-P. Malliavin. V, 456 pages. 1979.

Vol. 741: Algebraic Topology, Waterloo 1978. Proceedings. Edited by P. Hoffman and V. Snaith. XI, 655 pages. 1979.

Vol. 742: K. Clancey, Seminormal Operators. VII, 125 pages. 1979.

Vol. 743: Romanian-Finnish Seminar on Complex Analysis. Proceedings, 1976. Edited by C. Andreian Cazacu et al. XVI, 713 pages. 1979.

Vol. 744: I. Reiner and K. W. Roggenkamp, Integral Representations. VIII, 275 pages. 1979.

Vol. 745: D. K. Haley, Equational Compactness in Rings. III, 167 pages. 1979.

Vol. 746: P. Hoffman, τ-Rings and Wreath Product Representations. V, 148 pages. 1979.

Vol. 747: Complex Analysis, Joensuu 1978. Proceedings, 1978. Edited by I. Laine, O. Lehto and T. Sorvali. XV, 450 pages. 1979.

Vol. 748: Combinatorial Mathematics VI. Proceedings, 1978. Edited by A. F. Horadam and W. D. Wallis. IX, 206 pages. 1979.

Vol. 749: V. Girault and P.-A. Raviart, Finite Element Approximation of the Navier-Stokes Equations. VII, 200 pages. 1979.

Vol. 750: J. C. Jantzen, Moduln mit einem höchsten Gewicht. III, 195 Seiten. 1979.

Vol. 751: Number Theory, Carbondale 1979. Proceedings. Edited by M. B. Nathanson. V, 342 pages. 1979.

Vol. 752: M. Barr, *-Autonomous Categories. VI, 140 pages. 1979.

Vol. 753: Applications of Sheaves. Proceedings, 1977. Edited by M. Fourman, C. Mulvey and D. Scott. XIV, 779 pages. 1979.

Vol. 754: O. A. Laudal, Formal Moduli of Algebraic Structures. III, 161 pages. 1979.

Vol. 755: Global Analysis. Proceedings, 1978. Edited by M. Grmela and J. E. Marsden. VII, 377 pages. 1979.

Vol. 756: H. O. Cordes, Elliptic Pseudo-Differential Operators – An Abstract Theory. IX, 331 pages. 1979.

Vol. 757: Smoothing Techniques for Curve Estimation. Proceedings, 1979. Edited by Th. Gasser and M. Rosenblatt. V, 245 pages. 1979.

Vol. 758: C. Năstăsescu and F. Van Oystaeyen; Graded and Filtered Rings and Modules. X, 148 pages. 1979.

Vol. 759: R. L. Epstein, Degrees of Unsolvability: Structure and Theory. XIV, 216 pages. 1979.

Vol. 760: H.-O. Georgii, Canonical Gibbs Measures. VIII, 190 pages. 1979.

Vol. 761: K. Johannson, Homotopy Equivalences of 3-Manifolds with Boundaries. 2, 303 pages. 1979.

Vol. 762: D. H. Sattinger, Group Theoretic Methods in Bifurcation Theory. V, 241 pages. 1979.

Vol. 763: Algebraic Topology, Aarhus 1978. Proceedings, 1978. Edited by J. L. Dupont and H. Madsen. VI, 695 pages. 1979.

Vol. 764: B. Srinivasan, Representations of Finite Chevalley Groups. XI, 177 pages. 1979.

Vol. 765: Padé Approximation and its Applications. Proceedings, 1979. Edited by L. Wuytack. VI, 392 pages. 1979.

Vol. 766: T. tom Dieck, Transformation Groups and Representation Theory. VIII, 309 pages. 1979.

Vol. 767: M. Namba, Families of Meromorphic Functions on Compact Riemann Surfaces. XII, 284 pages. 1979.

Vol. 768: R. S. Doran and J. Wichmann, Approximate Identities and Factorization in Banach Modules. X, 305 pages. 1979.

Vol. 769: J. Flum, M. Ziegler, Topological Model Theory. X, 151 pages. 1980.

Vol. 770: Séminaire Bourbaki vol. 1978/79 Exposés 525–542. IV, 341 pages. 1980.

Vol. 771: Approximation Methods for Navier-Stokes Problems. Proceedings, 1979. Edited by R. Rautmann. XVI, 581 pages. 1980.

Vol. 772: J. P. Levine, Algebraic Structure of Knot Modules. XI, 104 pages. 1980.

Vol. 773: Numerical Analysis. Proceedings, 1979. Edited by G. A. Watson. X, 184 pages. 1980.

Vol. 774: R. Azencott, Y. Guivarc'h, R. F. Gundy, Ecole d'Eté de Probabilités de Saint-Flour VIII-1978. Edited by P. L. Hennequin. XIII, 334 pages. 1980.

Vol. 775: Geometric Methods in Mathematical Physics. Proceedings, 1979. Edited by G. Kaiser and J. E. Marsden. VII, 257 pages. 1980.

Vol. 776: B. Gross, Arithmetic on Elliptic Curves with Complex Multiplication. V, 95 pages. 1980.

Vol. 777: Séminaire sur les Singularités des Surfaces. Proceedings, 1976-1977. Edited by M. Demazure, H. Pinkham and B. Teissier. IX, 339 pages. 1980.

Vol. 778: SK1 von Schiefkörpern. Proceedings, 1976. Edited by P. Draxl and M. Kneser. II, 124 pages. 1980.

Vol. 779: Euclidean Harmonic Analysis. Proceedings, 1979. Edited by J. J. Benedetto. III, 177 pages. 1980.

Vol. 780: L. Schwartz, Semi-Martingales sur des Variétés, et Martingales Conformes sur des Variétés Analytiques Complexes. XV, 132 pages. 1980.

Vol. 781: Harmonic Analysis Iraklion 1978. Proceedings 1978. Edited by N. Petridis, S. K. Pichorides and N. Varopoulos. V, 213 pages. 1980.

Vol. 782: Bifurcation and Nonlinear Eigenvalue Problems. Proceedings, 1978. Edited by C. Bardos, J. M. Lasry and M. Schatzman. VIII, 296 pages. 1980.

Vol. 783: A. Dinghas, Wertverteilung meromorpher Funktionen in ein- und mehrfach zusammenhängenden Gebieten. Edited by R. Nevanlinna and C. Andreian Cazacu. XIII, 145 pages. 1980.

Vol. 784: Séminaire de Probabilités XIV. Proceedings, 1978/79. Edited by J. Azéma and M. Yor. VIII, 546 pages. 1980.

Vol. 785: W. M. Schmidt, Diophantine Approximation. X, 299 pages. 1980.

Vol. 786: I. J. Maddox, Infinite Matrices of Operators. V, 122 pages. 1980.

Vol. 787: Potential Theory, Copenhagen 1979. Proceedings, 1979. Edited by C. Berg, G. Forst and B. Fuglede. VIII, 319 pages. 1980.

Vol. 788: Topology Symposium, Siegen 1979. Proceedings, 1979. Edited by U. Koschorke and W. D. Neumann. VIII, 495 pages. 1980.

Vol. 789: J. E. Humphreys, Arithmetic Groups. VII, 158 pages. 1980.

Vol. 790: W. Dicks, Groups, Trees and Projective Modules. IX, 127 pages. 1980.

Vol. 791: K. W. Bauer and S. Ruscheweyh, Differential Operators for Partial Differential Equations and Function Theoretic Applications. V, 258 pages. 1980.

Vol. 792: Geometry and Differential Geometry. Proceedings, 1979. Edited by R. Artzy and I. Vaisman. VI, 443 pages. 1980.

Vol. 793: J. Renault, A Groupoid Approach to C*-Algebras. III, 160 pages. 1980.

Vol. 794: Measure Theory, Oberwolfach 1979. Proceedings 1979. Edited by D. Kölzow. XV, 573 pages. 1980.

Vol. 795: Séminaire d'Algèbre Paul Dubreil et Marie-Paule Malliavin. Proceedings 1979. Edited by M. P. Malliavin. V, 433 pages. 1980.

Vol. 796: C. Constantinescu, Duality in Measure Theory. IV, 197 pages. 1980.

Vol. 797: S. Mäki, The Determination of Units in Real Cyclic Sextic Fields. III, 198 pages. 1980.

Vol. 798: Analytic Functions, Kozubnik 1979. Proceedings. Edited by J. Ławrynowicz. X, 476 pages. 1980.

Vol. 799: Functional Differential Equations and Bifurcation. Proceedings 1979. Edited by A. F. Izé. XXII, 409 pages. 1980.

Vol. 800: M.-F. Vignéras, Arithmétique des Algèbres de Quaternions. VII, 169 pages. 1980.

Vol. 801: K. Floret, Weakly Compact Sets. VII, 123 pages. 1980.

Vol. 802: J. Bair, R. Fourneau, Etude Géometrique des Espaces Vectoriels II. VII, 283 pages. 1980.

Vol. 803: F.-Y. Maeda, Dirichlet Integrals on Harmonic Spaces. X, 180 pages. 1980.

Vol. 804: M. Matsuda, First Order Algebraic Differential Equations. VII, 111 pages. 1980.

Vol. 805: O. Kowalski, Generalized Symmetric Spaces. XII, 187 pages. 1980.

Vol. 806: Burnside Groups. Proceedings, 1977. Edited by J. L. Mennicke. V, 274 pages. 1980.

Vol. 807: Fonctions de Plusieurs Variables Complexes IV. Proceedings, 1979. Edited by F. Norguet. IX, 198 pages. 1980.

Vol. 808: G. Maury et J. Raynaud, Ordres Maximaux au Sens de K. Asano. VIII, 192 pages. 1980.

Vol. 809: I. Gumowski and Ch. Mira, Recurences and Discrete Dynamic Systems. VI, 272 pages. 1980.

Vol. 810: Geometrical Approaches to Differential Equations. Proceedings 1979. Edited by R. Martini. VII, 339 pages. 1980.

Vol. 811: D. Normann, Recursion on the Countable Functionals. VIII, 191 pages. 1980.

Vol. 812: Y. Namikawa, Toroidal Compactification of Siegel Spaces. VIII, 162 pages. 1980.

Vol. 813: A. Campillo, Algebroid Curves in Positive Characteristic. V, 168 pages. 1980.

Vol. 814: Séminaire de Théorie du Potentiel, Paris, No. 5. Proceedings. Edited by F. Hirsch et G. Mokobodzki. IV, 239 pages. 1980.

Vol. 815: P. J. Slodowy, Simple Singularities and Simple Algebraic Groups. XI, 175 pages. 1980.

Vol. 816: L. Stoica, Local Operators and Markov Processes. VIII, 104 pages. 1980.

Vol. 817: L. Gerritzen, M. van der Put, Schottky Groups and Mumford Curves. VIII, 317 pages. 1980.

Vol. 818: S. Montgomery, Fixed Rings of Finite Automorphism Groups of Associative Rings. VII, 126 pages. 1980.

Vol. 819: Global Theory of Dynamical Systems. Proceedings, 1979. Edited by Z. Nitecki and C. Robinson. IX, 499 pages. 1980.

Vol. 820: W. Abikoff, The Real Analytic Theory of Teichmüller Space. VII, 144 pages. 1980.

Vol. 821: Statistique non Paramétrique Asymptotique. Proceedings, 1979. Edited by J.-P. Raoult. VII, 175 pages. 1980.

Vol. 822: Séminaire Pierre Lelong–Henri Skoda, (Analyse) Années 1978/79. Proceedings. Edited by P. Lelong et H. Skoda. VIII, 356 pages, 1980.

Vol. 823: J. Král, Integral Operators in Potential Theory. III, 171 pages. 1980.

Vol. 824: D. Frank Hsu, Cyclic Neofields and Combinatorial Designs. VI, 230 pages. 1980.

Vol. 825: Ring Theory, Antwerp 1980. Proceedings. Edited by F. van Oystaeyen. VII, 209 pages. 1980.

Vol. 826: Ph. G. Ciarlet et P. Rabier, Les Equations de von Kármán. VI, 181 pages. 1980.

Vol. 827: Ordinary and Partial Differential Equations. Proceedings, 1978. Edited by W. N. Everitt. XVI, 271 pages. 1980.

Vol. 828: Probability Theory on Vector Spaces II. Proceedings, 1979. Edited by A. Weron. XIII, 324 pages. 1980.

Vol. 829: Combinatorial Mathematics VII. Proceedings, 1979. Edited by R. W. Robinson et al.. X, 256 pages. 1980.

Vol. 830: J. A. Green, Polynomial Representations of GL_n. VI, 118 pages. 1980.

Vol. 831: Representation Theory I. Proceedings, 1979. Edited by V. Dlab and P. Gabriel. XIV, 373 pages. 1980.

Vol. 832: Representation Theory II. Proceedings, 1979. Edited by V. Dlab and P. Gabriel. XIV, 673 pages. 1980.

Vol. 833: Th. Jeulin, Semi-Martingales et Grossissement d'une Filtration. IX, 142 Seiten. 1980.

Vol. 834: Model Theory of Algebra and Arithmetic. Proceedings, 1979. Edited by L. Pacholski, J. Wierzejewski, and A. J. Wilkie. VI, 410 pages. 1980.

Vol. 835: H. Zieschang, E. Vogt and H.-D. Coldewey, Surfaces and Planar Discontinuous Groups. X, 334 pages. 1980.

Vol. 836: Differential Geometrical Methods in Mathematical Physics. Proceedings, 1979. Edited by P. L. García, A. Pérez-Rendón, and J. M. Souriau. XII, 538 pages. 1980.

Vol. 837: J. Meixner, F. W. Schäfke and G. Wolf, Mathieu Functions and Spheroidal Functions and their Mathematical Foundations Further Studies. VII, 126 pages. 1980.

Vol. 838: Global Differential Geometry and Global Analysis. Proceedings 1979. Edited by D. Ferus et al. XI, 299 pages. 1981.

Vol. 839: Cabal Seminar 77 – 79. Proceedings. Edited by A. S. Kechris, D. A. Martin and Y. N. Moschovakis. V, 274 pages. 1981.

Vol. 840: D. Henry, Geometric Theory of Semilinear Parabolic Equations. IV, 348 pages. 1981.

Vol. 841: A. Haraux, Nonlinear Evolution Equations- Global Behaviour of Solutions. XII, 313 pages. 1981.

Vol. 842: Séminaire Bourbaki vol. 1979/80. Exposés 543–560. IV, 317 pages. 1981.

Vol. 843: Functional Analysis, Holomorphy, and Approximation Theory. Proceedings. Edited by S. Machado. VI, 636 pages. 1981.